Birkhäuser

Progress in Mathematics
Volume 324

More information about this series at http://www.springer.com/series/4848

Denis Auroux • Ludmil Katzarkov • Tony Pantev
Yan Soibelman • Yuri Tschinkel
Editors

Algebra, Geometry, and Physics in the 21st Century

Kontsevich Festschrift

Birkhäuser

Editors
Denis Auroux
Department of Mathematics
University of California
Berkeley, CA, USA

Ludmil Katzarkov
Institut für Mathematik
Universität Wien
Wien, Austria

Tony Pantev
Department of Mathematics
University of Pennsylvania
Philadelphia, PA, USA

Yan Soibelman
Department of Mathematics
Kansas State University
Manhattan, KS, USA

Yuri Tschinkel
Courant Institute of Mathematical Sciences
New York University
New York, NY, USA

ISSN 0743-1643 ISSN 2296-505X (electronic)
Progress in Mathematics
ISBN 978-3-319-59938-0 ISBN 978-3-319-59939-7 (eBook)
DOI 10.1007/978-3-319-59939-7

Library of Congress Control Number: 2017947499

Mathematics Subject Classification (2010): 14J33, 53D37, 18G60, 53C55

Printed on acid-free paper

This book is published under the trade name Birkhäuser, www.birkhauser-science.com
The registered company is Springer International Publishing AG
The registered company address is: Gewerbestrasse 11, 6330 Cham, Switzerland

Contents

Photo courtesy of M. Kontsevich

Preface

This volume is a tribute to Maxim Kontsevich, one of the most original and influential mathematicians of our time.

Maxim's vision has inspired major developments in countless areas of mathematics, ranging all the way from probability theory to motives over finite fields, and brought forth a paradigm shift at the interface of modern geometry and mathematical physics. Kontsevich integrals (in knot theory), Kontsevich complex (in Hodge theory), Kontsevich propagator (in deformation quantization), Kontsevich conjecture (in combinatorics) are now firmly part of the *lingua franca* of modern mathematics. Many of his papers have opened completely new directions of research. The list of Maxim's spectacular achievements starts already with his Ph.D. thesis on the proof of Witten's conjecture. Soon thereafter, his formulation of Homological Mirror Symmetry became a catalyst for major developments in symplectic geometry, algebraic geometry, and string theory alike; and his work on deformation quantization has fundamentally changed the way mathematicians think about the subject.

Many of his lectures became turning points in the respective subjects. To mention just two such examples: motivic integration originated in his 1995 talk at Université Paris-Sud in Orsay, and Fukaya–Seidel categories were introduced in his 1998 lectures proposing a vast extension of the framework for mirror symmetry. More recent examples abound, some of them still waiting to be fully digested by the mathematical community; Maxim continues to open vast new mathematical landscapes for the rest of us to explore.

As a student of Israel Gelfand, Maxim inherited his ideas about the unity of mathematics. Undoubtedly, Maxim's work is itself a brilliant illustration of the unity of modern mathematics, as well as its deep connections to theoretical physics.

We have been fortunate to have interacted with Maxim on various mathematical projects. We have marveled many times at the avalanche of ideas that he generates in response to a question, even seemingly far-removed from his main research interests; the facility with which he resolves, instantaneously, difficult technical problems; his vision; his generosity, curiosity, and kindness. We offer this volume to him, in friendship and admiration.

August 2016

Denis Auroux, Ludmil Katzarkov
Tony Pantev, Yan Soibelman
Yuri Tschinkel

Progress in Mathematics, Vol. 324, 1–29

Adiabatic Limits of Co-associative Kovalev–Lefschetz Fibrations

Simon Donaldson

To Maxim Kontsevich, on his 50th birthday

Abstract. We study co-associative fibrations of G_2-manifolds. We propose that the adiabatic limit of this structure should be given locally by a maximal submanifold in a space of indefinite signature and set up global versions of the constructions.

Mathematics Subject Classification (2010). 53C26.

Keywords. Exceptional holonomy, maximal submanifolds, co-associative submanifolds , $K3$ surfaces.

1. Introduction

This article makes the first steps in what we hope will be a longer project, investigating seven-dimensional G_2-manifolds from the point of view of co-associative fibrations, and in particular the "adiabatic limit", when the diameters of the fibres shrink to zero. To set up the background, recall that there is a notion of a "positive" exterior 3-form on an oriented 7-dimensional real vector space E, and that such a form defines a Euclidean structure on E (definitions are given in Section 2 below). Thus there is a notion of a positive 3-form ϕ on an oriented 7-manifold M, which defines a Riemannian metric g_ϕ and hence a 4-form $*_\phi\phi$ (writing $*_\phi$ for the Hodge $*$-operator of g_ϕ). A *torsion-free G_2-structure* can be defined to be a positive form ϕ such that ϕ and $*_\phi\phi$ are both closed. (We refer to [14] for the terminology and background.) We can also consider the weaker condition of a *closed G_2-structure* in which we just require that the 3-form ϕ is closed. We have then various fundamental questions, such as the following.

1. Given a compact oriented 7-manifold and a de Rham cohomology class $C \subset \Omega^3(M)$, does C contain a positive form? In other words, is there a non-empty subset $C^+ \subset C$ of positive forms?

2. If so, is there a torsion free G_2-structure in C^+, i.e., can we solve the equation $d *_\phi \phi = 0$ for $\phi \in C^+$?
3. The *Hitchin functional* H on the space of positive 3-forms is just the volume

$$H(\phi) = \mathrm{Vol}(M, g_\phi).$$

   The equation $d *_\phi \phi = 0$ is the Euler–Lagrange equation for this functional restricted to C^+ – in fact critical points are local maxima [11]. It is then interesting to ask if H is bounded above on C^+ and, if a critical point exists, whether it is a global maximum.
4. Bryant [4] defined a "Laplacian" flow on C^+:

$$\frac{\partial \phi}{\partial t} = d\left(*_\phi d *_\phi \phi\right),$$

   whose fixed points are torsion-free G_2 structures. It is interesting to ask if this flow converges as $t \to \infty$ to a fixed point.

While these are obvious and natural questions, any kind of systematic answers seem far out of reach at present. In this article we will study 7-manifolds with extra structure given by a "Kovalev–Lefschetz fibration" $\pi : M \to S^3$ with co-associative fibres. These fibres will be $K3$ surfaces, either smooth or with nodal singularities. The idea underlying our discussion is that there should be an "adiabatic limit" of the torsion-free G_2-condition, and more generally of each of the questions above. Leaving for the future any precise statement along those lines we will write down precise definitions which we propose as limiting geometric objects and we will see that questions (1)–(4) have sensible analogs in this setting.

Co-associative fibrations were studied by Gukov, Yau and Zaslow [9] and there is overlap between our discussion in Section 2 and theirs. More recent work by Baraglia [2] in the case of torus fibrations is especially relevant to our constructions in this paper (see also the brief comments in 5.1 below) In particular Baraglia made the connection with maximal submanifolds in spaces with indefinite signature, which is the fundamental idea that we use here. The notion of a Kovalev–Lefschetz fibration was essentially introduced by Kovalev in [16]. In other directions there is a substantial literature on adiabatic limits of various kinds of structures in Riemannian geometry. Examples include constant scalar curvature Kähler metrics [5] and many papers by different authors on "collapsing" Calabi–Yau manifolds, such as [8], [18]. We will not attempt to give a survey of existence results for compact G_2-manifolds here, but we mention the landmark results of Joyce [13], Kovalev [15] and Corti, Haskins, Nordström, Pacini [6].

2. Co-associative fibrations

We review some standard multi-linear algebra background. Let E be an oriented 7-dimensional real vector space and let ϕ be a 3-form in $\Lambda^3 E^*$. We have then a quadratic form on E with values in the oriented line $\Lambda^7 E^*$ defined by

$$G_\phi(v) = (v \rightharpoonup \phi)^2 \wedge \phi. \tag{1}$$

We say that ϕ is a *positive* 3-form if G_ϕ is positive definite. In this case, any choice of (oriented) volume form makes G_ϕ into a Euclidean structure on E and there is a unique choice of the volume form such that $|\phi|^2 = 7$. So we now have a Euclidean structure G_ϕ canonically determined by a positive 3-form ϕ.

For (E, ϕ) as above we say that a 4-dimensional subspace $V \subset E$ is *co-associative* if the restriction of ϕ to V vanishes. A standard model for such a triple (E, ϕ, V) can be built as follows. We start with $\mathbf{R}^4$ with co-ordinates $x_0, \ldots, x_3$ and standard metric and orientation. Let Λ^2_+ be the space of *self-dual 2-forms.* This has a standard basis $\omega_1, \omega_2, \omega_3$ given by

$$\omega_i = dx_0 dx_i + dx_j dx_k$$

where i, j, k run over cyclic permutations of $1, 2, 3$. Let t_1, t_2, t_3 be the co-ordinates on Λ^2_+ corresponding to this basis and set

$$\phi_0 = \sum_i \omega_i dt_i - dt_1 dt_2 dt_3, \tag{2}$$

an element of $\Lambda^3(E_0)^*$ where $E_0 = \mathbf{R}^4 \oplus (\Lambda^2_+)^*$. This is a positive 3-form if E_0 is given the orientation

$$-dt_1 dt_2 dt_3 dx_0 dx_1 dx_2 dx_3 dx_4 \tag{3}$$

and the corresponding Euclidean structure is the standard one. The subspace $\mathbf{R}^4 \subset E_0$ is co-associative and any co-associative triple (E, ϕ, V) is equivalent to the model $(E_0, \phi_0, \mathbf{R}^4)$ by an oriented linear isomorphism from E to E_0.

Somewhat more generally, let V be an oriented 4-dimensional real vector space. The wedge product gives a quadratic form on 2-forms up to a positive factor. Let H be a 3-dimensional vector space and $\underline{\omega} \in H^* \otimes \Lambda^2 V^*$. We say that $\underline{\omega}$ is *hypersymplectic* if, regarded as a linear map from H to $\Lambda^2 V^*$, it is an injection to a maximal positive subspace for the wedge product form. There is an induced orientation of H and, with a suitable sign convention, for any oriented volume form $\mathrm{vol}_H \in \Lambda^3 H^*$ the 3-form

$$\phi = \underline{\omega} + \mathrm{vol}_H \tag{4}$$

in $H^* \otimes \Lambda^2 V^* \oplus \Lambda^3 H^* \subset \Lambda^3 (V \oplus H)^*$ is positive. Moreover the general positive 3-form on $V \oplus H$ such that V is co-associative and H is the orthogonal complement of V with respect to the induced Euclidean structure has the shape above. Notice that a hypersymplectic $\underline{\omega}$ defines a conformal structure on V – the unique structure such that the image of H is the self-dual subspace.

Next we consider an oriented 7-manifold M. As stated in the introduction, a *closed* G_2*-structure* on M is a closed 3-form ϕ which is positive at each point and the structure is *torsion-free* if in addition $d *_\phi \phi = 0$. If ϕ is any positive 3-form a *co-associative submanifold* is a 4-dimensional submanifold $X \subset M$ such that each point $p \in X$ the tangent space TX_p is a co-associative subspace of TM_p with respect to $\phi(p)$ [10]. The form ϕ induces a standard orientation of X, for example by declaring that $*_\phi \phi$ is positive on X. As background, we recall that for general positive forms the co-associative condition is over-determined, but for closed forms ϕ there is an elliptic deformation theory and compact co-associative submanifolds are stable under small perturbations of the 3-form [17]. They are parametrized by a moduli space of dimension $b^+(X)$ (the dimension of a maximal positive subspace for the intersection form). If ϕ is torsion-free then co-associative submanifolds are examples of Harvey and Lawson's *calibrated submanifolds*[10], with calibrating form $*_\phi\phi$.

We pause here to note a useful identity.

Lemma 1. *If ϕ is a closed positive form then for any vector v in TM we have*

$$(v \rightharpoondown \phi) \wedge d *_\phi \phi = (v \rightharpoondown d\phi) \wedge *_\phi \phi.$$

This can be derived from the diffeomorphism invariance of the Hitchin functional $H(\phi) = \mathrm{Vol}(M, g_\phi)$. The first variation under a compactly supported variation is

$$\delta H = \frac{1}{3} \int_M (\delta\phi) \wedge *_\phi \phi.$$

If v is a compactly supported vector on M and $\delta\phi$ is the Lie derivative $L_v\phi$ then clearly the first variation is zero. Writing $L_v\phi = d(v \rightharpoondown \phi) + v \rightharpoondown d\phi$ and integrating by parts we get the identity

$$\int_M (v \rightharpoondown d\phi) \wedge *_\phi \phi - (v \rightharpoondown \phi) \wedge d *_\phi \phi = 0,$$

which implies the result.

We now move on to our central topic – fibrations $\pi : M \to B$ of a 7-manifold over a 3-dimensional base with co-associative fibres. Later we will consider "fibrations" in the algebraic geometers sense, with singular fibres, but in this section we will consider a genuine locally trivial bundle, with fibre a 4-manifold X. For definiteness we take X to be the smooth 4-manifold underlying a K3 surface, with the standard orientation. We start with a fixed C^∞ fibration $\pi : M \to B$ where M and B are oriented. We want to work with differential forms on the total space and to this end we recall some background.

Let $V \subset TM$ be the tangent bundle along the fibres, so we have an exact sequence

$$0 \to V \to TM \to \pi^*(TB) \to 0.$$

For $p, q \geq 0$ let $\Lambda^{p,q}$ be the vector bundle

$$\Lambda^p(\pi^* T^* B) \otimes \Lambda^q V^*$$

over M and write $\Omega^{p,q}$ for the space of sections of $\Lambda^{p,q}$. The exterior derivative on the fibres is a natural differential operator, for $p, q \geq 0$:

$$d_f : \Omega^{p,q} \to \Omega^{p,q+1},$$

and there is a canonical filtration of Ω^*_M with quotients the $\Omega^{*,*}$.

Now suppose that we have a *connection* H on $\pi : M \to B$. That is, a sub-bundle $H \subset TM$ with $TM = V \oplus H$. Then we get a direct sum decomposition

$$\Omega^*_M = \bigoplus_{p,q\geq 0} \Omega^{p,q}.$$

The exterior derivative $d : \Omega^*_M \to \Omega^*_M$ has three components with respect to this decomposition.

1. The exterior derivative along the fibres d_f, as above.
2. The coupled exterior derivative in the horizontal direction
$$d_H : \Omega^{p,q} \to \Omega^{p+1,q}.$$
This can be defined in various ways. One way is to work at a given point $b \in B$ and to choose a local trivialization of the fibration compatible with H on $\pi^{-1}(b)$. Then, over this fibre, d_H is given by the horizontal component of the exterior derivative in this trivialization. From another point of view we can regard H as a connection on a principal bundle with structure group $\mathrm{Diff}(X)$. Then d_H is the usual coupled exterior derivative on an infinite-dimensional associated vector bundle.
3. The curvature term. The curvature of the connection H is a section F_H of the bundle $V \otimes \pi^*\Lambda^2 T^*B$ defined by the equation
$$[\xi_1, \xi_2]_V = F_H(\xi_1, \xi_2),$$
for any sections ξ_1, ξ_2 of H, where $[\ ,\ ]_V$ denotes the vertical component of the Lie bracket. (This agrees with ordinary notion of curvature if we take the point of view of a $\mathrm{Diff}(X)$ bundle.) The tensor F_H acts algebraically on $\Omega^{p,q}$ by wedge product on the horizontal term and contraction on the vertical term, yielding a map (which we denote by the same symbol)
$$F_H : \Lambda^{p,q} \to \Lambda^{p+2,q-1}.$$

The fundamental differential-geometric formula for the exterior derivative on the total space is then

$$d = d_f + d_H + F_H : \Omega^{p,q} \to \Omega^{p,q+1} + \Omega^{p+1,q} + \Omega^{p+2,q-1}. \tag{5}$$

We say that $\underline{\omega} \in \Omega^{1,2}$ is a hypersymplectic element if it is hypersymplectic at each point, as defined above.

Proposition 1. *Given $\pi : M \to B$ as above, a closed G_2 structure on M such that the fibres of π are co-associative, with orientation compatible with those given on M and B, is equivalent to the following data.*

1. *A connection H.*
2. *A hypersymplectic element $\underline{\omega} \in \Omega^{1,2}$ with $d_f\underline{\omega} = 0, d_H\underline{\omega} = 0$.*
3. *A tensor $\underline{\lambda} \in \Omega^{3,0}$ with*
$$d_f\underline{\lambda} = -F_H\underline{\omega} \tag{6}$$
and such that the value of $\underline{\lambda}$ at each point is positive, regarded as an element of $\Lambda^3 T^*B$.

This proposition follows immediately from the algebraic discussion at the beginning of this section and the description of the exterior derivative above. The corresponding positive 3-form ϕ is of course $\underline{\omega} + \underline{\lambda}$.

Note that if we have *any* 3-form ϕ which vanishes on the fibres there is a well-defined component $\underline{\omega}$ of ϕ in $\Omega^{1,2}$, independent of a connection. This just reflects the naturality of the filtration of Ω^*_M. If ϕ is closed then $d_f\underline{\omega} = 0$. At this point it is convenient to switch to a local discussion (in the base), so we suppose for now that B is a 3-ball with co-ordinates t_1, t_2, t_3. Suppose we have any closed 3-form ϕ on M which vanishes on the fibres. Then the de Rham cohomology class of ϕ is zero and we can write $\phi = d\eta$ for a 2-form η on M. By assumption, the restriction of $d\eta$ to each fibre is zero so we get a map

$$h : B \to H^2(X), \tag{7}$$

taking a point to the cohomology class of η in the fibre over that point. Changing the choice of η changes h by the addition of a constant in $H^2(X)$. As discussed above, the form ϕ has a well-defined component $\underline{\omega} \in \Omega^{1,2}$ which we can write

$$\underline{\omega} = \sum_{i=1}^{3} \omega_i dt_i,$$

where the ω_i are closed 2-forms on the fibres. If we fix a trivialization of the bundle then the ω_i become families of closed 2-forms on a fixed 4-manifold X, parametrized by (t_1, t_2, t_3). A moments thought shows that the de Rham cohomology class $[\omega_i]$ is

$$[\omega_i] = \frac{\partial h}{\partial t_i}. \tag{8}$$

This formula makes intrinsic sense, since the cohomology groups of the different fibres are canonically identified. We say that a map $h : B \to H^2(X, \mathbf{R})$ is *positive* if it is an immersion and the image of its derivative at each point is a positive subspace with respect to the cup product form. Then it is clear that a necessary condition for the existence of a hypersymplectic element $\underline{\omega}$ representing the derivative is that h is positive.

We now change point of view and ask how we can build up a form ϕ. We start with a positive map $h : B \to H^2(X)$ and suppose that we have chosen a hypersymplectic element $\underline{\omega} = \sum \omega_i dt_i \in \Omega^{1,2}$ representing the derivative of h.

Lemma 2. *Given $\underline{\omega}$ as above there is a connection H such that $d_H\underline{\omega} = 0$.*

We work in a local trivialization so that the ω_i are regarded as $\underline{t}$-dependent 2-forms on the fixed 4-manifold X. The connection is represented by $\underline{t}$-dependent vector fields v_1, v_2, v_3 on X and the condition that $d_H\underline{\omega} = 0$ is

$$\frac{\partial \omega_i}{\partial t_j} - \frac{\partial \omega_j}{\partial t_i} + (L_{v_j}\omega_i - L_{v_i}\omega_j) = 0 \tag{9}$$

By assumption, ω_i represents the cohomology class $\frac{\partial h}{\partial t_i}$ and this implies that the cohomology class of $\frac{\partial \omega_i}{\partial t_j} - \frac{\partial \omega_j}{\partial t_i}$ is zero. So we can choose $\underline{t}$-dependent 1-forms a_1, a_2, a_3 such that

$$\frac{\partial \omega_i}{\partial t_j} - \frac{\partial \omega_j}{\partial t_i} = da_k,$$

for i, j, k cyclic. Since the ω_i are closed we can write (9) as

$$d(v_j \rightharpoonup \omega_i - v_i \rightharpoonup \omega_j) = da_k,$$

which is certainly satisfied if

$$v_j \rightharpoonup \omega_i - v_i \rightharpoonup \omega_j = a_k \tag{10}$$

Let

$$S : V \oplus V \oplus V \to V^* \oplus V^* \oplus V^*$$

be the map which takes a triple of tangent vectors v_1, v_2, v_3 to a triple of co-tangent vectors $v_j \rightharpoonup \omega_i - v_i \rightharpoonup \omega_j$. What we need is the elementary linear algebra statement that S is an isomorphism. To see this observe that it is a pointwise statement and by making a linear change of co-ordinates t_i we can reduce to the case when ω_i are the standard basis of the self-dual forms on $\mathbf{R}^4$. Using the metric to identify tangent vectors and cotangent vectors, S becomes the map which takes a triple v_1, v_2, v_3 to the triple

$$Jv_3 - Kv_2 \ , \ Kv_1 - Iv_3 \ , \ Iv_2 - Jv_1. \tag{11}$$

Here we make the standard identification of $\mathbf{R}^4$ with the quaternions and use quaternion multiplication. A short calculation in quaternion linear algebra shows that S is an isomorphism.

Once we have chosen $\underline{\omega}$ and H, as above, the remaining data we need to construct a closed G_2 structure with co-associative fibres on M is $\underline{\lambda} \in \Omega^{3,0}$ which we can write in co-ordinates as $-\lambda dt_1 dt_2 dt_3$ for a function λ on M. The identity $d^2 = 0$ on Ω^*_M implies that

$$d_f(F_H \underline{\omega}) = 0.$$

Since $H^1(X) = 0$ we can solve the equation

$$d_f \underline{\lambda} = -F_H \underline{\omega}.$$

The solution is unique up to the addition of a lift of a 3-form from B, i.e., up to changing λ by the addition of a function of $\underline{t}$. By adding a sufficiently positive function of $\underline{t}$ we can arrange that λ is positive. We conclude that any choice of hypersymplectic element representing the derivative of h can be extended to a closed G_2 structure with co-associative fibres.

We stay with the local discussion and go on to consider torsion-free G_2 structures. Suppose we have 2-forms ω_i spanning a maximal positive subspace of the vertical tangent space V. Let χ be an arbitrary volume form on V, so that $(\omega_i \wedge \omega_j)/\chi$ are real numbers. We form the determinant $\det((\omega_i \wedge \omega_j)/\chi)$ and the 4-form

$$(\det((\omega_i \wedge \omega_j)/\chi)))^{1/3} \chi.$$

It is clear that this is independent of the choice of χ and we just write it as $\det^{1/3}(\omega_i \wedge \omega_j)$.

Lemma 3. *If $\phi = \underline{\omega} + \underline{\lambda} \in \Omega^{1,2} + \Omega^{3,0}$ is a positive 3-form on M, as considered above, then*

$$*_\phi \phi = \underline{\Theta} + \underline{\mu}$$

where

1. $\underline{\Theta} = \mathcal{F}_1(\underline{\omega}, \underline{\lambda}) \in \Omega^{2,2}$ *is given by* $\sum_{\text{cyclic}} \Theta_i dt_j dt_k$ *where* Θ_i *are forms in the linear span of* $\omega_1, \omega_2, \omega_3$ *determined uniquely by the condition that*
$$\Theta_i \wedge \omega_j = \delta_{ij}(\lambda^{1/3}\det{}^{1/3}(\omega_a \wedge \omega_b)); \tag{12}$$
2. $\underline{\mu} = \mathcal{F}_2(\underline{\omega}, \underline{\lambda}) \in \Omega^{4,0}$ *is given by*
$$\det{}^{1/3}(\omega_a \wedge \omega_b)\lambda^{-2/3}. \tag{13}$$

This is a straightforward algebraic exercise using the definitions. The upshot of our local discussion is then the following statement.

Proposition 2. *A torsion-free G_2 structure on M with co-associative fibres is given by a map h (defined up to a constant), a hypersymplectic element $\underline{\omega}$ representing the derivative of h, a connection H and $\underline{\lambda} \in \Omega^{3,0}$ with $\underline{\lambda} > 0$, satisfying the five equations*

1. $d_H\underline{\omega} = 0$,
2. $d_f\underline{\lambda} = -F_H\underline{\omega}$,
3. $d_H\underline{\mu} = 0$,
4. $d_f\underline{\Theta} = -F_H\underline{\mu}$,
5. $d_H\underline{\Theta} = 0$.

Here $\underline{\mu}, \underline{\Theta}$ are determined algebraically from $\underline{\omega}, \underline{\lambda}$ by the formulae (12), (13).

Note that the equation (3) in Proposition 2 has a clear geometric meaning. It states that the structure group of the connection reduces to the volume preserving diffeomorphisms of X. It is also equivalent to the statement that the fibres are minimal submanifolds, with respect to the metric g_ϕ.

We will not attempt to analyse these equations further in this form, except to note one identity.

Lemma 4. *Suppose that $h, \underline{\omega}, \underline{\lambda}, H$ satisfy equations* (1), (2), (3), (4) *in Proposition* 2. *Then $d_H\underline{\Theta} = G dt_1 dt_2 dt_3$ where G is an anti-self dual harmonic form on each fibre with respect to the conformal structure determined by $\underline{\omega}$.*

First, the identity $d^2 *_\phi \phi = 0$ implies that, under the stated conditions $d_f(d_H\underline{\Theta}) = 0$ so $d_H\Theta = G dt_1 dt_2 dt_3$ where G is a closed 2-form on each fibre. Now we apply Lemma 1, taking v to be a horizontal vector. Since $d\phi = 0$ we have $(v \rightharpoonup \phi) \wedge d *_\phi \phi = 0$. The contraction $v \rightharpoonup \phi$ has terms in $\Omega^{0,2}$ and $\Omega^{2,0}$ but only the first contributes to the wedge product with $d *_\phi \phi$ and the result follows immediately.

Let us now move back to consider the "global" case, with a K3 fibration $\pi : M \to B$ over a general oriented 3-manifold B. We have a flat bundle $\mathcal{H} \to B$ with fibre $H^2(X; \mathbf{R})$ induced from an $H^2(X; \mathbf{Z})$ local system, so there is an intrinsic integer lattice in each fibre. Let ϕ be a closed 3-form on M, vanishing on the fibres. If we cover B by co-ordinate balls B_α we can analyse ϕ on each $\pi^{-1}(B_\alpha)$ as above, starting with a choice of a smooth section h_α of $\mathcal{H}|_{B_\alpha}$. Then on $B_\alpha \cap B_\beta$ the difference

$$\chi_{\alpha\beta} = h_\alpha - h_\beta \tag{14}$$

is a constant section of $\mathcal{H}$. This defines a Cěch cocycle, and hence a cohomology class

$$\chi \in H^1(B; \underline{\mathcal{H}}),$$

where $\underline{\mathcal{H}}$ is the sheaf of locally constant sections of the flat bundle $\mathcal{H}$. This class χ is determined by the de Rham cohomology class of ϕ. In fact the Leray–Serre spectral sequence gives us an exact sequence

$$0 \to H^3(B) \to H^3(M) \xrightarrow{b} H^1(B; \underline{\mathcal{H}}) \to 0, \tag{15}$$

and $[\phi] \in H^3(M)$ maps to $\chi \in H^1(B; \underline{\mathcal{H}})$. It is convenient to express this in terms of an "affine" variant of $\mathcal{H}$. Let G be the group of affine isometries of $H^2(X)$, an extension

$$1 \to H^2(X) \to G \to O(H^2(X)) \to 1.$$

A class $\chi \in H^1(B; \underline{\mathcal{H}})$ defines a flat bundle $\mathcal{H}_\chi$ with fibre $H^2(X)$ and structure group G over B. One way of defining this is to fix a cover and Cěch representative $\chi_{\alpha\beta}$ and then decree that locally constant sections of $\mathcal{H}_\chi$ are given by constant sections h_α over B_α satisfying (14). In these terms the conclusion is that if we start with a class $\Phi \in H^3(M)$, map to a class $\chi = b(\Phi)$ and form the bundle $\mathcal{H}_\chi$, then any representative $\phi \in \Omega^3(M)$ for Φ which vanishes on the fibres defines a canonical smooth section h of the flat bundle $\mathcal{H}_\chi$. Moreover if ϕ is a positive 3-form then h is a *positive* section of $\mathcal{H}_\xi$, using the obvious extension of the definition in the local situation.

2.1. The adiabatic limit

We have seen that the torsion-free condition for a positive 3-form involves six equations for a triple $(\underline{\omega}, \underline{\lambda}, H)$

$$\begin{aligned} d_f\underline{\omega} &= 0, & d_H\underline{\omega} &= 0, & d_f\underline{\lambda} &= -F_H\underline{\omega}, \\ d_H\underline{\mu} &= 0, & d_f\underline{\Theta} &= -F_H\underline{\mu}, & d_H\underline{\Theta} &= 0. \end{aligned}$$

where $\underline{\Theta} = \mathcal{F}_1(\underline{\omega}, \underline{\lambda})$ and $\underline{\mu} = \mathcal{F}_2(\underline{\omega}, \underline{\lambda})$ are determined algebraically by $\underline{\omega}, \underline{\lambda}$. We introduce a positive real parameter ϵ and write $\underline{\omega} = \epsilon\underline{\tilde{\omega}}$,

$$\underline{\tilde{\Theta}} = \mathcal{F}_1(\underline{\tilde{\omega}}, \lambda) = \epsilon\underline{\Theta}, \quad \underline{\tilde{\mu}} = \mathcal{F}_2(\underline{\tilde{\omega}}, \lambda) = \epsilon^2\underline{\mu}.$$

Then the six equations, written for $(\underline{\tilde{\omega}}, \underline{\lambda}, H)$ become

$$\begin{aligned} d_f\underline{\tilde{\omega}} &= 0, & d_H\underline{\tilde{\omega}} &= 0, & d_f\underline{\lambda} + \epsilon F_H\underline{\tilde{\omega}} &= 0, \\ d_H\underline{\tilde{\mu}} &= 0, & d_f\underline{\tilde{\Theta}} + \epsilon F_H\underline{\tilde{\mu}} &= 0, & d_H\underline{\tilde{\Theta}} &= 0. \end{aligned} \tag{16}$$

Here $\tilde{\mu}, \underline{\tilde{\Theta}}$ are related to $\underline{\tilde{\omega}}, \underline{\lambda}$ in just the same way as $\underline{\mu}, \underline{\Theta}$ were to $\underline{\omega}, \underline{\lambda}$. Geometrically, in the metric determined by $\underline{\tilde{\omega}}, \underline{\lambda}$ the volume of the fibre is ϵ^2 times that in the metric determined by $\underline{\omega}, \underline{\lambda}$. For non-zero ϵ these equations are completely equivalent to the original set; the fundamental idea we want to pursue is to pass to the "adiabatic limit", setting $\epsilon = 0$ so that the terms involving the curvature F_H drop out. At this stage we simplify notation by dropping the tildes, writing $\underline{\omega}, \underline{\Theta}, \underline{\mu}$ for $\underline{\tilde{\omega}}, \underline{\tilde{\Theta}}, \underline{\tilde{\mu}}$.

We say that $\underline{\omega} \in \Omega^{1,2}$ is a *hyperkähler element* if, in local co-ordinates on the base, $\underline{\omega} = \sum \omega_i dt_i$ with

$$\omega_i \wedge \omega_j = a_{ij}\nu,$$

where ν is a volume form along the fibres and a_{ij} is a positive definite matrix *constant* along the fibres.

Lemma 5. *Suppose that a pair $(\underline{\omega}, \underline{\lambda})$ satisfies the five equations*

$$d_f\underline{\omega} = 0, \quad d_H\underline{\omega} = 0, \quad d_f\underline{\lambda} = 0,$$
$$d_H\underline{\mu} = 0, \quad d_f\underline{\Theta} = 0.$$

Then $\underline{\omega}$ is a hyperkähler element and $\underline{\lambda}$ is the lift of a positive 3-form on B. Conversely given a hyperkähler element $\underline{\omega}$ and a positive 3-form on B there is a unique connection H which yields a solution to these equations.

In one direction, suppose that we have $\underline{\omega}, \underline{\lambda}$ satisfying these equations and work locally, writing $\underline{\omega} = \sum \omega_i dt_i, \underline{\lambda} = -\lambda dt_1 dt_2 dt_3, \underline{\Theta} = \sum \Theta_i dt_j dt_k$. The equation $d_f\underline{\lambda} = 0$ immediately gives that $\underline{\lambda}$ is a lift from B. We know that $\underline{\Theta}_i$ takes values in the self-dual space Λ^+ spanned by $\omega_1, \omega_2, \omega_3$ so the equation $d_f\underline{\Theta} = 0$ says that the Θ_i are self-dual harmonic forms. Since $b^+(X) = 3$ we have $\Theta_i = \sum g_{ij}\omega_j$ where the functions g_{ij} are constant on the fibres. Now

$$\Theta_i \wedge \omega_k = \lambda^{1/3}\ \delta_{ik}\ \det{}^{1/3}(\omega_a \wedge \omega_b),$$

so

$$\sum_j g_{ij}\omega_j \wedge \omega_k = \lambda^{1/3}\ \delta_{ik}\ \det{}^{1/3}(\omega_a \wedge \omega_b)$$

which implies that

$$\omega_j \wedge \omega_k = g^{jk}\lambda^{1/3}\det{}^{1/3}(\omega_a \wedge \omega_b)$$

where g^{jk} is the inverse matrix, also constant on the fibres. Hence $\underline{\omega}$ is hyperkähler.

For the converse, suppose that we have a hyperkähler element $\underline{\omega}$ and volume form on B, which we may as well take to be $-dt_1dt_2dt_3$ in local co-ordinates. Then we have a volume form $\underline{\mu}$ along the fibres given by $\det^{1/3}(\omega_i \wedge \omega_j)$ and we have to show that there is a unique connection H such that $d_H\underline{\omega} = 0$ and $d_H\underline{\mu} = 0$. For this we go back to the proof of Lemma 2. We can choose a local trivialization such that the volume form along the fibres is constant. For any choice of 1-forms a_1, a_2, a_3 with

$$\frac{\partial \omega_i}{\partial t_j} - \frac{\partial \omega_j}{\partial t_i} = da_k$$

we showed there that there is a unique choice of connection H. But we are free to change the 1-forms a_k by the derivatives of functions. It is easy to check that the condition that the connection is volume preserving is exactly that the 1-forms a_k satisfy $d^* a_k = 0$, with respect to the induced metric on the fibres. By Hodge theory, this gives the existence and uniqueness of the connection H. (In fact the argument shows that for any volume form along the fibres we can choose H to satisfy this additional condition, provided that the volume of the fibres is constant.)

We can clearly choose the solution in Lemma 5 so that the $\underline{\mu}$-volume of the fibres is 1, and we fix this normalisation from now on.

It remains to examine the last equation, $d_H\underline{\Theta} = 0$. Recall that $H^2(X, \mathbf{R})$ carries an intrinsic quadratic from of signature $(3, 19)$. This defines a volume form on any submanifold and hence an Euler–Lagrange equation defining stationary submanifolds. In our context we are interested in 3-dimensional stationary submanifolds whose tangent spaces are positive subspaces and these are called *maximal submanifolds.* A section h of $\mathcal{H}_\xi$ can be represented locally by an embedding of the base in $H^2(X)$ and we say that h is a maximal section if the image is a maximal submanifold. More generally, for any section h the *mean curvature* $m(h)$ is defined as a section of the vector bundle $(\mathrm{Im}\ dh)^\perp \subset \mathcal{H}$ which is the normal bundle of the submanifold.

Lemma 6. *Suppose that h is a positive section of $\mathcal{H}_X$, $\underline{\omega}$ is a hyperkähler element representing the derivative of h, and that H, $\underline{\lambda}$ are chosen as above. Then*

$$d_H\underline{\Theta} = m(h)\underline{\lambda}, \tag{17}$$

where we use the metric on the fibres defined by $\underline{\omega}$ to identify $(\mathrm{Im}\ dh)^\perp$ *with the anti-self-dual 2-forms along the fibres and use the result of Lemma 4. In particular, $d_H\underline{\Theta} = 0$ if and only if h is a maximal positive section.*

In local co-ordinates t_i let $g_{ij} = \int_X \omega_i \wedge \omega_j$ and set $J = \det(g_{ij})$. The volume of the image is given by

$$\int J^{1/2} dt_1 dt_2 dt_3$$

and the mean curvature is

$$m = J^{-1/2}\left(J^{1/2} g^{ij} h_{,j}\right)_{,i},$$

where g^{ij} is the inverse matrix and $_{,i}$ denote partial differentiation. That is

$$m = J^{-1/2} \sum \frac{\partial \Theta_i^*}{\partial t_i},$$

where $\Theta_i^* = J^{1/2} g^{ij} [\omega_j]$. Computing in a trivialization compatible with H at a given point we have

$$d_H\underline{\Theta} = \left(\sum \frac{\partial \Theta_i}{\partial t_i}\right) dt_1 dt_2 dt_3.$$

Now Θ_i is defined by the identity

$$\Theta_i \wedge \omega_j = \det{}^{1/3} \lambda^{1/3} \delta_{ij},$$

so

$$\int_X \Theta_i \wedge \omega_j = \lambda^{1/3} \int_X \det{}^{1/3} \delta_{ij},$$

whereas

$$\int_X \Theta_i^* \wedge \omega_j = J^{1/2}.$$

The condition that the μ-volume of the fibres is 1 is

$$\lambda^{2/3} = \int_X \det{}^{1/3}.$$

On the other hand one has

$$\int_X \det{}^{1/3} = J^{1/3}.$$

This implies that $\Theta_i^* = [\Theta_i]$ and $J^{-1/2} dt_1 dt_2 dt_3 = \underline{\lambda}$ and the result follows.

One can also prove this lemma more synthetically, comparing the Hitchin 7-dimensional volume functional with the 3-dimensional volume of the image of h.

2.2. Formal power series solutions

The conclusion that we are lead to in the preceding discussion is that the adiabatic limit of the torsion-free G_2-structure condition should be the maximal submanifold equation $m(h) = 0$. To give more support to this idea we show that a solution of the latter equation can be a extended to a formal power series solution of system of equations (16). For simplicity we work locally, so we take the base B to be a ball.

Proposition 3.

1. *Suppose that the quadruple $(h, \underline{\omega}, H, \underline{\lambda})$ satisfies the five equations of Lemma 5. Then there are formal power series*

$$\tilde{\underline{\omega}}^\epsilon = \underline{\omega} + \sum_{j=1}^{\infty} \underline{\omega}^{(j)} \epsilon^j, \quad H^\epsilon = H + \sum_{j=1}^{\infty} \zeta^{(j)} \epsilon^j, \quad \underline{\lambda}^\epsilon = \underline{\lambda} + \sum_{j=1}^{\infty} \underline{\lambda}^{(j)} \epsilon^j$$

which give a formal power series solution of the first five equations of (16). *We can choose $\underline{\omega}^{(j)}$ to be exact, so that $\tilde{\underline{\omega}}^\epsilon$ represents the derivative of the fixed map h.*

2. *Suppose that in addition h satisfies the maximal submanifold equation $m(h) = 0$ and is smooth up to the boundary of the ball B. Then we can find a formal power series*

$$h^\epsilon = h + \sum_{j=1}^{\infty} h^{(j)} \epsilon^j$$

and power series $\underline{\tilde{\omega}}^\epsilon, H^\epsilon, \underline{\mu}^\epsilon$ as above but now with $\underline{\tilde{\omega}}^\epsilon$ representing the derivative of h^ϵ and such that the quadruple $(h^\epsilon, \underline{\tilde{\omega}}^\epsilon, H^\epsilon, \underline{\lambda}^\epsilon)$ is a formal power series solution of the equations (16).

The proof follows standard lines once we understand the linearization of the equations, and for this we to develop some background. Let ω_i be a standard hyperkähler triple on X. Consider an infinitesimal variation $\omega_i + \eta_i$ with $d\eta_i = 0$ and write $\eta_i = \eta_i^+ + \eta_i^-$ in self-dual and anti-self-dual parts. So $\eta_i^+ = \sum S_{ij}\omega_j$ for some co-efficients S_{ij}. A short calculation shows that the variation in $d_f\underline{\Theta}$ is, to first order, $\sum dT_{ij} \wedge \omega_j$ where

$$T_{ij} = S_{ij} + S_{ji} - \frac{2}{3}\mathrm{Tr}(S)\delta_{ij}. \tag{18}$$

(I.e., T is obtained from S by projecting to the symmetric, trace-free component.) We want to vary ω_i in fixed cohomology classes, by da_i. The resulting first-order variation in $d\Theta_i$ is given by a linear map $\mathcal{L}_1$ from $\Omega^1_X \otimes \mathbf{R}^3$ to $\Omega^3_X \otimes \mathbf{R}^3$. So the calculation above shows that this linear map is a composite

$$\Omega^1_X \otimes \mathbf{R}^3 \overset{d^+\otimes 1}{\to} \Omega^{2,+}_X \otimes \mathbf{R}^3 \overset{A}{\to} \Omega^{2,+}_X \otimes \mathbf{R}^3 \overset{d\otimes 1}{\to} \Omega^3_X \otimes \mathbf{R}^3, \tag{19}$$

where A is the map taking $\sum S_{ij}\omega_j$ to $\sum T_{ij}\omega_j$, with T given by (18) above. Thus the image of A is the subspace of $\Omega^{2,+}_X \otimes \mathbf{R}^3$ consisting of tensors with values in $s^2_0(\mathbf{R}^3)$ – symmetric and trace-free. Here of course we are using repeatedly the fact that the bundle of self-dual forms is identified with $\mathbf{R}^3$ by the sections ω_i.

From another point of view, write S_+, S_- for the spin bundles over X and write S^p_+, S^p_- for their pth symmetric powers. Of course in our situation S_+ is a trivial flat bundle. The tensors with values in $s^2_0(\mathbf{R}^3)$ can be identified with sections of S^4_+. (More precisely, S^4_+ has a real structure and we should consider real sections, but in the discussion below we do not need to distinguish between the real and complex cases.) There is a contraction map from $\Omega^3_X \otimes \mathbf{R}^3$ to Ω^3_X given by

$$P(\psi_1, \psi_2, \psi_3) = \sum (*\psi_i) \wedge \omega_i \tag{20}$$

The kernel of this contraction map can be identified with $S^3_+ \otimes S_-$, corresponding to the decomposition:

$$(S_+ \otimes S_-) \otimes S^2_+ = (S_+ \otimes S_-) \oplus (S^3_+ \otimes S_-).$$

Starting with

$$d \otimes 1 : \Omega^{2,+}_X \otimes \mathbf{R}^3 \to \Omega^3_X \otimes \mathbf{R}^3,$$

we first restrict to sections of S^4_+ and then project to sections of $S^3_+ \otimes S_-$. This defines a differential operator

$$\delta_1 : \Gamma(S^4_+) \to \Gamma(S^3_+ \otimes S_-).$$

On the other hand, starting with

$$d : \Omega^3_X \otimes \mathbf{R}^3 \to \Omega^4_X \otimes \mathbf{R}^3,$$

we restrict to sections of $S^3_+ \otimes S_-$ to define

$$\delta_2 : \Gamma(S^3_+ \otimes S_-) \to \Gamma(S^2_+).$$

So we have

$$\Gamma(S^4_+) \xrightarrow{\delta_1} \Gamma(S^3_+ \otimes S_-) \xrightarrow{\delta_2} \Gamma(S^2_+).$$

The key fact we need is that this is exact at the middle term, i.e., that

$$\text{Ker } \delta_2 = \text{Im } \delta_1. \tag{21}$$

In fact, calculation shows that $\delta_2 + \delta_1^*$ can be identified with the Dirac operator coupled with S^3_+, i.e.,

$$D : \Gamma(S^3_+ \otimes S_-) \to \Gamma(S^3_+ \otimes S_+)$$

using the isomorphism

$$S^3_+ \otimes S_+ = S^+_4 \oplus S^+_2.$$

Since S_+ is trivial this Dirac operator is a sum of copies of the ordinary Dirac operator over the K3 surface and has no kernel in the S_- term. This implies (21). We can say a bit more. The kernel of δ_1 consists of the constant sections of S^4_+ so for any $\rho \in S^+_3 \otimes S_-$ we can solve the equation $\delta_1 T = \rho$ with T in the L^2-orthogonal complement of the constants. But this means that we can solve the equation $(d^+ \otimes 1)a = T$.

To sum up we have shown:

Lemma 7. *If $\rho \in \Omega^3_X \otimes \mathbf{R}^3$ satisfies the two conditions*
- *$P(\rho) = 0$ where P is the contraction* (20);
- *$(d \otimes 1)(\rho) = 0$ in $\Omega^4_X \otimes \mathbf{R}^3$;*

then ρ is in the image of $\mathcal{L}_1 : \Omega^1_X \otimes \mathbf{R}^3 \to \Omega^3_X \otimes \mathbf{R}^3$.

We now turn to the proof of the first item in Proposition 3. At stage k we suppose that we have found $\underline{\omega}^{(i)}, \underline{\lambda}^{(i)}, \zeta^{(i)}$ for $i \leq k$ so that the finite sums

$$\underline{\omega}^{[k]} = \underline{\omega} + \sum_{i=1}^{k} \underline{\omega}^{(k)} \quad \text{etc.},$$

satisfy the first five equations of (16) modulo ϵ^{k+1}. We want to choose $\underline{\omega}^{(k+1)}, \underline{\lambda}^{(k+1)}$ and $\zeta^{(k+1)}$ so that if we define

$$\underline{\omega}^{[k+1]} = \underline{\omega}^{[k]} + \epsilon^{k+1} \underline{\omega}^{(k+1)} \quad \text{etc.}$$

we get solutions of the first five equations of (16) modulo ϵ^{k+2}. The essential point is that this involves solving *linear* equations for $\underline{\omega}^{(k+1)}, \underline{\lambda}^{(k+1)}, \zeta^{(k+1)}$, and moreover these equations are essentially defined fibrewise.

Step 1 Let $E_{3,1}$ be the "error term"

$$d_f \underline{\lambda}^{[k]} + \epsilon F_{H[k]} \underline{\omega}^{[k]} = E_{3,1} \epsilon^{k+1} + O(\epsilon^{k+2}).$$

(Here the $O(\ )$ notation means in the sense of formal power series expansion.) The equation we need to solve to correct this error term is

$$d_f \underline{\lambda}^{(k+1)} = -E_{3,1},$$

involving only $\underline{\lambda}^{(k+1)}$. The identity $d^2 = 0$ on M implies that

$$d_f F_{H[k]} \underline{\omega}^{[k]} = -d_H d_{H[k]} \underline{\omega}^{[k]}$$

which is $O(\epsilon^{k+1})$. This means that $d_f E_{3,1} = 0$ so we can solve the equation for $\underline{\lambda}^{(k+1)}$. The solution is unique up to the addition of a term which is constant on the fibres.

Step 2 Now let $E_{2,3}$ be the error term

$$d_f \underline{\Theta}^{[k]} + \epsilon F_{H[k]} \underline{\mu}^{[k]} = E_{2,3}\epsilon^{k+1} + O(\epsilon^{k+2}).$$

The equation we need to solve to correct this error involves the first-order variation of $d_f \underline{\Theta}^{[k]}$ with respect to changes in $\underline{\omega}$ and $\underline{\lambda}$. For the first, we write $\omega^{(k+1)}$ as the derivative of $a \in \Omega^1 \otimes \mathbf{R}^3$ as above and the first-order variation is given by $\mathcal{L}_1(a)$. Denote the variation with respect to $\underline{\lambda}$ by $\mathcal{L}_2$. The equation we need to solve is

$$\mathcal{L}_1(a) + \mathcal{L}_2(\underline{\lambda}^{(k+1)}) = -E_{2,3}.$$

The same argument as before shows that $d_f E_{2,3} = 0$. We do not need to write our $\mathcal{L}_2$ explicitly because it is easy to see that $\mathcal{L}_2(\underline{\lambda}^{(k+1)})$ only involves the derivative along the fibre, so is unchanged if we change $\underline{\lambda}^{(k+1)}$ by a term which is constant on the fibres. Thus $\mathcal{L}_2(\underline{\lambda}^{(k+1)})$ is fixed by Step 1 and we have to solve the equation

$$\mathcal{L}_1(a) = \rho,$$

for a, where $\rho = -E_{2,3} - \mathcal{L}_2(\underline{\lambda}^{(k+1)})$. The same argument as before shows that $d_f E_{2,3} = 0$ and it is clear that $d_f \circ \mathcal{L}_2 = 0$. Thus $d_f \rho = 0$. We claim that, on each fibre, $P(\rho) = 0$. Given this claim, Lemma 7 implies that we can find the desired solution a.

To verify the claim we use the identity of Lemma 1. We take v to be a vertical vector and ϕ to be the 3-form

$$\phi = \omega^{[k]} + \underline{\lambda}^{[k+1]}.$$

By what we have arranged, the $(3,1)$ component of $d\phi$ is $O(\epsilon^{k+2})$ and this means that $(v \rightharpoondown \underline{\omega}^{[k]}) \wedge \rho$ is $O(\epsilon)$. In turn this implies that $(v \rightharpoondown \underline{\omega}) \wedge \rho$ vanishes for all vertical vectors v, and it is easy to check that this is equivalent to the condition $P(\rho) = 0$.

Step 3 At this stage we have chosen $\underline{\omega}^{(k+1)}$. We still have the ambiguity in $\underline{\lambda}^{(k+1)}$ up to constants along the fibres. We fix this by decreeing that the fibres have volume $1 + O(\epsilon^{k+2})$ with respect to the form $\underline{\mu}^{[k+1]}$ defined by $\underline{\lambda}^{[k+1]}, \underline{\omega}^{[k+1]}$.

Step 4 Now we choose $H^{[k+1]}$ so that $d_{H[k+1]} \underline{\omega}^{[k+1]}$ and $d_{H[k+1]} \underline{\mu}^{[k+1]}$ are $O(\epsilon^{k+2})$.

This completes the proof of the first item in Proposition 3.

For the second item, we take the power solution above for an input map h and write

$$d_{H^\epsilon}\Theta^\epsilon = M(\epsilon, h) = \left(M_0(h) + \epsilon M_1(h) + \epsilon^2 M_2(h) + \cdots\right)\underline{\lambda},$$

where $M_i(h)$ are sections of the normal bundle of the image of h in $H^2(X)$. We know that $M_0(h)$ is the mean curvature $m(h)$, whose linearization with respect to normal variations is a Jacobi operator J_h say. Locally, working over a ball, we can find a right inverse G_h to the Jacobi operator. Then starting with a solution h_0 of the equation $m(h_0) = 0$ we can construct a formal power series solution $h^\epsilon = h_0 + \sum \xi_k \epsilon^k$ to the equation $M(\epsilon, h^\epsilon) = 0$ in a standard fashion. For example the first term is given by

$$h^\epsilon = h_0 - \epsilon G_{h_0}(M_1(h_0)) + O(\epsilon^2).$$

2.3. The Torelli Theorem for K3 surfaces

The question of which compact 4-manifolds admit hypersymplectic structures is an interesting open problem. But for hyperkähler structures there is a complete classification. The underlying manifold must be a K3 surface or a 4-torus and we restrict here to the former case. We recall the fundamental *global Torelli Theorem*, for which see [3], Chapter VIII for example.

Proposition 4. *Let X be the oriented 4-manifold underlying a K3 surface. Write $\mathcal{C}$ for the set of classes $\delta \in H^2(X, \mathbf{Z})$ with $\delta.\delta = -2$ and write* Diff$_0$ *for the group of diffeomorphisms of X which act trivially on $H^2(X)$. A hyperkähler structure on X determines a self-dual subspace $H^+ \subset H^2(X, \mathbf{R})$ (the span of $[\omega_i]$) which has the two properties:*

- *H^+ is a maximal positive subspace for the cup product form.*
- *There is no class $\delta \in \mathcal{C}$ orthogonal to H^+.*

Conversely, given a subspace $H^+ \subset H^2(X; \mathbf{R})$ satisfying these two conditions there is a hyperkähler structure with volume 1 realizing H^+ as the self-dual subspace and the structure is unique up to the action of Diff$_0 \times SL(3, \mathbf{R})$.

(In the last sentence, the hyperkähler structure is regarded as an element of $\mathbf{R}^3 \otimes \Omega^2(X)$ and $SL(3, \mathbf{R})$ acts on $\mathbf{R}^3$.)

Let h be a positive map from the 3-ball to $H^2(X; \mathbf{R})$. We say that h *avoids -2 classes* if for each point in B there is no δ in $\mathcal{C}$ orthogonal to the image of the derivative. This is an open condition on h. The definition extends in the obvious way to the sections of the flat bundle $\mathcal{H}_\chi$ defined by a fibration $X \to M \to B$ as considered above. Then a hyperkähler element $\underline{\omega}$ defines a positive section of $\mathcal{H}_\chi$ which avoids -2 classes. A complete discussion of the converse brings in topological questions about the group Diff$_0$ which we do not want to go into here. At least in the local situation, when B is a ball, we can say that a positive section which avoids -2 classes defines a hyperkähler element.

3. Kovalev–Lefschetz fibrations

Baraglia [2] shows that there are essentially no co-associative fibrations over a compact base manifold B, so to get an interesting global problem we must allow singularities. The singularities that we consider here are those introduced by Kovalev [16], where the fibres are allowed to develop ordinary double points.

Definition 1. A differentiable Kovalev–Lefschetz (KL) fibration consists of data $(M, B, L, \tilde{L}, \pi)$ of the following form.

- M is a smooth oriented compact 7-manifold, B is a smooth oriented compact 3-manifold and $\pi : M \to B$ is a smooth map.
- $L \subset B$ is a 1-dimensional submanifold (i.e., a link).
- $\tilde{L} \subset M$ is a 1-dimensional submanifold and the restriction of π gives a diffeomorphism from $\tilde{L}$ to L.
- π is a submersion outside $\tilde{L}$ and for any point $p \in \tilde{L}$ there are oriented charts around $p, \pi(p)$ in which π is represented by a map
$$(z_1, z_2, z_3, t) \mapsto (z_1^2 + z_2^2 + z_3^2 + E(\underline{z}, t), t),$$
from $\mathbf{R}^7 = \mathbf{C}^3 \times \mathbf{R}$ to $\mathbf{R}^3 = \mathbf{C} \times \mathbf{R}$, where $E(\underline{z}, t)$ and its first and second partial derivatives vanish at $\underline{z} = 0$.

Thus the fibre $\pi^{-1}(p)$ of a KL fibration is a smooth 4-manifold if p is not in L, and if p is in L there is just one singular point, modeled on the quadric cone $\{z_1^2 + z_2^2 + z_3^2 = 0\}$ in $\mathbf{C}^3$, which is homeomorphic to the quotient $\mathbf{C}^2/\pm 1$. Now we define a torsion-free (respectively closed) co-associative KL fibration to be a differentiable one, as above, together with a smooth positive 3-form ϕ on M which is torsion-free (respectively closed) and such that the non-singular fibres are co-associative. We also require the orientations to be compatible with the form ϕ.

A good reason for focusing on these kind of singularities is that Kovalev has outlined a programme that would produce many examples of torsion-free co-associative fibrations of this shape [16]. (Kovalev's programme seems geometrically plausible, but there are some analytical difficulties in the available 2005 preprint [16]). One can certainly consider other kinds of singularities, but we will not do so here.

Note that a compact manifold with holonomy the full group G_2 has finite fundamental group [14]. Thus, up to finite coverings (and using the solution of the Poincaré conjecture), the essential case of interest is when the base B is the 3-sphere.

We can now adapt our previous discussion (for genuine fibrations) to the KL situation. We will only consider the case when the smooth fibres are diffeomorphic to a $K3$ surface. Let $\pi : M \to B$ be a differentiable KL fibration and γ be a small loop around one of the components of L. Then we can consider the *monodromy* around γ of the co-homology of the fibre. It is a well-known fact that this monodromy is the reflection in a "vanishing cycle" δ which is a class in $H^2(X; \mathbf{Z})$ with

$\delta.\delta = -2$. (In particular, this monodromy has order 2, so we do not need to discuss the orientation of γ.)

To build in this monodromy we consider B as an orbifold, modeled on $(\mathbf{R} \times \mathbf{C})/\sigma$ at points of L, where σ acts as -1 on $\mathbf{C}$ and $+1$ on $\mathbf{R}$. Recall that an orbifold atlas is a covering of B by charts $U_\alpha \subset B$ such that if U_α intersects L it is provided with an identification with a quotient $\tilde{U}_\alpha/\sigma$ for a σ-invariant open set $\tilde{U}_\alpha \subset \mathbf{R}\times\mathbf{C}$. A *flat orbifold vector bundle over* B with fibre a vector space V can be defined with reference to such an atlas as having local trivializations of the form $U_\alpha \times V$ over the charts which do not meet L and $(\tilde{U}_\alpha \times V, \tau_\alpha)$ over those which do, where τ_α is an involution of V. These trivializations satisfy obvious compatibility conditions on the overlaps of charts. In our situation the cohomology of the fibres provides such a flat orbifold vector bundle $\mathcal{H}$ over B with the involutions τ_α given by the reflections in the vanishing cycles. We also want to consider *flat orbifold affine bundles*. These are defined in just the same way except that the τ_α are replaced by affine involutions of V.

Let $\mathcal{V}$ be any flat orbifold vector bundle over B. We can define a sheaf $\underline{\mathcal{V}}$ of locally constant sections of $\mathcal{V}$. In a chart U_α which intersect L such a section is, by definition, given by a τ_α-invariant element of V. Any flat orbifold affine bundle determines a flat orbifold vector bundle via the natural map from the affine group of $\mathbf{R}^n$ to $GL(n, \mathbf{R})$. We will say that the affine bundle is a *lift* of the vector bundle. Extending the standard theory for flat bundles recalled in Section 2 above; if $\mathcal{V}$ is a flat orbifold vector bundle over B then there is a natural 1-1 correspondence between equivalence classes of flat orbifold affine bundles lifting $\mathcal{V}$ and elements of the cohomology group $H^1(B; \underline{\mathcal{V}})$.

Write $\mathcal{V}_\chi$ for the affine bundle determined by a flat orbifold vector bundle $\mathcal{V}$ and a class $\chi \in H^1(\underline{\mathcal{V}})$. We define the sheaf of smooth sections of $\mathcal{V}_\chi$ in the standard fashion; given over a chart $(\tilde{U}_\alpha, \tau_\alpha)$ by smooth maps $f_\alpha : \tilde{U}_\alpha \to V$ satisfying the equivariance condition

$$f_\alpha(\sigma(x)) = \tau_\alpha(f_\alpha(x)).$$

We will want to consider *orthogonal* vector and affine bundles. That is, we suppose that the model vector space V has a nondegenerate quadratic form of signature (p, q) and all the bundle data, and in particular each involution τ_α, is compatible with this. For our purposes we restrict to the case when $p = 3$. Away from L our bundle $\mathcal{V}_\chi$ is an ordinary flat affine bundle so we are in the situation considered before and we have a notion of a *positive section* – i.e., the image of the derivative is a maximal positive subspace of the fibre. We want to extend this notion over L. To discuss this we will now fix attention on the case when each τ_α is a reflection in a negative vector. That is we assume that

$$\tau_\alpha(v) = v + (\delta_\alpha, v)\delta_\alpha + \lambda_\alpha \delta_\alpha,$$

where $\delta_\alpha \in V$ with $(\delta_\alpha, \delta_\alpha) = -2$ and $\lambda_\alpha \in \mathbf{R}$. Let $f_\alpha : \tilde{U}_\alpha \to V$ be an equivariant map, where $\tilde{U}_\alpha$ is some neighbourhood of the origin in $\mathbf{R} \times \mathbf{C}$. Without real loss of generality, and to simplify notation, we can suppose that $\lambda_\alpha = 0$ and that $f_\alpha(0) = 0$. We write $Df_\alpha, D^2 f_\alpha$ for the first and second derivatives at 0.

Definition 2. We say that the f_α is a *branched positive map* at $(0,0)$ if

1. Df_α vanishes on $\mathbf{C}$, so the image of DF_α is spanned by a single vector $v_0 \in V$.
2. The restriction of $D^2 f_\alpha$ to $\mathbf{C}$ can be written as $v_1 \mathrm{Re} z^2 + v_2 \mathrm{Im} z^2$ where $v_1, v_2 \in V$.
3. The vectors v_0, v_1, v_2 span a maximal positive subspace in $\delta^\perp \subset V$.

In terms of local co-ordinates (w,t) on B, where $w = z^2$, this can be expressed by saying that we have multi-valued function

$$(w,t) \mapsto v_0 t + v_1 \mathrm{Re} w + v_2 \mathrm{Im} w + O(w^{3/2}, t^2),$$

with an ambiguity of sign in the component along $\delta \in V$.

With this definition in place we can define a *positive section* h of our flat affine orbifold bundle $\mathcal{V}_\chi$ to be a section which is positive away from L and represented by a branched positive map at points of L. We can define the 3-volume $\mathrm{Vol}_3(h)$ of the section in the obvious way. Finally we can define a *maximal positive section* to be a positive section which is maximal in the previous sense away from L. These definitions imply that near points of L such a section yields a branched solution of the maximal submanifold equation, in the usual sense of the literature.

3.1. Proposed adiabatic limit problem

Return now to a KL fibration $\pi : M \to B$, the flat orbifold vector bundle $\mathcal{H}$ over B and the corresponding sheaf $\underline{\mathcal{H}}$. It is straightforward to check that $\underline{\mathcal{H}}$ can be identified with the second derived direct image of the constant sheaf $\mathbf{R}$ on X and the first derived direct image is zero. So the Leray spectral sequence gives an exact sequence

$$0 \to H^3(B;\mathbf{R}) \to H^3(M;\mathbf{R}) \xrightarrow{b} H^1(B;\underline{\mathcal{H}}) \to 0.$$

Proposition 5. *Suppose that ϕ_0 is a closed positive 3-form on M with respect to which $\pi : M \to B$ is a co-associative KL fibration and set $\chi = b([\phi_0]) \in H^1((B;\underline{\mathcal{H}})$. Then ϕ_0 determines a branched positive section of $\mathcal{H}_\chi$.*

The proof is not hard but we will not go into it here.

We move on to the central point of this article. We say that a section h of a flat affine bundle $\mathcal{H}_\chi$ *avoids excess -2 classes* if the following two conditions hold.

- The restriction to $B \setminus L$ avoids -2 classes, in the sense defined before.
- At a point of L, and using the representation as in Definition 2 above, the only elements of $\mathcal{C}$ orthogonal to v_0, v_1, v_2 are $\pm\delta_\alpha$.

Write $[B] \in H^3(X)$ for the pull-back of the fundamental class of B. We make a two-part conjecture.

Conjecture 1. *Fix* $[\phi_0] \in H^3(M)$ *and* $\chi = b([\phi_0])$.

1. *If there is a positive section* h *of* $\mathcal{H}_\chi$ *which avoids excess* -2 *classes, then for sufficiently large* R *there is a closed positive* 3*-form on* M *in the class* $[\phi_0] + R[B]$ *with respect to which* $\pi : M \to B$ *has co-associative fibres.*
2. *If, in addition, the section* h *can be chosen to be maximal then, for sufficiently large* R*, we can choose the positive* 3*-form to define a torsion free* G_2*-structure.*

These are precise conjectures which express part of a more general – but more vague – idea that for large R the whole 7-dimensional theory should be expressed in terms of positive sections of $\mathcal{H}_\chi$. For example we would hope that there be should be converse statements, going from the existence of the structures on M, for large R, to sections of $\mathcal{H}_\chi$.

We can extend the discussion to Bryant's Laplacian flow for closed positive 3-forms

$$\frac{\partial \phi}{\partial t} = dd^* \phi.$$

The discussion in Section 2 suggests that the adiabatic limit should be the mean curvature flow for positive sections

$$\frac{\partial h}{\partial t} = m(h), \tag{22}$$

just as Hitchin's volume functional relates to the volume functional $\mathrm{Vol}_3 h$.

Proving precise statements such as in Conjecture 1 will clearly involve substantial analysis. But we can leave that aside and formulate questions, within the framework of sections of flat orbifold bundles, which can be studied independently. Here we can take any orthogonal flat affine orbifold bundle $\mathcal{V}_\chi$ with orbifold structure defined by reflections, as above.

1. Does $\mathcal{V}_\chi$ admit positive sections?
2. If so, does it admit a maximal positive section?
3. If $\mathcal{V}_\chi$ admits positive sections, is the volume functional Vol_3 bounded above?
4. Does the mean curvature flow (22) exist and converge to a maximal positive section?

Naturally we hope that these questions should have some relation with the corresponding questions for positive 3-forms at the beginning of this article.

All of these questions should be viewed as tentative, provisional, formulations. We include two remarks in that direction.

- The condition of "avoiding excess -2 classes" seems very unnatural from the point of view of sections of the flat orbifold bundle $\mathcal{H}_\chi$. One could envisage that there could be cases where there are positive sections of $\mathcal{H}_\chi$ which avoid excess -2 classes but where there is a maximal positive section which does not satisfy this condition. This might correspond to a torsion-free G_2 structure on a singular space obtained by collapsing subsets of M.

- In the definition of a Kovalev–Lefschetz fibration we required that there be at most one singular point in each fibre and the critical values form a union of disjoint circles. This should be the "generic" situation, but one could envisage that in generic 1-parameter families one should allow two critical points in a fibre. In that vein one could imagine that the long-time definition of the mean curvature flow (22) would require modifying the link L by allowing two components to cross, under suitable conditions on the monodromy of the flat bundle. Thus the right notion would not be a single KL fibration but an equivalence class generated by these operations.

4. More adiabatic analogues

4.1. Negative definite cup product

Let M be any compact 7-manifold with a torsion-free G_2 structure defined by a form ϕ. Suppose also that the holonomy is the full group G_2, which implies that $H^1(M;\mathbf{R}) = 0$ [14]. Then it is known via Hodge Theory that the quadratic form on $H^2(M;\mathbf{R})$ defined by

$$Q_{[\phi]}(\alpha) = \langle \alpha \cup \alpha \cup [\phi], [M] \rangle,$$

is negative definite [14]. Now suppose that M has a C^∞ Kovalev–Lefschetz fibration $\pi : M \to B$ with $K3$ fibres. The vanishing of $H^1(M)$ requires that $H^1(B;\mathbf{R}) = 0$. As in the previous section, let $[\phi_0]$ be any fixed class in $H^3(M)$ and $\phi_R = [\phi_0] + R[B]$. Then the limit of $R^{-1}Q_{[\phi_R]}$ as R tends to infinity is given by restricting to a smooth fibre X and taking the cup product on the fibre. Poincaré duality on B gives $H^2(B;\mathbf{R}) = 0$ and then the Leray spectral sequence shows that $H^2(M;\mathbf{R}) = H^0(\underline{\mathcal{H}})$, which can be identified with the subspace of $H^2(X)$ preserved by the monodromy of the fibration.

Now we have a related result in the context of positive sections.

Proposition 6. *Let $\mathcal{V}$ be a flat orbifold bundle over a compact base B and $\mathcal{V}_\chi$ be a lift to an affine orbifold bundle. Suppose that $\mathcal{V}_\chi$ admits a positive section and that $H^1(B,\mathbf{R}) = 0$. If σ is a non-trivial global section of $\underline{\mathcal{V}}$ then $\sigma.\sigma < 0$ at each point of B.*

We can write $\mathcal{V} = \mathbf{R}\,\sigma \oplus \mathcal{V}'$. The condition that $H^1(B;\mathbf{R}) = 0$ implies that χ is trivial in the $\mathbf{R}\,\sigma$ component, so if h is the positive section of $\mathcal{V}_\chi$ then there is a well-defined continuous function $\sigma.s$ on B which attains a maximum at some point $p \in B$. Consider first the case when p is not in L. The maximality implies that $\sigma(p)$ is orthogonal to the image of dh at p. By the definition of a positive section, this image is a maximal positive subspace which means that $\sigma(p).\sigma(p) < 0$. The definition of a positive section at points of L means that the same argument works when $p \in L$.

4.2. Associative submanifolds

Recall that a 3-dimensional submanifold $P \subset M$ in a 7-manifold with a positive 3-form ϕ is called *associative* if the restriction of ϕ to the orthogonal complement of TP in TM vanishes at each point of P. This implies that, with the right choice of orientation, the restriction of ϕ to TP is the volume form. These associative submanifolds are further examples of Harvey and Lawson's calibrated submanifolds [10], with calibrating form *phi* and there is an elliptic deformation theory [17]. They are interesting geometrical objects to investigate in G_2-manifolds: in particular there is a potential connection with moduli problems, because if a non-zero homology class $\pi \in H_3(M)$ can be represented by a compact associative submanifold then we have $\langle [\phi], \pi \rangle > 0$.

Let X be a K3 surface with a hyperkähler triple $\omega_1, \omega_2, \omega_3$. The cohomology classes $[\omega_i]$ span a positive subspace $H^+ \subset H^2(X;\mathbf{R})$. Each non-zero vector $\tau = (\tau_1, \tau_2, \tau_3)$ defines a complex structure I_τ on X and a symplectic form $\omega_\tau = \sum \tau_i \omega_i$. Write $H^\perp_\tau$ for the orthogonal complement of $[\omega_\tau]$ in H^+. If Σ is a smooth complex curve in X, for the complex structure I_τ, then the product of Σ and the line $R\tau$ is an associative submanifold in $X \times \mathbf{R}^3$ with the positive 3-form $\sum \omega_i dt_i - dt_1 dt_3 dt_3$. In particular, suppose that $c \in H^2(X;\mathbf{Z})$ is a class with $c^2 = -2$ and c is orthogonal $H^\perp_\tau$; then standard theory tells us that there is a unique such curve Σ in the class c, and Σ is a 2-sphere. This leads to a natural candidate for an "adiabatic description" of certain associative submanifolds in a Kovalev–Lefschetz fibration. We begin with the local picture, away from the critical set $L \subset B$. Suppose that we have a positive section given by a map h into $H^2(X;\mathbf{R})$ and fix a class c as above. Let $\gamma(s)$ be an embedded path in B so $h \circ \gamma$ is a path in $H^2(X;\mathbf{R})$ and for each s the derivative of $h \circ \gamma$ defines a complex structure on X and a 2-dimensional subspace $H^\perp(s) \subset \mathrm{Im}\, dh \subset H^2(X;\mathbf{R})$. If c is orthogonal to $H^\perp(s)$ for each s then we have a complex 2-sphere in the fibre over $\gamma(s)$ and the union of these 2-spheres, as s varies, is our candidate for an adiabatic limit of associative submanifolds. The condition on the derivative of $h \circ \gamma$ has a simple geometric meaning. Up to parametrization of γ, it is just the condition that γ be a *gradient path* for the function $\underline{t} \mapsto c.h(\underline{t})$ with respect to the metric on B induced by h.

To bring in the critical set $L \subset B$, recall that if $p \in B \setminus L$ is a point close to L there is a vanishing cycle Σ in the fibre X_p over p which is a 2-sphere bounding a "Lefschetz thimble" in M, and the homology class of the vanishing cycle is a class c as above. Let $\gamma : [0,1] \to B$ be a path with $\gamma(0)$ and $\gamma(1)$ in L and $\gamma(s) \in B \setminus L$ for $s \in (0,1)$. For small $\epsilon > 0$ there are vanishing cycles (defined up to isotopy) in the fibres over $\gamma(\epsilon), \gamma(1-\epsilon)$ and we say that γ is a C^∞ *matching path* if these agree under parallel transport along γ. In this case we can construct an embedded 3-sphere in M by capping off the 2-sphere fibration over $(\epsilon, 1-\epsilon)$ with the two Lefschetz thimbles. We say that γ is a *homological matching path* if the homology classes of the two vanishing cycles agree, a condition which is determined by the flat orbifold bundle $\mathcal{H}$. For the present discussion we will not distinguish between the homological and C^∞ conditions. Now we say that γ is a *matching gradient path* if it is a homological matching path and if over the interior $(0,1)$ it is a gradient path of the function $h.c$ (interpreted using parallel transport of c along γ). These are plausible candidates for the adiabatic limits of certain associative spheres. More precisely, one can hope that, provided γ is a "nondegenerate" solution of the matching gradient path condition in a suitable sense, then there will be an associative 3-sphere in the G_2 manifold for large values of R. On the other hand, the matching gradient paths are described entirely in terms of $\mathcal{H}_\chi$. The definition extends immediately to any flat orbifold bundle and they are objects which can be studied independent of any connection with G_2 geometry. This proposed description of associative submanifolds is in the same vein as constructions of Lagrangian spheres in symplectic topology [1] and of tropical descriptions of holomorphic curves in Calabi–Yau manifolds [7].

5. Variants

5.1. Torus fibres, special Lagrangian fibrations

The local discussion of co-associative G_2-fibrations in Section 2 applies with little change when the fibres are 4-tori rather than $K3$ surface. The "adiabatic limit data" is for a maximal submanifold in $\mathbf{R}^{3,3} = H^2(T^4)$. The difference in this case is that such a maximal submanifold yields a genuine exact solutions of the equations for non-zero ϵ. This was the case treated by Baraglia in [2]. The source of this difference is that the curvature of the connection H vanishes in this situation – the horizontal subspaces are tangent to sections of the fibration.

Suppose now that $M = S^1 \times N$ for a 6-manifold N, and consider G_2-structures on M induced from $SU(3)$ structures on N. Then we have

$$\phi = \rho + \Omega d\theta,$$

where $d\theta$ is the standard 1-form on S^1 and ρ, Ω are respectively a 3-form and a 2-form on N. Here we consider the case of torus fibrations $M \to B$ induced from *special Lagrangian fibrations* $\pi_N : N \to B$ by taking a product of the fibres with a circle. Now

$$H^2(T^3 \times S^1) = H^2(T^3) \oplus H^1(T^3)$$

and the quadratic form on $H^2(T^3 \times S^1)$ is induced from the dual pairing between $H^1(T^3), H^2(T^3)$. In other words the reduction to the product situation yields a natural splitting

$$\mathbf{R}^{3,3} = \mathbf{R}^3 \oplus \left(\mathbf{R}^3\right)^*.$$

The dual pairing also defines a standard symplectic form on $\mathbf{R}^{3,3}$ and the product structure implies that the image of the map $h : B \to \mathbf{R}^{3,3}$ is Lagrangian. It follows that this the image is the graph of the derivative of a function F on $\mathbf{R}^3$. The basic fact, observed by Hitchin [12], is that a Lagrangian submanifold of $\mathbf{R}^{3,3}$ defined as the graph of the derivative of a function F is a maximal submanifold if and only if F satisfies the Monge–Ampère equation

$$\det\left(\frac{\partial^2 F}{\partial x_i \partial x_j}\right) = C$$

for a constant C. This makes a bridge between the co-associative discussion and the extensive literature on special Lagrangian fibrations and "large complex structure" limits for Calabi–Yau 3-folds.

5.2. K3 fibrations of Calabi–Yau 3-folds

Suppose again that $M = S^1 \times N$ but now that $B = S^1 \times S^2$ and consider fibrations which are products with S^1 of $\varpi_N : N \to S^2$ with $K3$ fibres. More precisely, we suppose that ϖ_N is a Lefschetz fibration in the usual sense, with a finite set of critical values in S^2. The class $[\omega]$ gives a fixed class in the cohomology of the fibres so now we form a flat orbifold vector bundle $\mathcal{H}$ with fibre $R^{2,19}$. The discussion of positive sections goes over immediately to this 2-dimensional case. We could hope that this has a bearing on the question: when can a differentiable Lefchetz fibration with $K3$ fibres be given the structure of a holomorphic fibration of a Calabi–Yau 3-fold?

We can again make a bridge between this point of view and other, more standard, ones. Recall that the Grassmann manifold Gr^+ of positive oriented 2-planes in $\mathbf{R}^{2,19}$ can be identified with an open subset of a quadric $Q \subset \mathbf{CP}^{20}$. This is achieved by mapping the plane spanned by the orthonormal frame e_1, e_2 to the complex line $[e_1 + ie_2] \in \mathbf{CP}^{20}$. Let $\Sigma \subset \mathbf{R}^{2,19}$ be a surface with positive tangent space at each point. Then we have a Gauss map from Σ to Gr^+. Just as in the Euclidean case, the maximal condition implies that this is holomorphic with respect to the induced complex structure on Σ; hence the image is a complex curve. Conversely, start with a complex curve in $\mathrm{Gr}^+ \subset \mathbf{CP}^{20}$, lift it locally to $\mathbf{C}^{21}$ and parametrize it by a complex variable $z \in U \subset \mathbf{C}$. So we have holomorphic functions $f_0(z), \dots, f_{20}(z)$ with $Q(f_0, \dots, f_{20}) = 0$. Let H_i be the integrals, solving $\frac{dH_i}{dz} = f_i$. Then the map from U to $\mathbf{R}^{2,19}$ with components $\mathrm{Re} H_i$ has image a maximal surface in $\mathbf{R}^{2,19}$. This is the usual Weierstrass construction. In our context, the complex curve in Gr^+ amounts to a holomorphic description of a family of polarized $K3$ surfaces in terms of the Hodge data $H^{2,0} \subset H^2(X, \mathbf{C})$ and the Weierstrass construction makes the bridge with the maximal submanifold equation.

5.3. Cayley fibrations

Here we consider the local adiabatic problem for a fibration of an 8-dimensional manifold with holonomy Spin(7) by Cayley submanifolds. Recall that a torsion-free Spin(7) structure can be defined by a closed 4-form which is algebraically special at each point [14]. As the basic model we take two oriented 4-dimensional Euclidean spaces V_1, V_2 and a fixed isomorphism of the respective self-dual forms $\Lambda^2_+(V_i^*)$. Then we have a 4-form

$$\Omega_0 = d\mathrm{vol}_1 + \sum \omega_a \wedge \omega_a' + d\mathrm{vol}_2$$

on $V_1 \oplus V_2$, where $dvol_i$ are the volume forms in the two factors, ω_a runs over a standard orthonormal basis of $\Lambda^2_+(V_1^*)$ and ω'_a the corresponding basis for $\Lambda^2_+(V_2^*)$. A closed 4-form Ω on an 8-manifold M which is algebraically equivalent to Ω_0 at each point defines a torsion free Spin(7) structure. A 4-dimensional submanifold $X \subset M$ is a Cayley submanifold if at each point of X the triple $(TM_x, \Omega(x), TX_x)$ is algebraically equivalent to the model triple $(V_1 \oplus V_2, \Omega_0, V_1)$. These are, again, examples of calibrated submanifolds [10] and there is an elliptic deformation theory [17].

A fibration $\pi : M \to B$ with Cayley fibres can be expressed in the manner of Section 2. There is a connection H which furnishes a decomposition of the forms on M into (p,q) components, and a closed 4-form Ω with three components

$$\Omega = \Omega_{4,0} + \Omega_{2,2} + \Omega_{0,4}$$

The algebraic data is that at each point x of M we have oriented Euclidean structures on the horizontal and vertical subspaces, and an isomorphism between the spaces of self-dual forms, as above. The condition that Ω is closed can be viewed a set of coupled equations for the connection and the three components. A discussion parallel to that in the 7-dimensional case suggests that in the adiabatic limit the equations should de-couple, forcing the structure on the each fibre to be hyperkähler. This analysis leads us to formulate the following definition.

Definition 3. Let $\mathcal{H}$ be a real vector space with a quadratic form of signature $(3,q)$. *Adiabatic Cayley data* for $\mathcal{H}$ consists of an oriented Riemannian 4-manifold (B,g) and a closed $\mathcal{H}$-valued self-dual 2-form Ψ on B such that $\Psi^*\Psi$ is the identity endomorphism of Λ^2_+.

To explain the notation here, at each point $b \in B$, the value $\Psi(b)$ is regarded as a linear map from the fibre of Λ^2_+ to $\mathcal{H}$ (using the metric on Λ^2_+). Then $\Psi^*(b)$ is the adjoint, defined using the quadratic forms on each space, so the composite $\Psi^*\Psi$ is an endomorphism of Λ^2_+. For our immediate application we take $\mathcal{H}$ to be the 2-dimensional cohomology of a 4-torus or a $K3$ surface X. Given adiabatic Cayley data as above, the image of $\Psi(b)$ at each point b is a maximal positive subspace of $\mathcal{H}$ so defines a hyperkähler structure on X. (More precisely this is defined up to diffeomorphism, but that is not important for the local discussion here.) If $\omega_i, \omega_2, \omega_3$ is a standard orthonormal frame for $\Lambda^2_+(b)$ then we have a corresponding hyperkähler triple ω'_i of 2-forms on the fibre $\pi^{-1}(b)$. That is,

$$[\omega'_i] = \Psi(b)(\omega_i) \in H^2(\pi^{-1}(b)).$$

The sum

$$\Omega_{2,2} = \sum \omega_a \wedge \omega'_a$$

is independent of the choice of orthonormal frame and

$$d_f \Omega_{2,2} = 0.$$

Then is then a unique way to choose a connection H which is compatible with the volume form on the fibres and so that

$$d_H\Omega_{2,2} = 0.$$

We take $\Omega_{0,4}$ to be extension of the volume form on the fibres by this connection and $\Omega_{4,0}$ to be the lift of the volume form on the base. In this way we define a 4-form Ω on the total space of he correct algebraic type and with

$$d\Omega = F_H\Omega_{0,4} + F_H\Omega_{2,2},$$

terms which, as we argued before, should be suppressed in the adiabatic limit. (When the fibre is a 4-torus then, just as in Baraglia's case discussed above, the curvature F_H is zero, so we can write down genuine torsion-free Spin(7)-structures, given adiabatic Cayley data.)

One remark about Definition 3 is that the metric g is entirely determined by the $\mathcal{H}$-valued form Ψ. In other words suppose we are given an oriented 4-manifold B and an $\mathcal{H}$-valued 2-form Ψ on B. Then we can define $\Psi^*\Psi$, up to ambiguity by a positive scalar, using the wedge product form on 2-forms. Let us say that Ψ is *special* if at each point $\Psi^*\Psi$ is a projection onto a 3-dimensional positive subspace in Λ^2 (up to a factor). This subspace defines a conformal structure on B – the unique conformal structure so that this subspace is self-dual – and we fix a metric by by normalizing so that $\Psi^*\Psi$ is the identity on this subspace. In sum, an equivalent description of adiabatic Cayley data is a closed, special, $\mathcal{H}$-valued 2-form.

5.4. Ricci curvature

We recall first a standard result.

Proposition 7. *Let $V \subset \mathbf{R}^{p,q}$ be a p-dimensional maximal positive submanifold. Then the Ricci curvature of the induced Riemannian metric on V is non-negative.*

We review the proof. By elementary submanifold theory, the Riemann curvature tensor of V is expressed in terms of the second fundamental form S. Fix a point in V and orthonormal frames for the tangent and normal bundles so S has components S^a_{ij} where i, j label the tangent frame and a the normal frame. Then the Riemann curvature tensor is

$$R_{ijkl} = \sum_a S^a_{ik}S^a_{jl} - S^a_{jk}S^a_{il},$$

where we have an opposite sign to the familiar Euclidean case because the quadratic form is negative in the normal direction. This gives

$$R_{ik} = -\sum_{a,j}(S^a_{jj})S^a_{ik} + \sum_{a,j} S^a_{ij}S^a_{jk}.$$

The term $\sum_j S^a_{jj}$ is the mean curvature, which vanishes by assumption, and S is symmetric so we have

$$R_{ik} = \sum_{a,j} S^a_{ij}S^a_{kj},$$

which is non-negative. More precisely, the Ricci curvature in a tangent direction ξ to V is the square of the norm of the derivative of the Gauss map along ξ.

The relevance of this for us is an immediate consequence.

Corollary 1. *Suppose that $\mathcal{H}_\chi$ is a flat affine orbifold bundle over (B, L) which admits a maximal positive section. Then the induced Riemannian metric on $B \setminus L$ has non-negative Ricci curvature.*

We pass on to the higher-dimensional case, where the result is less standard.

Proposition 8. *Suppose that (B, g, Ψ) is a set of adiabatic Cayley data. Then the Ricci curvature of g is non-negative.*

We can regard Ψ as giving an isometric embedding of the bundle Λ^2_+ over B into the trivial bundle with fibre $\mathcal{H}$. There is thus a "second fundamental form" S of this subbundle, or equivalently the derivative of the map from B to the Grassmannian. Let $\mathcal{H}^\perp$ be the complementary bundle in $\mathcal{H}$. Then S is a tensor in $\Lambda^2_+ \otimes \Lambda^1 \otimes \mathcal{H}^\perp$. A little thought shows that the condition that $d\Psi = 0$ is equivalent to

- The connection on Λ^2_+ induced from the Levi-Civita connection is the same as the connection induced from the embedding $\Lambda^2_+ \subset \mathcal{H}$;
- The image of S in $\Lambda^3 \otimes \mathcal{H}^\perp$ induced the wedge product $\Lambda^1 \otimes \Lambda^2_+ \to \Lambda^3$ is zero.

The first item means that the Riemannian curvature of Λ^2_+ can be computed from S. Indeed it is given by $q_{\mathcal{H}^\perp}(S)$ where $q_{\mathcal{H}^\perp}$ is the quadratic map from $\Lambda^+ \otimes \Lambda^1 \otimes \mathcal{H}^\perp$ to $\Lambda^2 \otimes \Lambda^+$ by the tensor product of

- The skew-symmetric wedge product $\Lambda^1 \otimes \Lambda^1 \to \Lambda^2$;
- The skew-symmetric cross-product $\Lambda^2_+ \otimes \Lambda^2_+ \to \Lambda^2_+$;
- The symmetric inner product $\mathcal{H}^\perp \otimes \mathcal{H}^\perp \to \mathbf{R}$.

Of course to pin down signs it is crucial that the last form is negative definite, but otherwise it will not enter the picture. We define a quadratic map q from $\Lambda^+ \otimes \Lambda^1$ to $\Lambda^2 \otimes \Lambda^+$ by using the first two components above.

Next we want to recall how to compute the Ricci curvature of an oriented Riemannian 4-manifold (B, g) from the curvature tensor F of Λ^2_+. This curvature tensor F lies in $(\Lambda_+ \oplus \Lambda_-) \otimes \Lambda_+$ so has two components, say

$$F_+ \in \Lambda_+ \otimes \Lambda_+ \quad , \quad F_- \in \Lambda_- \otimes \Lambda_+ .$$

We work at a fixed point $b \in B$ and take a unit tangent vector e_0. This choice of e_0 induces an isomorphism between the self-dual and anti-self-dual forms at b. To fix signs, say that this isomorphism takes the self-dual form $dx_0 dx_1 + dx_2 dx_3$ to the anti-self-dual form $-dx_0 dx_1 + dx_2 dx_3$. Using this isomorphism we can define a trace

$$\mathrm{Tr}_{e_0} : \Lambda - \otimes \Lambda_+ \to \mathbf{R},$$

depending on e_0. Of course we also have an invariant trace:

$$\mathrm{Tr} : \Lambda_+ \otimes \Lambda_+ \to \mathbf{R}.$$

We define $T_{e_0} : \Lambda^2 \otimes \Lambda^+ \to \mathbf{R}$ to be the sum $\mathrm{Tr} + \mathrm{Tr}_{e_0}$ on the two factors. The formula we need is that

$$\mathrm{Ric}(e_0) = T_{e_0}(F).$$

This is easy to check, using the well-known decomposition of the curvature tensor of a 4-manifold, or otherwise. So in our situation

$$\mathrm{Ric}(e_0) = T_{e_0}(q_{\mathcal{H}^\perp}(S)).$$

So we need to show that for any $s \in \Lambda^+ \otimes \Lambda^1$ which is in the kernel of the wedge product $\Lambda^+ \otimes \Lambda^1 \to \Lambda^3$ we have $T_{e_0}(q(s)) \leq 0$. Calculate in an orthonormal frame e_i extending e_0. This induces a standard orthonormal frame ω_i for Λ_+. We can write $s = \sum_{i=1}^3 v_i \otimes \omega_i$, where $v_i = t_i e_0 + \sum_{i=1}^3 s_{ij} e_j$. One finds that the condition that s is in the kernel of the wedge product is that the the trace $\sum s_{ii}$ is zero and that the t_i are determined by the skew symmetric part of s_{ij}:

$$t_i = s_{jk} - s_{kj}$$

for i, j, k cyclic. On the other hand one finds that for any s in $\Lambda^1 \otimes \Lambda^+$

$$T_{e_0}(q(s)) = \sum_{\text{cyclic}} t_i (s_{jk} - s_{kj}).$$

So for s in the kernel of the wedge product this becomes $-\sum t_i^2$ which has the desired sign.

In fact this calculation shows that again $\mathrm{Ric}(e_0)$ is the square of the norm of the derivative of the map to the Grassmannian of 3-planes in $\mathcal{H}$, in the direction e_0.

References

[1] Auroux, D., Muñoz, V. and Presas, F. *Lagrangian submanifolds and Lefschetz pencils*, Jour. Symplectic Geom. 3 (2005) no. 2 171–219

[2] Baraglia, D. *Moduli of co-associative submanifolds and semi-flat G_2 structures*, Jour. Geometry and Physics 60 (2010) 1903–1918

[3] Barth, W., Peters, C. and van de Ven, A. *Compact complex surfaces*, Springer 1984

[4] Bryant, R. *Some remarks on G_2-structures*, Proc. Gokova Geometry and Topology conference (2005) 75–109

[5] Fine, J. *Constant scalar curvature Kähler metrics on fibred complex surfaces*, Jour. Differential Geometry 68 (2004) no. 3 397–432

[6] Corti, A., Haskins, M., Nordström, J., and Pacini, T. *G_2-manifolds and associative submanifolds via semi-Fano 3-folds*, Duke Math. J. 164 (2015) no. 10 1971–2092

[7] Gross, M. *Tropical geometry and mirror symmetry*, CBMS Regional Conference Series in Mathematics AMS 2011

[8] Gross, M. and Wilson, P. *Large complex structure limits of* K3 *surfaces*, Jour. Differential Geometry 55 (2000) no. 3 475–546

[9] Gukov, S. Yau, S.-T. and Zaslow, E. *Duality and fibrations on G_2 manifolds*, Turkish J. Math. 27 (2003) 61–97

[10] Harvey, R. and Lawson, H.B. *Calibrated geometries*, Acta Math. 148 (1982) 47–157

[11] Hitchin, N. *Stable forms and geometric structures*, in: Global Differential Geometry: the mathematical legacy of Alfred Gray, Contemporary Mathematics 288 70–89 Amer. Math. Soc. 2000

[12] Hitchin, N. *On the moduli space of special Lagrangian submanifolds*, Ann. Scuola Normale Sup. Pisa Cl Sci. 24 no. 4 (1998) 503–515

[13] Joyce, D.D. *Compact Riemannian 7-manifolds with holonomy G_2 I, II*, Jour. Differential Geometry 43 (1996) 291–375

[14] Joyce, D.D. *Compact manifolds with special holonomy*, Oxford U.P. 2000

[15] Kovalev, A. *Twisted connected sums and special Riemannian holonomy*, J. Reine Angew. Math. 565 (2003) 126–160

[16] Kovalev, A. *Co-associative $K3$ fibrations of compact G_2 manifolds*, arxiv 0511150

[17] McLean, R.C. *Deformations of calibrated submanifolds*, Commun. Analysis and Geometry 6 (1998) 705–747

[18] Tosatti, V. *Adiabatic limits of Ricci-flat Kähler metrics*, Jour. Differential Geometry 84 (2010) no. 2 427–453

Simon Donaldson
Simons Centre for Geometry and Physics
Stony Brook, NY 11794, USA
e-mail: `s.donaldson@imperial.ac.uk`

Progress in Mathematics, Vol. 324, 31–97

Ideal Webs, Moduli Spaces of Local Systems, and 3d Calabi–Yau Categories

A.B. Goncharov

Dedicated to Maxim Kontsevich, for his 50th birthday

Abstract. A *decorated surface* S is an oriented surface with punctures, and a finite set of marked points on the boundary, considered modulo isotopy. We assume that each boundary component has a marked point. We introduce *ideal bipartite graphs* on S. Each of them is related to a group G of type A_m or GL_m, and gives rise to cluster coordinate systems on certain moduli spaces of G-local systems on S. These coordinate systems generalize the ones assigned in [FG1] to ideal triangulations of S.

A bipartite graph W on S gives rise to a quiver with a canonical potential. The latter determines a triangulated 3d Calabi–Yau A_∞-category $\mathcal{C}_W$ with a *cluster collection* $\mathcal{S}_W$ – a generating collection of spherical objects of special kind [KS1].

Let W be an ideal bipartite graph on S of type G. We define an extension $\Gamma_{G,S}$ of the mapping class group of S, and prove that it acts by symmetries of the category $\mathcal{C}_W$.

There is a family of open CY threefolds over the *universal Hitchin base* $\mathcal{B}_{G,S}$, whose intermediate Jacobians describe Hitchin's integrable system [DDDHP], [DDP], [G], [KS3], [Sm]. We conjecture that the 3d CY category with cluster collection $(\mathcal{C}_W, \mathcal{S}_W)$ is equivalent to a full subcategory of the Fukaya category of a generic threefold of the family, equipped with a cluster collection of special Lagrangian spheres. For $G = SL_2$ a substantial part of the story is already known thanks to Bridgeland, Keller, Labardini-Fragoso, Nagao, Smith, and others, see [BrS], [Sm].

We hope that ideal bipartite graphs provide special examples of the Gaiotto–Moore–Neitzke spectral networks [GMN4].

Mathematics Subject Classification (2010). 14M99.

Keywords. Calabi–Yau categories, cluster coordinates, bipartite graphs, Fukaya categories.

1. Introduction and main constructions and results

1.1. Ideal webs and cluster coordinates on moduli spaces of local systems

Recall that a *decorated surface* S is an oriented topological surface with *punctures* and *marked points* on the boundary, considered modulo isotopy. The set of *special points* is the union of the set of punctures and marked points. We assume that each boundary component has a special point, the number n of special points is > 0, and that S is hyperbolic: if the genus $g(S) = 0$, then $n \geq 3$. Denote by Γ_S the mapping class group of S.

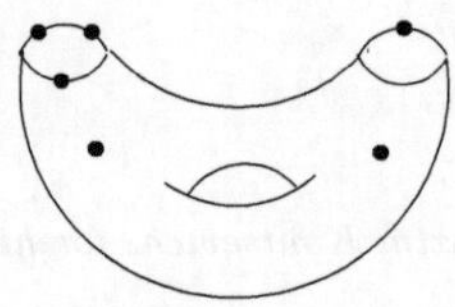

FIGURE 1. A decorated surface: punctures inside, special points on the boundary.

There is a pair of moduli spaces assigned to S, closely related to the moduli space of m-dimensional vector bundles with flat connections on S, defined in [FG1]:

$$\text{a } K_2\text{-moduli space } \mathcal{A}_{SL_m,S} \text{ and a Poisson moduli space } \mathcal{X}_{PGL_m,S}. \qquad (1)$$

As the names suggest, the space $\mathcal{A}_{SL_m,S}$ is equipped with a Γ_S-invariant class in K_2, and the space $\mathcal{X}_{PGL_m,S}$ is equipped with a Γ_S-invariant Poisson structure.[1]

An *ideal triangulation* of S is a triangulation with vertices at the special points. In [FG1] each of the spaces (1) was equipped with a collection of rational coordinate systems, called *cluster coordinate systems*, assigned to ideal triangulations of S. In particular this implies that these spaces are rational.[2] The ring of regular functions on the space $\mathcal{A}_{SL_m,S}$ is closely related to (and often coincides with) a cluster algebra of Fomin–Zelevinsky [FZI].

One of the crucial features of the cluster structure is that a single cluster coordinate system on a given space determines the cluster structure, and gives rise to a huge, usually infinite, collection of cluster coordinate systems, obtained by arbitrary sequences of *mutations.*

When G is of type A_1, the coordinate systems related to ideal triangulations provide most of the cluster coordinate systems. Any two ideal triangulations are related by a sequence of flips. A flip of an ideal triangulation gives rise to a cluster mutation of the corresponding cluster coordinate system. Therefore any two of them are related by a sequence of mutations.

When G is of type A_m, $m > 1$, the situation is dramatically different. First, the coordinate systems related to ideal triangulations form only a tiny part of

[1]There are similar pairs of moduli spaces for any split semisimple group G.

[2]For general S the moduli space of m-dimensional vector bundles with flat connections on S is not rational, and thus can not have a rational coordinate system.

all cluster coordinate systems. Second, although the coordinate systems assigned to any two ideal triangulations are still related by a sequence of mutations, the intermediate cluster coordinate systems are not related to any ideal triangulation.

In this paper we introduce a wider and much more flexible class of cluster coordinate systems on the moduli spaces (1). They are assigned to geometric data on S, which we call A_m*-webs on* S. We also introduce closely related and equally important A_m^**-webs on* S.

An example of an A_m-web. Let T be an ideal triangulation of S. Given a triangle of T, we subdivide it into $(m+1)^2$ small triangles by drawing three families of m equidistant lines, parallel to the sides of the triangle, as shown on the left of Figure 2.

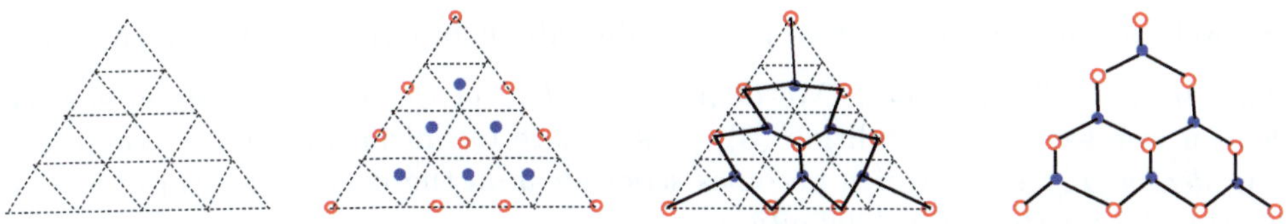

FIGURE 2. A bipartite graph associated with a 4-triangulation of a triangle.

For every triangle of T there are two kinds of small triangles: the "looking up" and "looking down" triangles. We put a $\bullet$-vertex into the center of each of the "down" triangles.

Let us color in red all "up" triangles. We consider the obtained red domains – some of them are unions of red triangles, and put a $\circ$-vertex into the center of each of them. Alternatively, the set of $\circ$-vertices is described as follows. Put a $\circ$-vertex into every special point, into the centers of little segments on the sides of t which do not contain the special points, and into the centers of interior "up" triangles, as shown on the middle picture on Figure 2.

A $\circ$-vertex and a $\bullet$-vertex are *neighbors* if the corresponding domains share an edge.

Recall that a *bipartite graph* is a graph with vertices of two kinds, so that each edge connects vertices of different kinds. So connecting the neighbors, we get a bipartite surface graph $\Gamma_{\mathrm{A}_m}(T)$, called the *bipartite* A_m*-graph* assigned to T, see Figure 2.

We associate to any bipartite surface graph a collection of its *zig-zag strands*. Namely, take the paths on the graph which go left at the $\bullet$-vertices, and right at the $\circ$ ones, and push them a bit off the vertices, getting a (green) web shown on Figure 3. The strands are oriented so that we go around $\bullet$-vertices clockwise. A zig-zag strand inside of a triangle t goes along one of its sides.

The strands associated to the bipartite graph $\Gamma_{\mathrm{A}_m}(T)$ enjoy the following properties:

1. The directions of the strands intersecting any given strand alternate.

2. The connected components of the complement to the union of strands are contractible.
3. The part of surface on the right of each strand is a disc, containing a single special point.

Webs and ideal webs. The alternation property is a well-known feature of the collection of zig-zag strands assigned to any bipartite graph on a surface. Adding contractibility of the domains, and axiomatizing this, one defines a *web* on a surface as a connected collection of oriented strands in generic position, considered modulo isotopy, satisfying conditions (1) and (2). So the collection of zig-zag strands assigned to a bipartite graph form a web, and it is well known that this way we get a bijection between webs and bipartite graphs on surfaces whose faces are discs – see [GK] or Section 2.1.

So webs are just an alternative way to think about bipartite graphs on surfaces.

The third condition is new. We define a *strict ideal web* on a decorated surface S as a web satisfying condition (3), plus a more technical minimality condition. An *ideal web* is a web which becomes strict ideal on the universal cover of S. See Section 2.3 for the precise definition.

In this paper we refer to webs rather then to bipartite graphs since the crucial third condition is formulated using the zig-zag strands.

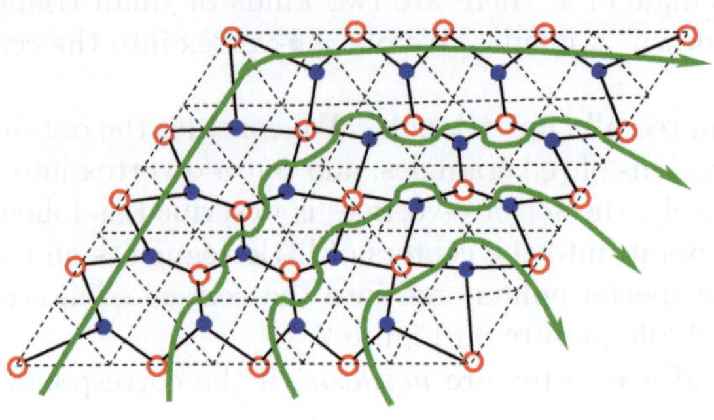

FIGURE 3. The strands of the web assigned to a 5-triangulation of a pair of adjacent triangles, which go around one of the four special points – the bottom right vertex of the quadrilateral.

Another example: the A^*_m-web assigned to ideal triangulations. Given an $m+1$-triangulation of T of S, let us define the *bipartite* A^*_m*-graph* $\Gamma_{A^*_m}(T)$ assigned to T, see Fig 4. Take a triangle t of the triangulation T. Assign the $\circ$-vertices to the centers of all "looking up" triangles of the $(m+1)$-triangulation of T, and the $\bullet$-vertices to the centers of all "looking down" triangles, as well as to the centers of $m+1$ little segments on each of the side of t. So every edge of the triangulation T carries $m+1$ $\bullet$-vertices. Connect them into a bipartite graph as shown on Figure 4. Notice that the special points of S are no longer among the vertices of the graph $\Gamma_{A^*_m}(T)$. This procedure creates $\bullet$-vertices on the boundary of the surface.

We remove the •-coloring from all external vertices. For example, if the surface S is just a triangle, we get a graph shown on Figure 5.

Shrinking the 2-valent •-vertices of the bipartite graph $\Gamma_{\mathrm{A}_m^*}(T)$ we get the graph $\Gamma_{\mathrm{A}_m}(T)$.

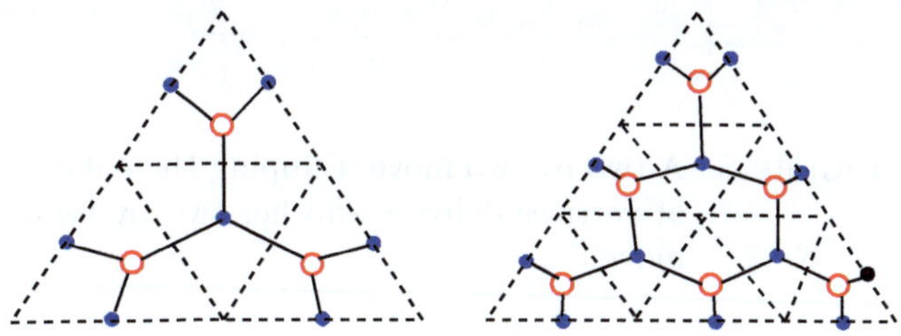

FIGURE 4. An A_1^*-graph, and an A_2^*-graph in an internal ideal triangle.

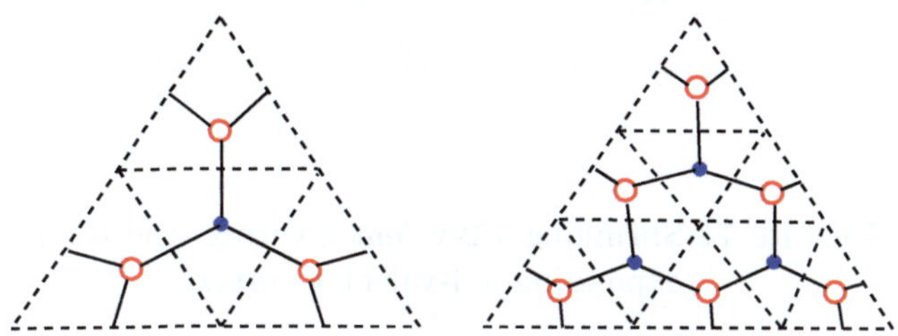

FIGURE 5. An A_1^*-graph, and an A_2^*-graph on a decorated surface given by a triangle.

Elementary moves. There are three elementary local transformations of bipartite surface graphs / webs. The crucial one is the two by two move.

Two by two move. It is shown on Figure 6. It affects four vertices of one color, and two of the other. It amounts to a transformation of webs, also known as a two by two move, see Figure 19.

Shrinking a 2-valent vertex. It is the first move shown on Figure 7. It amounts to resolving a bigon move of webs.

Expanding a k-valent vertex, $k > 3$. It is the second move on Figure 7. It amounts to creating a bigon move of webs.

A guide to classes of webs studied in the paper. We define several subclasses of webs:

$$\{\mathrm{A}_m \text{ and } \mathrm{A}_m^*\text{-webs}\} = \{\textit{rank } m+1 \textit{ ideal webs}\} \subset \{\textit{spectral webs}\}. \qquad (2)$$

- *Spectral webs* are defined on any oriented surface S. A spectral web $\mathcal{W}$ on S gives rise to a new surface $\Sigma_{\mathcal{W}}$, called the *spectral surface*, equipped with a ramified cover map $\Sigma_{\mathcal{W}} \longrightarrow S$, called the *spectral cover* – see Section 2.2. Its degree is the *rank* of $\mathcal{W}$.

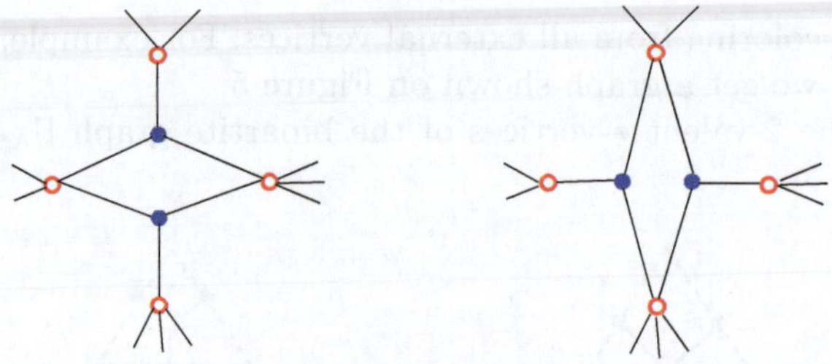

FIGURE 6. A two by two move. Flipping the colors of vertices delivers another two by two move.

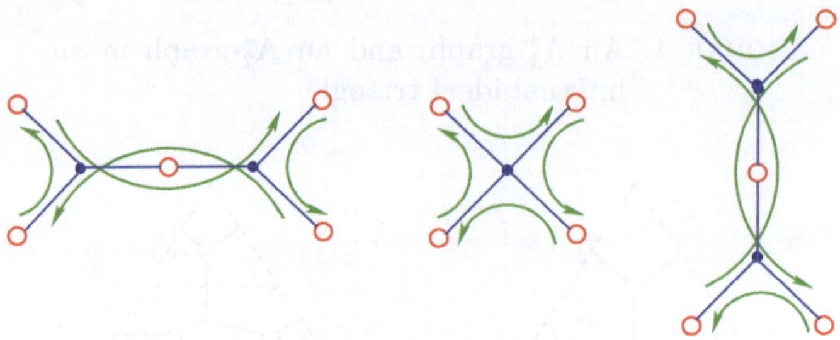

FIGURE 7. Shrinking a 2-valent ∘-vertex, and then expanding a 4-valent •-vertex.

- When S is a decorated surface, we introduce a subclass of spectral webs, called *ideal webs*, which play the central role in our story – see Section 2.3.
- There is an involution on the webs, whose orbits are *triple crossing diagrams*. It interchanges the A_m-web with A^*_m-webs – see Section 2.4.

Cluster coordinate systems assigned to $\mathbf{A}_{m-1}$-webs and $\mathbf{A}^*_{m-1}$-webs. An A_{m-1}-web $\mathcal{W}$ on S gives rise to a cluster coordinate system on each of the spaces (1). They generalize the ones assigned in [FG1] to ideal triangulations of S. Two by two moves amounts to cluster mutations.

Precisely, a face of a web is called *internal* if it does not intersect the boundary of S, and does not contain special points. Otherwise a face is called *external*. We define regular functions $\{A_F^{\mathcal{W}}\}$ on the space $\mathcal{A}_{SL_m,S}$, parametrized by the faces F of the web $\mathcal{W}$. They form a cluster coordinate system. The functions $\{A_F^{\mathcal{W}}\}$ corresponding to the external faces of $\mathcal{W}$ are the frozen cluster coordinates: they do not mutate.

We define rational functions $\{X_F^{\mathcal{W}}\}$ on the space $\mathcal{X}_{PGL_m,S}$ parametrized by the internal faces F of the web $\mathcal{W}$; they form a cluster Poisson coordinate system.

A *decorated flag* in an m-dimensional vector space V_m is a flag $F_0 \subset F_1 \subset \cdots \subset F_m$ where $\dim F_k = k$ equipped with non-zero vectors $v_i \in F_i/F_{i-1} - \{0\}$,

$i = 1, \ldots, m$. Denote by $\mathcal{A}_{GL_m}$ the moduli space of all decorated flags in V_m. So $\mathcal{A}_{GL_m} = GL_m/U$ where U is the upper triangular unipotent subgroup in GL_m.[3]

Pick a volume form Ω_m in V_m. Consider the moduli space $\mathrm{Conf}_n(\mathcal{A}^*_{SL_m})$ parametrizing n-tuples of decorated flags in V_m modulo the diagonal action of the group $SL_m = \mathrm{Aut}(V_m, \Omega_m)$:

$$\mathrm{Conf}_n(\mathcal{A}^*_{SL_m}) = (\mathcal{A}_{GL_m} \times \cdots \times \mathcal{A}_{GL_m})/\mathrm{Aut}(V_m, \Omega_m). \tag{3}$$

In the case when S is a disc with n special points on the boundary, an A^*_{m-1}-web $\mathcal{W}$ on S describes a cluster coordinate system on the moduli space (3). The external faces of $\mathcal{W}$ parametrize the frozen coordinates.

For any decorated surface S there is an analog of the space (3), denoted by $\mathcal{A}^*_{SL_m,S}$. It parametrizes twisted SL_m-local systems on S with an additional data: a choice of a decorated flag near each of the special points, invariant under the monodromy around the point. An A^*_{m-1}-web $\mathcal{W}$ on S describes a cluster coordinate system on the space $\mathcal{A}^*_{SL_m,S}$. Its external faces parametrize the frozen coordinates.

On the other hand, given a A^*_{m-1}-web $\mathcal{W}$ on S, we describe generic points of the spaces $\mathcal{A}_{GL_m,S}$ and $\mathcal{X}_{GL_m,S}$ via flat line bundles on the spectral surface $\Sigma_{S,\mathcal{W}}$. This description admits a generalization to non-commutative local systems on S [GKo].

1.2. Presenting an A_m-flip as a composition of two by two moves

Starting from now, we consider only triangulations without self-folded triangles.

We have seen that each ideal triangulation of S gives rise to a canonical A_m-web. Our next goal is to connect the A_m-webs assigned to two different triangulations by a sequence of elementary moves. Any two ideal triangulations without self-folded triangles can be connected by a sequence of flips that only involves ideal triangulations without self-folded triangles. Therefore it suffices to show that every flip in the sequence is a cluster transformation. So it is sufficient to do this for a single flip. We say that the A_m-webs related to an original ideal triangulation and a flipped one are related by an A_m-*flip*.

Figure 8 tells that an A_1-flip is the same thing as two by two move.

FIGURE 8. Two by two move corresponding to a flip of triangulations.

[3]There are two versions of decorated flags used in this paper: one of them includes the factor v_m in the data, the other does not. The first is decorated flag, that is a point of G/U, for G $= GL_m$, the second for G $= SL_m$.

Proposition 1.1 tells how to decompose an A_m-flip for any m as a composition of elementary moves. The existence of such a decomposition follows from Theorem 3.2. Here we present an A_m-flip as a specific composition of $\binom{m+2}{3}$ two by two moves, and shrinking / expanding moves.

Proposition 1.1. *There is a decomposition of an A_m-flip into a composition of elementary moves, including $\binom{m+2}{3}$ two by two moves, depicted on Figure 8 for $m = 1$, Figure 9 for $m = 2$, and on Figures 10–12 for $m = 3$.*

Proof. The pattern is clear from the $m = 3$ and $m = 4$ examples.

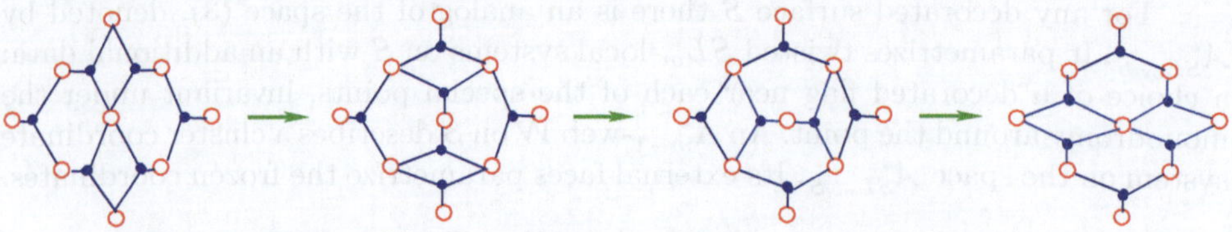

FIGURE 9. Presenting an A_2-flip as a sequence of elementary moves. We transform the A_2-web related to one triangulation into the A_2-web related to the other. We apply $2 \times 1 = 2$ two by two moves, shrink and expand the central vertex, and apply $1 \times 2 = 2$ two by two moves.

Figures 10–12 show how to decompose a transformation of A_3-webs arising from a flip of the vertical diagonal into a composition of $3 \times 1 + 2 \times 2 + 1 \times 3 = 3 + 4 + 3 = 10$ two by two moves.

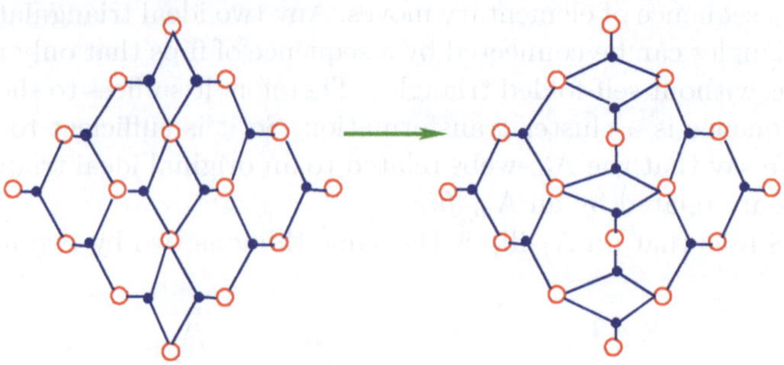

FIGURE 10. Step 1: Applying $3 \times 1 = 3$ two by two moves.

To go from Step 1 to Step 2 we shrink vertically and expand horizontally two of the "middle" $\circ$-vertices.
To go from Step 2 to Step 3 we shrink vertically and expand horizontally the other two "middle" $\circ$-vertices.

Similarly, a transformation of A_m-webs arising from a flip of a triangulation is a composition of $m \times 1 + (m-1) \times 2 + \cdots + 1 \times m = \binom{m+2}{3}$ two by two moves. Notice that the $\circ$-vertices do not change their position in the process. □

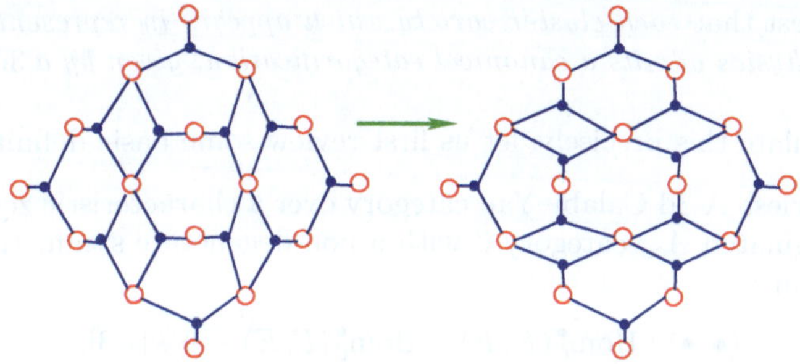

FIGURE 11. Step 2: Applying $2 \times 2 = 4$ two by two moves.

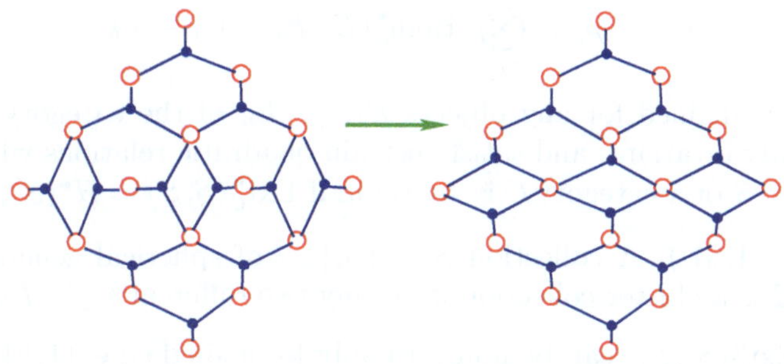

FIGURE 12. Step 3: Applying $1 \times 3 = 3$ two by two moves.

Remark. Given a positive integer m, we assigned in [FG1] to an ideal triangulation a quiver, called the A_m-quiver. A flip gives rise to a transformation of A_m-quivers, presented in [FG1], Section 10, as an explicit composition of cluster mutations. Proposition 1.1 tells that all intermediate quivers are described by A_m-webs, and the sequence of cluster mutations match the sequence of two by two moves. Each two by two move corresponds to gluing of an octahedron. This collection of two by two moves is parametrized by the octahedra of the *A_m-decomposition of a tetrahedra*, see *loc. cit.*

1.3. CY_3 categories categorifying cluster varieties

It was discovered more then a decade ago that many interesting rational varieties which appear in representation theory, geometry and physics come with a natural cluster structure of certain kind. Examples are given by Grassmannians [GSV], double Bruhat cells [BFZ], different versions of the classical Teichmüller space [GSV1], [FG1], and more generally different versions of moduli spaces of G-local systems on decorated surfaces, e.g., (1), [FG1].

So far cluster structures were used mostly to describe certain features of the relevant spaces, like Poisson / K_2-structures, canonical bases in the space of regular functions, non-commutative q-deformation, quantization, etc.

We suggest that *each cluster variety which appears in representation theory, geometry or physics admits a canonical categorification, given by a* 3d *Calabi–Yau category.*

To formulate this precisely, let us first review some basic definitions.

3d CY categories. A 3d Calabi–Yau category over a characteristic zero field k is a k-linear triangulated A_∞-category $\mathcal{C}$ with a non-degenerate symmetric degree -3 pairing on Hom's:

$$(\bullet, \bullet) : \mathrm{Hom}^\bullet_{\mathcal{C}}(E, F) \otimes \mathrm{Hom}^\bullet_{\mathcal{C}}(F, E) \longrightarrow k[-3]. \tag{4}$$

The A_∞-structure on a 3d CY category is encoded by the cyclic higher composition maps:

$$m_{E_1,\ldots,E_n} : \bigotimes_{1\leq i\leq n} \mathrm{Hom}^\bullet_{\mathcal{C}}(E_i, E_{i+1}[1]) \longrightarrow k. \tag{5}$$

These maps are defined for any objects $E_1, \ldots, E_n$ of the category $\mathcal{C}$. They are graded cyclically invariant, and satisfy certain quadratic relations which we omit.

An object S of a category $\mathcal{C}$ is *spherical* if $\mathrm{Ext}^\bullet_{\mathcal{C}}(S, S) = H^\bullet(S^3, k)$.

Definition 1.2 ([KS1]). A collection $\mathcal{S} = \{S_i\}_{i\in I}$ of spherical generators of a 3d CY category $\mathcal{C}$ is a cluster collection if for any two different $i, j \in I$ one has

$$\mathrm{Ext}^a_{\mathcal{C}}(S_i, S_j) \text{ can be non-zero only for a single } a \in \{1, 2\}. \tag{6}$$

A 3d CY category $\mathcal{C}$ with a cluster collection $\mathcal{S}$ gives rise to a quiver $Q_{\mathcal{C},\mathcal{S}}$. Its vertices are the isomorphism classes $[S_i]$ of objects of the cluster collection, and the number of arrows $[S_i] \to [S_j]$ equals $\mathrm{dimExt}^1_{\mathcal{C}}(S_i, S_j)$. Thus the quiver $Q_{\mathcal{C},\mathcal{S}}$ has no loops and length two cycles.

Let us assume that $m_1 = 0$. Then the cyclic composition maps, restricted to the cyclic product of $\mathrm{Ext}^1(E_i, E_{i+1})[1]$, determine a *potential* $P_{\mathcal{C},\mathcal{S}}$ of the quiver $Q_{\mathcal{C},\mathcal{S}}$, that is a linear functional on the vector spaces spanned by the cycles in the quiver. The original 3d CY category $\mathcal{C}$ with a cluster collection $\mathcal{S}$, considered modulo autoequivalences preserving the cluster collection $\mathcal{S}$, can be reconstructed from the quiver with potential $(Q_{\mathcal{C},\mathcal{S}}, P_{\mathcal{C},\mathcal{S}})$ [KS1].

A spherical object S of a CY category gives rise to the Seidel–Thomas [ST] reflection functor R_S, acting by an autoequivelence of the category:

$$R_S(X) := \mathrm{Cone}\Big(\mathrm{Ext}^\bullet(S, X) \otimes S \longrightarrow X\Big).$$

Given elements $0, i \in I$, we write $i < 0$ if $\mathrm{Ext}^1(S_i, S_0)$ is non-zero, and $i > 0$ otherwise.

Definition 1.3. A mutation of a cluster collection $\mathcal{S} = \{S_i\}_{i\in I}$ in a 3d CY category $\mathcal{C}$ at an object S_0 is a new spherical collection $\mathcal{S}' = \{S'_i\}$ in the same category $\mathcal{C}$ given by

$$S'_i = S_i,\ i < 0,\ S'_0 = S_0[-1],\ S'_i = R_{S_0}(S_i),\ i > 0,$$

Since the shift and reflection functors transform spherical objects to spherical ones, $\{S_i'\}_{i\in I}$ is a collection of spherical objects. Since the objects $\{S_i\}$ generate the triangulated category $\mathcal{C}$, the objects $\{S_i'\}$ are generators. However $\{S_i'\}_{i\in I}$ is not necessarily a cluster collection.

Mutations of cluster collections, being projected to $K_0(\mathcal{C})$, recover mutations of bases (42).

Mutations of cluster collections translate into *mutations of quivers with generic potentials* introduced earlier by Derksen, Weyman and Zelevinsky [DWZ].

Given a 3d CY category with a cluster collection $(\mathcal{C}, \mathcal{S})$, denote by $\mathrm{Auteq}(\mathcal{C}, \mathcal{S})$ the quotient of the group of autoequivelances of $\mathcal{C}$ by the subgroup preserving a cluster collection $\mathcal{S}$ in $\mathcal{C}$.

Categorification of cluster varieties. Returning to cluster varieties, we suggest the following metamathematical principle:

- *Each cluster variety $\mathcal{V}$ which appears in representation theory, geometry and physics, gives rise to a combinatorially defined 3-dimensional Calabi–Yau triangulated A_∞-category $\mathcal{C}_\mathcal{V}$.*

  The category $\mathcal{C}_\mathcal{V}$ comes with an array of cluster collections $\{\mathcal{S}_\alpha\}$, related by cluster mutations.

  There is a group $\Gamma_\mathcal{V}$ of symmetries of $\mathcal{C}_\mathcal{V}$, understood as a subgroup $\Gamma_\mathcal{V} \subset \mathrm{Auteq}(\mathcal{C}_\mathcal{V}, \mathcal{S})$.

  One should have the following additional features of this picture:

- A family of complex open Calabi–Yau threefolds $\mathcal{Y}_b$ over a base $\mathcal{B}$, which is smooth over $\mathcal{B}_0 \subset \mathcal{B}$, and a fully faithful functor

$$\varphi_\mathcal{V} : \mathcal{C}_\mathcal{V} \longrightarrow \mathcal{F}(\mathcal{Y}_b) \tag{7}$$

  to the Fukaya category $\mathcal{F}(\mathcal{Y}_b)$ of $\mathcal{Y}_b$, where $b \in \mathcal{B}_0$. It sends objects of cluster collections to the objects of the Fukaya category represented by special Lagrangian spheres.
- The group $\pi_1(\mathcal{B}_0, b)$ acts by symmetries of the category $\mathcal{F}(\mathcal{Y}_b)$. The equivalence $\varphi_\mathcal{V}$ intertwines the actions of the groups $\Gamma_\mathcal{V}$ and $\pi_1(\mathcal{B}_0, b)$:

$$\varphi_\mathcal{V} : (\mathcal{C}_\mathcal{V}, \Gamma_\mathcal{V}) \longrightarrow (\mathcal{F}(\mathcal{Y}_b), \pi_1(\mathcal{B}_0, b)). \tag{8}$$

A cluster collection $\mathcal{S}_\alpha$ in the category $\mathcal{C}_\mathcal{V}$ determines a canonical t-structure on $\mathcal{C}_\mathcal{V}$, whose heart is an abelian category denoted by $A(\mathcal{S}_\alpha)$.

An array of cluster collections $\mathcal{S} = \{\mathcal{S}_\alpha\}$ in the category $\mathcal{C}_\mathcal{V}$ determines an open domain in the space of stability conditions $\mathrm{Stab}(\mathcal{C}_\mathcal{V})$ on the category $\mathcal{C}_\mathcal{V}$:

$$\mathrm{Stab}_\mathcal{S}(\mathcal{C}_\mathcal{V}) \subset \mathrm{Stab}(\mathcal{C}_\mathcal{V}).$$

Namely, $\mathrm{Stab}_\mathcal{S}(\mathcal{C}_\mathcal{V})$ consists of stability conditions whose heart is one of the hearts $A(\mathcal{S}_\alpha)$. If any two of the cluster collections $\mathcal{S}_\alpha$ are related by a sequence of mutations, the domain $\mathrm{Stab}_\mathcal{S}(\mathcal{C}_\mathcal{V})$ is connected.

- One should have an embedding of $\mathcal{B}$ into the stack of stability conditions $\mathrm{Stab}(\mathcal{C}_\mathcal{V})$:

$$i : \mathcal{B} \hookrightarrow \mathrm{Stab}(\mathcal{C}_\mathcal{V}).$$

- Then the domain $\mathrm{Stab}_\mathcal{S}(\mathcal{C}_\mathcal{V}) \subset \mathrm{Stab}(\mathcal{C}_\mathcal{V})$ determines a subspace $\mathcal{B}_S \subset \mathcal{B}_0$:

$$\begin{array}{ccc} \mathcal{B}_S & \hookrightarrow & \mathrm{Stab}_\mathcal{S}(\mathcal{C}_\mathcal{V}) \\ \downarrow & & \downarrow \\ \mathcal{B}_0 & \hookrightarrow & \mathrm{Stab}(\mathcal{C}_\mathcal{V}) \end{array}$$

One should have an isomorphism

$$\Gamma_\mathcal{V} = \pi_1(\mathcal{B}_S).$$

Combined with the map induced by the embedding $\mathcal{B}_S \hookrightarrow \mathcal{B}_0$, it gives rise to a map $\Gamma_\mathcal{V} \longrightarrow \pi_1(\mathcal{B}_0, b)$, intertwined by the functor (8).

The pair $(\mathcal{C}_\mathcal{V}, \Gamma_\mathcal{V})$ is easily described combinatorially. Contrary to this, the Fukaya category $\mathcal{F}(\mathcal{Y}_b)$, whose definition we do not discuss, is a complicated object. The pair $(\mathcal{C}_\mathcal{V}, \Gamma_\mathcal{V})$ serves as a combinatorial model of a full subcategory of the Fukaya category $\mathcal{F}(\mathcal{Y}_b)$.

The homological mirror symmetry conjecture of M. Kontsevich predicts equivalence of a Fukaya category of a Calabi–Yau threefold $\mathcal{Y}$ with an A_∞-category of complexes of coherent sheaves on the mirror $\mathcal{Y}^\vee$.

The above picture adds a new ingredient: a combinatorial description of the A_∞-category. It can be viewed as a manifestation of a general principle:

- *All interesting categories are equivalent to Fukaya categories.*

The main message of our paper can be formulated as follows:

- *Ideal webs on decorated surfaces S play the same role for moduli spaces of G-local systems on S where G is of type A_m as ideal triangulations for the groups of type A_1. In particular ideal webs provide an explicit description of a "tame part" of the corresponding cluster atlas, canonical potentials for the related quivers, and therefore an explicitly defined 3d CY category with an array of cluster collections, together with its symmetry group.*

A_m-webs on S provide a minimal Γ_S-equivariant collection of cluster coordinate systems on each of the moduli spaces (1), connected in the following sense: any two of them are related by a sequence of elementary transformations involving only the coordinate systems from the collection. Similarly, A_m-webs are a minimal topological data on S which allow to define combinatorially a 3d CY category with a large symmetry group.

The fundamental geometric structures on surfaces generalizing ideal triangulations are the Gaiotto–Moore–Neitzke spectral networks, introduced in [GMN4], and studied further in [GMN5]. We discuss spectral networks in comparison with ideal webs in Section 1.4.

Our main example: G-local systems on decorated surfaces. In this example, which is the one we elaborate in this paper, the ingredients of the above picture look as follows:

- Varieties $\mathcal{V}$ are variants of the moduli spaces of G-local systems on a decorated surface S, where G is SL_m or GL_m, equipped with cluster variety structures defined in [FG1].
- The quivers assigned to ideal webs on S describe Γ_S-equivariant cluster structures of these moduli spaces. These quivers are equipped with canonical potentials.

  Therefore they give rise to a combinatorially defined 3d CY category with a $\Gamma_{G,S}$-equivariant array of cluster collections. The group $\Gamma_{G,S}$ is an extension of the mapping class group Γ_S by a generalized braid group, see Section 5.3.
- The base $\mathcal{B}$ is the base of the universal Hitchin integrable system for the pair (G, S).
- The family of open CY threefolds over $\mathcal{B}$ is the one related to the Hitchin integrable system of the pair (G, S) by [DDDHP], [DDP], [G], [KS3], [Sm].

The case $G = SL_2$. In this case, when S is sufficiently general, the existence of a fully faithful embedding $\varphi_{\mathcal{V}}$ is proved by Smith [Sm],[4] building on the works of Derksen, Weyman and Zelevinsky [DWZ], Gaiotto–Moore–Neitzke [GMN1]–[GMN3], Kontsevich and Soibelman [KS1], [KS3], Labardini–Fragoso [LF08], [LF12], Keller and Yang [KY], Bridgeland and Smith [BrS], and others. In particular:

Labardini-Fragoso [LF08], [LF12] introduced and studied canonical potentials related to surfaces with ideal triangulations as well as the tagged ideal triangulations [FST].

Bridgeland and Smith [BrS] proved that the spaces of stability conditions for the combinatorial categories provided by Labardini-Fragoso's work are naturally isomorphic to the spaces of quadratic differentials of certain kind on the corresponding Riemann surfaces.

K. Nagao in his lecture [N12] outlined a combinatorial action of a mapping class group.

Another example: dimer model integrable systems. The dimer integrable system is assigned to a Newton polygon $\mathbb{N}$, and was studied in [GK]. It gives rise to a cluster Poisson variety $\mathcal{X}_{\mathbb{N}}$ and a map given by the dimer model Hamiltonians

$$\mathcal{X}_{\mathbb{N}} \longrightarrow \mathcal{B}_{\mathbb{N}}.$$

The story goes as follows. Instead of ideal bipartite graphs on a decorated surface S we consider minimal bipartite graphs Γ on a torus T^2. Each of them is related to a convex polygon $\mathbb{N} \subset \mathbb{Z}^2$. The set of minimal bipartite graphs on a torus modulo isotopy with a given Newton polygon is finite, and any two of them are related by

[4] This requires an adjustment the definition of the Fukaya category of the threefold, incorporating the order two poles in the punctures, see [Sm].

the elementary moves. We assign to such a graph Γ the moduli space $\mathcal{L}_\Gamma$ of GL_1-local systems on Γ, and glue them via birational cluster Poisson transformations $\mathcal{L}_\Gamma \to \mathcal{L}_{\Gamma'}$ corresponding to two by two moves into a cluster variety $\mathcal{X}_\mathbb{N}$. The base $\mathcal{B}_\mathbb{N}$ is essentially the space of polynomials $P(x, y)$ with the Newton polygon $\mathbb{N}$.

Just as above, the bipartite graphs Γ give rise to quivers with canonical potentials related by two by two moves, and hence to 3d CY category $\mathcal{C}_\mathbb{N}$, determined just by the polygon $\mathbb{N}$, with an array of cluster collections. The corresponding CY open threefold is the toric threefold related to the cone over the Newton polygon $\mathbb{N}$. The geometry of these threefolds was extensively studied by the Physics and Math communities in the past.

In particular any convex polygon $\mathbb{N} \subset \mathbb{Z}^2$, considered up to translations, gives rise to a group $\Gamma_\mathbb{N}$, realized as the categorical symmetry group. This group is an extension of the cluster modular group of the cluster Poisson variety $\mathcal{X}_\mathbb{N}$ by a generalized braid group.

1.4. Concluding remarks

Bipartite graphs on decorated surfaces with abelian fundamental groups. Our coordinates on the moduli spaces (1) have many common features with Postnikov's cluster coordinates on the Grassmannians [P], see also [S] – the latter are given by a combinatorial data (directed networks on the disc) closely related to bipartite graphs on the disc. In particular, in both cases we have a subatlas of the cluster atlas closed under two by two moves, and the corresponding cluster coordinates are given simply by minors. Postnikov's story was partially generalized to the case of an annulus in [GSV]. Minimal bipartite graphs on decorated surfaces, with special attention to the minimal bipartite graphs on the torus, were studied in [GK].

Minimal bipartite graphs on decorated surfaces, considered modulo the two by two moves, admit a nice classification when the surface has an abelian fundamental group, i.e., either a disc, an annulus, or a torus. Namely, the set of equivalence classes is finite, and has an explicit description. For the disc this follows from [T] and [P], the annulus was considered in [GSV], and the torus in [GK].

For an arbitrary surface the classification is a wild problem. Ideal bipartite graphs provide an interesting subclass of bipartite surface graphs.

Gaiotto–Moore–Neitzke spectral networks. It would be very interesting to compare ideal webs with the Gaiotto–Moore–Neitzke spectral networks [GMN4].

SL_m-spectral networks describe a geometric object on a Riemann surface C related to a point of the Hitchin base, i.e., a collection $(t_2, \ldots, t_m)$ where $t_k \in \Omega^{\otimes k}(C)$ are holomorphic k-differentials on C. It is a remarkable geometric object, rather complicated, whose properties are not fully established yet.

Ideal webs are defined on topological surfaces. They parametrize cluster coordinate systems, and give rise to quivers with potentials, and therefore to 3d CY categories with cluster collections. It is not clear whether one can associate similar features to an arbitrary spectral network.

Ideal webs and spectral networks were developed independently, and relations between them are far from been clear. In [GMN5] the ideal webs assigned to ideal triangulations are related to spectral networks. I believe that all ideal webs are nice special examples of spectral networks.

Ideal webs and scattering amplitudes. Bipartite graphs on the disc, as well as slightly more general *on shell scattering diagrams* are in the center of the approach to the scattering amplitudes in the $N = 4$ super Yang–Mills theory developed in [ABCGPT]. We hope that similar constructions related to ideal webs on arbitrary surfaces lead to a physically meaningful story. The parallels between the two stories are highlighted below:

Scattering Amplitudes in the $N = 4$ SUYM	**Moduli spaces of local systems**
Disc with n special points on the boundary	An arbitrary decorated surface
Grassmannian $Gr(m, n) \simeq$ Configurations of n vectors in an m-dimensional space	Moduli space of GL_m-local systems on surfaces
On shell scattering diagrams	Ideal webs on surfaces
Scattering Amplitudes in the $N = 4$ SUYM	Similar integrals given by amalgamation

The structure of the paper. Sections 2–4 are entirely geometric: we study ideal webs and introduce the related cluster coordinates. They are completely independent from the other parts of the paper.

In Section 5.1 we present, for the convenience of the reader, the background material on quivers with potentials and 3d Calabi–Yau categories with cluster collections. In particular we recall the Kontsevich–Soibelman correspondence between these two objects ([KS1], Section 8). We only very briefly recall an alternative approach of Keller–Yang [KY] via Ginzburg algebras [G]. In Section 5.2 we define symmetry groups of 3d CY categories with cluster collections. The story is very similar to the definition of the cluster modular group [FG2]. In Section 5.3 we define an extended mapping class group. In Section 6.2 we formulate the conjecture relating the combinatorial 3d CY categories coming from ideal webs and Fukaya categories of open CY threefolds related to reductive groups of type A.

In Section 7, given a cluster Poisson variety $\mathcal{X}$, we define a space of cluster stability conditions. It is a complex manifold of complex dimension $\dim\mathcal{X}$ fibered over the positive part $\mathcal{X}^+(\mathbb{R}^t)$ of the space of real tropical points of $\mathcal{X}$, which is a piecewise linear space of real dimension $\dim\mathcal{X}$.

The MSRI lecture [G14] lecture can serve as an online introduction to the paper.

2. Ideal webs on decorated surfaces

2.1. Webs and bipartite graphs on decorated surfaces [GK]

In Section 2.1 we assume that S is an oriented surface with or without boundary. We do not assume that it is a decorated surface.

A *strand* on S is either an oriented loop, or an oriented path connecting two boundary points.

Definition 2.1. A **web** $\mathcal{W}$ on an oriented surface S is a union of strands $\{\gamma\}$ in generic position on S, considered modulo isotopy, such that:

1. As we move along a strand, the orientations of the crossing strands alternate.
2. The subset $\mathcal{W}$ is connected. The components of $S - \mathcal{W}$, called domains, are discs.

Our next goal is to show that webs on S can be encoded by bipartite graphs on S.

Any vertex of a graph is either *internal*, that is of valency ≥ 2, or *external.*

Definition 2.2. A *bipartite graph on a surface* S, possibly with a boundary, is a graph with external vertices at the boundary, whose internal vertices and internal edges form an abstract bipartite graph. The external vertices are not colored.

Lemma 2.3 is borrowed from [GK]. We provide a proof for the convenience of the reader.

Lemma 2.3. *Let S be an oriented surface, possibly with boundary. There is a bijection between the isotopy classes of webs and connected bipartite graphs, whose faces are discs, on S.*

Proof. A path on the boundary of a domain on an oriented surface is *positively oriented* if its orientation coincides with the orientation of the boundary, provided by the surface orientation. Otherwise it is *negatively oriented.* We picture positive orientation as clockwise.

A web $\mathcal{W}$ on S gives rise to a bipartite graph Γ on S as follows.

Condition 1) implies that there are three types of the domains:

1. $\bullet$-*domains*: the sides are oriented positively.
2. $\circ$-*domains*: the sides are oriented negatively.
3. *Faces*: the directions of the sides alternate.

Since each domain is a disc (Condition 2), assigning a $\circ$ (respectively $\bullet$) central point to each $\circ$ (respectively $\bullet$) domain, we get a bipartite graph Γ on S. Its edges correspond to the crossing points of the strands, see Figure 13. Indeed, since the strands are in generic position, every crossing point of $\mathcal{W}$ is an intersection of two arrows. So it determines the $\circ$ and the $\bullet$-domains.

Let us now go in the opposite direction, and assign to a bipartite graph Γ on S a web $\mathcal{W}$.

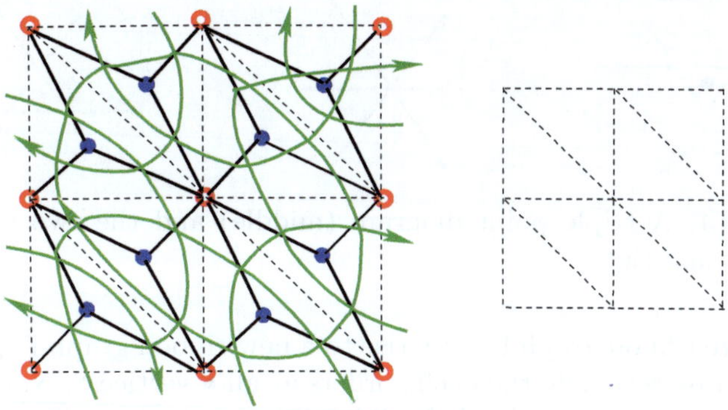

FIGURE 13. Zig-zag strands (green) for the bipartite graph assigned to a triangulation (shown by dotted lines) of a torus with four punctures.

A *zig-zag path* on a bipartite ribbon graph Γ is an oriented path on Γ which turns maximally left at $\bullet$-vertices, and maximally right at $\circ$-vertices. We isotope zig-zag paths slightly to strands going clockwise (respectively counterclockwise) around the $\bullet$ (respectively the $\circ$) vertices. Namely, at each vertex v on a zig-zag path γ we push the path γ a bit inside of the angle formed by the edges incident to v. These strands are called *zig-zag strands.*

A bipartite graph Γ on S gives rise to a web $\mathcal{W}$, given by the zig-zag strands of Γ. They satisfy Conditions 1 and 2. □

Webs and triple point diagrams. A web on a surface is closely related to a slight generalization of Dylan Thurston's *triple point diagram on the disc*:

Definition 2.4 ([T]). **A triple point diagram on the disc** is a collection of oriented strands in the disc whose intersection points are as in the middle of Figure 14, and the endpoints of the strands are distinct points on the boundary of the disc.

We consider triple point diagrams on an oriented surface with boundary rather then on a disc, and resolve all triple point crossings.

An important difference is that there are two non-equivalent ways to resolve triple point crossings, shown on the left and on the right on Figure 14. So there is a pair of webs associated with a triple point diagram, and vice verse. See Lemma-Construction 2.18 for more details. This is the reason we work with webs rather then with triple point diagrams.

The conjugate surface [GK]. A *ribbon graph* is a graph with a cyclic order of the edges at every vertex. A *face path* on a ribbon graph Γ is an oriented path on the graph which turns right at every vertex. A closed face path is called a *face loop.*

A graph embedded into an oriented surface has an induced ribbon structure. We assume that its external vertices are on the boundary of S.

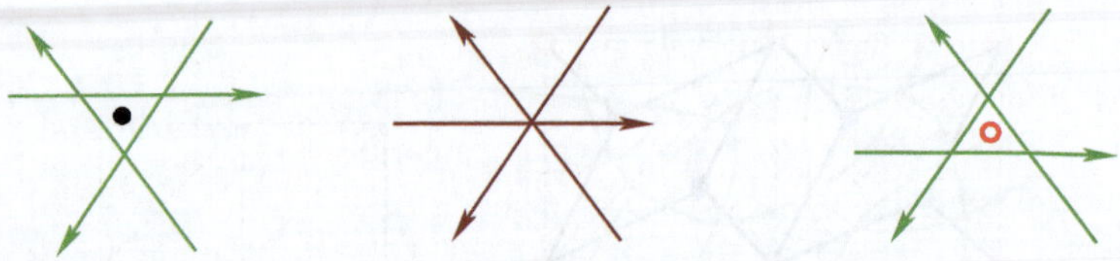

FIGURE 14. A triple point diagram (middle) and the two associated webs (left and right).

A bipartite ribbon graph Γ gives rise to a new ribbon graph Γ^*, the *conjugate graph*, obtained by reversing the cyclic orders at all $\bullet$-vertices[5]. Notice that Γ and Γ^* are the same graphs – only their ribbon structures are different. There is a bijection:

$$\{\text{zig-zag strands } \{\gamma_i\} \text{ on } \Gamma\} \longleftrightarrow \{\text{face paths on } \Gamma^*\}. \tag{9}$$

The orientations of zig-zag strands match the orientations of the face paths on Γ^*.

A ribbon graph Γ gives rise to a surface in a standard way – we replace the edges of Γ by thin ribbons. Let Σ be the topological surface with boundary corresponding to the ribbon graph Γ^*. It can be visualized by taking the surface glued from the ribbons assigned to the edges of Γ, cutting each of the ribbons in the middle, twisting its ends incident to black vertices by 180°, and gluing it back. Then bijection (9) gives rise to a bijection

$$\{\text{zig-zag strands } \{\gamma_i\} \text{ on } \Gamma\} \longleftrightarrow \{\text{boundary components of } \Sigma\}. \tag{10}$$

Orientations of zig-zag strands match orientations of boundary components on Σ.

2.2. Spectral webs and spectral covers

Let $\mathcal{W}$ be a web on a decorated surface S. Let $p : \widetilde{S} \to S$ be the universal cover of S. The surface $\widetilde{S}$ inherits a set of special points: the preimage of the ones on S. So it is a decorated surface. Then $\widetilde{\mathcal{W}} := p^{-1}(\mathcal{W})$ is a web on $\widetilde{S}$. Defining different types of webs $\mathcal{W}$ on S, like minimal webs, spectral webs, ideal webs, we work with the web $\widetilde{\mathcal{W}}$. We use the same strategy introducing cluster coordinates related to an ideal web on S. Since the web $\widetilde{\mathcal{W}}$ is $\pi_1(S)$-equivariant, all constructions lead to notions and objects related to the surface S itself.

Key example. Figure 15 illustrates how misleading could be the geometry of a web considered on S rather then its universal cover. Consider a torus with a single special point. Pick an ideal triangulation. The corresponding bipartite graph has a unique $\circ$-vertex, and two $\bullet$-vertices, as illustrated on the left of Figure 15. The web assigned to this bipartite graph consists of a single selfintersecting zig-zag path. However the universal cover, illustrated on the right of Figure 15, consists of nonselfintersecting zig-zag loops. Each of them surrounds a single special point, which in this particular case is also a single $\circ$-vertex inside of the loop. We show

[5]Reversing the cyclic order at $\circ$-vertices we get another ribbon graph which differs from Γ^* by the orientation.

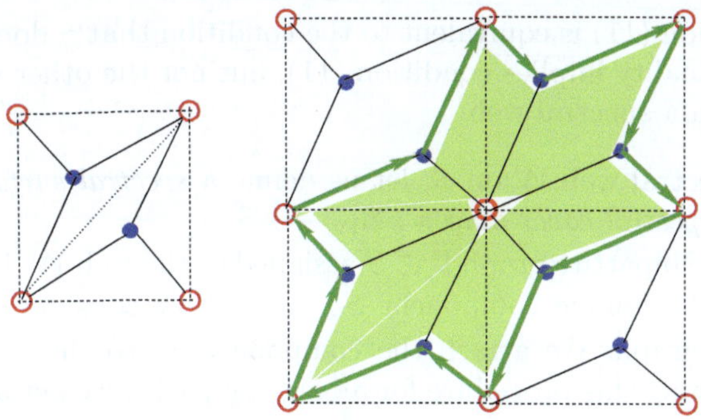

FIGURE 15. On the left: the bipartite graph for a punctured torus, triangulated by a diagonal. On the right: the bipartite graph on its universal cover with a (green) zig-zag loop γ. The domain S_γ is the interior of the filled 12-gon. It contains a single $\circ$-vertex.

in Figure 15 one of its zig-zag loops, surrounding a $\circ$-vertex. The corresponding domain is a 12-gon, filled green. Therefore this web on the torus is an ideal A_1-web for the definition given below.

Definition 2.5. A web $\mathcal{W}$ on a surface S is **minimal** if the web $\widetilde{\mathcal{W}}$ has the following property:

- Strands have no selfcrossings, and there are no parallel bigons, see Figure 16.

A bipartite graph on a surface S is minimal if the corresponding web is minimal.

FIGURE 16. Minimality: no selfcrossings and no parallel bigons on the universal cover.

Definition 2.6. A web $\mathcal{W}$ on a surface S is **spectral** if the web $\widetilde{\mathcal{W}}$ has the following property:

- Each strand γ of $\widetilde{\mathcal{W}}$ cuts the surface $\widetilde{S}$ into two components:

$$\widetilde{S} - \gamma = \widetilde{S}_\gamma \cup \widetilde{S}'_\gamma. \tag{11}$$

The connected components in (11) are orientated by the orientation of S. The domain $\widetilde{S}_\gamma$ is the one for which its orientation induces the original (clockwise on the pictures) orientation of γ.

Remark. Condition (11) is equivalent to the condition that γ does not have selfintersections. Minimality implies condition (11), but not the other way: we can have parallel bigons on a spectral web.

Given a spectral web $\mathcal{W}$ on S, let us define a *spectral surface* $\Sigma_{\mathcal{W}}$, together with a ramified *spectral cover* map $\pi : \Sigma_{\mathcal{W}} \longrightarrow S$.

Consider the bipartite graph $\widetilde{\Gamma} \subset \widetilde{S}$ assigned to the web $\widetilde{\mathcal{W}}$. Each zig-zag path γ on $\widetilde{\Gamma}$ bounds the unique component $\widetilde{S}_\gamma$ on $\widetilde{S}$. Now consider $\widetilde{\Gamma}$ as an abstract ribbon bipartite graph. We attach the component $\widetilde{S}_\gamma$ to the graph $\widetilde{\Gamma}$ along the strand γ. Performing this procedure for all zig-zag paths, we get a surface $\Sigma_{\widetilde{\mathcal{W}}}$. By construction, each component $\widetilde{S}_\gamma$ becomes a part of $\Sigma_{\widetilde{\mathcal{W}}}$, denoted by F_γ. So

$$\Sigma_{\widetilde{\mathcal{W}}} = \widetilde{\Gamma} \cup \cup_{\gamma_i} F_{\gamma_i}.$$

Using the tautological isomorphisms $F_\gamma \stackrel{\sim}{\rightarrow} \widetilde{S}_\gamma$, we get a map, called *spectral cover map*

$$\widetilde{\pi} : \Sigma_{\widetilde{\mathcal{W}}} = \widetilde{\Gamma} \cup \cup_{\gamma_i} F_{\gamma_i} \longrightarrow \widetilde{S}. \tag{12}$$

The fundamental group $\pi_1(S)$ acts on both $\widetilde{S}$ and $\Sigma_{\widetilde{\mathcal{W}}}$, and by the construction its action commutes with the projection π. So the group $\pi_1(S)$ acts by automorphisms of the cover. Set

$$\Sigma_{\mathcal{W}} := \Sigma_{\widetilde{\mathcal{W}}}/\pi_1(S).$$

Then we get a commutative diagram, where the horizontal maps are given by the factorisation by the free action of the group $\pi_1(S)$:

$$\begin{array}{ccc} \Sigma_{\widetilde{\mathcal{W}}} & \longrightarrow & \Sigma_{\mathcal{W}} := \Sigma_{\widetilde{\mathcal{W}}}/\pi_1(S) \\ \widetilde{\pi}\downarrow & & \downarrow \\ \widetilde{S} & \longrightarrow & S = \widetilde{S}/\pi_1(S) \end{array}$$

Therefore we arrive at a map, which we call the spectral cover map:

$$\pi : \Sigma_{\mathcal{W}} \longrightarrow S. \tag{13}$$

The following theorem summarizes its properties.

Theorem 2.7. *Let $\mathcal{W}$ be a spectral web on an oriented surface S. Then the spectral cover map* (13) *can be ramified only at the $\bullet$-vertices of the associated bipartite graph Γ. The ramification index of a $\bullet$-vertex is its valency minus two.*

Proof. This is a local question, so it is sufficient to argue for the graph $\widetilde{\Gamma}$ on the universal cover. The map $\widetilde{\pi}$ can be ramified only at the vertices of the graph $\widetilde{\Gamma}$. Take a vertex v. Let n be the valency of v. Then a little neighborhood U of v is a union of n sectors $C_1, \ldots, C_n$ formed by pairs of consecutive edges emanating from v. A zig-zag path γ passing through v goes along the boundary of one of the sectors C_i. If v is a $\bullet$-vertex, the disc $S_\gamma \cap U$ contains all these sectors but the one C_i. Thus the union of $S_\gamma \cap U$ over all zig-zag paths $\widetilde{\gamma}$ passing through an n-valent $\bullet$-vertex v covers $n-1$ times $U - v$. It is $1:1$ at the vertex. So the ramification index of the map π at a $\bullet$-vertex v is $n-1$. If v is a $\circ$-vertex, then $S_\gamma \cap U$ contains only one of the sectors, C_i. So the map π is unramified near a $\circ$-vertex. □

Definition 2.8. The **rank** $\mathrm{rk}(\mathcal{W})$ of a spectral web $\mathcal{W}$ is the degree of the map (13).

Lemma 2.9. *For an ideal web $\mathcal{W}$ on a decorated surface S one has*

$$\deg(\widetilde{\pi}) = \deg(\pi).$$

Proof. The degree $\deg(\widetilde{\pi})$ is the number of points in the fiber $\widetilde{\pi}^{-1}(x)$ for a generic $x \in \widetilde{S}$. The group $\pi_1(S)$ acts discretely, so the factorisation by the action of $\pi_1(S)$ does not change the cardinality of the fiber. □

Spectral surface = completed conjugate surface. Let us add to the conjugate ribbon graph Γ^* its faces F_{φ_i} bounding the face paths φ_i. Namely, if φ_i is a face loop, then F_{φ_i} is a disc bounded by this loop. If φ_i is an unclosed face path, then F_{φ_i} is a disc; the half of its boundary is identified with the path φ_i, and the rest becomes boundary of Σ. We arrive at a *completed conjugate surface* $\mathbf{\Sigma}$:

$$\mathbf{\Sigma} = \Gamma^* \cup \cup_{\varphi_i} F_{\varphi_i}. \tag{14}$$

The surface $\mathbf{\Sigma}$ still may have a boundary. Topologically it is obtained by filling the holes on Σ.

Lemma 2.10. *Let $\mathcal{W}$ be a spectral web on S. Then the spectral surface $\Sigma_{\mathcal{W},S}$ is canonically identified with the completed conjugate surface $\mathbf{\Sigma}$ of the bipartite ribbon graph Γ assigned to $\mathcal{W}$:*

$$\mathbf{\Sigma} = \Sigma_{\mathcal{W}}. \tag{15}$$

Proof. We construct the spectral surface $\Sigma_{\mathcal{W}}$ by attaching the domain S_γ to every zig-zag path γ on the bipartite graph Γ. In our case domains S_γ are discs. Furthermore, zig-zag paths γ on Γ are in bijection with the face paths on Γ^* – see (9). □

From now on we make no distinction between the spectral surface $\Sigma_{\mathcal{W}}$ of a spectral web and the completed conjugate surface $\mathbf{\Sigma}$, and keep the latter notation only.

The preimages on $\mathbf{\Sigma}$ of the special points on S are declared to be the special points on $\mathbf{\Sigma}$.

Let Γ be the bipartite graph assigned to the web $\mathcal{W}$. Let φ be a face path on the conjugate graph Γ^*. Let $\mathbf{\Sigma}_\varphi$ be the unique domain obtained by cutting $\mathbf{\Sigma}$ along φ, so that the orientation of the boundary of $\mathbf{\Sigma}_\varphi$ coincides with the orientation of the face path φ. Each face path on Γ^* surrounds just one special point on $\mathbf{\Sigma}$, providing a bijection

$$\{\text{face paths on } \Gamma^*\} \leftrightarrow \{\text{special points of } \mathbf{\Sigma}\}. \tag{16}$$

$$\{\text{face path } \varphi\} \longmapsto \{\text{the unique special point } s_\varphi \text{ inside of the domain } \mathbf{\Sigma}_\varphi\}.$$

2.3. Ideal webs on decorated surfaces

Let us introduce now the notion of an *ideal web*, which is the central new object in the paper. Ideal webs form a particular class of spectral webs.

From now on, S is a decorated surface with $n > 0$ special points $\{s_1, \ldots, s_n\}$. Set

$$S^{\times} = S - \{s_1, \ldots, s_n\}. \quad (17)$$

A bipartite graph on S can have vertices at special points. A web on S means a web on $S^{\times}$.

The double of a decorated surface with boundary. Given an oriented decorated surface S with boundary, its *double* $S_{\mathcal{D}}$ is an oriented decorated surface obtained by gluing S with its mirror S°, given by the surface S with the opposite orientation, into a surface without boundary, by identifying the corresponding components. The set of special points on $S_{\mathcal{D}}$ is inherited from S and S° as follows. Each special point on the boundary of S gives rise to a single special point on the double $S_{\mathcal{D}}$. Each special point s inside of S gives rise to two special points on $S_{\mathcal{D}}$, given by s and its mirror. This way we get all special points on the double.

A web $\mathcal{W}$ on S gives rise to a mirror web $\mathcal{W}^{\circ}$ on S°, given by the strands of the web $\mathcal{W}$ with the reversed orientations. Gluing the webs $\mathcal{W}$ and $\mathcal{W}^{\circ}$ we get a web $\mathcal{W}_{\mathcal{D}}$ on $S_{\mathcal{D}}$.

We can develop the whole story for decorated surfaces without boundary first, and then just require that the story on decorated surfaces with boundary is determined by the condition that everything going fine on its double. We, nevertheless, will spell some of the definitions, e.g., the minimality condition, for decorated surfaces with boundary.

Ideal webs

Definition 2.11. A strand γ on $S^{\times}$ is an **ideal strand** if $S - \gamma$ has two components, and

- The component S_{γ}, see (11), is a disc which contains a unique special point s.

We say that an ideal strand γ as above is *associated to* s.

Definition 2.12. A web $\mathcal{W}$ on a decorated surface S without boundary is **ideal** if

- The web $\widetilde{\mathcal{W}}$ is a minimal web of ideal strands on $\widetilde{S}$.

Observe that the strands of $\widetilde{\mathcal{W}}$ can not have selfintersections due to the condition from Definition 2.11 that $\widetilde{S}_{\gamma}$ is a disc. The minimality implies that parallel bigons are also prohibited.

Definition 2.13. A web $\mathcal{W}$ on a decorated surface S with boundary is **ideal** if the web $\mathcal{W}_{\mathcal{D}}$ on $S_{\mathcal{D}}$ is ideal.

We say that a bipartite graph Γ on S is ideal if the corresponding web is ideal.

A web on S is a *strict ideal web* if it is already a minimal web of ideal strands.

A web on a decorated surface S with boundary is ideal if it consists of ideal strands, minimal, and satisfies the following condition:

- There are no parallel half-bigons, see Figure 17.

FIGURE 17. No parallel half-bigons: the boundary of S is punctured.

Lemma 2.14. *A strict ideal web on a decorated surface S enjoys the following condition:*

- *The strands associated with a given special point do not intersect.*

Proof. Due to, say, the minimality condition, the strands have no selfintersections. So they are simple loops going clockwise around the corresponding special points. Two such loops going around the same special point do not intersect. Indeed, otherwise, since they have the same orientation, this will create a parallel bigon or half-bigon. □

Corollary 2.15. *Let $\mathcal{W}$ be a rank m ideal web on S. Then the strands of the web $\widetilde{\mathcal{W}}$ associated with a special point $s \in \widetilde{S}$ form m concentric (half-)circles on $\widetilde{S}$ going clockwise around s.*

Key example continued. The web in Figure 15 assigned to an ideal triangulation of a torus with a single special point is an ideal web, but it is not a strict ideal web.

Given an ideal web $\mathcal{W}$ on a decorated surface S, the *codistance* $\langle F, s\rangle$ between a face F of the web $\widetilde{\mathcal{W}}$ and a special point s on $\widetilde{S}$ is the number of strands associated with s containing F.

Corollary 2.16. *Let $\mathcal{W}$ be an ideal web on S. Then for any face F of the web $\widetilde{\mathcal{W}}$ one has*

$$\sum_s \langle F, s\rangle = \mathrm{rk}(\mathcal{W}).$$

In particular, the number of strands associated with a given special point equals $\mathrm{rk}(\mathcal{W})$.

Proof. Follows from Theorem 2.7. □

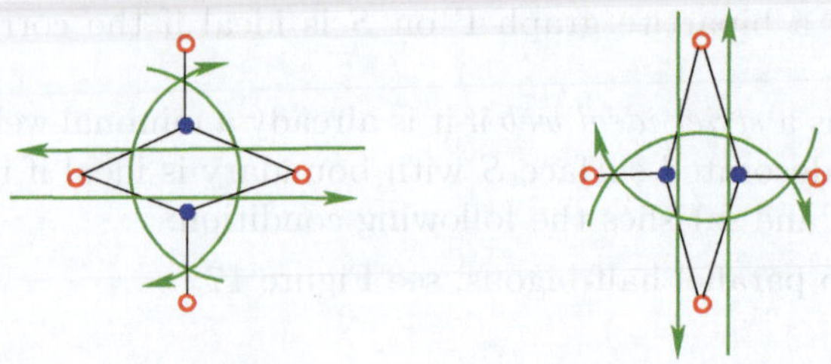

FIGURE 18. A two by two move of bipartite graphs.

Two by two moves. A two by two move of a web on S is a local transformation involving four zig-zag strands, shown in Figure 19, see [T], [GK]. It clearly takes an ideal web to another ideal web. It preserves the rank. Indeed, it decreases by one the codistances to two special points and increases by one the codistances to other two ones. A two by two move amounts to a transformation of bipartite graphs, also known as a two by two move, see Figure 18.

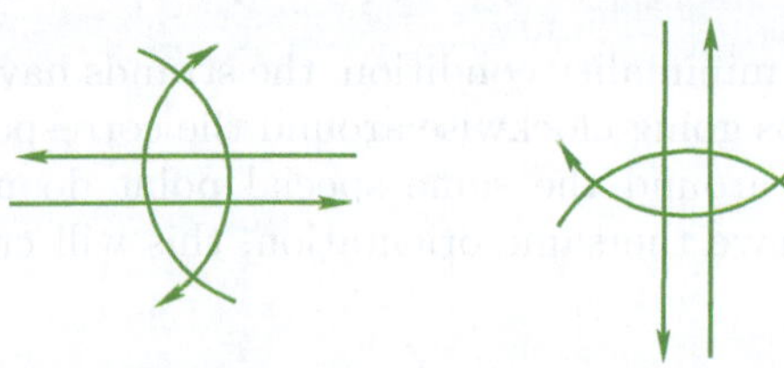

FIGURE 19. A two by two move of webs.

2.4. Ideal webs arising from ideal triangulations of decorated surfaces

Let S be a decorated surface. Given an integer $m \geq 1$, we have assigned to an ideal triangulation of T of S two bipartite graphs on S, the graphs $\Gamma_{\mathrm{A}_m}(T)$ and $\Gamma_{\mathrm{A}^*_m}(T)$. So there are webs associated to each of them. Below we define A_m-webs and A^*_m-webs. Then we show that the associated to these graphs webs are the A_m-webs and A^*_m-webs.

A_m-webs and A^*_m-webs

Definition 2.17. An ideal web $\mathcal{W}$ on S is an A_m**-web** (respectively A^*_m**-web**) if the web $\widetilde{\mathcal{W}}$ on the universal cover $\widetilde{S}$ has m (respectively $m+1$) strands around each special point, and its rank is $m+1$.

As we show below, A_m-webs are related to moduli spaces of A_m-local systems on S, while A^*_m-webs are related to moduli spaces of GL_{m+1}-local systems on S. The number of strands around each special point of an A_m / A^*_m-web is the rank of the corresponding reductive group.

Recall that there are two ways to resolve a triple point diagram, shown on Figure 14. So there is an involution acting on the webs. Lemma 2.18 tells that it interchanges A_m-webs with A_m^*-webs.

Lemma 2.18. *There is an involution $\mathcal{W} \to *\mathcal{W}$ on the set of isotopy classes of webs, which preserves ideal webs on a decorated surface, and interchanges the A_{m+1}-webs with the A_m^*-webs. The orbits of this involution are in bijection with the triple-point diagrams on S.*

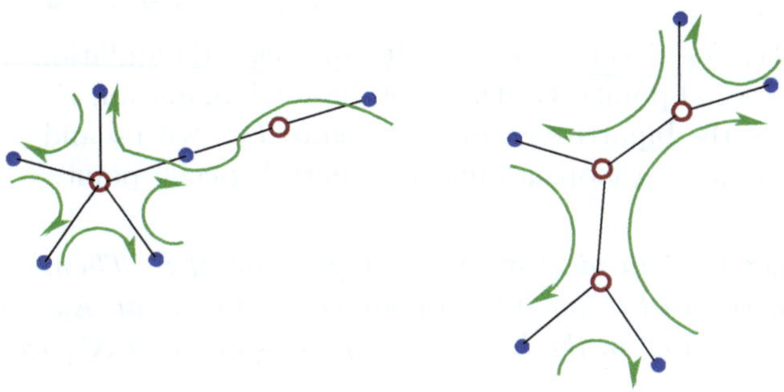

FIGURE 20. Breaking a 5-valent ∘-vertex into three 3-valent ones.

Proof. We replace an arbitrary bipartite graph by a bipartite graph with 3-valent ∘-vertices, so that the corresponding web of zig-zag strands does not change, see Figure 20. Namely, collapse the segments containing 2-valent ∘-vertices, collapsing the two •-vertices of such segments into •-vertices of higher valency, and removing the 2-valent ∘-vertices. Then replace each k-valent ∘-vertex, $k > 3$, by a pair of ∘-vertices of valencies 2 and $k-2$, by inserting a 2-valent •-vertex between them, and keep doing so till no more $k > 3$ valent ∘-vertices remains. Although the obtained bipartite graphs are different, their webs are isotopic.

Assuming all ∘-vertices are 3-valent, we alter the web at each ∘-vertex by moving one of the three strands surrounding this ∘-vertex across the intersection point of the other two. This way we get an involution $\mathcal{W} \to *\mathcal{W}$ on the set of webs, see Figure 14. The involution does not preserve the web rank. It interchanges the A_m-webs and A_m^*-webs.

Assuming that all ∘-vertices are 3-valent, and shrinking all ∘-domains of the corresponding web into points, we get a triple point diagram. Conversely, any triple point diagram can be resolved into a web in two different ways, so that each triple crossing is replaced by either ∘- or •-vertex. □

A_m-webs and A_{m-1}^*-webs assigned to ideal triangulations. Recall that an ideal triangulation T of S is regular if every triangle of the triangulation has three distinct sides.

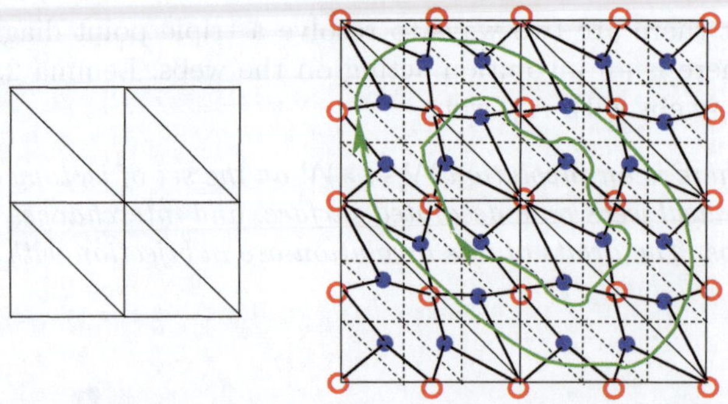

FIGURE 21. A_2-picture. On the left: an ideal triangulation of a torus with four special points. On the right: its 3-triangulation T_3 – shown by dotted lines; the bipartite graph assigned to T_3 – with • and ○- vertices; and the two zig-zag loops around the central special point.

Lemma 2.19. *Let T be an ideal regular triangulation of S. Then:*
i) *The web associated to the bipartite graph $\Gamma_{A_m}(T)$ is an A_m-web on S.*
ii) *The web associated to the bipartite graph $\Gamma_{A^*_m}(T)$ is a A^*_m-web on S.*

Proof. i) It is clear now from Figure 3 that zig-zag strands on the graph are ideal strands. Indeed, given a special point s of S, take the triangles $t_1, \ldots, t_k$ sharing s, ordered clockwise. In each triangle t_i consider the zig-zag strands parallel to the side opposite to s. They form m arcs going clockwise around s. Considering all triangles sharing the puncture, we get m strands associated with s. The rank of the obtained ideal web is $m + 1$.

ii) The proof if very similar to i), and thus is omitted. □

Lemma 2.20. *There is a bijection between A_1-webs and ideal triangulations of S, such that two by two moves of ideal webs correspond to flips of triangulations, see Figure* 8.

Proof. We can assume that all •-vertices are 3-valent. Given an A_1-web on a decorated surface S, its zig-zag strands are in the bijection with the set of special vertices of S. Since the rank of the web is two, each face of the web determines two special points, and defines a homotopy class of a path connecting them. Let us choose their representatives, and declare them the edges. Then each •-vertex provides us a triangle, and we arrive at an ideal triangulation of S. By the construction, the A_1-web assigned to it is the original web. Then ○-vertices are the special points. The second claim is clear from Figure 8. □

2.5. Quivers associated to bipartite graphs on decorated surfaces

We follow [GK, Section 4.1.3], with modifications for surfaces with boundaries.

Let S be a decorated surface. A face of a bipartite graph Γ on S is *external* if it intersects the boundary of S, or its interior contains a special point. It is called *internal* otherwise.

Below we use a slightly relaxed definition of a quiver, where we do not require the vectors $\{e_i\}$ to form a basis of Λ, replacing this by the condition that they generate the lattice. We call the vectors $\{e_i\}$ the *basic vectors.* Each basic vector is classified as either frozen or non-frozen.

Definition 2.21. A quasi-quiver is a data $\Big(\Lambda, \{e_i\}, (*,*)\Big)$, where Λ is a lattice, $\{e_i\}$ is a finite set of vectors of two types – non-frozen and frozen – generating the lattice Λ, and $(*,*)$ a skew-symmetric $\frac{1}{2}\mathbb{Z}$-valued bilinear form on Λ such that $(e_i, e_j) \in \mathbb{Z}$ unless both e_i and e_j are frozen.

All quivers in this paper are either traditional quivers, or quasi-quivers in the sense of Definition 2.21 where the basic vectors e_i satisfy just a single relation: $\sum_i e_i = 0$. Abusing terminology, we refer to them as quivers. In all examples in this paper $(e_i, e_j) \in \mathbb{Z}$ for any e_i, e_j.

A quiver for a decorated surface S without boundary. So S is a compact surface with a non-empty subset of special points. Let Γ be a bipartite graph on S.

Let us assume that special points can not be vertices of Γ. Then

$$\Gamma \subset S^\times := S - \{\text{special points}\}.$$

Set

$$\Lambda_\Gamma := \text{Ker}\Big(H_1(\Gamma;\mathbb{Z}) \longrightarrow H_1(S;\mathbb{Z})\Big). \tag{18}$$

$$\Lambda_\Gamma^\circ := \text{Ker}\Big(H_1(\Gamma;\mathbb{Z}) \longrightarrow H_1(S^\times;\mathbb{Z})\Big). \tag{19}$$

The inclusion $S^\times \subset S$ induces a map $H_1(S^\times;\mathbb{Z}) \longrightarrow H_1(S;\mathbb{Z})$. So there is a canonical embedding

$$\Lambda_\Gamma^\circ \subset \Lambda_\Gamma.$$

Recall the spectral surface Σ_Γ of Γ. It is the surface associated with the conjugate bipartite ribbon graph Γ^*. Therefore the surface Σ_Γ is homotopy equivalent to Γ^*. Since the abstract graphs Γ^* and Γ coincide, Σ_Γ is homotopy equivalent to Γ. Filling the faces on Σ_Γ by discs, we get the completed spectral surface $\mathbf{\Sigma}_\Gamma$. The inclusion $\Sigma_\Gamma \hookrightarrow \mathbf{\Sigma}_\Gamma$ provides an exact sequence

$$0 \longrightarrow Z_\Gamma \longrightarrow H_1(\Sigma_\Gamma;\mathbb{Z}) \longrightarrow H_1(\mathbf{\Sigma}_\Gamma;\mathbb{Z}) \longrightarrow 0. \tag{20}$$

The subgroup Z_Γ is generated by the zig-zag loops of Γ, which are the same as the face loops of Σ_Γ. They satisfy a single relation: the sum of all generators equals to zero. The intersection form on $H_1(\mathbf{\Sigma}_\Gamma;\mathbb{Z})$ provides a skewsymmetric bilinear form with the kernel Z_Γ:

$$H_1(\Sigma_\Gamma;\mathbb{Z}) \wedge H_1(\Sigma_\Gamma;\mathbb{Z}) \longrightarrow \mathbb{Z}. \tag{21}$$

Since Γ is homotopy equivalent to Σ_Γ, we get a form on $H_1(\Gamma;\mathbb{Z})$. Restricting it to the Λ_Γ we get a skewsymmetric bilinear form

$$(*,*)_\Gamma : \Lambda_\Gamma \wedge \Lambda_\Gamma \longrightarrow \mathbb{Z}. \tag{22}$$

Let $\mathbb{Z}[X]$ be the free abelian group generated by a set X. It has a canonical basis $\{x\}$, $x \in X$.

Since we assumed that the faces of Γ are discs on S, there are exact sequences

$$0 \longrightarrow H_2(S,\mathbb{Z}) \stackrel{r}{\longrightarrow} \mathbb{Z}[\{\text{Faces of } \Gamma\}] \longrightarrow H_1(\Gamma;\mathbb{Z}) \longrightarrow H_1(S;\mathbb{Z}) \longrightarrow 0. \quad (23)$$

$$0 \longrightarrow \mathbb{Z}[\{\text{Internal faces of } \Gamma\}] \longrightarrow H_1(\Gamma;\mathbb{Z}) \longrightarrow H_1(S^{\times};\mathbb{Z}) \longrightarrow 0. \quad (24)$$

So there are isomorphisms

$$\Lambda_\Gamma^\circ = \mathbb{Z}[\{\text{Internal faces of } \Gamma\}], \ \Lambda_\Gamma = \frac{\mathbb{Z}[\{\text{Faces of } \Gamma\}]}{r(H_2(S,\mathbb{Z}))}.$$

There is a basis $\{e_G\}$ of the lattice Λ_Γ° given by the internal faces G of Γ, and a collection of vectors $\{e_F\}$ given by all faces F of Γ, satisfying a single relation

$$\sum_{F:\text{ faces of } \Gamma} e_F = 0.$$

The quiver $\mathbf{q}_\Gamma$ assigned to a bipartite graph Γ on S is given by the triple

$$\Big(\Lambda_\Gamma, \{e_F\}, \langle *,* \rangle_\Gamma\Big).$$

The vectors e_G assigned to the internal faces of Γ are the non-frozen basic vectors.

A quiver for a decorated surface with boundary. Suppose now that a decorated surface S has boundary components. Let $S_{\mathcal{D}}$ be the topological double of S. It is an oriented surface without boundary obtained by gluing the surface S and its mirror S° along the matching boundary components. It inherits the special points from S.

Let Γ be any bipartite ribbon graph. Changing the color of each vertex, and the cyclic order of the half-edges at every vertex, we arrive at its mirror – a bipartite ribbon graph Γ°.

Let Γ be a bipartite ribbon graph on S. So its external vertices are on the boundary, and they are not colored. The mirror graph Γ° lives naturally on S°. We glue the graphs Γ and Γ° along the matching external vertices into a bipartite graph $\Gamma_{\mathcal{D}} \subset S_{\mathcal{D}}$. Let $S_{\mathcal{D}}^\circ$ be the surface $S_{\mathcal{D}}$ with the opposite orientation. There is a canonical involutive isomorphism $\sigma : S_{\mathcal{D}} \longrightarrow S_{\mathcal{D}}^\circ$ flipping S and S°. It induces an isomorphism of bipartite ribbon graphs

$$\sigma : \Gamma_{\mathcal{D}} \stackrel{\sim}{\longrightarrow} \Gamma_{\mathcal{D}}^\circ.$$

Notice that as abstract graphs, $\Gamma_{\mathcal{D}}$ coincides with $\Gamma_{\mathcal{D}}^\circ$. Moreover there is a canonical isomorphism of lattices $\Lambda_{\Gamma_{\mathcal{D}}^\circ} = \Lambda_{\Gamma_{\mathcal{D}}}$, and the isomorphism σ induces an involution

$$\sigma_* : \Lambda_{\Gamma_{\mathcal{D}}} \stackrel{\sim}{\longrightarrow} \Lambda_{\Gamma_{\mathcal{D}}^\circ} = \Lambda_{\Gamma_{\mathcal{D}}}.$$

Definition 2.22. Let Γ be a bipartite graph Γ on a decorated surface S with boundary.

Lattices $(\Lambda_\Gamma, \Lambda_\Gamma^\circ)$ are the coinvariants of the induced involution σ_* on the lattices $(\Lambda_{\Gamma_{\mathcal{D}}}, \Lambda_{\Gamma_{\mathcal{D}}}^\circ)$:

$$\Lambda_\Gamma := \Big(\Lambda_{\Gamma_{\mathcal{D}}}\Big)_{\sigma_*} := \Lambda_{\Gamma_{\mathcal{D}}}/(\sigma_*(l) - l), \ \Lambda_\Gamma^\circ := \Big(\Lambda_{\Gamma_{\mathcal{D}}}^\circ\Big)_{\sigma_*}.$$

Lemma 2.23. *Let Γ be a bipartite graph on a decorated surface S with a boundary. Then*

$$\Lambda_\Gamma = \frac{\mathbb{Z}[\{\text{Faces of } \Gamma\}]}{r(H_2(S_\mathcal{D};\mathbb{Z}))}, \quad \Lambda^\circ_\Gamma = \mathbb{Z}[\{\text{Internal faces of } \Gamma\}].$$

Proof. The lattice Λ_Γ has canonical vectors associated with the faces of Γ, internal or external. Namely, if F is an internal face of Γ, it is also a face $F_\mathcal{D}$ of $\Gamma_\mathcal{D}$. If F is an external face of Γ, gluing it to its mirror F° we get a single face $F_\mathcal{D}$ of $\Gamma_\mathcal{D}$. Projecting the vectors $e_{F_\mathcal{D}} \in \Lambda_{\Gamma_\mathcal{D}}$ to the coinvariants we get canonical vectors $e_F \in \Lambda_\Gamma$. □

Definition 2.24. The quiver assigned to a bipartite surface graph Γ is given by

$$\mathbf{q}_\Gamma := \Big(\Lambda_\Gamma, \{e_F\}, \langle *,*\rangle_\Gamma\Big).$$

The vectors $\{e_G\}$ assigned to the internal faces G of Γ are the non-frozen basic vectors.

The non-frozen subquiver is given by the sublattice Λ°_Γ with a basis $\{e_G\}$ and the induced form:

$$\mathbf{q}^\circ_\Gamma := \Big(\Lambda^\circ_\Gamma, \{e_G\}, \mathrm{Res}_{\Lambda^\circ_\Gamma}\langle *,*\rangle_\Gamma\Big).$$

Lemma 2.25. [GK, Lemma 4.5]. *A two by two move $\Gamma \longrightarrow \Gamma'$ centered at an internal face G leads to a mutation of quivers $\mathbf{q}_\Gamma \longrightarrow \mathbf{q}_{\Gamma'}$ in the direction of the vector e_G.*

A combinatorial description of the form $\langle *,*\rangle_\Gamma$. Let F be a face of a bipartite graph Γ on a decorated surface S, which may have a boundary. It gives rise to a face $F_\mathcal{D}$ of the bipartite graph $\Gamma_\mathcal{D}$. Its boundary, oriented by the orientation of $S_\mathcal{D}$, provides a zig-zag loop $\partial F_\mathcal{D}$ with its canonical orientation on the conjugate surface $\Sigma_{\Gamma_\mathcal{D}}$. Given a pair of faces F, G of Γ, the intersection index $\langle F, G\rangle_\Gamma$ of the loops $\partial F_\mathcal{D}$ and $\partial G_\mathcal{D}$ on the conjugate surface $\Sigma_{\Gamma_\mathcal{D}}$ can be calculated as follows.

Let E be an edge of the graph Γ. We set $\langle F, G\rangle^{(E)} = 0$ if E is not a common edge of F and G, $\langle F, G\rangle^{(E)} = +1$ if the (clockwise on our pictures) orientation of F induces $\bullet \to \circ$ orientation of E, and $\langle F, G\rangle^{(E)} = -1$ otherwise.

Lemma 2.26. [GK, Proposition 8.3] *One has*

$$\langle F, G\rangle_\Gamma = \sum_{E:\text{ edges of }\Gamma} \langle F, G\rangle^{(E)}.$$

Let Γ be a bipartite graph on a decorated surface S with boundary. Consider a lattice

$$\widetilde{\Lambda}_\Gamma := \mathbb{Z}[\{\text{Faces of } \Gamma\}].$$

It has a basis $\{e_F\}$ provided by the faces F of Γ. The skew-symmetric bilinear form $\langle *,*\rangle_\Gamma$ is well defined on the lattice $\widetilde{\Lambda}_\Gamma$. We define a traditional quiver $\widetilde{\mathbf{q}}_\Gamma$ as a triple

$$\widetilde{\mathbf{q}}_\Gamma = \Big(\widetilde{\Lambda}_\Gamma, \{e_F\}, \langle *,*\rangle_\Gamma\Big).$$

The quiver $\mathbf{q}_\Gamma$ from Definition 2.24 is its quotient by the rank one sublattice spanned by the vector $\sum_F e_F$ in the kernel of the form $\langle *, * \rangle_\Gamma$. External faces provide the frozen basis vectors.

Quivers for A_m / A_m^*-bipartite graphs. Let S be a decorated surface.

Let Γ be a graph on S. We assume that its vertices are disjoint from the special points of S. We say that a face F of Γ is *non-special* if it does not contain a special point of S.

Let Γ_{A_m} be an A_m-bipartite graph on S, and $\Gamma_{\mathrm{A}_m^*}$ the associated A_m^*-bipartite graph on S. Although the special points of S form a subset of the set of $\circ$-vertices of Γ_{A_m}, the $\Gamma_{\mathrm{A}_m^*}$ is a graph on the punctured surface $S^\times$. Moreover, one has

$$\{\text{Faces of the graph } \Gamma_{\mathrm{A}_m}\} = \{\text{Non-special faces of the graph } \Gamma_{\mathrm{A}_m^*}\}. \quad (25)$$

It is clear from the way we glue the spectral surface from the discs associated with the zig-zag loops that there is a canonical isomorphism of surfaces

$$\mathbf{\Sigma}_{\Gamma_{\mathrm{A}_m}} = \mathbf{\Sigma}_{\Gamma_{\mathrm{A}_m^*}}.$$

Lemma 2.27. *Let S be a decorated surface without boundary. Then there are canonical isomorphisms of lattices*

$$\mathbb{Z}[\{\text{Faces of the graph } \Gamma_{\mathrm{A}_m}\}] = \mathbb{Z}[\{\text{Non-special faces of the graph } \Gamma_{\mathrm{A}_m^*}\}] \quad (26)$$
$$= \mathrm{Ker}\Big(H_1(\Gamma_{\mathrm{A}_m^*};\mathbb{Z}) \longrightarrow H_1(S^\times;\mathbb{Z})\Big) = \mathrm{Ker}\Big(H_1(\Sigma_{\Gamma_{\mathrm{A}_m^*}};\mathbb{Z}) \stackrel{\pi_*}{\longrightarrow} H_1(S^\times;\mathbb{Z})\Big).$$

Proof. The first isomorphism follows from (25). The second is obvious since S has no boundary, and the external faces of $\Gamma_{\mathrm{A}_m^*}$ are the ones which have special points inside. The third follows since the embedding $\Gamma_{\mathrm{A}_m^*} \hookrightarrow \Sigma_{\Gamma_{\mathrm{A}_m^*}}$ is a homotopy equivalence, and followed by the spectral map, induces the embedding $\Gamma_{\mathrm{A}_m^*} \hookrightarrow S^\times$. □

Lemma 2.28. *There is an isomorphism of lattices, where π_* is induced by the spectral map:*

$$\mathrm{Ker}\Big(H_1(\Sigma_\Gamma;\mathbb{Z}) \stackrel{\pi_*}{\longrightarrow} H_1(S;\mathbb{Z})\Big) = \mathrm{Ker}\Big(H_1(\Gamma;\mathbb{Z}) \longrightarrow H_1(S;\mathbb{Z})\Big). \quad (27)$$

Proof. The conjugate surface for the bipartite surface graph Γ is identified with its spectral surface Σ_Γ. The conjugate surface of any bipartite ribbon graph Γ is homotopy equivalent to Γ. So $H_1(\Gamma;\mathbb{Z}) = H_1(\Sigma_\Gamma;\mathbb{Z})$. Using this identification, the map $H_1(\Gamma;\mathbb{Z}) \longrightarrow H_1(S;\mathbb{Z})$ is identified with the map $H_1(\Sigma_\Gamma;\mathbb{Z}) \stackrel{\pi_*}{\longrightarrow} H_1(S;\mathbb{Z})$. □

3. Ideal webs on polygons and configurations of (decorated) flags

3.1. Ideal webs on a polygon

Consider an example when the decorated surface S is a polygon P, whose vertices are the special points. Let us spell the definition of ideal webs in this case. Notice that in this case any ideal web is strict ideal – there are no non-trivial covers of the polygon.

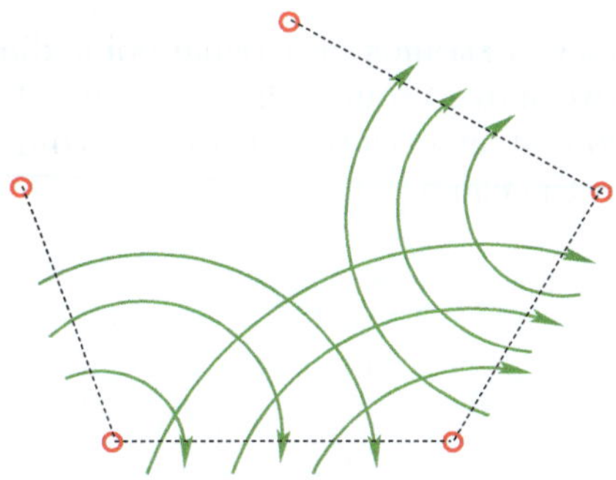

FIGURE 22. Strands of a rank 4 ideal web around three consecutive (red) vertices.

Definition 3.1. An ideal web in a polygon is a collection of oriented strands in the polygon with simple crossings, considered modulo isotopy, such that

1. As we move along a strand, the orientations of the crossing strands alternate.
2. The web is connected, the complement to the web is a disjoint union of discs.
3. Each strand surrounds exactly one vertex of the polygon.
4. Strands have no selfcrossings; there are no parallel bigons.
5. Strands have no parallel half-bigons.

Recall that two webs are equivalent if they are related by a sequence of two by two moves.

Theorem 3.2. *Any rank $m+1$ ideal web on a polygon is either an A_m-web, or a A_m^*-web.*

Any two A_m-webs are equivalent, and any two A_m^*-webs are equivalent

Proof. Recall that resolving the intersection points of a triple point diagram in one of the two standard ways, see Figure 14, we get a web, and this procedure can be reversed. The two by two moves of triple crossing diagrams, see Figure 23, match the ones of the webs.

Therefore an ideal web in a polygon gives rise to a minimal triple point diagram. Each strand has the "in" endpoint, and the "out" endpoint. The operations above do not change the matching of the "in" and "out" points on the boundary provided by the oriented strands, so that the strand is oriented from "in" to "out". □

FIGURE 23. A two by two move of triple point diagrams.

Let n be the number of strands of a triple point diagram on the disc. The endpoints alternate as we move around the boundary. Indeed, the orientations of the strands induce consistent orientations on the complementary regions, see Figure 24, which implies the claim.

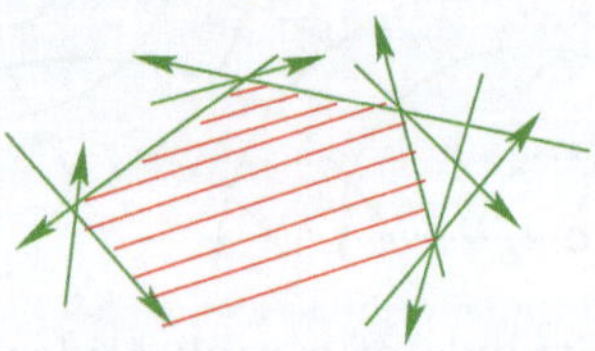

FIGURE 24. For any triple point diagram, the orientations of the strands induce consistent orientations on the complementary regions.

A triple point diagram on the disc is *minimal* if the number of its intersection points is not bigger than for any other diagram inducing the same matching on the boundary.

Given a collection of $2n$ alternatively oriented points on the boundary, a triple crossing diagram inducing such a collection provides a matching between the "in" and "out" points.

Theorem 3.3 ([T]).

1) *In a disc with* $2n$ *endpoints on the boundary, all* $n!$ *matchings of "in" endpoints with "out" endpoints are achieved by minimal triple point diagrams.*
2) *Any two minimal triple point diagrams on the disc with the same matching on the endpoints are equivalent, that is can be related by a sequence of* $2 \leftrightarrow 2$ *moves, shown on Figure* 23.

Let us return to the proof of Theorem 3.2. Each strand goes clockwise around a vertex v of the polygon. Its "in" and "out" endpoints are at the sides of the polygon shared by v.

Lemma 3.4. *Consider an ideal rank* m *web on a polygon. Then each side of the polygon supports* $2(m-1)$ *alternating "out" and "in" endpoints. It starts from an "out" endpoint if we count counterclockwise, see Figure* 22.

Proof. Due to the minimality there are no selfintersections or parallel bigons formed by the strands surrounding a given vertex of the polygon. This plus condition 5) implies (Corollary 2.15) that the strands surrounding a vertex v of the polygon form "concentric half-circles". They start at the side preceding v, and ends on the next side, following counterclockwise orientation.

Recall that for any triple point diagram on the disc, as we move along the boundary of the disc, the directions of the ends at the boundary alternate. Therefore for any web on the disc, the directions of the ends at the boundary alternate.

Finally, take the closest to v strand surrounding a vertex v. It surround at least two endpoints. Indeed, if it surrounds no points, it can be contracted, which contradicts to the minimality. The only other option compatible with the above properties is that it surrounds two endpoints. □

Corollary 3.5. *All rank m ideal webs on a polygon have the same "in" and "out" matching of the ends, induced by the strands of the web.*

Therefore thanks to Theorem 3.3 the triple point crossing diagrams assigned to any two rank m ideal webs on the polygon are equivalent. Thus there are exactly two equivalences classes of rank m ideal webs on the polygon.

The ideal A_m-web on the disc assigned to a triangulation of the polygon has the configuration of "in" and "out" endpoints and their matching as described in Lemma 3.4. A two by two move of an A_m-web is an A_m-web. The same is true for the ideal A_m^*-web on the disc assigned to a triangulation of the polygon. Theorem 3.2 is proved.

3.2. $\mathbf{A}_m$ / $\mathbf{A}_m^*$-webs on polygons and cluster coordinates

3.2.1. $\mathbf{A}_m$-webs and cluster coordinates on configurations of decorated flags. Let (V_m, Ω_m) be an m-dimensional vector space with a volume form Ω_m.

Definition 3.6.

i) A decorated flag in (V_m, Ω_m) is a data $(F_\bullet, \alpha_\bullet)$:

$$F_\bullet : \ 0 = F_0 \subset F_1 \subset \cdots \subset F_{m-1} \subset F_m = V_m, \quad \mathrm{dim} F_p = p, \ \alpha_p \in \det F_p - 0, \tag{28}$$

consisting of a flag $F_\bullet$ and non-zero p-vectors $\alpha_p \in \mathrm{det} F_p$ for $p = 1, \ldots, m-1$.

ii) A $*$-decorated flag in (V_m, Ω_m) is a similar data $(F_\bullet, \alpha_\bullet)$ where $\alpha_p \in \mathrm{det} F_p$ for $p = 1, \ldots, m$.

Instead of the p-vectors α_p one can choose non-zero vectors $v_p \in F_p/F_{p-1}$, $p = 1, \ldots, m-1$. Indeed, given the vectors, the volume forms are $\alpha_p = v_1 \wedge \cdots \wedge v_p$.

Denote by $\mathcal{A}_{SL_m}$ the moduli space of decorated flags in (V_m, Ω_m), and by $\mathrm{Conf}_n(\mathcal{A}_{SL_m})$ the moduli space of configurations of n decorated flags:

$$\mathrm{Conf}_n(\mathcal{A}_{SL_m}) := (\mathcal{A}_{SL_m})^n / \mathrm{Aut}(V_m, \Omega_m).$$

Take an oriented n-gon P_n whose vertices are numbered by $(1, \ldots, n)$ following the clockwise orientation. Consider configurations of decorated flags $(A_1, \ldots, A_n)$ in (V_m, Ω_m) labeled by the vertices of P_n. Let $\mathcal{W}$ be an A_{m-1}-web on the polygon P_n. Given a face F, let d_i be the codistance from F to the vertex v_i. Corollary 2.16 tells that

$$d_1 + \cdots + d_n = m. \tag{29}$$

Given an decorated flag $A = (F_\bullet, \alpha_\bullet)$, see (28), we use the convention

$$\alpha_0(A) = 1 \in \mathbb{C} = \det F_0. \tag{30}$$

So $\alpha_0(A)$ is the unit in the exterior algebra of V_m.

Definition 3.7. Given an A_{m-1}-web $\mathcal{W}$ on P_n, we assign to its face F a regular function $A_F^{\mathcal{W}}$ on the space $\mathrm{Conf}_n(\mathcal{A}_{SL_m})$ given by

$$A_F^{\mathcal{W}} := \langle \Omega_m, \alpha_{d_1}(A_1) \wedge \cdots \wedge \alpha_{d_n}(A_n) \rangle. \tag{31}$$

Lemma 3.8. *The functions $A_F^{\mathcal{W}}$ are invariant under the twisted cyclic shift*

$$\begin{aligned} t : \mathrm{Conf}_n(\mathcal{A}_{SL_m}) &\longrightarrow \mathrm{Conf}_n(\mathcal{A}_{SL_m}), \\ (A_1, \ldots, A_n) &\longmapsto ((-1)^{m-1} A_n, A_1, \ldots, A_{n-1}). \end{aligned}$$

Proof. Straightforward. □

3.2.2. A_m^*-webs and cluster coordinates on configurations of $*$-decorated flags. Recall the configuration space of n $*$-decorated flags for the group $SL_m = \mathrm{Aut}(V_m, \Omega_m)$:

$$\begin{aligned} \mathrm{Conf}_n(\mathcal{A}^*_{SL_m}) &:= \Big(\mathcal{A}_{GL_m} \times \cdots \times \mathcal{A}_{GL_m}\Big)/\mathrm{Aut}(V_m, \Omega_m), \\ \mathcal{A}_{GL_m} &:= GL_m/U. \end{aligned} \tag{32}$$

Given an A^*_{m-1}-web $\mathcal{W}_{\mathrm{A}^*_{m-1}}$ on an n-gon, just the same construction as above provides regular functions $A_F^{\mathcal{W}}$ on the space (32). We assign the coordinates to <u>all faces</u> of the A^*_{m-1}-web $\mathcal{W}_{\mathrm{A}^*_{m-1}}$. The number of the faces equals the dimension of the space (32).

The coordinates at the <u>internal</u> faces of $\mathcal{W}_{\mathrm{A}^*_{m-1}}$ are nothing else but the coordinates assigned to the associated A_{m-1}-web $\mathcal{W}_{\mathrm{A}_{m-1}}$: this web is obtained by shrinking all 2-valent $\bullet$-vertices, and creating as a result a $\circ$-vertex at each vertex of the polygon.

Finally, one has

$$\mathrm{Conf}_n(\mathcal{A}_{GL_m}) := \Big(\mathcal{A}_{GL_m} \times \cdots \times \mathcal{A}_{GL_m}\Big)/\mathrm{Aut}(V_m) = \mathrm{Conf}_n(\mathcal{A}^*_{SL_m})/GL(1). \tag{33}$$

Here the group $GL(1)$ acts diagonally. The action of an element $t \in GL(1)$ amounts to multiplying by t each of the coordinates $\{A_F^{\mathcal{W}}\}$ at the faces of an A^*_{m-1}-web $\mathcal{W}_{\mathrm{A}^*_{m-1}}$.

3.2.3. A_m-webs and cluster Poisson coordinates on configurations of flags. Let $\mathcal{B}_m$ be the space of flags in V_m. Consider the moduli space of configurations of n flags:

$$\mathrm{Conf}_n(\mathcal{B}_m) := (\mathcal{B}_m)^n/\mathrm{Aut}(V_m).$$

Let F, G be two faces of an A_{m-1}-web $\mathcal{W}$ on a polygon P_n. Recall the pairing $\langle F, G\rangle_{\mathcal{W}}$.

There is a projection $\mathcal{A}_{SL_m} \to \mathcal{B}_m$. Given a configuration of flags $(F_1, \ldots, F_n)$, choose a configuration $(A_1, \ldots, A_n)$ of decorated flags projecting onto it. Then for any face G of an ideal web $\mathcal{W}$ we have a function $A_G^{\mathcal{W}}$.

Definition 3.9. Given an A_{m-1}-web $\mathcal{W}$ on P_n, we assign to an internal face F of $\mathcal{W}$ a rational function $X_F^{\mathcal{W}}$ on the space $\mathrm{Conf}_n(\mathcal{B}_m)$, given by the following product over the faces G:

$$X_F^{\mathcal{W}} := \prod_G (A_G^{\mathcal{W}})^{\langle F,G\rangle_{\mathcal{W}}}. \tag{34}$$

Lemma 3.10. *The function $X_F^{\mathcal{W}}$ does not depend on the choice of configuration $(A_1, \ldots, A_n)$.*

Proof. Take a strand γ. It is associated with a special point s. Let d be the codistance from γ to s. Let us calculate how the coordinate X_F changes when we multiply α_d by a non-zero number λ. Clearly X_F might change only if γ intersects some edges of the face F. In this case, it must intersect a pair of consecutive edges E_1, E_2 sharing a vertex v, or may be several such pairs. Let us investigate what happens near v. There is a unique face G_i containing the edge E_i and different then F, see Figure 25. The faces G_1, G_2 share the vertex v. They are on the same side of the zig-zag strand γ. Thus their contribution to X_F, using notation from Lemma 2.26, is

$$(A_{G_1})^{\langle F,G_1\rangle^{(E_1)}}(A_{G_2})^{\langle F,G_1\rangle^{(E_2)}} = (A_{G_1}/A_{G_2})^{\pm 1}.$$

Therefore it does no change if we multiply α_d by λ. □

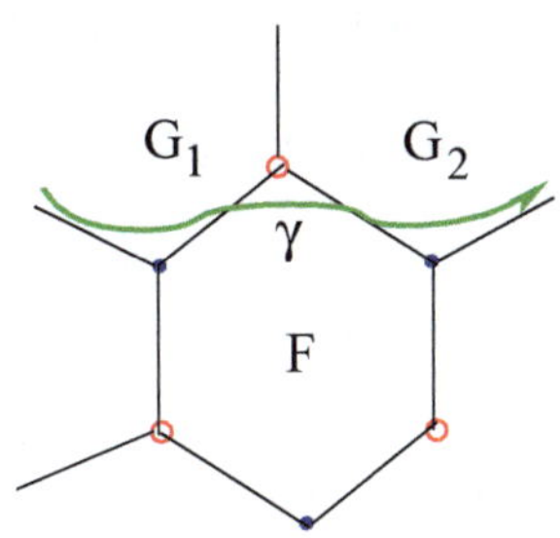

FIGURE 25. Contribution to X_F of the decorated flag associated with a strand γ.

3.3. Cluster nature of the coordinates assigned to A_m / A_m^*-webs

Recall that what we call a cluster / cluster Poisson coordinate system was called in [FG2] cluster $\mathcal{A}$ / cluster $\mathcal{X}$- coordinate system. A collection of cluster / cluster Poisson coordinate systems related by cluster / cluster Poisson transformations form a cluster / cluster Poisson atlas.

Theorem 3.11.

a) A_{m-1}*-webs on an n-gon provide a cluster atlas on* $\mathrm{Conf}_n(\mathcal{A}_{SL_m})$. *Precisely:*
  1. *The functions* $\{A_F^{\mathcal{W}}\}$ *at the faces F of a given* A_{m-1}*-web* $\mathcal{W}$ *form a regular coordinate system on the space* $\mathrm{Conf}_n(\mathcal{A}_{SL_m})$.
  2. *Any two* A_{m-1}*-webs are related by two by two and shrink / expand moves.*
  Any two by two move of webs $\mu_F : \mathcal{W} \to \mathcal{W}'$ *centered at a face F of* $\mathcal{W}$ *amounts to a cluster mutation at the face F. The shrink / expand moves do not affect the coordinates.*

b) *Internal faces of* A_{m-1}*-webs on an n-gon provide a cluster Poisson atlas on* $\mathrm{Conf}_n(\mathcal{B}_m)$*:*
  1. *The functions* $\{X_F^{\mathcal{W}}\}$ *assigned to the internal faces F of a given* A_{m-1}*-web* $\mathcal{W}$ *form a rational coordinate system on the space* $\mathrm{Conf}_n(\mathcal{B}_m)$.
  2. *Any two by two move* $\mu_F : \mathcal{W} \to \mathcal{W}'$ *amounts to a cluster Poisson mutation at the face F. The shrink / expand moves do not affect the coordinates.*

c) *The* A_{m-1}^**-webs on an n-gon provide a cluster atlas on* $\mathrm{Conf}_n(\mathcal{A}_{SL_m}^*)$*: The functions* $\{A_F^{\mathcal{W}}\}$ *at the faces F of a given* A_{m-1}^**-web* $\mathcal{W}$ *form a regular coordinate system on* $\mathrm{Conf}_n(\mathcal{A}_{SL_m}^*)$. *A two by two move* $\mu_F : \mathcal{W} \to \mathcal{W}'$ *amounts to a cluster mutation at F.*

d) *Given an* A_{m-1}^**-web* $\mathcal{W}$ *on an n-gon, the functions* $\{A_F^{\mathcal{W}}\}$ *provide a birational isomorphism* $\Lambda_{\mathcal{W}} \otimes_{\mathbb{Z}} \mathbb{G}_m \longrightarrow \mathrm{Conf}_n(\mathcal{A}_{GL_m})$.

Proof. Our construction recovers the cluster coordinates introduced in [FG1, Section 8].

Lemma 3.12. *Let* $\mathcal{W}_T$ *be the* A_m*-web assigned to an ideal triangulation T of the polygon. Then*

i) *For the web* $\mathcal{W}_T$*, the functions from Definitions* 3.7 *and* 3.9 *coincide with the coordinates assigned to T in loc. cit.*

ii) *The form* $\langle F, G\rangle$ *for the web* $\mathcal{W}_T$ *coincides with the form* ε_{ij} *on the set I parametrizing the coordinates considered in loc. cit.*

iii) *A two by two move of the web* $\mathcal{W}_T$ *corresponds to a mutation in loc. cit.*

Proof. i) For the cluster coordinates this is clear from the very definition. Then formula (37) is the standard formula relating the cluster and cluster Poisson coordinates.

ii) It is clear from the very definitions.

iii) Follows from Section 2.1. □

a) We proved in [FG1] that the coordinates assigned to an ideal triangulation T of the polygon are indeed coordinates on the moduli space $\mathrm{Conf}_n(\mathcal{A}_{SL_m})$. The part i) of Lemma 3.12 tells that those coordinates coincide with the ones arising from the web assigned to T. Finally and two A_{m-1}-webs on the polygon are related by two by two moves by Theorem 3.2. Thanks to the Plücker identity between the minors of a 2×4 matrix, and similar identities between higher order minors induced by it, a two by two move amounts to a cluster mutation of the $\mathcal{A}$-coordinates. The part a) is proved.

b) Follows from a) and Lemma 3.12 by using the standard relationship between the and cluster Poisson coordinates, see (34).

c) Deduced easily from a).

d) It is deduced from c). Indeed, $\mathrm{Conf}_n(\mathcal{A}_{GL_m}) = \mathrm{Conf}_n(\mathcal{A}^*_{SL_m})/\mathbb{G}_m$, and on the other hand the elements e_F generate the lattice $\Lambda_{\mathcal{W}}$ and satisfy a single relation $\sum e_F = 0$. □

4. Ideal webs and coordinates on moduli spaces of local systems

4.1. Cluster coordinates for moduli spaces of local systems on surfaces

Denote by $T'S$ the bundle of non-zero tangent vectors to the surface S. A *twisted SL_m-local system* on S is a local system on $T'S$ which has monodromy $(-1)^{m-1}$ around a loop rotating a tangent vector at any point by 360°.

Let G be either SL_m or GL_m. Let U be a maximal unipotent subgroup of G. Let $\mathcal{L}$ be a G-local system on a space. Set

$$\mathcal{L}_{\mathcal{A}} := \mathcal{L}/U.$$

When $\mathcal{L}$ is an SL_m-local system on a space, we also set

$$\mathcal{L}_{\mathcal{A}^*} := \mathcal{L} \times_{SL_m} (GL_m/U).$$

Given a PGL_m-local system $\mathcal{L}$ on S, set $\mathcal{L}_{\mathcal{B}} := \mathcal{L}/B$, where B is a Borel subgroup of PGL_m.

We remove a little disc around each special point of a decorated surface S, getting a surface S°. Let $\partial_{\mathrm{th}}S^\circ$ be the *thickened boundary of* S° – a small neighborhood of the boundary of S°.

Definition 4.1 ([FG1]). Let S be a decorated surface.

- Let G be either SL_m or GL_m. A decoration on a twisted G-local system $\mathcal{L}$ on S° is a flat section of restriction of $\mathcal{L}_{\mathcal{A}}$ to $T'(\partial_{\mathrm{th}}S^\circ)$. The moduli space $\mathcal{A}_{G,S}$ parametrizes decorated twisted G-local system on S°.
- A $*$-decoration on a twisted SL_m-local system $\mathcal{L}$ on S° is a flat section of the restriction of $\mathcal{L}_{\mathcal{A}^*}$ to $T'(\partial_{\mathrm{th}}S^\circ)$. The moduli space $\mathcal{A}^*_{SL_m,S}$ parametrizes $*$-decorated twisted SL_m-local systems on S°.
- A framing on a PGL_m-local system $\mathcal{L}$ on S° is a flat section of the restriction of $\mathcal{L}_{\mathcal{B}}$ to $\partial_{\mathrm{th}}S^\circ$. The moduli space $\mathcal{X}_{PGL_m,S}$ parametrizes framed G-local systems on S°.

Let $\mathcal{W}$ be an A_{m-1}-web on a decorated surface S, and $\widetilde{\mathcal{W}}$ its lift to the universal cover $\widetilde{S}$. We are going to assign the coordinates to the faces / internal faces of $\widetilde{\mathcal{W}}$. By the construction, the coordinates assigned to the faces which differ by the action of the deck transformation group $\pi_1(S)$ will be the same. This way we assign the coordinates to the faces of $\mathcal{W}$.

If $\mathcal{W}$ is a strict A_{m-1}-web, there is no need to go to the universal cover, and the coordinates are assigned just to the faces / internal faces of the web $\mathcal{W}$.

Let $\widetilde{\mathcal{L}}$ be the lift of a decorated SL_m-local system $\mathcal{L}$ to $\widetilde{S}$. Let s be a special point on $\widetilde{S}$. Denote by $A(s)$ the flat section of $\widetilde{\mathcal{L}}_{\mathcal{A}}$ providing the decoration $\widetilde{\mathcal{L}}$. Given a zig-zag strand γ of $\widetilde{\mathcal{W}}$ associated to s, we can extend $A(s)$ to a flat section of $\widetilde{\mathcal{L}}_{\mathcal{A}}$ on the punctured disc $\widetilde{S}_\gamma - s$.

Let F be a face of $\widetilde{\mathcal{W}}$. Choose a non-zero tangent vector v at a point inside of F. Connect it by a path in $T'\widetilde{S}$ with a non-zero tangent vector v_i at a point near the special point s_i. Transport the section $A(s_i)$ along this path from v_i to v. The resulting decorated flag $A_v(s_i)$ over v is well defined since $A(s_i)$ is a flat section of $\widetilde{\mathcal{L}}_{\mathcal{A}}$ over $\widetilde{S}_\gamma - s_i$. We get decorated flags $A_v(s_1), \ldots, A_v(s_n)$ over v. Given a face F, let d_{s_i} be the codistance from F to s_i. Corollary 2.16 implies that

$$d_{s_1} + \cdots + d_{s_n} = m. \tag{35}$$

Definition 4.2. Given an A_{m-1}-web $\mathcal{W}$ on S and a face F of $\widetilde{\mathcal{W}}$, we define a regular function $A_F^{\mathcal{W}}$ on the moduli space $\mathcal{A}_{SL_m,S}$ by setting

$$A_F^{\mathcal{W}} := \langle \Omega_m, \alpha_{d_1}(A_v(s_1)) \wedge \cdots \wedge \alpha_{d_n}(A_v(s_n)) \rangle. \tag{36}$$

The functions $A_F^{\mathcal{W}}$ are evidently $\pi_1(S)$-invariant, and so we may assume that F stands here for the $\pi_1(S)$-orbit in the space of faces of $\widetilde{\mathcal{W}}$.

Just the same way, given an A^*_{m-1}-web $\mathcal{W}$, we define regular functions $A_F^{\mathcal{W}}$ on the moduli space $\mathcal{A}^*_{SL_m,S}$ by formula (36). The only difference is that now we have extra coordinates at the external faces F containing special points of S.

Definition 4.3. Given an A_{m-1}-web $\mathcal{W}$ on S and an internal face F of $\widetilde{\mathcal{W}}$, we define a rational function $X_F^{\mathcal{W}}$ on the space $\mathcal{X}_{PGL_m,S}$ by the following product over the faces G:

$$X_F^{\mathcal{W}} := \prod_G (A_G^{\mathcal{W}})^{\langle F,G \rangle}, \tag{37}$$

The analog of Lemma 3.12 is valid. In particular:

Lemma 4.4. *For the* A_{m-1}*-web* $\mathcal{W}_T$ *assigned to an ideal triangulation* T *of* S*, the coordinates from Definitions* 4.2 *and* 4.3 *coincide with the coordinates assigned to* T *in* [FG1], *Section* 8.

Theorem 4.5. *Given a strict ideal* A_{m-1}*-web* $\mathcal{W}$ *on* S*:*

i) *The regular functions* $\{A_F^{\mathcal{W}}\}$ *at the faces* F *of* $\mathcal{W}$ *are cluster coordinates on* $\mathcal{A}_{SL_m,S}$

ii) *Rational functions* $\{X_F^{\mathcal{W}}\}$ *at the faces* F *of* $\mathcal{W}$ *are cluster Poisson coordinates on* $\mathcal{X}_{PGL_m,S}$.

Given a strict ideal A_{m-1}^**-web* $\mathcal{W}$ *on* S:

iii) *The regular functions* $\{A_F^{\mathcal{W}}\}$ *at the faces* F *of* $\mathcal{W}$ *are cluster coordinates on* $\mathcal{A}_{SL_m,S}^*$.

Proof. Given a strict ideal web $\mathcal{W}$, and picking an ideal triangulation T of S, the restriction $\mathcal{W}_t$ of $\mathcal{W}$ to each triangle t of T is a strict ideal web on t. Therefore by Theorem 3.11 it can be transformed by elementary transformations to the standard web of type either A_{m-1} or A_{m-1}^* on t. So the original web $\mathcal{W}$ is equivalent to one of the two standard webs $\mathcal{W}_{\mathrm{A}_{m-1},T}$ or $\mathcal{W}_{\mathrm{A}_{m-1}^*,T}$ related to the triangulation T, depending on the type of the web $\mathcal{W}$.

It was proved in [FG1] that the $\mathcal{A}$-functions assigned to an ideal triangulation T form a regular cluster coordinate system on the space $\mathcal{A}_{SL_m,S}$, and the $\mathcal{X}$-functions form a rational cluster Poisson coordinate system on the space $\mathcal{X}_{PGL_m,S}$.

A two by two move amounts to a cluster transformation of the $\mathcal{A}$-coordinates. This implies that a two by two move amounts to a cluster Poisson transformation of the $\mathcal{X}$-coordinates. So we get the parts i) and ii). The part iii) reduces easily to i). □

Remark. Any spectral web $\mathcal{W}$ on a decorated surface S satisfying the condition that

$$\text{each domain } S_\gamma \text{ is a disc} \tag{38}$$

produces a collection of functions $\{A_F\}$ assigned to the faces F of $\mathcal{W}$ as well as a collection of functions $\{X_F\}$ assigned to the internal faces F of $\mathcal{W}$. Indeed, the decoration / framing provides a flat section on the disc S_γ / punctured disc S_γ. The counting argument related to the degree of the spectral map, guarantees that we can use this flat sections to define the functions.

However the number of faces / internal faces of such a web is typically bigger then the dimension of the spaces $\mathcal{A}_{SL_m,S}$ and, respectively, $\mathcal{X}_{PGL_m,S}$. So non-ideal spectral webs satisfying condition (38) usually do not produce coordinate systems on these spaces.

A_{m-1}^*-bipartite graphs and $\mathcal{A}_{GL_m,S}$.

Theorem 4.6. *Let* Γ *be a strict* A_{m-1}^**-graph on a decorated surface* S *without boundary. Then there is a birational isomorphism*

$$\mathcal{A}_{GL_m,S} \stackrel{\sim}{\longrightarrow} \text{the moduli space of } GL(1)\text{-local system on } \Gamma.$$

Proof. Theorem 4.6 is equivalent to Theorem 4.5 iii). Indeed, one has

$$0 \longrightarrow \frac{\mathbb{Z}[\{\text{faces of } \Gamma\}]}{H_2(S,\mathbb{Z})} \longrightarrow H_1(\Gamma,\mathbb{Z}) \longrightarrow H_1(S,\mathbb{Z}) \longrightarrow 0.$$

Thus a line bundle with connection on the graph Γ is uniquely determined by the following data:

i) Monodromies around the faces of the graph Γ.

ii) The global monodromies on S.

Since $GL_m = SL_m \times \mathbb{G}_m$, and thanks to (32), the space $\mathcal{A}_{GL_m,S}$ is a product:

$$\mathcal{A}_{GL_m,S} = \mathcal{A}^*_{SL_m,S}/GL(1) \times \text{the moduli space of } GL(1)\text{-local system on } S. \quad (39)$$

The data ii) describes a $GL(1)$-local system on S. It is responsible for the second factor.

The data i) accounts for the description of the space $\mathcal{A}^*_{SL_m,S}$ in Theorem 4.5 iii) by the coordinates on the faces of Γ. The quotient by the $GL(1)$ in the first factor in (39) amounts to the quotient of $\mathbb{Z}[\{\text{faces of } \Gamma\}]$ by $H_2(S,\mathbb{Z}) = \mathbb{Z}$. □

4.2. Ideal loops and monodromies of framed local systems around punctures

Recall that, given a decorated surface S, a *puncture* is an internal special point on S. Recall

$$S^\times := S - \{\text{punctures of } S\}.$$

The monodromy of a framed PGL_m-local system on $S^\times$ around a puncture lies in the Cartan group H of PGL_m, isomorphic to $\mathbb{G}_m^{m-1}$. Monodromies around the punctures provide a map

$$\mathcal{X}_{PGL_m,S} \longrightarrow H^{\{\text{punctures of } S^\times\}}. \quad (40)$$

Its fibers are the generic symplectic leaves on $\mathcal{X}_{PGL_m,S}$. Monodromies around the punctures generate the center of the Poisson algebra of functions on $\mathcal{X}_{PGL_m,S}$ [FG1].

To each zig-zag loop γ we assign a function C_γ given by the product of the face coordinates assigned to the faces F sitting inside of the disc S_γ:

$$C_\gamma := \prod_{F \subset S_\gamma} X_F$$

Given a strict A_{m-1}-web $\mathcal{W}$ on S, for every puncture s there are $m-1$ ideal loops around s.

Theorem 4.7. *Let $\mathcal{W}$ be a strict A_{m-1}-web on S. Then*

i) *The functions C_γ are invariant under the two by two moves.*

ii) *The functions C_γ are Casimirs. Their product is equal to* 1, *and this is the only relation between them. They generate the center of the Poisson algebra of functions on $\mathcal{X}_{PGL_m,S}$.*

iii) *The map* (40) *is described by the collection of functions C_γ.*

Proof. i) Indeed, under a two by two moves the change of the face coordinates located inside of the zig-zag strand shown on Fig 26 is given by

$$A' = A(1+X^{-1})^{-1}, \ B' = B(1+X).$$

Thus $ABX = A'B'$, which proves the claim.

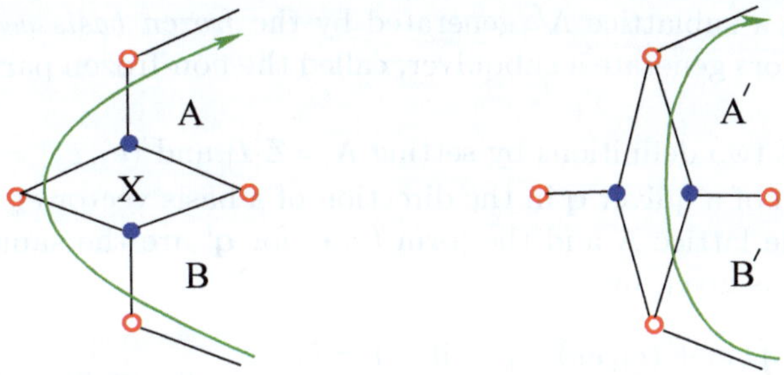

FIGURE 26. The action of a two by two move on the face coordinates.

ii) The zig-zag loops match the boundary components on the spectral surface Σ. This plus the exact sequence (20) implies all the claims since the Poisson bracket is given by the formula

$$\{X_F, X_G\} = \langle F, G\rangle X_F X_G.$$

iii) For the A_{m-1}-web related to a triangulation of S this is clear from the explicit description of the framed local system assigned to the coordinates which was given in [FG1]. So in general it follows from i). □

5. Webs, quivers with potentials, and 3d Calabi–Yau categories

5.1. Quivers with potentials and CY_3 categories with cluster collections

In Section 5.1 we collect for convenience of the reader some foundational material, including the relationship between quivers with potentials and 3d Calabi–Yau categories [KS1, Section 8.1].

1. Quivers and potentials. A quiver $\mathbf{q}$ without loops and 2-cycles is described by a set of vertices I, a subset $I^f \subset I$ of frozen vertices, and a skewsymmetric function $\varepsilon_{ij} : I \times I \to \mathbb{Z}$, encoding the number of oriented arrows $i \to j$ between the vertices.

Any element $k \in I - I^f$ provides a mutated in the direction k quiver $\mathbf{q}'$ described by the same set I and new function $\varepsilon'_{ij} = \mu_k(\varepsilon_{ij})$ defined by the Fomin–Zelevinsky formula [FZI]:

$$\varepsilon'_{ij} := \begin{cases} -\varepsilon_{ij} & \text{if } k \in \{i,j\}, \\ \varepsilon_{ij} & \text{if } \varepsilon_{ik}\varepsilon_{kj} \leq 0, \quad k \notin \{i,j\} \\ \varepsilon_{ij} + |\varepsilon_{ik}| \cdot \varepsilon_{kj} & \text{if } \varepsilon_{ik}\varepsilon_{kj} > 0, \quad k \notin \{i,j\}. \end{cases} \tag{41}$$

This procedure is involutive: the mutation of ε'_{ij} at the vertex k is the original function ε_{ij}.

There is a geometric description of a quiver as a data $\Big(\Lambda, \Lambda^f, \{e_i\}, (*,*)\Big)$, given by a lattice Λ, a basis $\{e_i\}_{i\in I}$ of Λ, a skew-symmetric $\mathbb{Z}$-valued bilinear form

$(*,*)$ on Λ, and a sublattice Λ^f generated by the *frozen basis vectors.* The non-frozen basis vectors generate a subquiver, called the non-frozen part of the original quiver.

One relates two definitions by setting $\Lambda = \mathbb{Z}[I]$ and $(e_i, e_j) = \varepsilon_{ij}$.

A mutation of a quiver $\mathbf{q}$ in the direction of a basis vector e_k is a new quiver $\mathbf{q}' = \mu_{e_k}(\mathbf{q})$. The lattice Λ and the form $(*,*)$ for $\mathbf{q}'$ are the same as for $\mathbf{q}$. The basis $\{e_i'\}$ for $\mathbf{q}'$ is given by

$$e_i' := \begin{cases} e_i + (e_i, e_k)_+ e_k & \text{if} \quad i \neq k, \\ -e_k & \text{if} \quad i = k, \end{cases} \qquad a_+ := \max(a, 0). \tag{42}$$

The composition of mutations $\mu_{e_k'} \circ \mu_{e_k}$ no longer acts as the identity on the original basis $\{e_k\}$. It is rather the reflection of the collection $\{e_i\}$ at e_k.

In the "simply-laced case" a cluster algebra [FZI], and more generally a cluster variety [FG2], are described by a collection of quivers related by quiver mutations (41).

Categorification requires a more elaborate version of a quiver, which we use throughout the paper. A quiver Q is given by a set of vertices $\{i\}_{i \in I}$ and a collection of finite-dimensional vector spaces $A(i,j)$, the *arrow spaces*, assigned to each pair of vertices (i,j). It is customary to choose a basis in each arrow space $\mathrm{A}(i,j)$, which we thought of as a collection of arrows $i \to j$.

To define the potential we consider the path algebra $\mathcal{P}(Q)$ of the quiver Q:

$$\mathcal{P}(Q) := \oplus_{n \geq 2} \bigoplus_{i_1, \ldots, i_n \in I} A(i_1, i_2) \otimes \cdots \otimes A(i_{n-1}, i_n).$$

The product is given by concatenation. Let us consider its cyclic envelope

$$\mathrm{HH}_0(\mathcal{P}(Q)) := \oplus_{n \geq 2} \bigoplus_{i_1, \ldots, i_n \in I} \Big(A(i_1, i_2) \otimes \cdots \otimes A(i_{n-1}, i_n) \otimes A(i_n, i_1)\Big)_{\mathbb{Z}/n\mathbb{Z}}. \tag{43}$$

Here the subscript $\mathbb{Z}/n\mathbb{Z}$ denotes the coinvariants of the cyclic shift.

Definition 5.1. A potential is a linear functional $P : \mathrm{HH}_0(\mathcal{P}(Q)) \longrightarrow k$.[6]

Informally, the space (43) is a vector space with a natural basis given by *cyclic words* of arrows of the quiver – the paths ending at their starting vertex. So a potential can be thought of as a formal linear sum, possibly infinite, of cyclic words.

---

[6]To motivate this definition observe that any element $P \in \mathrm{HH}_0(\mathcal{P}(Q))$ gives rise to a function F_P on the space of representations of the quiver Q: we assign to a cycle C in (43) a function F_C given by the trace of the product of operators assigned to the arrows of C, and extend by linearity: $F_{C_1+C_2} := F_{C_1} + F_{C_2}$.

2. An example: quivers with potentials from bipartite graphs. A bipartite graph Γ on S gives rise to a *quiver with a canonical potential* (Q_Γ, P_Γ). The quiver Q_Γ is given by the dual graph: its vertices are the faces of the graph Γ; its edges are dual to the edges E of Γ, oriented so that the $\bullet$-vertex of E is on the right. We call the oriented edges of the quiver *arrows.*

Each $\bullet$-vertex b of a bipartite graph Γ gives rise to a simple cyclic path C_b on Q_Γ given by the arrows going clockwise around the vertex b. Each $\circ$-vertex r of Γ gives rise to a similar simple loop C_r on Q_Γ given by the arrows going counterclockwise around the vertex r. Set

$$P_\Gamma := \sum_{b:\ \bullet\text{-vertices of } \Gamma} C_b - \sum_{r:\ \circ\text{-vertices of } \Gamma} C_r \in \mathrm{HH}_0(\mathcal{P}(Q_\Gamma)). \tag{44}$$

Quivers with potentials assigned to bipartite graphs were invented by physicists, see [FHKVW] and references there.

Below we briefly recall Kontsevich–Soibelman's construction [KS1], Section 8.

3. 3d Calabi–Yau categories and cluster collections. A 3d Calabi–Yau category over a characteristic zero field k is a (weakly unital) k-linear triangulated A_∞-category $\mathcal{C}$ with the following properties:

1. For any two objects E, F, the $\mathrm{Hom}^\bullet(E,F)$ is a $\mathbb{Z}$-graded finite-dimensional vector space:
$$\mathrm{Hom}^\bullet(E,F) = \oplus_n \mathrm{Hom}^n(E,F).$$
2. There is a non-degenerate symmetric pairing
$$(\bullet,\bullet) : \mathrm{Hom}^\bullet(E,F) \otimes \mathrm{Hom}^\bullet(F,E) \longrightarrow k[-3]. \tag{45}$$
3. For any objects $E_1, \ldots, E_{n+1}$, there are higher composition maps
$$m_n : \bigotimes_{1\leq i\leq n} \Big(\mathrm{Hom}^\bullet(E_i, E_{i+1})[1]\Big) \longrightarrow \mathrm{Hom}^\bullet(E_1, E_{n+1})[2] \tag{46}$$
satisfying the axioms of the A_∞-category, which we recall later on.
4. For any objects $E_1, \ldots, E_n$, $n \geq 2$, we use pairing (45) to produce a map $(E_{n+1} = E_1)$:
$$W_n : \bigotimes_{1\leq i\leq n} \Big(\mathrm{Hom}^\bullet(E_i, E_{i+1}[1])\Big) \longrightarrow k.$$
$$W_n(a_1, \ldots, a_n) := (m_{n-1}(a_1, \ldots, a_{n-1}), a_n). \tag{47}$$
The last condition is that this map must be cyclically invariant.

In particular, the map $m_1 : \mathrm{Hom}^\bullet(E_1,E_2) \longrightarrow \mathrm{Hom}^\bullet(E_1,E_2)[-1]$ is a differential: $m_1^2 = 0$. The Ext-groups are its cohomology:

$$\mathrm{Ext}^\bullet(E,F) := H^\bullet(\mathrm{Hom}^\bullet(E,F), m_1).$$

For any object E, the *total potential of the object* E, is a formal power series

$$W_E(\alpha) := \sum_n \frac{W_n(\alpha,\ldots,\alpha)}{n}, \quad \alpha \in \mathrm{Hom}^\bullet(E,E)[1].$$

Consider the cyclic tensor envelope of a graded vector space V:

$$\mathcal{C}\mathrm{T}(V) := \oplus_{n=0}^{\infty} (V^{\otimes n})_{\mathbb{Z}/n\mathbb{Z}}. \tag{48}$$

The total potential W_E lies in the dual to the cyclic tensor envelope of $\mathrm{Hom}^\bullet(E,E)[1]$:

$$\mathrm{Hom}\Big(\mathcal{C}\mathrm{T}(\mathrm{Hom}^\bullet(E,E)[1]), k\Big). \tag{49}$$

The inverse of the odd non-degenerate pairing (45) is an odd bivector, providing an odd non-commutative Poisson bracket $\{*,*\}$ on the graded vector space (49), see, e.g., [K92].

Let $\mathcal{C}$ be a non-triangulated 3d CY A_∞-category with finitely many objects $S_1, \ldots, S_n$. Set $\mathbb{S} := \oplus_{i=1}^n S_i$. It is the same thing as a 3d CY A_∞-algebra structure on the graded space $\mathrm{Hom}^\bullet(\mathbb{S},\mathbb{S})$ equipped with idempotents $p_i : \mathrm{Hom}^\bullet(\mathbb{S},\mathbb{S}) \to \mathrm{Hom}^\bullet(\mathbb{S}, S_i)$. Equivalently, it can be described by a potential $W_\mathbb{S}$ satisfying the classical master equation. Precisely, the total potential $W_\mathbb{S}$ of $\mathbb{S}$ is an element

$$W_\mathbb{S} \in \mathrm{Hom}\Big(\mathcal{C}\mathrm{T}(\mathrm{Hom}^\bullet(\mathbb{S},\mathbb{S})[1]), k\Big). \tag{50}$$

It satisfies the *classical master equation*

$$\{W_\mathbb{S}, W_\mathbb{S}\} = 0. \tag{51}$$

Vice versa, assuming conditions 1) and 2) above, any element W in (49) satisfying $\{W,W\} = 0$ determines a non-triangulated 3d CY A_∞-category $\mathcal{C}_W$ with the objects $S_1, \ldots, S_n$ such that W is the total potential of the object $\mathbb{S}$ of $\mathcal{C}_W$. The maps m_n are the components of the non-commutative Hamiltonian vector field $\{W,*\}$ provided by the Hamiltonian W [K92].

The non-triangulated 3d CY category $\mathcal{C}$ gives rise to its triangulated envelope, the category $\mathrm{Tw}(\mathcal{C})$ of twisted complexes of $\mathcal{C}$ [BK]. Its objects are given by direct sums $X_1[n_1] \oplus \ldots \oplus X_k[n_k]$ of shifted generators S_i of $\mathcal{C}$, equipped with a matrix $f = (f_{ij})_{i<j}$ of morphisms $f_{ij} : X_i \longrightarrow X_j$, $\deg(f_{ij}) = n_i - n_j + 1$, so that $f := \sum_{i<j} f_{ij}$ satisfies the Maurer–Cartan equation

$$\textstyle\sum_{l=1}^k m_l(f, \ldots, f) = 0.$$

4. A quiver with potential from a cluster collection [KS1]. A cluster collection $\mathcal{S}$ gives rise to a quiver $Q_{\mathcal{C},\mathcal{S}}$ whose vertices $\{i\}$ are the objects S_i, and the arrow spaces are

$$\mathrm{A}(i,j) := \mathrm{Ext}^1(S_i, S_j). \tag{52}$$

Alternatively, we get a quiver by considering a lattice $\Lambda := K_0(\mathcal{C})$, with a basis given by the classes $[S_i]$, and a skew symmetric form given by the negative of the Euler form:

$$\begin{aligned} &(*,*) : K_0(\mathcal{C}) \wedge K_0(\mathcal{C}) \longrightarrow \mathbb{Z}. \\ &(A,B) := -\textstyle\sum_a (-1)^a \dim \mathrm{Ext}^a(A,B). \end{aligned} \tag{53}$$

The total potential $W_{\mathbb{S}}$ of $\mathbb{S} = \oplus_{i=1}^{n} S_i$, restricted to tensor products of $\mathrm{Ext}^1(S_i, S_j)[1]$'s, provides a potential $P_{\mathcal{C},\mathcal{S}}$ of the quiver $Q_{\mathcal{C},\mathcal{S}}$:

$$P_{\mathcal{C},\mathcal{S}} : \mathrm{HH}_0(\mathcal{P}(Q_{\mathcal{C},\mathcal{S}})) \longrightarrow k.$$

This way a 3d CY A_∞-category with a cluster collection $(\mathcal{C}, \mathcal{S})$ gives rise to a quiver with potential $(Q_{\mathcal{C},\mathcal{S}}, P_{\mathcal{C},\mathcal{S}})$. The quiver does not have loops and cycles of length two, and the potential is *minimal*, i.e., it starts from the cubic terms.

Given a quiver with potential (Q, P), the group of automorphisms of the path algebra $\mathcal{P}(Q)$ of the quiver Q preserving the idempotents corresponding to the vertices of the quiver acts on the potentials. Two potentials are *gauge equivalent* if belong to the same orbit of this group.

5. The Kontsevich–Soibelman correspondence [KS1, Section 8]. It asserts that assigning to a 3d CY category with a cluster collection $(\mathcal{C}, \mathcal{S})$ its quiver with potential $(Q_{\mathcal{C},\mathcal{S}}, P_{\mathcal{C},\mathcal{S}})$ we get a 1-1 correspondence

$$\left\{\begin{matrix}\text{3d CY categories} \\ + \text{ cluster collections}\end{matrix} \Big/ \begin{matrix}\text{equivalences preserving CY str.} \\ \& \text{ cluster collections}\end{matrix}\right\}$$
$$\leftrightarrow \left\{\begin{matrix}\text{quivers without length one and two} \\ \text{cycles, with minimal potentials}\end{matrix} \Big/ \begin{matrix}\text{gauge} \\ \text{equivalence}\end{matrix}\right\}. \tag{54}$$

The inverse construction proceeds in two steps.

i). We assign to a quiver with potential a 3d CY non-triangulated A_∞-category $\widetilde{\mathcal{C}}$ with a set of objects $\{S_i\}_{i\in I}$ and $m_1 = 0$. The category is determined by the full potential $W_{\mathbb{S}}$ of the object $\mathbb{S} = \oplus_{i=1}^{n} S_i$, defined as a sum of two terms

$$W_{\mathbb{S}} := W'_{\mathbb{S}} + W_{\mathrm{can}} \in \mathrm{Hom}\Big(\mathcal{C}\mathrm{T}(\mathrm{Ext}^\bullet(\mathbb{S}, \mathbb{S})[1]), k\Big).$$

The $W'_{\mathbb{S}}$ is non-zero only on the cyclic products of $\mathrm{Ext}^1(\mathbb{S}, \mathbb{S})[1]$. It is given by the quiver potential.

The W_{can} is defined by postulating that its only non-zero components are the maps

$$\mathrm{Ext}^a(S_i, S_j)[1] \otimes \mathrm{Ext}^{3-a}(S_j, S_i)[1] \otimes \mathrm{Ext}^0(S_i, S_i)[1] \longrightarrow k, \;\; a = 0, 1, 2. \tag{55}$$

These maps are induced by the pairing $\mathrm{Ext}^a(S_i, S_j) \otimes \mathrm{Ext}^{3-a}(S_j, S_i) \longrightarrow k$ for $a = 0, 1$ and by its negative for $a = 2$. The term W_{can} can be defined by a formula. Namely, denote by x_{ij}^s a basis of coordinates on $\mathrm{Ext}^1(S_i, S_j)[1]$, by ξ_{ji}^s the dual basis for $\mathrm{Ext}^2(S_j, S_i)[1]$, and by α_i and a_i the natural coordinates on $\mathrm{Ext}^0(S_i, S_i)[1] = k[1]$ and $\mathrm{Ext}^3(S_i, S_i)[1] = k[-2]$. Then

$$W_{\mathrm{can}}(\alpha, x, \xi, a) := \textstyle\sum_{i=1}^{n} \alpha_i^2 a_i + \sum_{i,j=1}^{n} (\alpha_i x_{ij}^s \xi_{ji}^s - \alpha_i \xi_{ij}^s x_{ji}^s).$$

The potential W_{can} provides the graded vector space $\mathrm{Ext}^\bullet(\mathbb{S}, \mathbb{S})$ with a structure of a graded associative 3d CY algebra with a collection of idempotents $\{p_i\}_{i\in I}$ and a unit e:

$$p_i := 1 \in \mathrm{Ext}^0(S_i, S_i), \;\; e := \textstyle\sum_{i\in I} p_i.$$

The multiplication vanishes on the graded component

$$\mathrm{Ext}^1(\mathbb{S},\mathbb{S}) \otimes \mathrm{Ext}^1(\mathbb{S},\mathbb{S}) \to \mathrm{Ext}^2(\mathbb{S},\mathbb{S}).$$

It is given by a non-degenerate bilinear form on

$$\mathrm{Ext}^1(\mathbb{S},\mathbb{S}) \otimes \mathrm{Ext}^2(\mathbb{S},\mathbb{S}) \to \mathrm{Ext}^3(\mathbb{S},\mathbb{S}).$$

This implies that $\{W_{\mathrm{can}}, W_{\mathrm{can}}\} = 0$, which one can easily check directly. Clearly, $\{W'_{\mathbb{S}}, W'_{\mathbb{S}}\} = 0$. One proves that $\{W_{\mathrm{can}}, W'_{\mathbb{S}}\} = 0$. Thus $\{W_{\mathbb{S}}, W_{\mathbb{S}}\} = 0$. Therefore we get an A_∞ CY category $\widetilde{\mathcal{C}}_W$ with the objects $S_1, \ldots, S_n$, which by construction form a cluster collection.

ii). The desired 3d CY category $\mathcal{C}$ is the category $\mathrm{Tw}(\widetilde{\mathcal{C}})$ of twisted complexes of $\widetilde{\mathcal{C}}$.

The category $\mathrm{Tw}(\widetilde{\mathcal{C}})$ can be described as the DG category of finite-dimensional DG modules over the Ginzburg algebra related to the quiver with potential $(Q_{\mathcal{C},\mathcal{S}}, P_{\mathcal{C},\mathcal{S}})$. Although we are not using this, for convenience of the reader we recall the definitions.

6. The Ginzburg algebra of a quiver with potential [G]. The *Ginzburg algebra* [G] associated to a quiver with potential (Q,P) is a DG-algebra $A^\bullet_{Q,P}$ concentrated in non-positive degrees, with the following generators in the degrees $0, -1, -2$:

- Projectors $\{e_i\}$, $e_i^2 = e_i$, $\sum_i e_i = 1$, corresponding to vertices $\{i\}$ of the quiver Q.
- Degree 0 generators $\{a_E\}$, and degree -1 generators $\{a_E^*\}$, assigned to arrows $\{E\}$ of Q.
- Degree -2 generators $\{t_i\}$ assigned to vertices $\{i\}$ of the quiver Q.
- The differential d acts on the generators by

$$da_E = 0, \quad da_E^* = \frac{\partial P}{\partial a_E}, \quad dt_i = \sum_{E \,\mid\, t(E)=i} a_E^* a_E - \sum_{E \,\mid\, h(E)=i} a_E a_E^*. \tag{56}$$

Here $\frac{\partial P}{\partial a_E}$ is the non-commutative derivative. These definitions can be written using only arrow spaces. Indeed, choose a basis $\{a_E\}$ in each arrow space; let $\{a_E^*\}$ be the dual basis in the dual space. Then to write the differential we employ the identity elements

$$\textstyle\sum_E a_E \otimes a_E^* \in \mathrm{A}(i,j) \otimes \mathrm{A}(i,j)^*, \qquad \sum_E a_E^* \otimes a_E \in \mathrm{A}(i,j)^* \otimes \mathrm{A}(i,j).$$

The *Jacobian algebra* $H^0(A^\bullet_{Q,P})$ is generated by the generators e_v, a_E satisfying the relations $\frac{\partial P}{\partial a_E} = 0$ for each arrow E.

Denote by $\mathcal{F}_{Q,P}$ the DG-category of DG-modules over the Ginzburg algebra $A^\bullet_{Q,P}$, whose homology are finite-dimensional, nilpotent modules over $H^0(A^\bullet_{Q,P})$. The subcategory of degree zero $A^\bullet_{Q,P}$-modules is an abelian heart of this category. It is the category $H^0(A^\bullet_{Q,P})$-modules.

Lemma 5.2. *The DG category $\mathcal{F}_{Q,P}$ is canonically equivalent to the category of twisted complexes over the CY A_∞-category $\widetilde{\mathcal{C}}_{Q,P}$ defined above. The cluster collection $\{S_i\}$ in $\widetilde{\mathcal{C}}_{Q,P}$ is identified with the one-dimensional $A^\bullet_{Q,P}$-modules corresponding to the vertices $\{i\}$ of Q.*

There is another 3d Calabi–Yau category $\mathcal{P}_{Q,P}$, the category of perfect complexes over the Ginzburg algebra $A^\bullet_{Q,P}$. It is Koszul dual to the category $\mathcal{F}_{Q,P}$. Namely, the category $\mathcal{F}_{Q,P}$ is equivalent to the category of DG-functors from the category $\mathcal{P}_{Q,P}$ to the category $\mathrm{Vect}^\bullet$ of bounded complexes of vector spaces with finite-dimensional cohomology, and vice versa:

$$\mathcal{F}_{Q,P} \sim \mathrm{Funct}_{dg}(\mathcal{P}_{Q,P}, \mathrm{Vect}^\bullet), \qquad \mathcal{P}_{Q,P} \sim \mathrm{Funct}_{dg}(\mathcal{F}_{Q,P}, \mathrm{Vect}^\bullet).$$

7. Mutations of cluster collections [KS1]. A spherical object S of a CY category gives rise to the Seidel–Thomas [ST] reflection functor R_S, acting by an autoequivelence of the category:

$$R_S(X) := \mathrm{Cone}\Big(\mathrm{Ext}^\bullet(S, X) \otimes S \longrightarrow X\Big).$$

Given elements $0, i \in I$, we write $i < 0$ if $\mathrm{Ext}^1(S_i, S_0)$ is non-zero, and $i > 0$ otherwise.

Definition 5.3. A mutation of a cluster collection $\mathcal{S} = \{S_i\}_{i\in I}$ in a 3d CY category $\mathcal{C}$ at an object S_0 is a new spherical collection $\mathcal{S}' = \{S'_i\}$ in the same category $\mathcal{C}$ given by

$$S'_i = S_i, \ i < 0, \ S'_0 = S_0[-1], \ S'_i = R_{S_0}(S_i), \ i > 0,$$

Since the shift and reflection functors transform spherical objects to spherical ones, $\{S'_i\}_{i\in I}$ is a collection of spherical objects. Since the objects $\{S_i\}$ generate the triangulated category $\mathcal{C}$, the objects $\{S'_i\}$ are generators. However $\{S'_i\}_{i\in I}$ is not necessarily a cluster collection.

Mutations of cluster collections, being projected to $K_0(\mathcal{C})$, recover mutations of bases (42).

Mutations of cluster collections translate into *mutations of quivers with generic potentials* introduced earlier by Derksen, Weyman and Zelevinsky [DWZ]. Keller and Yang [KY] promoted mutations of quivers with generic potentials $(Q, P) \to (Q', P')$ to *mutation functors* – explicit equivalences $\mathcal{F}_{Q,P} \longrightarrow \mathcal{F}_{Q',P'}$ between the corresponding 3d CY categories.

In the Kontsevich–Soibelman picture a mutation does not affect the 3d CY category, altering only the cluster collection. It is build on the cluster collection mutation ansatz, which a posteriori implies the Derksen–Weyman–Zelevinsky mutation rule for quivers with potentials. The ansatz itself is a categorification of the geometric form of quiver mutation (42), and related to tiltings of t-structures [Br1].

In symplectic topology we are given a 3d CY category – the Fukaya category, while construction of a cluster collection, realized by special Lagrangian spheres, requires some choices.

To produce a 3d CY category combinatorially, we need as an input a quiver with potential. However there is no preferred quiver. Furthermore, there is an issue:

- *A mutation of a cluster collection may deliver a non-cluster collection.*
- *Mutations of quivers with potentials are defined only for quivers with generic potentials.*

Given a quiver with potential, it is hard to determine whether after a finite number of mutations we will get a quiver with potential which does not allow mutations in some directions.

8. Cluster varieties and quivers with canonical potentials. In general, there is no canonical potential assigned to an arbitrary quiver. The combinatorial part of our proposal in Section 1.3 boils down to the following:

- Every cluster variety $\mathcal{V}$ which appears in representation theory, geometry and physics admits a "tame" collection of quivers, equipped with canonical potentials.

  The corresponding 3d CY category is the combinatorial category $\mathcal{C}_{\mathcal{V}}$.
- Any two quivers with potentials from this collection can be connected by mutations inside of the collection; elements of the group $\Gamma_{\mathcal{V}}$ are realized by compositions of mutations.
- The equivalence (8) identifies the cluster collections in $\mathcal{C}_{\mathcal{V}}$ with the cluster collections provided by special Lagrangian spheres in the CY threefold $\mathcal{Y}_b$.

Recall that collections of bipartite surface graphs related by two by two moves give rise to a cluster Poisson variety [GK]. The quiver of a bipartite surface graph comes with a canonical potential (44). We suggest that this is a universal source of cluster varieties in the examples related to simple groups of type A.

5.2. The symmetry group of a CY$_3$ category with a cluster collection

Let $\mathcal{S} = \{S_i\}$ be a cluster collection in CY$_3$ category $\mathcal{C}$. Consider the quotient group

$$\mathrm{Auteq}(\mathcal{C};\mathcal{S}) := \frac{\text{Autoequivalences of the category } \mathcal{C}}{\text{Autoequivalences preserving objects of the cluster collection } \mathcal{S}}. \tag{57}$$

Our first goal is to define a group of symmetries of a pair $(\mathcal{C};\mathcal{S})$:

$$\Gamma_{\mathcal{C},\mathcal{S}} \subset \mathrm{Auteq}(\mathcal{C};\mathcal{S}).$$

1. The braid group of a cluster collection.

Definition 5.4. Given a cluster collection $\mathcal{S} = \{S_i\}_{i\in I}$ in $\mathcal{C}$, the reflection functors R_{S_i} and their inverses generate a subgroup, called the *braid group of the cluster collection* $\mathcal{S}$:

$$\mathrm{Br}_{\mathcal{C},\mathcal{S}} \subset \mathrm{Auteq}(\mathcal{C};\mathcal{S}).$$

By Propositions 2.12–2.13 in [ST], the functors R_{S_i} satisfy the "braid relations":

$$\mathrm{R}_{S_a}\mathrm{R}_{S_b} = \mathrm{R}_{S_b}\mathrm{R}_{S_a} \text{ if } \mathrm{Ext}^1(S_a, S_b) = 0.$$
$$\mathrm{R}_{S_a}\mathrm{R}_{S_b}\mathrm{R}_{S_a} = \mathrm{R}_{S_b}\mathrm{R}_{S_a}\mathrm{R}_{S_b} \text{ if } \mathrm{dimExt}^1(S_a, S_b) = 1.$$

2. Categorified cluster modular groupoid

Definition 5.5. The categorified cluster modular groupoid $\mathcal{C}\mathrm{Mod}$ is the following groupoid:

- Its objects are pairs $(\mathcal{C}, \mathcal{S})$, where $\mathcal{S}$ is a cluster collection in a 3d CY category $\mathcal{C}$.
- Morphisms $\psi : (\mathcal{C}, \mathcal{S}) \longrightarrow (\mathcal{C}', \mathcal{S}')$ are given by certain pairs (α, β), where $\alpha : \mathcal{C} \longrightarrow \mathcal{C}'$ is an equivalence of categories, and $\beta : \mathcal{S} \to \mathcal{S}'$ is a bijection. Precisely:

  The morphisms are generated by the following elementary ones:

  i) mutations $\mu_{S_k} : (\mathcal{C}, \mathcal{S}) \longrightarrow (\mathcal{C}, \mu_{S_k}(\mathcal{S}))$, where $\mathcal{C} \to \mathcal{C}$ is the identity functor, and $\mu_{S_k}(\mathcal{S})$ is the mutation of the cluster collection $\mathcal{S}$ in the direction S_k;

  ii) equivalences $\varphi : (\mathcal{C}_1, \mathcal{S}_1) \xrightarrow{\sim} (\mathcal{C}_2, \varphi(\mathcal{S}_1))$, where $\varphi : \mathcal{C}_1 \longrightarrow \mathcal{C}_2$ is an any equivalence.

  Compositions of morphisms are defined in the obvious way.
- Relations: two compositions of generating morphisms $\psi_1, \psi_2 : (\mathcal{C}_1, \mathcal{S}_1) \longrightarrow (\mathcal{C}_2, \mathcal{S}_2)$ are equal if the underlying equivalences coincide, and $\beta_1(S_i) = \beta_2(S_i)$, $\forall i \in I$.

We stress that mutations do not affect the category, changing the cluster collections only.

Lemma 5.6. *In the groupoid* $\mathcal{C}\mathrm{Mod}$ *the square of a mutation is the reflection functor:*

$$\mu_{S[-1]} \circ \mu_S = \mathrm{R}_S \ \forall S \in \mathcal{S}. \tag{58}$$

Proof. The composition $\mu_{S[-1]} \circ \mu_S$ is given by a pair (α, β) where α is the identity functor on $\mathcal{C}$, and β transforms the cluster collection $\mathcal{S}$ to the one $\mathrm{R}_S(\mathcal{S})$. The reflection functor R_S is an autoequivalence of the category $\mathcal{C}$, which has the same effect on the cluster collection. This just means that we have (58). □

Let $\mathrm{Aut}_{\mathcal{C}\mathrm{Mod}}(\mathcal{C}, \mathcal{S})$ be the automorphism group of the object $(\mathcal{C}, \mathcal{S})$ in the groupoid $\mathcal{C}\mathrm{Mod}$.

Lemma 5.7. *There is a canonical injective homomorphism*

$$\mathrm{Aut}_{\mathcal{C}\mathrm{Mod}}(\mathcal{C}, \mathcal{S}) \longrightarrow \mathrm{Auteq}(\mathcal{C}; \mathcal{S}). \tag{59}$$

Proof. Any element of $\mathrm{Aut}_{\mathcal{C}\mathrm{Mod}}(\mathcal{C}, \mathcal{S})$ is given by a sequence of generating morphisms

$$(\mathcal{C}, \mathcal{S}) \stackrel{(\alpha_0,\beta_0)}{\longrightarrow} (\mathcal{C}_1, \mathcal{S}_1) \stackrel{(\alpha_1,\beta_1)}{\longrightarrow} (\mathcal{C}_2, \mathcal{S}_2) \stackrel{(\alpha_2,\beta_2)}{\longrightarrow} \ldots \stackrel{(\alpha_{n-1},\beta_{n-1})}{\longrightarrow} (\mathcal{C}_n, \mathcal{S}_n) \stackrel{(\alpha_n,\beta_n)}{\longrightarrow} (\mathcal{C}, \mathcal{S}).$$

The projection of the composition $\alpha_n \circ \cdots \circ \alpha_1 \circ \alpha_0 \in \text{Auteq}(\mathcal{C})$ to the group $\text{Auteq}(\mathcal{C}, \mathcal{S})$ provides a group homomorphism (59). It is injective since, by definition, two sequences resulting in the same element of the group $\text{Auteq}(\mathcal{C}; \mathcal{S})$ determine the same morphism of the groupoid $\mathcal{C}$Mod. □

3. Comparing groupoids $\mathcal{C}$Mod and Mod. The categorified cluster modular groupoid $\mathcal{C}$Mod is similar to the cluster modular groupoid Mod defined in [FG2].

The objects of the groupoid Mod are quivers $(\Lambda, \{e_i\}_{i\in I}, (*,*))$. Its morphisms are generated by the elementary ones of two types: quiver mutations (42) and isomorphisms of quivers. Similarly to Definition 5.5, quiver mutations do not affect the lattice with the form $(\Lambda, (*,*))$, changing bases only. Two compositions of elementary morphisms are equal if they induce the same cluster transformation of the corresponding quantum cluster variety [FG2].

Theorem 5.8. *There is a canonical functor, see (53):*

$$\mathcal{C}\text{Mod} \longrightarrow \text{Mod}, \ \ (\mathcal{C}, \mathcal{S}) \longrightarrow (K_0(\mathcal{C}), \{[S_i]\}, [*,*]). \tag{60}$$

Proof. The cluster collection mutation μ_{S_k} induces in $K_0(\mathcal{C})$ the half-reflection (42). Therefore the assignment (60) is defined on the generating morphisms. The fact that it sends the relations between the generating morphisms $\mathcal{C}$Mod to the relations in the cluster modular groupoid Mod is implied by [K11, Theorem 5.2]. □

The difference between the two groupoids is that in the cluster modular groupoid the square of a mutation is the identity: $\mu_{-e_k}\mu_{e_k} = \text{Id}$, while in $\mathcal{C}$Mod it is the reflection functor:

Lemma 5.9. *The group* $\text{Aut}_{\mathcal{C}\text{Mod}}(\mathcal{C}, \mathcal{S})$ *is an extension of the cluster modular group by the cluster braid group:*

$$1 \longrightarrow \text{Br}_{\mathcal{C},\mathcal{S}} \longrightarrow \text{Aut}_{\mathcal{C}\text{Mod}}(\mathcal{C}, \mathcal{S}) \longrightarrow \text{Aut}_{\text{Mod}}(K_0(\mathcal{C}), [\mathcal{S}], (*,*)) \longrightarrow 1. \tag{61}$$

Proof. Thanks to relation (58), the cluster braid group $\text{Br}_{\mathcal{C},\mathcal{S}}$ is a subgroup of $\text{Aut}_{\mathcal{C}\text{Mod}}(\mathcal{C}, \mathcal{S})$. □

5.3. A_m-webs and extended mapping class groups

1. Categorified two by two moves. A web $\mathcal{W}$ on S gives rise to a quiver with potential $(Q_\mathcal{W}, P_\mathcal{W})$, and therefore to a 3d CY category $\mathcal{C}_\mathcal{W} := \mathcal{C}_{Q_\mathcal{W}, P_\mathcal{W}}$ with a cluster collection $\mathcal{S}_\mathcal{W}$, whose objects are parametrized by the faces of the web $\mathcal{W}$.

A two by two move $\mu_F : \mathcal{W} \longrightarrow \mathcal{W}'$ of webs, centered at a face F of $\mathcal{W}$, gives rise to a quiver with potential $(Q_{\mathcal{W}'}, P_{\mathcal{W}'})$, and hence to a 3d CY category with a cluster collection $(\mathcal{C}_{\mathcal{W}'}, \mathcal{S}_{\mathcal{W}'})$.

Proposition 5.10. *The quiver with potential* $(Q_{\mathcal{W}'}, P_{\mathcal{W}'})$ *describes the mutated cluster collection* $\mu_F(\mathcal{S}_\mathcal{W})$ *in the category* $\mathcal{C}_\mathcal{W}$. *In particular the categories* $\mathcal{C}_{\mathcal{W}'}$ *and* $\mathcal{C}_\mathcal{W}$ *are equivalent.*

Proposition 5.10 is proved in Section 5.4

Therefore we arrived at the categorified two by two move:

$$\mu_F : (\mathcal{C}_\mathcal{W}, \mathcal{S}_\mathcal{W}) \longrightarrow (\mathcal{C}_\mathcal{W}, \mu_F(\mathcal{S}_\mathcal{W})) \sim (\mathcal{C}_{\mathcal{W}'}, \mathcal{S}_{\mathcal{W}'}).$$

Let us explain now how we construct symmetries of the category with cluster collection related to the group PGL_m and a decorated surface S.

2. Groupoid of regular ideal triangulations. An ideal triangulation of a decorated surface S has triangles of two kinds: *regular triangles*, with three distinct sides, and *folded triangles*, where two sides are glued together. Some or all vertices of a regular ideal triangle can be glued together. An example is given by an ideal triangulation of a punctured torus.

Below we consider only ideal triangulations which do not have folded triangles – we call them *regular ideal triangulations*. We define the *groupoid* Tr_S *of regular ideal triangulations of* S:

Definition 5.11. Let S be a decorated surface.

- The objects of the groupoid Tr_S are regular ideal triangulations of S.
- The morphisms are generated by isomorphisms of decorated surfaces and flips:
  i) Any isomorphism $i : S \to S'$ gives rise to morphisms $i : (S, T) \to (S', i(T))$.
  ii) Any internal edge E of a regular ideal triangulation T provides a morphism $\mu_E : T \to T_E$ where T_E is the unique modulo isotopy ideal triangulation obtained by flipping the edge E.
- The relations between the morphisms are generated by the following:
  a) Isomorphisms commute with the flips: $i \circ \mu_E = \mu_{i(E)} \circ i$.
  b) The square relations: flips at disjoint edges commute.
  c) The pentagon relations.

The mapping class group Γ_S is isomorphic to the automorphism group $\mathrm{Aut}_{\mathrm{Tr}_S}(T)$ of the object of groupoid $\mathcal{T}_S$ given a regular ideal triangulation T. Indeed, for any $g \in \Gamma_S$, there is a sequence of flips of regular ideal triangulations $T \to T_1 \to \cdots \to g(T)$ connecting T with $g(T)$. Combining with the isomorphism $g^{-1} : g(T) \to T$, we get an automorphism $a_{g,T} \in \mathrm{Aut}_{\mathcal{T}_S}(T)$, given by the composition

$$T \to T_1 \to \cdots \to g(T) \xrightarrow{g^{-1}} T.$$

Different sequences of flips connecting T with $g(T)$ can be homotoped to each other by using the square and pentagon relations. So the element $a_{g,T}$ is well defined. The map $g \longmapsto a_{g,T}$ is a group homomorphism. It is known to be an isomorphism.

3. The extended mapping class group $\Gamma_{G,S}$. We assigned to a regular ideal triangulation T of S an ideal A_{m-1}-bipartite graph, see Figure 2, and hence an ideal A_{m-1}-web $\mathcal{W}_T$. So we get a quiver with potential, and thus a 3d CY category with cluster collection $(\mathcal{C}^T_{G,S}, \mathcal{S}^T_{G,S})$, $G = PGL_m$.

Definition 5.12. Let S be a decorated surface and $G = PGL_m$. Then

$$\widetilde{\Gamma}_{G,S} := \mathrm{Aut}_{\mathcal{C}\mathrm{Mod}}(\mathcal{C}^T_{G,S}, \mathcal{S}^T_{G,S})$$

is the automorphism group of the object $(\mathcal{C}^T_{G,S}, \mathcal{S}^T_{G,S})$ of the groupoid $\mathcal{C}$Mod.

Lemma 5.13. *The isomorphism class of the group $\widetilde{\Gamma}_{G,S}$ does not depend on the choice of T.*

Proof. Any two regular ideal triangulations of S are related by a sequence of flips, A flip $T \to T'$ gives rise to a sequence of $(m-1)^2$ two by two moves, and shrink / expand moves, transforming the A_{m-1}-web $\mathcal{W}_T$ to $\mathcal{W}_{T'}$, see Section 1.2. Proposition 5.10 categorifies the two by two moves to mutations of cluster collections. Their composition is a morphism in the groupoid $\mathcal{C}$Mod:

$$f^m_T : (\mathcal{C}^T_{G,S}, \mathcal{S}^T_{G,S}) \longrightarrow (\mathcal{C}^{T'}_{G,S}, \mathcal{S}^{T'}_{G,S}). \qquad (62)$$

So all objects $(\mathcal{C}^T_{G,S}, \mathcal{S}^T_{G,S})$ belong to the same connected component of the groupoid $\mathcal{C}$Mod. □

The cluster braid subgroup $\mathrm{Br}_{G,S} \subset \mathrm{Aut}_{\mathcal{C}\mathrm{Mod}}(\mathcal{C}^T_{G,S}, \mathcal{S}^T_{G,S})$ provides an extension

$$1 \longrightarrow \mathrm{Br}_{G,S} \longrightarrow \widetilde{\Gamma}_{G,S} \longrightarrow \widetilde{\Gamma}_{G,S}/\mathrm{Br}_{G,S} \longrightarrow 1. \qquad (63)$$

Denote by $\overline{\mathcal{C}\mathrm{Mod}}$ the groupoid obtained from $\mathcal{C}$Mod by imposing conditions $\mu_{S_k[-1]} \circ \mu_{S_k} = \mathrm{Id}$. Lemma 5.6 implies that $\mathrm{Aut}_{\overline{\mathcal{C}\mathrm{Mod}}}(\mathcal{C}, \mathcal{S}) = \widetilde{\Gamma}_{G,S}/\mathrm{Br}_{G,S}$.

Theorem 5.14. *There is a morphism of groupoids $\tau : \mathrm{Tr}_S \longrightarrow \overline{\mathcal{C}\mathrm{Mod}}$ providing a group map*

$$\tau : \Gamma_S \longrightarrow \widetilde{\Gamma}_{G,S}/\mathrm{Br}_{G,S}. \qquad (64)$$

Proof. We define a functor τ on objects by setting $\tau : T \longmapsto (\mathcal{C}^T_{G,S}, \mathcal{S}^T_{G,S})$. Its action on the generating morphisms given by by the flips is given by $\tau(T \to T') := f^m_T$. It remains to show that τ transforms the compositions of flips related to the square and pentagon relations to the identity. This is obvious for the square relations, but it is a rather non-trivial task for the pentagon relations. We prove the pentagon relation in Section 5.4. First, we prove the basic pentagon relation in the cluster set-up. Then we use a geometric result of [DGG], translated from the language of octahedral moves into the language of web mutations to deduce the general pentagon relation to the basic pentagon relation. □

Definition 5.15. The group $\Gamma_{G,S}$ is the pull back via map (64) of extension (63):

$$\begin{array}{ccccccccc}
1 & \longrightarrow & \mathrm{Br}_{G,S} & \longrightarrow & \widetilde{\Gamma}_{G,S} & \longrightarrow & \widetilde{\Gamma}_{G,S}/\mathrm{Br}_{G,S} & \longrightarrow & 1 \\
 & & \uparrow = & & \uparrow & & \uparrow \pi & & \\
1 & \longrightarrow & \mathrm{Br}_{G,S} & \longrightarrow & \Gamma_{G,S} & \longrightarrow & \Gamma_S & \longrightarrow & 1
\end{array}$$

We conclude that the two by two moves of ideal A_{m-1}-webs on a decorated surface S give rise to an action of the group $\Gamma_{G,S}$ by symmetries of the category $\mathcal{C}^T_{G,S}$, i.e., to a homomorphism

$$\Gamma_{G,S} \longrightarrow \mathrm{Aut}_{\mathcal{C}\mathrm{Mod}}(\mathcal{C}^T_{G,S}, \mathcal{S}^T_{G,S}). \tag{65}$$

Theorem 5.16. *Let $G = SL_m$. Then the homomorphism* (65) *is injective.*

This follows from a general result of Keller.

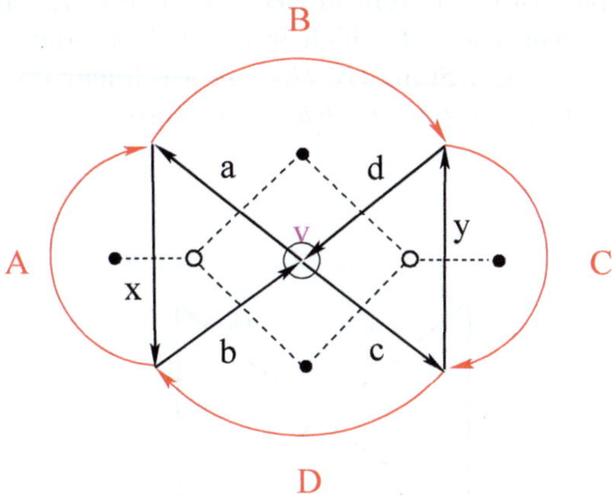

FIGURE 27. The potential $W = xab + Ax + ycd + yC + Bad + Dcb + W'$.

5.4. The potentials and two by two moves

Denote by $h(a)$ the head of an arrow a, and by $t(a)$ the tale. We write ab for the composition of two arrows b and a, where b is followed by a, so that $h(b) = t(a)$. Recall that given a potential W on a quiver Q, a mutation of the pair (Q, W) at a vertex v is performed as follows [DWZ].

- For each arrow p incident to v: reverse its order to the opposite one, getting an arrow p^*.
- For each loop α passing through the vertex v: for each consecutive pair of arrows p, q such that $h(p) = t(q) = v$, so that $\alpha = Aqp$, alter the loop α to the loop $A[qp]$.
- For each pair of arrows p, q such that $h(p) = t(q) = v$: add new arrow $[qp]$ to the graph, and add a new loop $[qp]p^*q^*$ to the potential.

We can alter a quiver with potential, getting an *equivalent quiver with potential,* in two ways:

1. By applying to it an automorphism of the path algebra of the quiver.
2. Presenting a quiver with potential as a direct sum of the two quivers with potentials

$$(Q, W) = (Q_1, W_1) \oplus (Q_2, W_2),$$

where W_1 is a sum of two-edge loops: $W_1 = \sum_i a_i b_i$. The "direct sum" means that the a_i, b_i do not enter to the W_2. Then the quiver with potential (Q, W) is said to be equivalent to (Q_2, W_2).

Consider a two by two move on a bipartite graph $\Gamma \to \Gamma'$ centered at a vertex v.

The potential of the original quiver Q_Γ can be written as follows, see Figure 27:

$$W = xab + Ax + ycd + yC + Bad + Dcb + W'.$$

Here the W' is the part of the potential free from a, b, c, d, x, y. The A is a linear combination of the paths each of which start at $h(x)$ and end at $t(x)$, so the composition Ax makes sense. Similarly B, C, D are linear combinations of paths such that the compositions Bad, Cy, Dcb make sense.

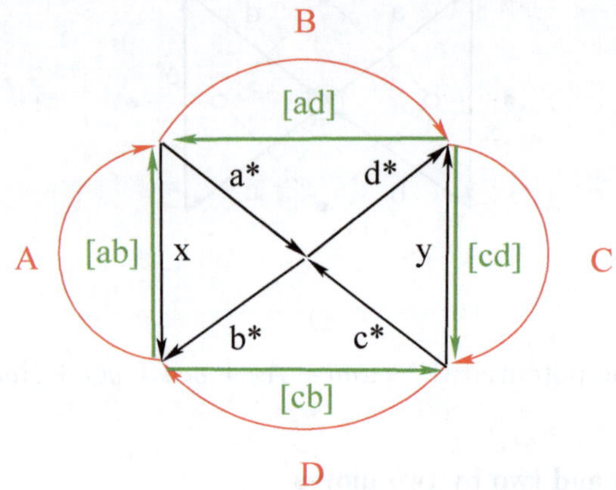

FIGURE 28. The mutated potential $\widetilde{W}$.

Let us mutate the quiver with potential W at the central vertex v of the quiver Q_Γ, shown by a circle on Figure 27. We get a new quiver with a potential $\widetilde{W}$:

$$\widetilde{W} = x[ab] + [cd]y + Ax + B[ad] + yC + D[cb] + [cd]d^*c^* \\ + [ad]d^*a^* + [cb]b^*c^* + [ab]b^*a^* + W'.$$

Consider an automorphism of the path algebra which acts on the generators identically except on the following ones:

$$[ab] \longmapsto [ab] - A,\ x \longmapsto x - b^*a^*,\ [cd] \longmapsto [cd] - C,\ y \longmapsto y - d^*c^*.$$

It sends the potential $\widetilde{W}$ to the following one:

$$\begin{aligned}(x - b^*a^*)([ab] - A) + ([cd] - C)(y - d^*c^*) + A(x - b^*a^*) + B[ad] \\ + (y - d^*c^*)C + D[cb] + ([cd] - C)d^*c^* + [ad]d^*a^* + [cb]b^*c^* \\ + ([ab] - A)b^*a^* + W' \\ = x[ab] + [cd]y + B[ad] + D[cb] - Cd^*c^* + [ad]d^*a^* + [cb]b^*c^* - Ab^*a^* + W'.\end{aligned} \tag{66}$$

This potential is a direct sum of the potential $x[ab] + [cd]y$ and the potential

$$\widetilde{W}^* = B[ad] + D[cb] - Cd^*c^* - Ab^*a^* + [ad]d^*a^* + [cb]b^*c^* + W'.$$

Let us consider the automorphism of the path algebra which acts on the generators

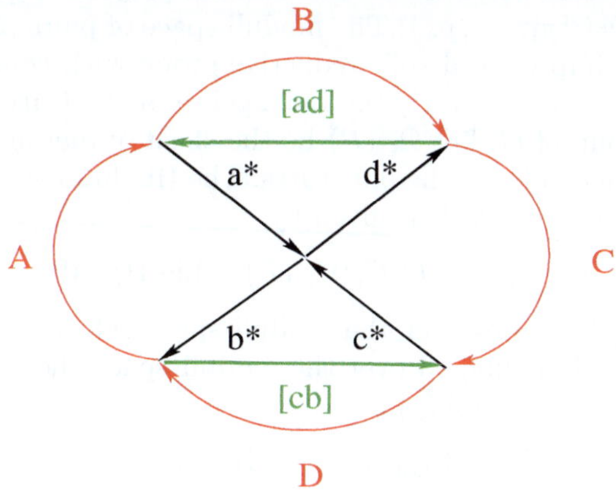

FIGURE 29. The equivalent quiver with potential W^*.

identically except the following ones:

$$b^* \longmapsto -b^*,\ d^* \longmapsto -d^*,$$

It transforms the potential $\widetilde{W}^*$ to the following one:

$$W^* = B[ad] + D[cb] + Cd^*c^* + Ab^*a^* - [ad]d^*a^* - [cb]b^*c^* + W'.$$

But this is nothing else but the potential for the quiver $Q_{\Gamma'}$ for the bipartite graph Γ' obtained from Γ by the two my two move at the vertex v, see Figure 30.

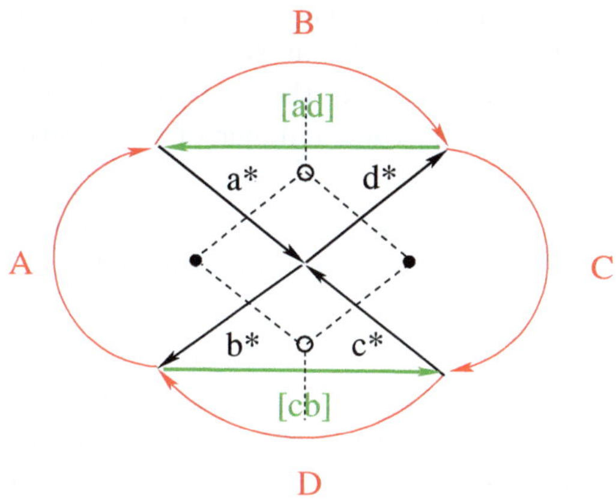

FIGURE 30. The mutated potential W^* is the potential for the quiver $Q_{\Gamma'}$.

6. 3d Calabi–Yau categories from ideal webs and open Calabi–Yau threefolds

6.1. Geometry of the Hitchin fibration

1. The universal Hitchin base. Let C be a smooth projective genus g curve with a set of punctures $P := \{p_1, \ldots, p_n\}$. The moduli space of pairs (C, P) is denoted by $\mathcal{M}_{g,n}$. Let G be a simply laced split reductive group with connected center, e.g., PGL_m or GL_m. Denote by Lie(H) the Lie algebra of the Cartan group of G, and by W the Weyl group of G. Let $\Omega_C(P)$ be the sheaf of meromorphic differentials on C with logarithmic poles at the punctures. The Hitchin base $\mathrm{B}_{G,C,P}$ for a group G and a punctured curve (C, P) is given by

$$\mathrm{B}_{G,C,P} := H^0(C, (\Omega_C(P) \otimes \mathrm{Lie(H)})/W).$$

When the pair (C, P) varies over the moduli space $\mathcal{M}_{g,n}$, we get the *universal Hitchin base* $\mathrm{B}_{G,g,n}$. It is fibered over the moduli space the $\mathcal{M}_{g,n}$ with the fiber $\mathrm{B}_{G,C,P}$ over a point $(C, P) \in \mathcal{M}_{g,n}$:

$$\begin{array}{ccc} \mathrm{B}_{G,C,P} & \hookrightarrow & \mathrm{B}_{G,g,n} \\ \downarrow & & \downarrow \\ (C,P) & \hookrightarrow & \mathcal{M}_{g,n} \end{array}$$

2. The Hitchin base and the spectral cover for PGL_m. A point t of the Hitchin's base of (C, P) is given by the following data:

$$t = (C, P; t_2, t_3, \ldots, t_m), \quad t_k \in \Omega_C(P)^{\otimes k}.$$

The associated spectral curve is a curve in T^*C given by solutions of the polynomial equation:

$$\Sigma_t := \{\lambda \in T^*C \mid \lambda^m + t_2\lambda^{m-2} + \cdots + t_{m-1}\lambda + t_m = 0\} \subset T^*C. \qquad (67)$$

The projection $T^*C \to C$ induces the spectral cover $\pi_t : \Sigma_t \to C$. It is an $m : 1$ ramified cover of C. For generic t it has only simple ramification points. Their projections are precisely the zeros of the discriminant $\Delta_m = (\prod_{1 \leq i<j \leq m}(\lambda_i - \lambda_j))^2$ of equation (67), where $\{\lambda_i\}_{i=1,\ldots,m}$ are the roots. For example, $\Delta_3 = 4t_2^3 - 27t_3^2$. So $\Delta_m \in \Omega_C(P)^{m(m-1)}$. So the total number of ramification points is $m(m-1)(2g-2+n)$.

3. A family of open CY threefolds over the universal Hitchin base. It is a family $\mathrm{Y}_{G,C,P}$ of open Calabi–Yau complex algebraic threefolds over the Hitchin base $\mathrm{B}_{G,C,P}$:

$$\mathrm{Y}_{G,C,P} \longrightarrow \mathrm{B}_{G,C,P}. \qquad (68)$$

When the set of punctures is empty, it was studied in [DDDHP], [DDP], [G]. Then the intermediate Jacobians of the fibers of fibration (68) provide Hitchin's integrable system related to G [DDP]. A general construction for an arbitrary decorated surface, possibly with boundary components carrying special points, is given [KS3, Section 8].

When the pair (C, P) varies over $\mathcal{M}_{g,n}$, we get the universal family of open threefolds:

$$\mathrm{Y}_{G,g,n} \longrightarrow \mathrm{B}_{G,g,n}. \tag{69}$$

Example. Let $G = PGL_2$. Then Hitchin's base parametrizes pairs (C, P, t), where t is a quadratic differential on C with the order two poles at the punctures. Consider a threefold

$$\mathrm{Y}^t_{C,P} := \{x \in C - P; \alpha_1, \alpha_2, \alpha_3 \in \Omega^1_x(C) \mid \alpha_1^{\otimes 2} + \alpha_2^{\otimes 2} + \alpha_3^{\otimes 2} = t\}. \tag{70}$$

The four-dimensional variety of the quadruples $(x; \alpha_1, \alpha_2, \alpha_3)$ has a canonical volume form. Indeed, the determinant of its cotangent space is identified with $T_x(C)^{\otimes 3} \otimes T^*_x(C) = T_x(C)^{\otimes 2}$. The quadratic differential t trivializes it. Therefore the hypersurface given by equation (70) inherits a volume form. This way we get an open Calabi–Yau threefold.

4. The Hitchin integrable system for $G = PGL_m$. A Higgs bundle on (C, P) is a holomorphic G-bundle $\mathcal{E}$ on C with an operator $\Phi : \mathcal{E} \longrightarrow \mathcal{E} \otimes \Omega_C(P)$. A Higgs bundle $(\mathcal{E}, \Phi)$ on C is the same thing as a coherent sheaf $\mathcal{L}_{\mathcal{E}}$ on the surface $\Omega_C(P)$. Informally $\mathcal{L}_{\mathcal{E}}$ is given by the eigenvalues of the Higgs field operator $\Phi : \mathcal{E} \longrightarrow \mathcal{E} \otimes \Omega_C(P)$. Let $\pi : \Omega_C(P) \longrightarrow C$ be the canonical projection. Then $\pi_* \mathcal{L}_{\mathcal{E}} = \mathcal{E}$. Denote by $\mathcal{H}_{G,g,n}$ the moduli space of Higgs G-bundles over the universal curve (C, P) over $\mathcal{M}_{g,n}$. Then there is the Hitchin integrable system

$$p : \mathcal{H}_{G,g,n} \longrightarrow \mathrm{B}_{G,g,n} \; (\mathcal{E}, \Phi) \longmapsto (\mathrm{Tr}\Phi^2, \mathrm{Tr}\Phi^3, \ldots, \mathrm{Tr}\Phi^m).$$

Let $\mathrm{B}^\times_{G,g,n}$ be the complement to the discriminant, that is the open part of the universal Hitchin base formed by the points over which the map p is non-singular. Then the fiber is an abelian variety A_t. It is described as follows. The spectral cover $\pi_t : \Sigma_t \longrightarrow C$ gives rise to a map of Jacobians $\pi_{t*} : J_{\Sigma_t} \longrightarrow J_C$. One has

$$A_t = \mathrm{Ker}(\pi_{t*} : J_{\Sigma_t} \longrightarrow J_C).$$

There is a local system of lattices over $\mathrm{B}^\times_{G,g,n}$ with the fibers

$$\Lambda_t := H_1(A_t, \mathbb{Z}) = \mathrm{Ker}(H_1(\Sigma_t, \mathbb{Z}) \longrightarrow H_1(C, \mathbb{Z})).$$

The space $\mathrm{B}^\times_{G,g,n}$ is fibered over the moduli space $\mathcal{M}_{g,n}$ with a fiber $\mathrm{B}^\times_{G,C,P}$, the complement to the discriminant in the Hitchin base. The fundamental group $\pi_1(\mathrm{B}^\times_{G,g,n}, t)$ is an extension

$$1 \longrightarrow \pi_1(\mathrm{B}^\times_{G,C,P}) \longrightarrow \pi_1(\mathrm{B}^\times_{G,g,n}, t) \longrightarrow \pi_1(\mathcal{M}_{g,n}, (C, P)) \longrightarrow 1.$$

The fundamental group $\pi_1(\mathrm{B}^\times_{G,g,n}, t)$ acts on the lattice Λ_t.

5. Framed Hitchin base. We also need a $W:1$ cover $\widetilde{\mathcal{B}}_{G,S}$ of the universal Hitching base:

$$\widetilde{\mathcal{B}}_{G,S} \longrightarrow \widetilde{\mathrm{B}}_{G,S}.$$

When $G = PGL_m$, a point of $\widetilde{\mathcal{B}}_{G,S}$ is given by a point $t \in \widetilde{\mathrm{B}}_{G,S}$ plus for each puncture p on S a choice of an ordering of the points of the spectral curve over the point p. Equivalently, it is given by an ordering of the roots $\lambda_1, \ldots, \lambda_m$ of the polynomial (67) near the point p. The λ_i's are 1-forms. So we get an ordered set of their residues $c_i := \mathrm{Res}_p \lambda_i$ at p.

We call the $\widetilde{\mathcal{B}}_{G,S}$ the framed universal Hitchin base. Dealing with the moduli space $\mathcal{X}_{G,S}$ of framed G-local systems on S we always deal with the framed universal Hitchin base $\widetilde{\mathcal{B}}_{G,S}$.

6.2. CY$_3$ categories from ideal webs and open Calabi–Yau threefolds

1. Geometry of the topological spectral cover. Denote by S a genus g oriented topological surface with n punctures. Let $\mathcal{W}$ be an ideal A_{m-1}-web on S. Denote by $\Gamma_{\mathcal{W}}$ the A_{m-1}-bipartite graph on S corresponding to $\mathcal{W}$. The spectral surface $\Sigma_{\mathcal{W}}$ related to the bipartite graph $\Gamma_{\mathcal{W}}$ is a surface with mn punctures. The spectral map is a ramified $m:1$ map

$$\pi_{\mathcal{W}} : \Sigma_{\mathcal{W}} \longrightarrow S. \tag{71}$$

It is unramified over punctures: there are m punctures on $\Sigma_{\mathcal{W}}$ over each puncture on S.

Lemma 6.1. *For a generic t, the spectral cover $\pi_t : \Sigma_t \to C$ has the same number of ramification points and the same genus as the topological spectral cover $\pi_{\mathcal{W}} : \Sigma_{\mathcal{W}} \to S$.*

Proof. The ramification points of the spectral map $\pi_{\mathcal{W}}$ are the $\bullet$-vertices of the web $\mathcal{W}$ on S of valency ≥ 3: a $\bullet$-vertex of valency v has ramification index $v-2$. To count them, take the ideal A_{m-1}-web corresponding to an ideal triangulation of S. Then the number of 3-valent $\bullet$-vertices in a single triangle is $\binom{m}{2}$, while the number of ideal triangles is $2(2g-2+n)$. So the number 3-valent $\bullet$-vertices is $\binom{m}{2}2(2g-2+n) = m(m-1)(2g-2+n)$. This matches the number of ramification points of the spectral map π_t, see Section 6.1. The degree of each of the covers is m. So the surface $\Sigma_{\mathcal{W}}$ has the same genus and the same number of punctures, mn, as the Σ_t. □

2. A conjectural combinatorial description of Fukaya categories via ideal webs. The fiber over a point $t \in \mathrm{B}^{\times}_{G,g,n}$ of the complement to discriminant of fibration (69) is a smooth open Calabi–Yau threefold $\mathcal{Y}_t$. Its Fukaya category $\mathcal{F}(\mathcal{Y}_t)$ is a Calabi–Yau category. We do not discuss its definition here. We arrive at a family of Calabi–Yau categories over $\mathrm{B}^{\times}_{G,g,n}$. The fundamental group $\pi_1(\mathrm{B}^{\times}_{G,g,n}, t)$ acts by symmetries of the Fukaya category $\mathcal{F}(\mathcal{Y}_t)$. There is a map

$$\pi_1(\mathrm{B}^{\times}_{G,g,n}, t) \longrightarrow \pi_1(\mathcal{M}_{g,n}; (C, P)) = \Gamma_S. \tag{72}$$

So there are two kinds of 3d CY categories, coming with certain groups of symmetries:

- The combinatorial category $\mathcal{C}_{\mathcal{W}}$ assigned to an ideal A_{m-1}-web $\mathcal{W}$ on S.
  The group $\Gamma_{G,S}$ from Definition 5.15 acts by its symmetries.
- The Fukaya category $\mathcal{F}(\mathcal{Y}_t)$ of a complex smooth open CY threefold $\mathcal{Y}_t$, for $G = PGL_m$, over a generic point $t \in \mathrm{B}^{\times}_{G,g,n}$ of the universal Hitchin base.
  The fundamental group $\pi_1(\mathrm{B}^{\times}_{G,g,n})$ acts by its symmetries.

We conjecture that the easily defined category $\mathcal{C}_{\mathcal{W}}$ provides a transparent combinatorial model of a full subcategory of the Fukaya category.

Conjecture 6.2. *Let $\mathcal{W}$ be an A_{m-1}-web on S. Then there is a fully faithful functor*

$$\varphi : \mathcal{C}_{\mathcal{W}} \longrightarrow \mathcal{F}(\mathcal{Y}_t).$$

It transforms cluster collections in $\mathcal{C}_{\mathcal{W}}$ to cluster collections in the Fukaya category provided by special Lagrangian spheres. The functor φ intertwines the action of the symmetry group $\Gamma_{G,S}$ with the action of $\pi_1(\mathrm{B}^{\times}_{G,g,n})$ on the Fukaya category, providing a commutative diagram

$$\begin{array}{ccc} \Gamma_{G,S} & \longrightarrow & \Gamma_S \\ \downarrow & & \downarrow = \\ \pi_1(\mathrm{B}^{\times}_{G,g,n}) & \longrightarrow & \pi_1(\mathcal{M}_{g,n}). \end{array} \tag{73}$$

One should have a similar story when S has special points on the boundary.

Let us summarize how the combinatorial data provided by the web should match the one on the Fukaya category side.

Complex geometry:	**Topology:**
Hitchin system for PGL_m	**Moduli space of framed PGL_m-local systems**
A Riemann surface C	A punctured oriented surface S
A domain $\mathcal{B}_{\mathcal{W}}$ in the complement to the discriminant of the universal Hitchin base	An A_{m-1}-web $\mathcal{W}$ on S = its bipartite graph $\Gamma_{\mathcal{W}}$
Spectral curve Σ_t, where $t \in \mathcal{B}_{\mathcal{W}}$, and the spectral cover $\pi_t : \Sigma_t \longrightarrow C$	Spectral curve $\Sigma_{\mathcal{W}}$, and the spectral cover $\pi_{\mathcal{W}} : \Sigma_{\mathcal{W}} \longrightarrow S$
Ramification points of the spectral cover	$\bullet$-points of the graph $\Gamma_{\mathcal{W}}$
Lattice Λ_t $:= \mathrm{Ker}(H_1(\Sigma_t;\mathbb{Z}) \to H_1(C;\mathbb{Z}))$	Lattice $\Lambda_{\mathcal{W}}$ $:= \mathrm{Ker}(H_1(\Sigma_{\mathcal{W}};\mathbb{Z}) \to H_1(S;\mathbb{Z}))$
Gauss–Manin connection on lattices Λ_t	Two by two moves of webs
Fukaya category of the open Calabi–Yau threefold $\mathcal{Y}_t$ related to the Hitchin system	3d CY A_∞-category $\mathcal{C}_{\mathcal{W}}$ of the quiver with potential assigned to $\mathcal{W}$
A collection of special Lagrangian spheres generating Fukaya category	A cluster collection of spherical objects in $\mathcal{C}_{\mathcal{W}}$, parametrized by internal faces of $\Gamma_{\mathcal{W}}$

7. A complex manifold of cluster stability conditions

Below we construct a manifold whose fundamental group should realize the generalized braid group.

1. The positive and negative tropical domains. Let $\mathcal{X}$ be a cluster Poisson variety. A cluster coordinate system $\{X_i^{\mathbf{q}}\}$ assigned to a quiver $\mathbf{q}$ gives rise to a tropicalized cluster Poisson coordinate system $\{x_i^{\mathbf{q}}\}$ on the space of real tropical points $\mathcal{X}(\mathbb{R}^t)$. It determines a cone in $\mathcal{X}(\mathbb{R}^t)$ consisting of the points with non-negative coordinates in the coordinate system $\{x_i^{\mathbf{q}}\}$:

$$\mathcal{X}_{\mathbf{q}}^+(\mathbb{R}^t) := \{l \in \mathcal{X}(\mathbb{R}^t) \mid x_i^{\mathbf{q}}(l) \geq 0\} \subset \mathcal{X}(\mathbb{R}^t).$$

It was conjectured in [FG4] and proved in [GHKK] that when $\mathbf{q}$ runs through all quivers obtained from a given one by mutations, the internal parts of the domains $\mathcal{X}_{\mathbf{q}}^+(\mathbb{R}^t)$ are disjoint. The domains $\mathcal{X}_{\mathbf{q}}^+(\mathbb{R}^t)$ meet along the walls where one of the coordinates become zero. The wall $x_k^{\mathbf{q}} = 0$ separates the domain $\mathcal{X}_{\mathbf{q}}^+(\mathbb{R}^t)$ and $\mathcal{X}_{\mathbf{q}'}^+(\mathbb{R}^t)$ where the quivers $\mathbf{q}$ and $\mathbf{q}'$ are related by a mutation in the direction k. The domains $\mathcal{X}_{\mathbf{q}}^+(\mathbb{R}^t)$ usually do not cover the space $\mathcal{X}(\mathbb{R}^t)$.

For example, the space $\mathcal{X}_{PGL_2,T}(\mathbb{R}^t)$ for a punctured torus T is illustrated in [FG4, Fig 1]. For any punctured surface S, the complement $\mathcal{X}_{PGL_2,S}(\mathbb{R}^t) - \mathcal{X}_{PGL_2,S}^+(\mathbb{R}^t)$ is of measure zero.

Definition 7.1. The union of the domains $\mathcal{X}_{\mathbf{q}}^+(\mathbb{R}^t)$ is called the *positive tropical domain*:

$$\mathcal{X}^+(\mathbb{R}^t) \subset \mathcal{X}(\mathbb{R}^t).$$

Evidently, it is a connected domain. Similarly, there is a *negative tropical domain*

$$\mathcal{X}^-(\mathbb{R}^t) \subset \mathcal{X}(\mathbb{R}^t).$$

It is the union of the domains $\mathcal{X}_{\mathbf{q}}^-(\mathbb{R}^t) := \{l \in \mathcal{X}(\mathbb{R}^t) \mid x_i^{\mathbf{q}}(l) \leq 0\} \subset \mathcal{X}(\mathbb{R}^t)$.

The cluster modular group Γ acts by automorphisms of the domains $\mathcal{X}^{\pm}(\mathbb{R}^t)$.

Given any mutation-connected set $\mathcal{S}$ of quivers, there are subdomains

$$\mathcal{X}_{\mathcal{S}}^{\pm}(\mathbb{R}^t) := \coprod_{\mathbf{q}\in\mathcal{S}} \mathcal{X}_{\mathbf{q}}^{\pm}(\mathbb{R}^t) \subset \mathcal{X}(\mathbb{R}^t). \tag{74}$$

2. The complex domain $\mathcal{U}_{\mathcal{X}}$ of cluster stability conditions. Our goal is to define a Γ-equivariant complex domain $\mathcal{U}_{\mathcal{X}}$ equipped with a Γ-equivariant surjective projection

$$p : \mathcal{U}_{\mathcal{X}} \longrightarrow \mathcal{X}^+(\mathbb{R}^t).$$

It sits in the space of stability conditions $\mathrm{Stab}_{\mathcal{X}}/\mathrm{Br}_{\mathcal{X}}$ for the 3d CY category related to the cluster variety $\mathcal{X}$ with a generic potential, i.e., there is a Γ-equivariant embedding:

$$\mathcal{U}_{\mathcal{X}} \subset \mathrm{Stab}_{\mathcal{X}}/\mathrm{Br}_{\mathcal{X}}.$$

So each point $u \in \mathcal{U}_{\mathcal{X}}$ has a neighborhood isomorphic to an open subset in $\mathrm{Hom}(\Lambda, \mathbb{C})$.

Take an upper half-plane

$$\mathcal{H} := \{r\exp(i\varphi) \mid r > 0, 0 < \varphi \leq \pi\}.$$

A quiver $\mathbf{q} = (\Lambda, \{e_i\}_{i\in \mathrm{I}}, (*,*))$ gives rise to a complex domain $\mathcal{U}_\mathbf{q}$ parametrizing homomorphisms of abelian groups $f : \Lambda \to \mathbb{C}$ (the central charges) with $f(e_i) \in \mathcal{H}$:

$$\mathcal{U}_\mathbf{q} := \{f : \Lambda \longrightarrow \mathbb{C} \mid f(a+b) = f(a) + f(b),\ f(e_i) \in \mathcal{H},\ \forall i \in \mathrm{I}\}. \quad (75)$$

So the domain $\mathcal{U}_\mathbf{q}$ is the product of the domains $\mathcal{H}$ over the set I of vertices of $\mathbf{q}$:

$$\mathcal{U}_\mathbf{q} := \prod_{j\in \mathrm{I}} \mathcal{H}_j, \qquad \mathcal{H}_j := \{z_j = r_j\exp(i\varphi_j) \mid r_j > 0, 0 < \varphi_j \leq \pi\}. \quad (76)$$

Recall that we assign to a quiver $\mathbf{q}$ a set of tropical cluster Poisson coordinates $\{x_i\}_{i\in \mathrm{I}}$. The imaginary part map $\mathrm{Im} : \mathcal{H} \longrightarrow \mathbb{R}_{\geq 0}$, $z \longmapsto \mathrm{Im}(z)$ provides a projection

$$\mathrm{Im} : \mathcal{U}_\mathbf{q} \longrightarrow \mathcal{X}^+_\mathbf{q}(\mathbb{R}^t),\ \{z_i\} \longmapsto \{x_i := \mathrm{Im}(z_i)\}. \quad (77)$$

So we assign to a quiver $\mathbf{q}$ a triple

$$\Big(\mathcal{U}_\mathbf{q},\ \mathcal{X}^+_\mathbf{q}(\mathbb{R}^t),\ \mathrm{Im} : \mathcal{U}_\mathbf{q} \longrightarrow \mathcal{X}^+_\mathbf{q}(\mathbb{R}^t)\Big). \quad (78)$$

An isomorphism of quivers $\mathbf{q} \longrightarrow \mathbf{q}'$ gives rise to a unique isomorphism of the triples.

Let us glue the triples (78) assigned to the quivers obtained by mutations of a quiver $\mathbf{q}$.

Mutations. Denote by $\overline{\mathcal{U}}_\mathbf{q}$ the partial closure of the domain $\mathcal{U}_\mathbf{q}$ defined by using the inequalities $0 \leq \varphi_j \leq \pi$ in (76) while keeping $r > 0$. The domain $\overline{\mathcal{U}}_\mathbf{q}$ has boundary walls of real codimension one given by setting either $\varphi_j = \pi$ or $\varphi_j = 0$ for some $j \in \mathrm{I}$.

Given a mutation of quivers $\mathbf{q} \stackrel{k}{\longrightarrow} \mathbf{q}'$ in the direction k, let us glue the domain $\overline{\mathcal{U}}_\mathbf{q}$ to the domain $\overline{\mathcal{U}}_{\mathbf{q}'}$ along the two real codimension one walls $\varphi_k = \pi$ and $\varphi_k = 0$:

$$\begin{aligned} G_{\mathbf{q}\to\mathbf{q}'} &: \{\text{the wall } \varphi_k = \pi \text{ in } \overline{\mathcal{U}}_\mathbf{q}\} \stackrel{\text{glued}}{\longrightarrow} \{\text{the wall } \varphi'_k = 0 \text{ in } \overline{\mathcal{U}}_{\mathbf{q}'}\}, \\ G_{\mathbf{q}\to\mathbf{q}'} &: \{\text{the wall } \varphi_k = 0 \text{ in } \overline{\mathcal{U}}_\mathbf{q}\} \stackrel{\text{glued}}{\longrightarrow} \{\text{the wall } \varphi'_k = \pi \text{ in } \overline{\mathcal{U}}_{\mathbf{q}'}\}. \end{aligned} \quad (79)$$

Namely, let $\{z_i\}_{i\in \mathrm{I}}$ be the coordinates in the domain $\mathcal{U}_\mathbf{q}$, and $\{z'_i\}_{i\in \mathrm{I}}$ the ones in $\mathcal{U}_{\mathbf{q}'}$. We define the gluing maps as follows:

$$z'_i := (G_{\mathbf{q}\to\mathbf{q}'})^*(z_i) := \begin{cases} -z_k & \text{if } i = k, \\ z_i + [\varepsilon_{ik}]_+ z_k & \text{if } i \neq k. \end{cases} \quad (80)$$

Observe that if $\mathrm{Im}(z_k) = 0$, and $\mathrm{Im}(z_i) > 0$ for $i \neq k$, then $\mathrm{Im}(z'_k) = 0$ and $\mathrm{Im}(z'_i) > 0$ for $i \neq k$. Furthermore, if $\varphi_k = 0$ then $\varphi'_k = \pi$, and vice versa, if $\varphi_k = \pi$ then $\varphi'_k = 0$.

Here is an equivalent way to look at the gluing of the domains $\overline{\mathcal{U}}_\mathbf{q}$ and $\overline{\mathcal{U}}_{\mathbf{q}'}$.

Recall that the basis $\{e'_i\}$ of the quiver $\mathbf{q}'$ is related to the basis $\{e_i\}$ of the quiver $\mathbf{q}$ by

$$e'_i := \begin{cases} -e_k & \text{if } i = k, \\ e_i + [\varepsilon_{ik}]_+ e_k & \text{if } i \neq k. \end{cases} \tag{81}$$

Both domains $\mathcal{U}_{\mathbf{q}}$ and $\mathcal{U}_{\mathbf{q}'}$ sit in the space $\mathrm{Hom}(\Lambda, \mathbb{C})$ via (75). It is clear from (81) that they are glued by the map (80) precisely the way they intersect there; the result of gluing is identified with the union of domains

$$\mathcal{U}_{\mathbf{q}} \cup \mathcal{U}_{\mathbf{q}'} \subset \mathrm{Hom}(\Lambda, \mathbb{C}).$$

In particular the $\mathcal{U}_{\mathbf{q}} \cup \mathcal{U}_{\mathbf{q}'}$ is a manifold.

Denote by $\{x'_i\}$ the tropical cluster Poisson coordinates assigned to $\mathbf{q}'$. They are related to the ones $\{x_i\}_{i \in \mathrm{I}}$ for the quiver $\mathbf{q}$ by the formula

$$\mu_k^*(x'_i) := \begin{cases} -x_k & \text{if } i = k \\ x_i - \varepsilon_{ik} \min\{0, -\mathrm{sgn}(\varepsilon_{ik}) x_k\} & \text{otherwise.} \end{cases} \tag{82}$$

Lemma 7.2. *The projection* (77) *is compatible with mutations. So we get a projection*

$$\mathcal{U}_{\mathbf{q}} \cup \mathcal{U}_{\mathbf{q}'} \longrightarrow \mathcal{X}^+_{\mathbf{q}}(\mathbb{R}^t) \cup \mathcal{X}^+_{\mathbf{q}'}(\mathbb{R}^t). \tag{83}$$

Proof. If $x_k \geq 0$, the mutation formula (82) for tropical cluster Poisson coordinates reduces to:

$$\mu_k^*(x'_i) := \begin{cases} -x_k & \text{if } i = k \\ x_i + [\varepsilon_{ik}]_+ x_k & \text{otherwise.} \end{cases} \tag{84}$$

So taking the imaginary part of the gluing map (80) we recover (84). □

The result of gluing of the two domains corresponding to a pair of quivers $\mathbf{q} \stackrel{k}{\longleftrightarrow} \mathbf{q}'$ is a domain which is homotopy equivalent to a circle. Indeed, it is identified with

$$\mathcal{U}_{\mathbf{q}} \cup \mathcal{U}_{\mathbf{q}'} = \mathbb{C}^* \times \prod_{j \neq k} \mathcal{H}_j.$$

Consider now the composition of two mutations in the same direction k:

$$\mathbf{q} \stackrel{k}{\longrightarrow} \mathbf{q}' \stackrel{k}{\longrightarrow} \mathbf{q}''.$$

It is described by mutations of the basis $\{e_i\}$ defining the quiver $\mathbf{q}$:

$$\{e_i\} \stackrel{k}{\longrightarrow} \{e'_i\} \stackrel{k}{\longrightarrow} \{e''_i\}.$$

The basis $\{e''_i\}$ is usually different than the one $\{e_i\}$. However quivers $\mathbf{q}''$ and $\mathbf{q}$ are canonically isomorphic: there is a unique isomorphism of lattices $\varphi : \Lambda \longrightarrow \Lambda$ preserving the form $(*, *)$ such that $\varphi(e_i) = e''_i$. It defines an isomorphism of quivers, and hence domains:

$$\varphi : \mathcal{U}_{\mathbf{q}} \longrightarrow \mathcal{U}_{\mathbf{q}''}.$$

It is well known that the cones $\mathcal{X}^+_{\mathbf{q}}(\mathbb{R}^t)$ and $\mathcal{X}^+_{\mathbf{q}''}(\mathbb{R}^t)$ coincide. So we get a commutative diagram

$$\begin{array}{ccc} \mathcal{U}_{\mathbf{q}} & \xrightarrow{\varphi} & \mathcal{U}_{\mathbf{q}''} \\ \downarrow{\scriptstyle \mathrm{Im}_{\mathbf{q}}} & & \downarrow{\scriptstyle \mathrm{Im}_{\mathbf{q}''}} \\ \mathcal{X}^+_{\mathbf{q}}(\mathbb{R}^t) & \xrightarrow{=} & \mathcal{X}^+_{\mathbf{q}''}(\mathbb{R}^t) \end{array}$$

Cluster transformations. Consider a tree T_n whose edges incident to a given vertex are parametrized by the set I. We assign to each vertex a of T_n a quiver $\mathbf{q}_a$ considered up to an isomorphism, so that the quivers $\mathbf{q}_a$ and $\mathbf{q}_b$ assigned to the vertices of an edge labeled by an element $k \in \mathrm{I}$ are related by the mutation μ_k:

$$\mathbf{q}_b = \mu_k(\mathbf{q}_a), \ \mathbf{q}_a = \mu_k(\mathbf{q}_b). \tag{85}$$

We assign to each vertex a of the tree T_n the domain $\mathcal{U}_{\mathbf{q}_a}$. Gluing the pairs of domains $\mathcal{U}_{\mathbf{q}_a}$ and $\mathcal{U}_{\mathbf{q}_b}$ assigned to each edge of the tree T_n as explained above, we get a manifold denoted $\widetilde{\mathcal{U}}_{\mathcal{X}}$. Lemma 7.2 implies that projections (77) are glued into a projection

$$\mathrm{Im} : \widetilde{\mathcal{U}}_{\mathcal{X}} \longrightarrow \mathcal{X}(\mathbb{R}^t). \tag{86}$$

For any pair of vertices a, b of the tree T_n, there is a unique path $\mathbf{i}$ on the tree connecting them:

$$\mathbf{i}: \ a = a_0 \xrightarrow{k_1} a_1 \xrightarrow{k_2} \cdots \xrightarrow{k_{m-1}} a_{m-1} \xrightarrow{k_m} a_m = b. \tag{87}$$

Moreover, there is a unique automorphism of trees $i_{a\to b} : \mathrm{T}_n \longrightarrow \mathrm{T}_n$ which sends a to b and preserves the decoration of the edges by the set I.

Now take a pair of vertices a and b of the tree T_n such that

- The quivers $\mathbf{q}_a$ and $\mathbf{q}_b$ are isomorphic.

Then for any vertex c, the automorphism $i_{a\to b} : \mathrm{T}_n \longrightarrow \mathrm{T}_n$ induces an isomorphism of the quivers assigned to the vertices c and $i_{a\to b}(c)$. Therefore it induces an automorphism

$$\varphi_{a\to b} : \widetilde{\mathcal{U}}_{\mathcal{X}} \longrightarrow \widetilde{\mathcal{U}}_{\mathcal{X}}.$$

On the other hand, it induces an automorphism of the cluster Poisson variety:

$$\Phi_{a\to b} : \mathcal{X} \longrightarrow \mathcal{X}. \tag{88}$$

Its tropicalisation induces an automorphism of the space of real tropical points:

$$\Phi^t_{a\to b} : \mathcal{X}(\mathbb{R}^t) \longrightarrow \mathcal{X}(\mathbb{R}^t).$$

We have a commutative diagram:

$$\begin{array}{ccc} \widetilde{\mathcal{U}}_{\mathcal{X}} & \xrightarrow{\varphi_{a\to b}} & \widetilde{\mathcal{U}}_{\mathcal{X}} \\ \downarrow{\scriptstyle \mathrm{Im}} & & \downarrow{\scriptstyle \mathrm{Im}} \\ \mathcal{X}^+(\mathbb{R}^t) & \xrightarrow{\Phi^t_{a\to b}} & \mathcal{X}^+(\mathbb{R}^t) \end{array}$$

Now take a pair of vertices a and b of the tree T_n such that

- The quivers $\mathbf{q}_a$ and $\mathbf{q}_b$ are isomorphic. The map $\Phi^t_{a\to b}$ is the identity.

The corresponding automorphisms $\varphi_{a\to b}$ of the manifold $\widetilde{\mathcal{U}}_{\mathcal{X}}$ form a group denoted by Γ_{triv}.

Definition 7.3. The orbifold $\mathcal{U}_{\mathcal{X}}$ is the quotient of the manifold $\widetilde{\mathcal{U}}_{\mathcal{X}}$ by the action of the group Γ_{triv}:

$$\mathcal{U}_{\mathcal{X}} := \widetilde{\mathcal{U}}_{\mathcal{X}}/\Gamma_{\text{triv}}.$$

Then, by the definition, the projection (86) induces a canonical projection

$$\text{Im} : \mathcal{U}_{\mathcal{X}} \longrightarrow \mathcal{X}^+(\mathbb{R}^t). \tag{89}$$

By the definition, the automorphisms (88) generate the cluster modular group Γ. Evidently, the projection (89) is Γ-equivariant. The complex manifold $\mathcal{U}_{\mathcal{X}}$ has a Γ-equivariant Poisson structure $\{*, *\}$ which in any cluster coordinate system $\{z_i\}$ is given by

$$\{z_i, z_j\} = (e_i, e_j).$$

The real tropical space also has a Poisson structure given by the same formula $\{x_i, x_j\} = (e_i, e_j)$ in any tropical cluster coordinate system $\{x_i\}$. The projection (89) is evidently Poisson.

We summarize the main result of this Section.

Theorem 7.4. *Let $\mathcal{X}$ be a cluster Poisson variety. Then the domains $\mathcal{U}_{\mathbf{q}}$ are glued into a complex Poisson orbifold $\mathcal{U}_{\mathcal{X}}$. The cluster modular group Γ acts on $\mathcal{U}_{\mathcal{X}}$. The projections* (77) *give rise to a Γ-equivariant Poisson projection on the space of positive real tropical points of $\mathcal{X}$:*

$$p : \mathcal{U}_{\mathcal{X}} \longrightarrow \mathcal{X}^+(\mathbb{R}^t).$$

There is a Γ-equivariant embedding of the orbifold $\mathcal{U}_{\mathcal{X}}$ into the space of stability conditions $\text{Stab}_{\mathcal{X}}/\text{Br}_{\mathcal{X}}$ *on the* 3d *Calaby–Yau category related to a generic potential of a quiver* $\mathbf{q}$*:*

$$\mathcal{U}_{\mathcal{X}} \subset \text{Stab}_{\mathcal{X}}/\text{Br}_{\mathcal{X}}. \tag{90}$$

Proof. The only claim which was not discussed yet is the Γ-equivariance of the projection p. Let $\widetilde{\Gamma}$ be the group of automorphisms $i_{a\to b}$ of the tree. Denote by Γ'_{triv} the subgroup of the automorphisms $i_{a\to b}$ which induce the trivial automorphisms $\Phi_{a\to b}$ of the cluster variety. The cluster modular group is given by $\Gamma := \widetilde{\Gamma}/\Gamma'_{\text{triv}}$. However the group which acts naturally on $\widetilde{\mathcal{U}}_{\mathcal{X}}$ is $\widetilde{\Gamma}/\Gamma_{\text{triv}}$. Evidently $\Gamma'_{\text{triv}} \subset \Gamma_{\text{triv}}$: if a cluster Poisson transformation is trivial, its tropicalisation is also trivial. The crucial claim that in fact $\Gamma'_{\text{triv}} = \Gamma_{\text{triv}}$ is given by Theorem 7.5.

Theorem 7.5. *Let $\sigma : \mathbf{q} \to \mathbf{q}'$ be a cluster transformation which induces the identity transformation on the set of the positive tropical points. Then the quivers $\mathbf{q}$ and $\mathbf{q}'$ are isomorphic, and the corresponding quantum cluster transformation $\Phi(\sigma)$ is the identity.*

Theorem 7.5 is derived in [GS16, Section 2] from a theorem of Keller [K11], which was stated using *c-vectors* and *C-matrices* rather than the tropical points of cluster Poisson varieties. □

There is a similar complex Poisson manifold $\mathcal{U}_{\mathcal{X}}^{-}$ equipped with a Γ-equivariant projection

$$p: \mathcal{U}_{\mathcal{X}}^{-} \longrightarrow \mathcal{X}^{-}(\mathbb{R}^t).$$

An example. Consider the cluster Poisson variety $\mathcal{X}_{\mathrm{A}_2}$ of type A_2. Its tropicalisation $\mathcal{X}_{\mathrm{A}_2}(\mathbb{R}^t)$ is identified with the real plane $\mathbb{R}^2$ decomposed into the union of five cones, denoted by $C_1, \ldots, C_5$, numbered clockwise. We assign to each cone C_i a complex domain $\mathcal{U}_i$, isomorphic to $\mathcal{H} \times \mathcal{H}$. Then we glue these domains in the cyclic order:

$$\mathcal{U}_1 \longleftrightarrow \mathcal{U}_2 \longleftrightarrow \mathcal{U}_3 \longleftrightarrow \mathcal{U}_4 \longleftrightarrow \mathcal{U}_5 \longleftrightarrow \mathcal{U}_1.$$

The mapping class group is isomorphic to $\mathbb{Z}/5\mathbb{Z}$. Its generator acts by shifting $\mathcal{U}_i$ to $\mathcal{U}_{i+1}$.

Acknowledgement

The geometric part of the paper, Sections 2–4, describing ideal webs and cluster coordinates was written at the IHES during the Summer 2011. I am grateful to Maxim Kontsevich for useful discussions, and to the IHES for the hospitality and support. I am grateful to Efim Abrikosov for useful discussions and comments on the paper. I was supported by the NSF grants DMS-1059129 and DMS-1301776.

The whole story was discussed at talks at the Mirror Symmetry conference at Miami (January 2012) and Cluster Algebras conference at Northeastern University (April 2012). I am grateful to participants for useful discussions.

References

[ABCGPT] Arkani-Hamed N., Bourjaily J., Cachazo F., Goncharov A., Postnikov A., Trnka J.: *Scattering Amplitudes and the Positive Grassmannian.* arXiv: 1212.5605.

[BFZ] Berenstein A., Fomin S., Zelevinsky A.: *Cluster algebras. III Upper bounds and double Bruhat cells.* Duke Math. J. 126 (2005) no. 1, 1–52.

[BK] Bondal A., Kapranov M.: *Enhanced triangulated categories.* USSR Sbornik. Tom 181 (1990).

[Br] Bridgeland T.: *Stability conditions on triangulated categories.* arXiv:math/ 0212237.

[Br1] Bridgeland T.: *t-structures on some local Calabi–Yau varieties.* J. Algebra 289 (2005), no. 2, 453–483.

[BrS] Bridgeland T., Smith I.: *Quadratic differentials as stability conditions.* arXiv: 1302.7030.

[DWZ] Derksen H., Weyman J., Zelevinsky A.: *Quivers with potentials and their representations I: Mutations.* arXiv:0704.0649.

[DDDHP] Diaconescu, D.-E.; Dijkgraaf, R.; Donagi, R.; Hofman, C.; Pantev, T.: *Geometric transitions and integrable systems.* Nuclear Phys. B 752 (2006), no. 3, 329–390. arXiv:hep-th/0506196.

[DDP] Diaconescu D.-E., Donagi R., Pantev, T.: *Intermediate Jacobians and ADE Hitchin Systems.* Math. Res. Lett. 14 (2007), no. 5, 745–756. arXiv:hep-th/0607159.

[DGG] Dimofte T., Gabella M., Goncharov A.B.: K-Decompositions and 3d Gauge Theories. arXiv:1301.0192.

[FG1] Fock V.V., Goncharov A.B.: *Moduli spaces of local systems and higher Teichmüller theory.* Publ. Math. IHES, no. 103 (2006) 1–212. arXiv:math.AG/0311149.

[FG2] Fock V., Goncharov A.B.: *Cluster ensembles, quantization and the dilogarithm.* Ann. Sci. L'Ecole Norm. Sup. (2009). arXiv:math.AG/0311245.

[FG3] Fock V., Goncharov A.B.: *The quantum dilogarithm and representations quantum cluster varieties.* arXiv:math/0702397.

[FG4] Fock V., Goncharov A.B.: *Cluster varieties at infinity.* arXiv:1104.0407. To appear in Selecta Math.

[FST] Fomin S., Shapiro M., Thurston D.: *Cluster algebras and triangulated surfaces. Part I: Cluster complexes.* arXiv:math/0608367.

[FT] Fomin S., Thurston D.: *Cluster algebras and triangulated surfaces. Part II: Lambda lengths.* arXiv:1210.5569.

[FZI] Fomin S., Zelevinsky A.: *Cluster algebras. I.* JAMS. 15 (2002), no. 2, 497–529.

[FHKVW] S. Franco, A. Hanany, K.D. Kennaway, D. Vegh, B. Wecht: *Brane dimers and quiver gauge theories.* hep-th/0504110.

[GMN1] Gaiotto D., Moore G., Neitzke A.: *Wall-crossing, Hitchin Systems, and the WKB Approximation.* arXiv:0907.3987.

[GMN2] Gaiotto D., Moore G., Neitzke A.: *Framed BPS States.* arXiv:1006.0146.

[GMN3] Gaiotto D., Moore G., Neitzke A.: *Wall-Crossing in Coupled* 2d-4d *Systems.* arXiv:1103.2598.

[GMN4] Gaiotto D., Moore G., Neitzke A.: *Spectral networks.* arXiv:1204.4824.

[GMN5] Gaiotto D., Moore G., Neitzke A.: *Spectral Networks and Snakes.* arXiv: 1209.0866.

[GSV] Gekhtman M., Shapiro M., Vainstein A.: *Cluster algebras and Poisson geometry.* Mathematical Surveys and Monographs, Volume 167, AMS, 2010.

[GSV1] Gekhtman M., Shapiro M., Vainstein A.: *Cluster algebras and Weil–Petersson forms.* arXiv:math/0309138.

[G] Ginzburg V.: *Calabi–Yau algebras.* arXiv:math/0612139.

[G14] Goncharov A.B.: A lecture in MSRI, https://www.msri.org/workshops/708/schedules/19165.

[GK] Goncharov A.B., Kenyon R.: *Dimers and cluster integrable systems.* arXiv: 1107.5588. Ann. Sci. ENS, 2013.

[GKo] Goncharov A.B., Kontsevich M.: *Non commutative local systems.* To appear.

[GS13] Goncharov A.B., Shen L.: *Geometry of canonical bases and mirror symmetry.* arXiv:1309.5922

[GS16] Goncharov A.B., Shen L.: *Donaldson–Thomas transformations for moduli spaces of local systems.* arXiv:1602.06479.

[GHKK] Gross M., Hacking P., Keel S., Kontsevich M.: *Canonical bases for cluster algebras.* arXiv:1411.1394.

[HM] Hubbard J., Masur H.A.: *Quadratic Differentials and Foliations.* Acta Math. 142 (1979), 221–274.

[K11] Keller B.: *On the cluster theory and quantum dilogarithm identity.* arXive:1102.4148.

[K12] B. Keller: *Cluster algebras and derived categories.* Derived categories in algebraic geometry, 123–183, EMS Ser. Congr. Rep., Eur. Math. Soc., Zürich, 2012. arXiv:1202.4161.

[KY] Keller B., Yang D.: *Derived equivalences from mutations of quivers with potential.* Adv. in Math 26 (2011), 2118–2168. arXiv:0906.0761.

[K92] Kontsevich M., *Formal (non)-commutative symplectic geometry.* The Gelfand Mathematical seminars, 1990–1992. Birkhäuser. L. Corwin, I. Gelfand, J. Lepowsky eds., 1993, 173–188.

[KS1] Kontsevich M., Soibelman Y.: *Stability structures, motivic Donaldson–Thomas invariants and cluster transformations.* arXiv:0811.2435.

[KS2] Kontsevich M., Soibelman Y.: *Cohomological Hall algebra, exponential Hodge structures and motivic Donaldson–Thomas invariants.* arXiv:1006.2706.

[KS3] Kontsevich M., Soibelman Y.: *Wall-crossing structures in Donaldson–Thomas invariants, integrable systems and Mirror Symmetry.* arXiv:1303.3253

[LF08] Labardini-Fragoso D. *Quivers with potentials associated to triangulated surfaces.* arXiv:0803.1328.

[LF12] Labardini-Fragoso D.: *Quivers with potentials associated to triangulated surfaces, part IV: Removing boundary assumptions.* arXiv:1206.1798.

[LF13] Labardini-Fragoso D.: *On triangulations, quivers with potentials and mutations.* arXiv:1302.1936.

[N10] Nagao K.: *Donaldson–Thomas invariants and cluster transformations.* arXiv: 1002.4884.

[N12] Nagao K.: A lecture at Cluster algebras conference at the Northeastern Univ., April 2012.

[P] Postnikov A.: *Total positivity, Grassmannians, and networks.* arXiv:math/0609764.

[S] Scott J.: *Grassmannians and Cluster Algebras.* arXiv:math/0311148.

[ST] Seidel P., Thomas R.: *Braid group actions on derived categories of coherent sheaves.* arXiv:math/0001043.

[Sm] Smith I.: *Quiver algebras as Fukaya categories.* arXiv:1309.0452.

[T] Thurston D.: *From dominos to hexagons.* arxiv:math/0405482.

A.B. Goncharov
Department of Mathematics
Yale University
10 Hillhouse ave
New Haven CT 06511, USA
e-mail: `alexander.goncharov@yale.edu`

Progress in Mathematics, Vol. 324, 99–129

Spectral Sequences for Cyclic Homology

D. Kaledin

To Maxim Kontsevich, for his 50th birthday

Abstract. We prove the non-commutative Hodge-to-de Rham Degeneration Conjecture of Kontsevich and Soibelman.

Mathematics Subject Classification (2010). 14F99.

Keywords. Non-commutative, Hodge-to-de Rham, degeneration, conjugate spectral sequence.

Introduction

In modern non-commutative algebraic geometry as formalized for example by M. Kontsevich and Y. Soibelman [KS], algebraic varieties are replaced by associative unital DG algebras $A_\bullet$ considered up to a derived Morita equivalence. The role of differential forms is played by Hochschild homology classes of $A_\bullet$, and de Rham cohomology corresponds to periodic cyclic homology $HP_\bullet(A_\bullet)$. Periodic cyclic homology is related to Hochschild homology by a standard spectral sequence. If $A_\bullet$ is derived Morita-equivalent to a smooth algebraic variety X, then the spectral sequence reduces to the classical Hodge-to-de Rham spectral sequence of P. Deligne. Because of this, one also calls it the *Hodge-to-de Rham* spectral sequence in the general case.

If the smooth variety X is also proper, and the base ring is a field of characteristic 0, then the Hodge-to-de Rham spectral sequence degenerates. This is Deligne's reformulation of the classical Hodge theory, and there is also an alternative purely algebraic proof due to Deligne and L. Illusie [DI].

Motivated by this, Kontsevich and Soibelman conjectured in [KS] that if a general DG algebra $A_\bullet$ over a field of characteristic 0 is homologically smooth and homologically proper, then the Hodge-to-de Rham spectral sequence degenerates.

Partially supported by RSF, grant 14-50-00005.

This conjecture has been largely proved in [Ka2], by an adaptation of the method of Deligne and Illusie. However, the argument in that paper suffers from two related drawbacks:

- One has to impose an additional assumption that the DG algebra $A_\bullet$ is question is concentrated in non-negative homological degrees (that is, $A_i = 0$ for $i > 0$).
- To pass form algebras to DG algebras, one uses Dold–Kan equivalence and simplicial methods. This is what forces one to impose the assumption above, and this is what makes a large part of the argument very hard to understand.

In this paper, we revisit the subject, and we give a proof of Kontsevich–Soibelman Degeneration Conjecture free from any additional technical assumptions.

Our method is still the same – it is based on reduction to positive characteristic and adapting the approach of [DI]. In principle, one could remove the assumption (•) by using simplicial-cosimplicial objects as, e.g., in [Ka3]. However, this would make the argument even more opaque. Instead, we opt for an earlier approach tried in the unpublished preprint [Ka1]. This is much closer to [DI], in that it uses explicitly two different spectral sequences that exist in positive characteristic – the Hodge-to-de Rham spectral sequence, on one hand, and the so-called *conjugate spectral sequence* on the other hand. The spectral sequences are completely different, but they have the same first page and the same last page. So, what one actually proves is that the conjugate spectral sequence degenerates, under some assumptions; the Hodge-to-de Rham sequence then degenerates for dimension reasons.

For associative algebras of finite homological dimension, it is relatively easy to construct a non-commutative version of the conjugate spectral sequence, and this has been essentially done in [Ka1]. However, for general DG algebras, the question is much more delicate. It took a while to realize that the sequence in question simply does not exist – or rather, it does exist, but does not converge to periodic cyclic homology. What it converges to is a completely new additive invariant of DG algebras and DG categories constructed in [Ka5] under the name of *co-periodic cyclic homology* and denoted $\overline{HP}_\bullet(A_\bullet)$. In retrospect, this state of affairs has been also suggested by Kontsevich 10 years ago [Ko1, Ko2], but the major push for actually developing the theory has been given by recent works of A. Beilinson [Be] and B. Bhatt [Bh]. Whatever the origins of the theory are, now we know that co-periodic cyclic homology does exist, it has a conjugate spectral sequence converging to it, and for a homologically smooth and homologically bounded DG algebra $A_\bullet$, there is a comparison theorem providing a canonical isomorphism $HP_\bullet(A_\bullet) \cong \overline{HP}_\bullet(A_\bullet)$. Therefore a Deligne–Illusie type of argument for degeneration should still be possible. This is what the present paper provides.

A couple of words about the organization of the paper. Out of necessity, a large part of it is a recapitulation of my earlier papers. In particular, Section 1 and Section 2 contain the relevant results of [Ka5]. Section 1 is concerned with general results about co-periodic cyclic homology, summarized in Theorem 1.3, and Section 2 contains the construction of the conjugate spectral sequence. Section 3 starts with some general results on Tate (co)homology of finite groups. Everything is completely standard but we do not know any good references (there are some intersections with [Ka6, Subsection 6.3]). Then we apply the results to define a relative version of Tate cohomology, and use it to prove results about co-periodic cyclic homology. In particular, this includes a degeneration criterion (Proposition 3.6). At this point, we can already prove all out degeneration results; however, we make a detour and use the opportunity to correct one fault of [Ka5] – namely, we construct the conjugate spectral sequence in characteristic 2, the case excluded in [Ka5] for reasons explained in [Ka5, Subsection 5.5]. This is the subject of Section 4. The technology used is a combination of some splitting results of [Ka6, Section 6] and the notion of a trace functor of [Ka3]. Finally, in Section 5, we prove our degeneration results. This includes Theorem 5.4 equivalent to the Degeneration Conjecture of [KS].

1. Co-periodic cyclic homology

In this section, we recall main facts about co-periodic cyclic homology introduced in [Ka5], together with some terminology and notation.

1.1. Mixed complexes

A *mixed complex* $\langle V_{\bullet}, B\rangle$ in an abelian category $\mathcal{E}$ is a complex $V_{\bullet}$ in $\mathcal{E}$ equipped with a map of complexes $B : V_{\bullet} \to V_{\bullet}[-1]$ such that $B^2 = 0$ (we will drop B from notation when it is clear from the context). The *periodic expansion* $\operatorname{Per}(V_{\bullet})$ of a mixed complex $V_{\bullet}$ is the complex

$$\operatorname{Per}(V_{\bullet}) = V_{\bullet}((u)),$$

with the differential $d + Bu$, where d is the differential in the complex $V_{\bullet}$, u is a formal generator of cohomological degree 2, and $V_{\bullet}((u))$ is shorthand for "formal Laurent power series in u with coefficients in $V_{\bullet}$". Analogously, *co-periodic* and *polynomial periodic* expansions $\overline{\operatorname{Per}}(V_{\bullet})$, $\operatorname{per}(V_{\bullet})$ are given by

$$\overline{\operatorname{Per}}(V_{\bullet}) = V_{\bullet}((u^{-1})), \qquad \operatorname{per}(V_{\bullet}) = V_{\bullet}[u, u^{-1}],$$

again with the differential $d + uB$. By definition, the space of Taylor power series $V_{\bullet}[[u]] \subset V_{\bullet}((u))$ is a subcomplex in the periodic expansion $\operatorname{Per}(V_{\bullet})$; the *expansion* $\operatorname{Exp}(V_{\bullet})$ is the quotient complex

$$\operatorname{Exp}(V_{\bullet}) = V_{\bullet}[u^{-1}] = V_{\bullet}((u))/uV_{\bullet}[[u]].$$

Multiplication by u induces an invertible degree-2 endomorphism of the complexes $\operatorname{Per}(V_{\bullet})$, $\overline{\operatorname{Per}}(V_{\bullet})$, $\operatorname{per}(V_{\bullet})$ and a non-invertible periodicity map $u : \operatorname{Exp}(V_{\bullet}) \to \operatorname{Exp}(V_{\bullet})[2]$. We have

$$\operatorname{Per}(V_{\bullet}) \cong \lim_{\overset{u}{\leftarrow}} \operatorname{Exp}(V_{\bullet}). \tag{1.1}$$

Since a Laurent polynomial in u is a Laurent power series both in u and in u^{-1}, we have natural functorial maps

$$\operatorname{Per}(V_\bullet) \longleftarrow \operatorname{per}(V_\bullet) \longrightarrow \overline{\operatorname{Per}}(V_\bullet) \tag{1.2}$$

for any mixed complex $V_\bullet$. If $V_\bullet$ is concentrated in a finite range of degrees, then both maps are isomorphisms, but in general, they are not.

Example 1.1. Assume given a module E over a ring R, and assume that a cyclic group $\mathbb{Z}/n\mathbb{Z}$ of some order n acts on E, with $\sigma : E \to E$ being the generator. Then the length-2 complex

$$E \xrightarrow{\mathsf{id}-\sigma} E \tag{1.3}$$

has a natural structure of a mixed complex, with the map B given by

$$B = \mathsf{id} + \sigma + \cdots + \sigma^{n-1} : E_\sigma \to E^\sigma.$$

The expansion of the complex (1.3) is the standard homology complex

$$C_\bullet(\mathbb{Z}/n\mathbb{Z}, E),$$

and the periodic expansion is the standard Tate homology complex

$$\check{C}_\bullet(\mathbb{Z}/n\mathbb{Z}, E).$$

1.2. Small categories

In a sense, the standard example of a mixed complex that appears in nature combines the complexes of Example 1.1 for all integers $n \geq 1$. To package the data, it is convenient to use the technology of homology of small categories. For any small category I and ring R, we denote by $\operatorname{Fun}(I, R)$ the abelian category of functors from I to R-modules, with the derived category $\mathcal{D}(I, R)$, and for any functor $\gamma : I \to I'$ between small categories, we denote by $\gamma^* : \operatorname{Fun}(I', R) \to \operatorname{Fun}(I, R)$ the pullback functor. The functor γ^* is exact, and it has a left and a right adjoint that we denote by $\gamma_!, \gamma_* : \operatorname{Fun}(I, R) \to \operatorname{Fun}(I', R)$. For any $E \in \operatorname{Fun}(I, R)$, the homology $H_\bullet(I, E)$ of the category I with coefficients in E is by definition given by

$$H_i(I, E) = L^i \tau_! E,$$

where $\tau : I \to \mathsf{pt}$ is the tautological projection to the point category pt.

The specific small category that we need is A. Connes' cyclic category Λ of [C]. We do not reproduce here the full definition (see, e.g., [L]), but we do recall that objects of Λ correspond to cellular decompositions of the circle S^1, and morphisms correspond to homotopy classes of cellular maps of a certain type. We call 0-cells *vertices*, and we call 1-cells *edges*. For any $n \geq 1$, there is exactly one decomposition with n vertices and n edges. The corresponding object in Λ is denoted $[n]$, and we denote by $V([n])$ the set of its vertices. Any map $f : [n'] \to [n]$ in Λ induces a map $f : V([n']) \to V([n])$. For any $v \in V([n])$, the preimage $f^{-1}(v) \subset V([n'])$ carries a natural total order. We have $\mathsf{Aut}([n]) = \mathbb{Z}/n\mathbb{Z}$, the cyclic group, so that for any $E \in \operatorname{Fun}(\Lambda, R)$, $E([n])$ is naturally an $R[\mathbb{Z}/n\mathbb{Z}]$-module. Moreover, we have a natural embedding $j : \Delta^o \to \Lambda$, where as usual, Δ is the category of finite non-empty totally ordered sets, and Δ^o is the opposite category. To keep notation consistent with the embedding j, we denote by $[n] \in \Delta$ the set with n elements.

By definition, the category $\mathrm{Fun}(\Delta^o, R)$ is category of simplicial R-modules, and for any $E \in \mathrm{Fun}(\Delta^o, R)$, we have the standard chain complex $CH_\bullet(E)$ with terms $CH_i(E) = E([i+1])$, $i \geq 0$, and the differential $d_i : CH_i(E) \to CH_{i-1}(E)$ given by

$$d_i = \sum_{0 \leq j \leq i} \delta^i_j, \tag{1.4}$$

where $\delta^\bullet_\bullet$ are the face maps. Moreover, we also have another complex $CH'_\bullet(E)$ with the same terms as $CH_\bullet(E)$, but with the differential given by (1.4) with the summation extended over j from 0 to $i-1$ (that is, we drop the last term). Then the complex $CH_\bullet(E)$ computes the homology $H_\bullet(\Delta^o, E)$, while the complex $CH'_\bullet(E)$ is acyclic – in fact, canonically contractible.

Now, for any object $E \in \mathrm{Fun}(\Lambda, R)$, we have the simplicial object $j^*E \in \mathrm{Fun}(\Delta^o, R)$, and it is well known that the complexes (1.3) for $E([n])$, $n \geq 1$ fit together into a single bicomplex

$$CH'_\bullet(E) \xrightarrow{\mathsf{id} - \sigma^\dagger} CH_\bullet(E), \tag{1.5}$$

where for any $[n] \in \Lambda$, $\sigma \in \mathsf{Aut}([n])$ is the generator of the cyclic group $\mathbb{Z}/n\mathbb{Z}$, and $\sigma^\dagger = (-1)^{n+1}\sigma$. We denote by $CH_\bullet(E)$ the total complex of the bicomplex (1.5). Furthermore, it is also well known that the maps B of Example 1.1 fit together into a single map $B : CH_\bullet(E) \to CH_\bullet(E)[-1]$ that turns $CH_\bullet(E)$ into a mixed complex.

It will be convenient to recast this construction in a slightly different way. For any $[n] \in \Lambda$, denote by $\mathbb{K}_\bullet([n])$ the standard cellular chain complex computing the homology of the circle S^1 with respect to the decomposition corresponding to $[n]$. Then it turns out that $K_\bullet$ is functorial with respect to morphisms in Λ, so that we obtain an exact sequence

$$0 \longrightarrow \mathbb{Z} \xrightarrow{\kappa_1} \mathbb{K}_1 \longrightarrow \mathbb{K}_0 \longrightarrow \mathbb{Z} \xrightarrow{\kappa_0} 0 \tag{1.6}$$

in the category $\mathrm{Fun}(\Lambda, \mathbb{Z})$. For any $E \in \mathrm{Fun}(\Lambda, R)$, denote $\mathbb{K}_\bullet(E) = \mathbb{K}_\bullet \otimes E$. Then $\mathbb{K}_\bullet(E)$ is a mixed complex in $\mathrm{Fun}(\Lambda, R)$, with the map B given by

$$B = (\kappa_1 \circ \kappa_0) \otimes \mathsf{id}. \tag{1.7}$$

Now for any $E \in \mathrm{Fun}(\Lambda, R)$, denote by $cc_\bullet(E)$ the cokernel of the natural map (1.5). Then one can show that we have natural identifications

$$CH_\bullet(j^*E) \cong cc_\bullet(\mathbb{K}_0(E)), \qquad CH'_\bullet(j^*E) \cong cc_\bullet(\mathbb{K}_1(E)),$$

and the mixed complex $CH_\bullet(E)$ is then given by

$$CH_\bullet(E) \cong cc_\bullet(\mathbb{K}_\bullet(E)), \tag{1.8}$$

with the map B induced by the map B of (1.7).

Finally, observe that if we are given a complex $E_\bullet$ in the category $\operatorname{Fun}(\Lambda, R)$, then we can apply $cc_\bullet(-)$ and $CH_\bullet(-)$ to $E_\bullet$ termwise. We denote by $cc_\bullet(E_\bullet)$, $CH_\bullet(E_\bullet)$ the sum-total complexes of the resulting bicomplexes. Explicitly, we have

$$cc_j(E_\bullet) = \bigoplus_{n\geq 1} E_{j+n}([n])_{\sigma^\dagger}, \qquad j \in \mathbb{Z}, \tag{1.9}$$

with the differential induced by the differential (1.4) and the differential in the complex $E_\bullet$.

1.3. Cyclic homology

We can now define periodic and co-periodic cyclic homology.

Definition 1.2. Assume given a ring R and a complex $E_\bullet$ in the category $\operatorname{Fun}(\Lambda, R)$. Then the *cyclic homology complex* $CC_\bullet(E_\bullet)$, the *periodic cyclic homology complex* $CP_\bullet(E_\bullet)$, the *co-periodic cyclic homology complex* $\overline{CP}_\bullet(E_\bullet)$, and the *polynomial periodic cyclic homology complex* $cp_\bullet(E_\bullet)$ are given by

$$\begin{aligned} CC_\bullet(E_\bullet) &= \operatorname{Exp}(CH_\bullet(E_\bullet)), & CP_\bullet(E_\bullet) &= \operatorname{Per}(CH_\bullet(E_\bullet)), \\ \overline{CP}_\bullet(E_\bullet) &= \overline{\operatorname{Per}}(CH_\bullet(E_\bullet)), & cp_\bullet(E_\bullet) &= \operatorname{per}(CH_\bullet(E_\bullet)). \end{aligned} \tag{1.10}$$

The *periodic* resp. *co-periodic* cyclic homology $HP_\bullet(E_\bullet)$ resp. $\overline{HP}_\bullet(E_\bullet)$ is the homology of the complexes $CP_\bullet(E_\bullet)$ resp. $\overline{CP}_\bullet(E_\bullet)$.

We note that the first line in (1.10) is completely standard; it is the second line that defines new theories introduced in [Ka5]. The homology of the complex $CC_\bullet(E_\bullet)$ is the usual cyclic homology $HC_\bullet(E_\bullet)$, and it is well known that we have a natural identification

$$HC_\bullet(E_\bullet) \cong H_\bullet(\Lambda, E_\bullet), \tag{1.11}$$

where $E_\bullet$ in the right-hand side is understood as the corresponding object in the derived category $\mathcal{D}(\Lambda, R)$. One can combine this with (1.1) to express $HP_\bullet(-)$. Co-periodic cyclic homology functor $\overline{HP}_\bullet(-)$ does not admit such a homological expression, and in fact, it is not true that quasiisomorphic complexes have isomorphic $\overline{HP}_\bullet$. For any complex $E_\bullet$, we do have functorial maps

$$CP_\bullet(E_\bullet) \longleftarrow cp_\bullet(E_\bullet) \longrightarrow \overline{CP}_\bullet(E_\bullet) \tag{1.12}$$

induced by the maps (1.2), but in general, these maps are not quasiisomorphisms. We also note that we have a natural functorial map

$$\alpha : CC_\bullet(E_\bullet) \to cc_\bullet(E_\bullet), \tag{1.13}$$

and in general, this map is not a quasiisomorphism either. One example where it is a quasiisomorphism is $E_\bullet = \mathbb{K}_\bullet(E'_\bullet)$ for some complex $E'_\bullet$ in $\operatorname{Fun}(\Lambda, R)$. In this case, by [Ka5, Lemma 3.11], α induces a natural isomorphism

$$HC_\bullet(\mathbb{K}_\bullet(E'_\bullet)) \cong HH_\bullet(E'_\bullet), \tag{1.14}$$

where $HH_\bullet(E'_\bullet)$ is the homology of the complex $CH_\bullet(E'_\bullet)$ (or equivalently, of the complex $CH_\bullet(j^*E'_\bullet)$).

Assume now given a Noetherian commutative ring k and a DG algebra $A_\bullet$ termwise-flat over k. Then one defines a complex $A_\bullet^\natural$ in $\mathrm{Fun}(\Lambda, k)$ as follows. For any $[n] \in \Lambda$, $A_\bullet^\natural([n]) \cong A_\bullet^{\otimes_k n}$, with terms of the product numbered by vertices $v \in V([n])$. For any map $f : [n'] \to [n]$, the corresponding map $A_\bullet^\natural(f)$ is given by

$$A_\bullet^\natural(f) = \bigotimes_{v \in V([n])} m_{f^{-1}(v)}, \tag{1.15}$$

where for any finite set S, $m_S : A_\bullet^{\otimes_k S} \to A_\bullet$ is the multiplication map in the DG algebra $A_\bullet$. Then by definition, cyclic homology and periodic cyclic homology of the DG algebra $A_\bullet$ are the cyclic homology and the periodic cyclic homology of the complex $A_\bullet^\natural$, and we define co-periodic and polynomial periodic cyclic homology $\overline{HP}_\bullet(A_\bullet)$, $hp_\bullet(A_\bullet)$ in the same way: we set

$$\overline{HP}_\bullet(A_\bullet) = \overline{HP}_\bullet(A_\bullet^\natural), \qquad hp_\bullet(A_\bullet) = hp_\bullet(A_\bullet^\natural).$$

Here are, then, the main two results about $\overline{HP}_\bullet(A_\bullet)$ proved in [Ka5].

Theorem 1.3.

(i) *Co-periodic cyclic homology functor $\overline{HP}_\bullet(-)$ extends to an additive invariant of small DG categories (in particular, a quasiisomorphism of DG algebras induces an isomorphism of their co-periodic cyclic homology groups).*

(ii) *Assume that $k \otimes \mathbb{Q} = 0$ and $A_\bullet$ is homologically smooth and homologically bounded over k. Then the maps (1.12) for $E_\bullet = A_\bullet^\natural$ are quasiisomorphisms, so that we have*

$$HP_\bullet(A_\bullet) \cong hp_\bullet(A_\bullet) \cong \overline{HP}_\bullet(A_\bullet).$$

Proof. The first statement (i) is [Ka5, Theorem 6.6], and the second statement (ii) is contained in [Ka5, Theorem 6.7]. □

In Theorem 1.3 (ii), *homologically smooth* as usual means that $A_\bullet$ is perfect as an $A_\bullet$-bimodule (that is, $A_\bullet^o \otimes_k A_\bullet$-module). If k is a field, then *homologically bounded* simply means that the complex $A_\bullet$ has a finite number of non-trivial homology groups. In the general case, this must hold for $A_\bullet \otimes \overline{k}$, where $\overline{k}$ is any residue field of the ring k.

2. Conjugate spectral sequence

For any mixed complex $\langle V_\bullet, B\rangle$ in an abelian category $\mathcal{E}$, the u-adic filtration on $\mathrm{Per}(V_\bullet) \cong V_\bullet((u))$ induces a convergent spectral sequence

$$H_\bullet(V_\bullet)((u)) \Rightarrow H_\bullet(\mathrm{Per}(V_\bullet)),$$

where $H_\bullet(-)$ stands for homology objects. In particular, for any ring R and complex $E_\bullet \in \mathrm{Fun}(\Lambda, R)$, we have a convergent spectral sequence

$$HH_\bullet(E_\bullet)((u)) \Rightarrow HP_\bullet(E_\bullet). \tag{2.1}$$

If $E_\bullet = A_\bullet^\natural$ for a DG algebra $A_\bullet$ over a commutative ring k, $HH_\bullet(A_\bullet^\natural)$ is naturally identified with the Hochschild homology $HH_\bullet(A_\bullet)$ of the DG algebra $A_\bullet$, so that (2.1) reads as

$$HH_\bullet(A_\bullet)((u)) \Rightarrow HP_\bullet(A_\bullet). \tag{2.2}$$

This is the *Hodge-to-de Rham spectral sequence*.

For co-periodic cyclic homology, no analog of (2.1) is currently known, but under some assumptions, we do have a version of (2.2). This was introduced in [Ka5] under the name of the *conjugate spectral sequence*. In this section, we briefly recall the construction.

2.1. Filtrations

The main technical tool used in [Ka5] for studying co-periodic cyclic homology is the use of filtrations and filtered derived categories. Filtrations are decreasing and indexed by all integers – that is, a *filtered complex* in an abelian category $\mathcal{E}$ is a complex $E_\bullet$ equipped with a collection of subcomplexes $F^iE_\bullet \subset E_\bullet$, $i \in \mathbb{Z}$ such that $F^iE_\bullet \subset F^jE_\bullet$ for $i \geq j$. A filtration $F^\bullet$ is *termwise-split* if for any i and j, the embedding $F^iE_j \to E_j$ admits a one-sided inverse $E_j \to F^iE_j$. The stupid filtration $F^\bullet$ on a complex $E_\bullet$ is obtained by setting $F^jE_i = E_i$ if $i + j \leq 0$ and 0 otherwise; it is tautologically termwise-split. For any filtration $F^\bullet$ and any integer $n \geq 1$, the *nth rescaling* $F^\bullet_{[n]}$ of $F^\bullet$ is given by $F^i_{[n]} = F^{in}$, $i \in \mathbb{Z}$, and the *shift by n* $F^\bullet_n$ is given by $F^i_n = F^{i+n}$. A map $\langle E'_\bullet, F'_\bullet\rangle \to \langle E_\bullet, F_\bullet\rangle$ of filtered complexes is a *filtered quasiisomorphism* if for any i, the induced map $\operatorname{gr}^i_{F'} E'_\bullet \to \operatorname{gr}^i_F E_\bullet$ of associated graded quotients is a quasiisomorphism (there is no requirement on the map $E'_\bullet \to E_\bullet$ nor on the maps $F^iE'_\bullet \to F^iE_\bullet$). Inverting filtered quasiisomorphisms in the category of filtered complexes, we obtain the *filtered derived category* $\mathcal{DF}(\mathcal{E})$.

The *completion* $\widehat{E}_\bullet$ of a filtered complex $\langle E_\bullet, F^\bullet\rangle$ is given by

$$\widehat{E}_\bullet = \lim_{\overset{i}{\leftarrow}} \lim_{\overset{j}{\rightarrow}} F^jE_\bullet/F^iE_\bullet \cong \lim_{\overset{j}{\rightarrow}} \lim_{\overset{i}{\leftarrow}} F^jE_\bullet/F^iE_\bullet,$$

where the limit is over all integers $i \geq j$, with i going to ∞ and j to $-\infty$. If a map $E'_\bullet \to E_\bullet$ is a filtered quasiisomorphism, the induced map $\widehat{E}'_\bullet \to \widehat{E}_\bullet$ of completions is a quasiisomorphism. The converse is not true: two very different filtrations can have the same completion. Specifically, say that two filtrations $F^\bullet_1$, $F^\bullet_2$ on the same complex $E_\bullet$ are *commensurable* if for any integers i, j, there exist integers $j_1 \leq j \leq j_2$ such that $F_2^{j_2}E_i \subset F_1^jE_i \subset F_2^{j_1}E_i$ and $F_1^{j_2}E_i \subset F_2^jE_i \subset F_1^{j_1}E_i$. Then two commensurable filtrations on a complex $E_\bullet$ obviously give the same completion $\widehat{E}_\bullet$.

Every shift $F^\bullet_n$ and every rescaling $F^\bullet_{[n]}$ of a filtration $F^\bullet$ is obviously commensurable to $F^\bullet$. For a more non-trivial example, assume given a filtered complex $\langle E_\bullet, F^\bullet\rangle$, and define its *filtered truncations* by setting

$$\begin{aligned} \tau^F_{\geq n}E_i &= d^{-1}(F^{n+1-i}E_{i-1}) \cap F^{n-i}E_i \subset E_i, \\ \tau^F_{\leq n}E_i &= E_i/(F^{n+1-i}E_i + d(F^{n-i}E_{i+1})) \end{aligned} \tag{2.3}$$

for any integer n. Denote also $\tau^F_{[n,m]} = \tau^F_{\geq n}\tau^F_{\leq m}$ for any integers $n \leq m$. Then the subcomplexes $\tau^F_{\geq n}E_\bullet \subset E_\bullet$ for varying n give a filtration $\tau^\bullet$ on $E_\bullet$, and this filtration is commensurable to $F^\bullet$.

Example 2.1. Assume that $E_\bullet$ is the sum-total complex of a bicomplex $E_{\bullet,\bullet}$, and let $F^\bullet$ be the stupid filtration with respect to the first coordinate. Then $\tau^F_{\geq n}E_\bullet$, $\tau^F_{\leq n}E_\bullet$ are canonical truncations with respect to the second coordinate.

In the general case, the truncation functors τ^F are also related to the canonical filtrations: for any integers i, j, we have natural isomorphisms

$$\mathsf{gr}^j_F\,\tau^F_{\geq i}E_\bullet \cong \tau_{\geq i-j}\,\mathsf{gr}^j_F\,E_\bullet, \qquad \mathsf{gr}^j_F\,\tau^F_{\leq i}E_\bullet \cong \tau_{\leq i-j}\,\mathsf{gr}^j_F\,E_\bullet,$$

where $\tau_{\geq n}$, $\tau_{\leq n}$ are the usual canonical truncations. In particular, the functors τ^F preserve filtered quasiisomorphisms and descend to the filtered derived category $\mathcal{DF}(\mathcal{E})$, where they become truncation functors with respect to the well-known t-structure of [BBD] (the heart of this t-structure is the category of complexes in $\mathcal{E}$). However, it will be useful to have the truncation functors already on the level of filtered complexes.

We note that since the filtrations $F^\bullet$ and $\tau^\bullet$ are commensurable, $\mathsf{gr}^i_\tau\,E_\bullet$ is complete with respect to $F^\bullet$ for any integer i, so that for any filtered quaiisomorphism $f : \langle E'_\bullet, F'_\bullet\rangle \to \langle E_\bullet, F_\bullet\rangle$ is also a filtered quasiisomorphism with respect to the filtrations τ. Thus sending $\langle E_\bullet, F^\bullet\rangle$ to $\langle E_\bullet, \tau^\bullet\rangle$ descends to an endofunctor

$$\mathcal{DF}(\mathcal{E}) \to \mathcal{DF}(\mathcal{E}) \tag{2.4}$$

of the filtered derived category $\mathcal{DF}(\mathcal{E})$. We also note that for any integer i and filtered complex $\langle E_\bullet, F^\bullet\rangle$, $\mathsf{gr}^i_\tau\,E_\bullet$ is quasiisomorphic (but not isomorphic) to $\tau^F_{[i,i]}E_\bullet$.

Now, for any ring R and any complex $E_\bullet$ in the category $\operatorname{Fun}(\Lambda, R)$ equipped with a termwise-split filtration $F^\bullet$, define the *standard filtration* on the complex $cc_\bullet(E_\bullet)$ by setting

$$F^i cc_j(E_\bullet) = \bigoplus_{n\geq 1} F^{i+n}E_{j+n}([n])_{\sigma^\dagger}, \qquad i,j \in \mathbb{Z}, \tag{2.5}$$

where we use the decomposition (1.9) of the complex $cc_\bullet(E_\bullet)$. By virtue of (1.8), the standard filtration extends to the complex $CH_\bullet(E_\bullet)$ and then to its periodic expansions $\overline{CP}_\bullet(E_\bullet)$, $cp_\bullet(E_\bullet)$.

Lemma 2.2 ([Ka5, Lemma 3.8]). *Equip a complex $E_\bullet$ in $\operatorname{Fun}(\Lambda, R)$ with the stupid filtration $F^\bullet$. Then the co-periodic complex $\overline{CP}_\bullet(E_\bullet)$ is isomorphic to the completion of the polynomial periodic complex $cp_\bullet(E_\bullet)$ with respect to the standard filtration.*

By virtue of this result, one can reduce the study of $\overline{CP}_\bullet(E_\bullet)$ to the study of $cp_\bullet(E_\bullet)$ equipped with the standard filtration.

Let us now do the following. Fix an integer $p \geq 1$, and for any complex $E_\bullet$ in $\operatorname{Fun}(\Lambda, R)$ equipped with the stupid filtration, denote by $cp_\bullet(E_\bullet)^{[p]}$ the complex $cp_\bullet(E_\bullet)$ equipped with the pth rescaling of the standard filtration of (2.5).

Definition 2.3. The *conjugate filtration* $V^{\bullet}$ on $cp_{\bullet}(E_{\bullet})$ is given by

$$V^n cp_{\bullet}(E_{\bullet}) = \tau_{\geq 2n-1} cp_{\bullet}(E_{\bullet})^{[p]}, \qquad n \in \mathbb{Z}. \tag{2.6}$$

Note that since the complex $cp_{\bullet}(E_{\bullet})$ is by definition 2-periodic, the conjugate filtration is periodic: we have

$$V^n cp_{\bullet}(E_{\bullet}) \cong V^0 cp_{\bullet}(E_{\bullet})[2n] \tag{2.7}$$

for any integer n. By definition, the conjugate filtration is a shift of a rescaling of the filtration $\tau^{\bullet}$, so that it is commensurable to the pth rescaling of the standard filtration on $cp_{\bullet}(E_{\bullet})$. This is in turn commensurable to the standard filtration itself. Therefore by Lemma 2.2, the co-periodic cyclic homology complex $\overline{CP}_{\bullet}(E_{\bullet})$ is isomorphic to the completion of the complex $cp_{\bullet}(E_{\bullet})$ with respect to the conjugate filtration (2.6). We then have a convergent spectral sequence

$$H_{\bullet}(\mathrm{gr}^0_V cp_{\bullet}(E_{\bullet}))((u^{-1})) \Rightarrow \overline{HP}_{\bullet}(E_{\bullet}), \tag{2.8}$$

where as before, $H_{\bullet}(-)$ stands for homology objects, u is a formal generator of cohomological degree 2, and we have used the identifications (2.7).

2.2. Edgewise subdivision

In general, the spectral sequence (2.8) does not seem to be particularly useful, since its initial term is rather obscure. However, under some assumptions, it can be computed explicitly. The first step in this computation is the so-called *edgewise subdivision*.

Recall that for any integer l, the category Λ has a cousin Λ_l corresponding to the l-fold cover $S^1 \to S^1$. Objects in Λ_l are objects $[nl] \in \Lambda$, $n \geq 1$, equipped with the order-l automorphism $\tau = \sigma^n : [nl] \to [nl]$, and morphisms are morphisms in Λ that commute with the automorphism τ. It is convenient to number objects in Λ_l by positive integers, so that $[n] \in \Lambda_l$ corresponds to $[nl] \in \Lambda$. We have the forgetful functor $i_l : \Lambda_l \to \Lambda$, $[n] \mapsto [nl]$, and we also have a natural projection $\pi_l : \Lambda_l \to \Lambda$ that sends $[nl]$ to the object $[n]$ considered as an induced cellular decomposition of $S^1 \cong S^1/\tau$. The functor $\pi_l : \Lambda_l \to \Lambda$ is a Grothendieck bifibration with fiber pt_l, the groupoid with one object with automorphism group $\mathbb{Z}/l\mathbb{Z}$. The functor i_l induces the pullback functor $i_l^* : \mathrm{Fun}(\Lambda, R) \to \mathrm{Fun}(\Lambda_l, R)$, classically known as "edgewise subdivision functor", and the functor π_l induces functors $\pi_{l!}, \pi_{l*} : \mathrm{Fun}(\Lambda_l, R) \to \mathrm{Fun}(\Lambda, R)$. We will need a slightly more complicated functor $\pi_{l\flat}$ that sends complexes in $\mathrm{Fun}(\Lambda_l, R)$ to complexes $\mathrm{Fun}(\Lambda, R)$. It is given by

$$\pi_{l\flat} E_{\bullet} = \mathsf{per}(\pi_{l!}(i_l^* \mathbb{K}_{\bullet} \otimes E_{\bullet})). \tag{2.9}$$

Equivalently, one can use π_{l*} – we have a natural trace map $\pi_{l!} \to \pi_{l*}$, and when evaluated on the complex $i_l^* \mathbb{K}_{\bullet} \otimes E_{\bullet}$, this map is an isomorphism. The functors i_l^* and $\pi_{l\flat}$ extend to filtered complexes in the obvious way.

Now assume that our base ring R is annihilated by a prime p, and restrict our attention to the p-fold cover Λ_p. Then we have the following result.

Proposition 2.4 ([Ka5, Proposition 4.4]). *Assume given a complex $E_\bullet$ in* $\operatorname{Fun}(\Lambda, R)$ *equipped with some filtration $F^\bullet$, and denote by $E_\bullet^{[p]}$ the same complex equipped with the pth rescaling $F^\bullet_{[p]}$ of the filtration. Then there exists a functorial map*

$$\nu_p : cc_\bullet(\pi_{p\flat} i_p^* E_\bullet^{[p]}) \to cp_\bullet(E_\bullet)^{[p]},$$

and this map is a filtered quasiisomorphism with respect to the standard filtrations.

For the next step, we need to impose a condition on the complex $E_\bullet$. Recall that we have assumed $pR = 0$. Therefore the cohomology algebra $H^\bullet(\mathbb{Z}/p\mathbb{Z}, R)$ of the cyclic group $\mathbb{Z}/p\mathbb{Z}$ is given by

$$H^\bullet(\mathbb{Z}/p\mathbb{Z}, R) \cong R[u]\langle\varepsilon\rangle, \tag{2.10}$$

where u is a generator of degree 2, and ε is a generator of degree 1. The relations are: u commutes with ε, and

$$\varepsilon^2 = \begin{cases} u, & p = 2, \\ 0, & p \text{ is odd.} \end{cases} \tag{2.11}$$

In particular, for any $R[\mathbb{Z}/p\mathbb{Z}]$-module E and any integer i, we have a natural map

$$\varepsilon_i : \check{H}_i(\mathbb{Z}/p\mathbb{Z}, E) \to \check{H}_{i-1}(\mathbb{Z}/p\mathbb{Z}, E), \tag{2.12}$$

where $\check{H}_\bullet(\mathbb{Z}/p\mathbb{Z}, -)$ is the Tate homology of $\mathbb{Z}/p\mathbb{Z}$. Since the Tate homology of the cyclic group is 2-periodic, we have $\varepsilon_i = \varepsilon_{i+2n}$ for any n, so that ε_i only depends on the parity of the integer i.

Definition 2.5. An $R[\mathbb{Z}/p\mathbb{Z}]$-module E is *tight* if ε_1 is an isomorphism. A complex $E_\bullet$ of $R[\mathbb{Z}/p\mathbb{Z}]$-modules is *tight* if E_i is a tight $R[\mathbb{Z}/p\mathbb{Z}]$-module when i is divisible by p and a projective $R[\mathbb{Z}/p\mathbb{Z}]$-module otherwise. A complex $E_\bullet$ in the category $\operatorname{Fun}(\Lambda_p, R)$ is *tight* if for any $[n] \in \Lambda_p$, $E_\bullet([n])$ is a tight complex with respect to the action of the group $\mathbb{Z}/p\mathbb{Z}$ generated by $\tau \in \mathsf{Aut}([n])$.

Note that if p is odd, then (2.11) shows that for a tight $R[\mathbb{Z}/p\mathbb{Z}]$-module E, we have $\varepsilon_0 = 0$. Conversely, if $p = 2$, ε_0 is also an isomorphism, and in fact the tightness condition is always satisfied (both for an object and for a complex).

Lemma 2.6.

(i) *Assume given a tight complex $E_\bullet$ of $R[\mathbb{Z}/p\mathbb{Z}]$-modules, and denote by $E_\bullet^{[p]}$ the complex $E_\bullet$ equipped with the pth rescaling of the stupid filtration. Denote by $\check{C}_\bullet(\mathbb{Z}/p\mathbb{Z}, E_\bullet^{[p]})$ the sum-total complex of the Tate homology bicomplex of the group $\mathbb{Z}/p\mathbb{Z}$ with coefficients in $E_\bullet$ equipped with the filtration $F^\bullet$ induced by the filtration on $E_\bullet^{[p]}$. Let*

$$\mathsf{I}(E_\bullet) = \tau^F_{[0,0]}\check{C}_\bullet(\mathbb{Z}/p\mathbb{Z}, E_\bullet^{[p]}).$$

Then the induced filtration $F^\bullet$ on $\mathsf{I}(E_\bullet)$ is the stupid filtration, and for any integer i, we have a natural filtered isomorphism

$$\tau^F_{[i,i]}\check{C}_\bullet(\mathbb{Z}/p\mathbb{Z}, E_\bullet^{[p]}) \cong \mathsf{I}(E_\bullet)[i]. \tag{2.13}$$

(ii) *Assume given a tight complex $E_\bullet$ in* $\mathrm{Fun}(\Lambda_p, R)$, *and denote by $E_\bullet^{[p]}$ the complex $E_\bullet$ equipped with the pth rescaling of the stupid filtration. Consider the induced filtration on the complex $\pi_{p\flat}E_\bullet^{[p]}$, and let $\mathsf{I}(E_\bullet) = \tau^F_{[0,0]}\pi_{p\flat}E_\bullet^{[p]}$. Then for any $[n] \in \Lambda$, we have $\mathsf{I}(E_\bullet)([n]) \cong \mathsf{I}(E_\bullet([n]))$, and for any integer $[i]$, the isomorphisms* (2.13) *induce an isomorphism*

$$\tau^F_{[i,i]}\pi_{p\flat}E_\bullet^{[p]} \cong \mathsf{I}(E_\bullet)[i]. \tag{2.14}$$

Proof. Almost all of the statements are obvious; the non-obvious ones are [Ka5, Lemma 5.3]. □

Explicitly, the isomorphisms (2.13) can be described as follows. By periodicity, we have an isomorphism

$$u : \check{C}_\bullet(\mathbb{Z}/p\mathbb{Z}, E_\bullet^{[p]}) \cong \check{C}_\bullet(\mathbb{Z}/p\mathbb{Z}, E_\bullet^{[p]})[2]$$

corresponding to the action of the generator u of the cohomology algebra (2.10). Twists by powers of u provide isomorphisms (2.13) for even i. To obtain the isomorphisms for odd i, one considers the action of the generator ε. This gives natural maps

$$\varepsilon_i : \tau^F_{[i,i]}\check{C}_\bullet(\mathbb{Z}/p\mathbb{Z}, E_\bullet^{[p]}) \to \tau^F_{[i-1,i-1]}\check{C}_\bullet(\mathbb{Z}/p\mathbb{Z}, E_\bullet^{[p]}), \tag{2.15}$$

a filtered refinement of (2.12). Since $E_\bullet$ is tight, ε_i is a filtered isomorphism for any odd i.

All of this works relatively over the category Λ; in particular, we have natural maps

$$\varepsilon_i : \tau^F_{[i,i]}\pi_{p\flat}E_\bullet^{[p]} \to \tau^F_{[i-1,i-1]}\pi_{p\flat}E_\bullet^{[p]} \tag{2.16}$$

for any tight complex $E_\bullet$ in $\mathrm{Fun}(\Lambda_p, R)$, and these maps are invertible for odd i.

2.3. Localized conjugate filtration

Now, as it turns out, for a complex $E_\bullet$ in $\mathrm{Fun}(\Lambda, R)$ with tight edgewise subdivision $i_p^*E_\bullet$, we can localize the conjugate filtration (2.6) with respect to the category Λ and express it in terms of the complex $\pi_{p\flat}i_p^*E_\bullet$. Namely, introduce the following.

Definition 2.7. For any complex $E_\bullet$ in $\mathrm{Fun}(\Lambda_p, R)$, the *conjugate filtration* $V^\bullet$ on the complex $\pi_{p\flat}E_\bullet$ is given by

$$V^n\pi_{p\flat}E_\bullet = \tau^F_{\geq 2n}\pi_{p\flat}E_\bullet^{[p]}, \tag{2.17}$$

where $E_\bullet^{[p]}$ stands for $E_\bullet$ equipped with the pth rescaling of the stupid filtration.

We can then take a complex $E_\bullet$ in $\mathrm{Fun}(\Lambda, R)$, consider the corresponding complex $\pi_{p\flat}i_p^*E_\bullet$ equipped with the conjugate filtration (2.17), and apply to it the cyclic homology complex functor $CC_\bullet(-)$. Since the functor $CC_\bullet(-)$ is exact, the conjugate filtration induces a filtration $V^\bullet$ on $CC_\bullet(\pi_{p\flat}i_p^*E_\bullet)$. We denote by $\widehat{CC}_\bullet(\pi_{p\flat}i_p^*E_\bullet)$ the completion of the complex $CC_\bullet(\pi_{p\flat}i_p^*E_\bullet)$ with respect to the filtration $V^\bullet$.

Consider now the composition

$$CC_{\bullet}(\pi_{p\flat} i_p^* E_{\bullet}) \xrightarrow{\alpha} cc_{\bullet}(\pi_{p\flat} i_p^* E_{\bullet}) \xrightarrow{\nu_p} cp_{\bullet}(E_{\bullet}) \tag{2.18}$$

of the map α of (1.13) and the natural map ν_p of Proposition 2.4.

Lemma 2.8.

(i) *Assume that the complex $i_p^* E_{\bullet}$ in $\mathrm{Fun}(\Lambda_p, R)$ is tight in the sense of Definition 2.5. Then the map (2.18) extends to a quasiisomorphism*

$$\widehat{CC}_{\bullet}(\pi_{p\flat} i_p^* E_{\bullet}) \cong \overline{CP}_{\bullet}(E_{\bullet}).$$

(ii) *Moreover, assume that the prime p is odd. Then (2.18) itself is a filtered quasiisomorphism with respect to the filtrations $V^{\bullet}$ of (2.17) resp. (2.6).*

(iii) *In addition, still assuming that p is odd, let $\mathsf{I}(i_p^* E_{\bullet})$ be canonical complex of Lemma 2.6. Then we have a natural isomorphism*

$$\mathsf{gr}_V^0 \, \pi_{p\flat} i_p^* E_{\bullet} \cong \mathbb{K}(\mathsf{I}(i_p^* E_{\bullet})) \tag{2.19}$$

in the derived category $\mathcal{D}(\Lambda, R)$.

Proof. (i) is [Ka5, Lemma 5.19], (ii) is [Ka5, Lemma 5.9], and (iii) is [Ka5, Lemma 5.7]. □

By virtue of this result, for an odd prime p, we can rewrite the spectral sequence (2.8) as

$$CC_{\bullet}(\mathsf{gr}_V^0 \, \pi_{p\flat} i_p^* E_{\bullet})((u^{-1})) \Rightarrow \overline{HP}_{\bullet}(E_{\bullet}), \tag{2.20}$$

where we have used the obvious counterpart of the periodicity isomorphism (2.7) for the filtration (2.17). Moreover, by (1.11), (2.19), and (1.14), we can further rewrite (2.21) as

$$HH_{\bullet}(\mathsf{I}(i_p^* E_{\bullet}))((u^{-1})) \Rightarrow \overline{HP}_{\bullet}(E_{\bullet}), \tag{2.21}$$

where as in (2.8), u is a formal generator of cohomological degree 2.

2.4. DG algebras

To make the spectral sequence (2.21) useful, it remains to compute the complex $\mathsf{I}(i_p^* E_{\bullet})$. In order to do this, we need to assume further that $E_{\bullet}$ comes from a DG algebra $A_{\bullet}$. We thus assume given a commutative ring k annihilated by an odd prime p. We denote by $k^{(1)}$ the Frobenius twist of k – that is, k considered as a module over itself via the absolute Frobenius map $k \to k$. For any flat k-module V, we denote $V^{(1)} = V \otimes_k k^{(1)}$.

Proposition 2.9 ([Ka5, Proposition 6.10]). *Assume that the ring k is Noetherian. Then for any complex $V_{\bullet}$ of flat k-modules, the complex $V_{\bullet}^{\otimes_k p}$ of $k[\mathbb{Z}/p\mathbb{Z}]$-modules is tight in the sense of Definition 2.5, and we have a natural identification*

$$\mathsf{I}(V_{\bullet}^{\otimes_k p}) \cong V_{\bullet}^{(1)},$$

where $\mathsf{I}(-)$ is the canonical complex provided by Lemma 2.6.

Corollary 2.10. *Assume given a DG algebra $A_\bullet$ termwise-flat over a commutative ring k annihilated by an odd prime p. Then the complex $i_p^* A_\bullet^\natural$ in the category* $\mathrm{Fun}(\Lambda_p, k)$ *is tight in the sense of Definition* 2.5, *and we have a natural identification*

$$\mathsf{I}(i_p^* A_\bullet^\natural) \cong \left(A_\bullet^\natural\right)^{(1)}.$$

Proof. This is [Ka5, Lemma 6.19]. □

By virtue of this corollary, we have a natural identification

$$HH_\bullet(\mathsf{I}(i_p^* A_\bullet^\natural)) \cong HH_\bullet^{(1)}(A_\bullet),$$

where for any complex $E_\bullet$ in $\mathrm{Fun}(\Lambda, k)$ termwise-flat over k, $HH_\bullet^{(1)}(E_\bullet)$ denotes the homology of the Frobenius twist $CH_\bullet(E_\bullet)^{(1)}$ of the Hochschild homology complex $CH_\bullet(E_\bullet)$. If the commutative ring k is finitely generated and regular, so that $k^{(1)}$ is flat over k, then we have

$$HH_\bullet^{(1)}(A_\bullet) \cong HH_\bullet(A_\bullet^{(1)}),$$

where $A_\bullet^{(1)}$ is the Frobenius twist of the DG algebra $A_\bullet$. Then (2.21) takes its final form – what we have is a spectral sequence

$$HH_\bullet(A_\bullet^{(1)})((u^{-1})) \Rightarrow \overline{HP}_\bullet(A_\bullet). \tag{2.22}$$

This is the *conjugate spectral sequence* for the DG algebra $A_\bullet$.

We finish the section with an alternative description of the complex $i_p^* A_\bullet^\natural$ in terms of the p-tensor power algebra $A_\bullet^{\otimes_k p}$; this goes back to [Ka2, Subsection 2.2], and we will need it later in Section 3.

For any small category $\mathcal{C}$, sending an object $[n] \in \Lambda$ to $\mathcal{C}^{V([n])}$ defines a contravariant functor from Λ to the category of small categories, and the Grothendieck construction [G] associates a category fibered over Λ to this functor. We will denote the category by $\mathcal{C} \wr \Lambda$, and we will denote the fibration by $\pi : \mathcal{C} \wr \Lambda \to \Lambda$. Then for any DG algebra $A_\bullet$ in the category $\mathrm{Fun}(\mathcal{C}, k)$ – or equivalently, for any functor from $\mathcal{C}$ to DG algebras over k – the construction of the complex $A_\bullet^\natural$ of Subsection 1.3 admits a straightforward refinement that produces a complex $A_\bullet^\natural$ in the category $\mathrm{Fun}(\mathcal{C} \wr \Lambda, k)$.

Consider the case $\mathcal{C} = \mathsf{pt}_p$, the groupoid with one object with automorphism group $G = \mathbb{Z}/p\mathbb{Z}$. Then a DG algebra in $\mathrm{Fun}(\mathsf{pt}_p, k)$ is simply a DG algebra $A_\bullet$ over k equipped with an action of the group G, and what we obtain is a complex $A_\bullet^\natural$ in the category $\mathrm{Fun}(\mathsf{pt}_p \wr \Lambda, k)$. The categories Λ and $\mathsf{pt}_p \wr \Lambda$ have the same objects, and for any such object $[n]$, we have

$$A_\bullet^\natural([n]) = A_\bullet^{\otimes_k n},$$

with the natural action of G^n. By base change [Ka2, Lemma 1.7], we have

$$\pi_* A_\bullet^\natural([n]) \cong \left(A_\bullet^\natural([n])\right)^{G^n} = \overline{A}_\bullet^\natural([n]), \tag{2.23}$$

where we denote by $\overline{A}_\bullet = A_\bullet^G \subset A_\bullet$ the subalgebra of G-invariants. Taken together, these identifications produce an isomorphism

$$\pi_* A_\bullet^\natural \cong \overline{A}_\bullet^\natural. \tag{2.24}$$

Now note that we also have a natural embedding

$$\lambda : \Lambda_p \to \mathsf{pt}_p \wr \Lambda \tag{2.25}$$

that induces the diagonal embedding $\mathsf{pt}_p \to \mathsf{pt}_p^n$ on the fiber over any object $[n] \in \Lambda$. Then for any DG algebra $A_\bullet$ over k, we have natural isomorphism

$$i_p^* A_\bullet^\natural \cong \lambda^* \left(A_\bullet^{\otimes_k p}\right)^\natural, \tag{2.26}$$

where in the right-hand side, the pth power $A_\bullet^{\otimes_k p}$ is equipped with the natural G-action by the longest permutation.

3. Tate cohomology

To analyze the conjugate spectral sequence (2.22), we need a more invariant definition of the functor $\pi_{p\flat}$ of (2.9). The relevant conceptual formalism is that of Tate cohomology.

3.1. Relative Tate cohomology

Assume given a ring k, a finite group G, and a bounded complex $E_\bullet$ of $k[G]$-modules. Recall that the *Tate cohomology* of the group G with coefficients in $E_\bullet$ is given by

$$\check{H}^\bullet(G, E_\bullet) = \operatorname{Ext}^\bullet_{\mathcal{D}(k[G])/\mathcal{D}^{pf}(k[G])}(k, E_\bullet), \tag{3.1}$$

where k is the trivial $k[G]$-module, and $\mathcal{D}(k[G])/\mathcal{D}^{pf}(k[G])$ is the quotient of the derived category $\mathcal{D}(k[G])$ of all $k[G]$-modules by its full subcategory $\mathcal{D}^{pf}(k[G]) \subset \mathcal{D}(k[G])$ spanned by perfect complexes of $k[G]$-modules. In order to compute Tate cohomology, it is convenient to introduce the following (we use the same notation and terminology as in [Ka6, Subsection 6.3]).

Definition 3.1.

(i) *Resolution data* for a finite group G is a pair $\nu = \langle P_\bullet, I^\bullet \rangle$ of a left free resolution $P_\bullet$ and a right free resolution $I^\bullet$ of the trivial $\mathbb{Z}[G]$-module $\mathbb{Z}$.

(ii) For any associative unital ring k, any bounded complex $E_\bullet$ of $k[G]$-modules, and any resolution data ν, the *Tate cohomology complex* of G with coefficients in $E_\bullet$ is given by

$$\check{C}^\bullet(G, \nu, E_\bullet) = \left(E_\bullet \otimes \overline{P}_\bullet \otimes I^\bullet\right)^G, \tag{3.2}$$

where $\overline{P}_\bullet$ is the cone of the augmentation map $P_\bullet \to \mathbb{Z}$.

(iii) For any associative unital ring k, any bounded complex $E_{\bullet}$ of $k[G]$-modules, and any resolution data ν, the *reduced Tate cohomology complex* of G with coefficients in $E_{\bullet}$ is given by

$$\check{C}^{\bullet}_{red}(G,\nu,E_{\bullet}) = \left(E_{\bullet} \otimes \widetilde{P}_{\bullet}\right)^G, \tag{3.3}$$

where $\widetilde{P}_{\bullet}$ is the cone of the natural map $P_{\bullet} \to \mathbb{Z} \to I^{\bullet}$.

Remark 3.2. Here as well as elsewhere, we use homological and cohomological indexes for complexes interchangeably, with the convention that $E_i = E^{-i}$.

Resolution data form a category in the obvious way, and this category is connected (for example, every ν is obviously connected by a chain of maps to a functorial resolution data set ν' obtained by taking bar resolutions). For any resolution data ν, we have a natural map $\widetilde{P}_{\bullet} \to \overline{P}_{\bullet} \otimes I^{\bullet}$, and the induced map

$$\check{C}^{\bullet}_{red}(G,\nu,E_{\bullet}) \to \check{C}^{\bullet}(G,\nu,E_{\bullet})$$

of Tate cohomology complexes is a quasiisomorphism for any bounded complex $E_{\bullet}$. Moreover, a map of resolution data induces a map of Tate complexes, and these maps are also quasiisomorphisms. Therefore we can drop ν from notation and obtain a well-defined object $\check{C}^{\bullet}(G,E_{\bullet}) \cong \check{C}^{\bullet}_{red}(G,E_{\bullet})$ in the derived category of k-modules. Its cohomology modules are then canonically identified with Tate cohomology groups $\check{H}^{\bullet}(G,E_{\bullet})$.

It is obvious from (3.1) that $\check{H}^{\bullet}(G,k)$ is an algebra, and for any bounded complex $E_{\bullet}$, $\check{H}^{\bullet}(G,E_{\bullet})$ is a module over $\check{H}^{\bullet}(G,k)$. To lift it to the derived category level, one chooses resolution data $\nu = \langle P_{\bullet}, I^{\bullet}\rangle$ that are *multiplicative* in the following sense: both $I^{\bullet}$ and $\overline{P}_{\bullet}$ are DG algebras, and the natural maps $\mathbb{Z} \to I^{\bullet}$, $\mathbb{Z} \to \overline{P}_{\bullet}$ are algebra maps. Multiplicative resolution data exist. For each multiplicative ν, $\check{C}^{\bullet}(G,\nu,k)$ is a DG algebra, and $\check{C}^{\bullet}(G,\nu,E_{\bullet})$ is a module over this DG algebra. Passing to the derived category, we obtain well-defined multiplication maps

$$\check{C}^{\bullet}(G,k) \overset{\mathsf{L}}{\otimes} \check{C}^{\bullet}(G,E_{\bullet}) \to \check{C}^{\bullet}(G,E_{\bullet})$$

that do not depend on the choice of the resolution data ν.

Assume now given small categories $\mathcal{C}$, $\mathcal{C}'$ and a functor $\pi : \mathcal{C}' \to \mathcal{C}$. Assume further that π is a Grothendieck bifibration in the sense of [G], and that the fibers of this bifibrations are equivalent to pt_G, the groupoid with one object with automorphism group G. By base change [Ka2, Lemma 1.7], for any object $c' \in \mathcal{C}'$ with $c = \pi(c') \in \mathcal{C}$, and any functor $E \in \operatorname{Fun}(\mathcal{C}',k)$, we have natural isomorphisms

$$\pi_! E(c') \cong E(c')_G, \qquad \pi_* E(c) \cong E(c')^G, \tag{3.4}$$

where G acts on $E(c')$ via the embedding $\mathsf{pt}_G \to \mathcal{C}'$ of the fiber over $c \in \mathcal{C}$. Then one can obviously make the constructions above work "relatively over $\mathcal{C}$". Namely, one defines *resolution data* for π as pairs $\nu = \langle P_{\bullet}, I^{\bullet}\rangle$ of a left and a right resolution of the constant functor $\mathbb{Z} \in \operatorname{Fun}(\mathcal{C}',\mathbb{Z})$ such that for any $c \in \mathcal{C}$, ν restricted to the fiber $\mathsf{pt}_G \subset \mathcal{C}'$ over c gives resolution data in the sense of Definition 3.1. One shows easily that resolution data exist (e.g., take the bar resolutions) and that the category of resolution data is connected. Then for any resolution data ν, one defines

$$\pi_\nu^\flat(E) = \pi_*(E \otimes \overline{P}_\bullet \otimes I^\bullet), \qquad \pi_\nu^{\flat\flat}(E) = \pi_*(E \otimes \widetilde{P}_\bullet), \tag{3.5}$$

where $\overline{P}_\bullet$ is as in (3.2), and $\widetilde{P}_\bullet$ is as in (3.3). By (3.4), we have a natural quasiisomorphism $\pi_\nu^\flat(E) \cong \pi_\nu^{\flat\flat}(E)$, and both complexes do not depend on the choice of ν. All in all, we obtain a well-defined object

$$\pi^\flat(E) \in \mathcal{D}(\mathcal{C}, k)$$

in the derived category $\mathcal{D}(\mathcal{C}, k)$. For any object $c' \in \mathcal{C}'$ with $c = \pi(c') \in \mathcal{C}$, (3.4) gives a natural identification

$$\pi^\flat(E)(c) \cong \check{C}^\bullet(G, E(c')).$$

Moreover, as in [Ka6, Subsection 6.3], one can choose multiplicative resolution data for π. This shows that $\pi^\flat(k)$ is an algebra object in $\mathcal{D}(\mathcal{C}, k)$, and for any E, we have natural action maps

$$\pi^\flat(k) \overset{\mathsf{L}}{\otimes} \pi^\flat(E) \to \pi^\flat(E) \tag{3.6}$$

that turn $\pi^\flat(E)$ into a module object over the algebra $\pi^\flat(k)$.

Finally, we observe that all of the above can be repeated *verbatim* for a bounded complex $E_\bullet$ instead of a single object E in $\operatorname{Fun}(\mathcal{C}', k)$. Moreover, assume given a filtered complex $\langle E_\bullet, F^\bullet \rangle$ in $\operatorname{Fun}(\mathcal{C}', k)$, and assume that the filtration $F^\bullet$ is termwise-split, and $\mathsf{gr}_F^i E_\bullet$ is bounded for any integer i. Then again, (3.5) gives an object

$$\pi^\flat E_\bullet \in \mathcal{D}F(\mathcal{C}, k) \tag{3.7}$$

in the filtered derived category $\mathcal{D}F(\mathcal{C}, k)$, and as such, it does not depend on the choice of the resolution data ν.

3.2. Conjugate filtration

We now want to apply relative Tate cohomology to the study of the conjugate filtration of Definition 3.2. We assume that the base ring k is commutative and annihilated by a prime p. For simplicity, we also assume right away that k is finitely generated and regular, so that the absolute Frobenius map $k \to k$ is finite and flat. Now, consider the natural projection

$$\pi_p : \Lambda_p \to \Lambda.$$

This is a Grothendieck bifibration, and its fiber is pt_p, the groupoid with one object with automorphism group $\mathbb{Z}/p\mathbb{Z}$. Therefore for any filtered complex $\langle E_\bullet, F^\bullet \rangle$ in $\operatorname{Fun}(\Lambda_p, k)$ such that the filtration $F^\bullet$ is termwise-split and $\mathsf{gr}_F^i E_\bullet$ is bounded for any i, (3.7) provides a natural object

$$\pi_p^\flat E_\bullet \in \mathcal{D}F(\Lambda, k). \tag{3.8}$$

We note that we have a natural isomorphism

$$\pi_{\nu}^{\flat\flat}E_{\bullet} \cong \pi_{p\flat}E_{\bullet}[1], \tag{3.9}$$

where $\pi_{p\flat}E_{\bullet}$ is the complex of (2.9), and ν is the pair of resolutions of $\mathbb{Z}$ obtained by periodization of the complex $\mathbb{K}_{\bullet}$ of (1.6). Therefore the object (3.8) coincides with $\pi_{p\flat}E_{\bullet}$ up to a homological shift.

In particular, let $E_{\bullet}$ be an arbitrary complex in $\operatorname{Fun}(\Lambda_p, k)$, and let $E^{[p]}_{\bullet}$ be $E_{\bullet}$ equipped with the pth rescaling of the stupid filtration. Then all the assumptions on the filtration are satisfied, so that we have a well-defined object $\pi_p^{\flat}E_{\bullet}^{[p]}$ in $\mathcal{D}F(\Lambda, k)$. We also have the action map

$$\pi_p^{\flat}k \overset{\mathsf{L}}{\otimes} \pi_p^{\flat}E_{\bullet}^{[p]} \to \pi_p^{\flat}E_{\bullet}^{[p]} \tag{3.10}$$

induced by the map (3.6).

Lemma 3.3 ([Ka4, Lemma 3.2]). *We have a canonical isomorphism*

$$\pi_{p\flat}k \cong \mathbb{K}_{\bullet}(k)((u)) = \bigoplus_{i\in\mathbb{Z}}\mathbb{K}_{\bullet}(k)[2i] \tag{3.11}$$

of complexes in $\operatorname{Fun}(\Lambda, k)$.

By (3.9), this gives an isomorphism $\tau_{[i,i]}\pi_p^{\flat}k \cong k[i]$ for any integer i. Since the filtered truncation functors (2.3) are obviously compatible with the tensor products, the map (3.10) then induces a map

$$\varepsilon_i : \tau_{[1,1]}\pi_p^{\flat}k \otimes \tau^F_{[i,i]}\pi_p^{\flat}E_{\bullet}^{[p]} \cong \tau^F_{[i,i]}\pi_p^{\flat}E_{\bullet}^{[p]}[1] \to \tau^F_{[i+1,i+1]}\pi_p^{\flat}E_{\bullet}^{[p]}$$

for any integer i. These are exactly the maps (2.16). In particular, ε_i only depends on the parity of i, and if $E_{\bullet}$ is tight in the sense of Definition 2.5, then ε_1 is an isomorphism. In this case, for any integer j, (3.10) also induces an isomorphism

$$\mathsf{I}(E_{\bullet})[j] = \tau_{[j,j]}\pi_{\flat}^{p}k \otimes \tau^F_{[0,0]}\pi_p^{\flat}E_{\bullet}^{[p]} \cong \tau^F_{[j,j]}\pi_p^{\flat}E_{\bullet}^{[p]}, \tag{3.12}$$

a version of the isomorphism (2.13).

If p is odd, tightness of $E_{\bullet}$ further implies that $\varepsilon_0 = 0$. One immediate corollary of this is a short construction of the quasiisomorphism (2.19) of Lemma 2.8, for odd p. Indeed, since $\varepsilon_2 = \varepsilon_0 = 0$ and the filtered truncations are multiplicative, the map (3.10) induces a natural map

$$\begin{aligned}\mathbb{K}(\mathsf{I}(E_{\bullet}))[1] = \mathbb{K}_{\bullet}[1] \otimes \mathsf{I}(E_{\bullet}) &\cong \tau_{[1,2]}\pi_p^{\flat}k \otimes \tau^F_{[0,0]}\pi_p^{\flat}E_{\bullet}^{[p]}\\ &\to \tau^F_{[1,2]}\pi_p^{\flat}E_{\flat}^{[p]} \cong \mathsf{gr}^0_V\,\pi_{p\flat}E_{\bullet}[1],\end{aligned}$$

and since the maps (3.12) are isomorphisms, this map is a quasiisomorphism.

However, we will need another corollary. Namely, keep the assumption that $E_{\bullet}$ is tight, let

$$\overline{\pi}_p^{\flat}E_{\bullet} = \tau^F_{\geq 0}\pi_p^{\flat}E_{\bullet}^{[p]}, \tag{3.13}$$

and consider this complex as an object in the filtered derived category $\mathcal{DF}(\Lambda, k)$ by equipping it with the filtration $\tau^{\bullet}$, via the functor (2.4). We then have the augmentation map

$$a : \overline{\pi}^{\flat}_p E_{\bullet} \to \tau_{[0,0]} \pi^{\flat}_p E^{[p]}_{\bullet} = \mathsf{I}(E_{\bullet}) \tag{3.14}$$

in $\mathcal{DF}(\Lambda, k)$, where $\mathsf{I}(E_{\bullet})$ is placed in filtered degree 0.

Lemma 3.4. *Assume that the map a of* (3.14) *admits a one-sided inverse s : $\mathsf{I}(E_{\bullet}) \to \overline{\pi}^{\flat}_p E_{\bullet}$, $s \circ a = \mathsf{id}$ in the filtered derived category $\mathcal{DF}(\Lambda, k)$. Then the spectral sequence* (2.21) *degenerates.*

Proof. By definition, we have a natural embedding $\overline{\pi}^{\flat}_p E_{\bullet} \to \pi^{\flat}_p E_{\bullet}$. Composing it with s, we obtain a map

$$\mathsf{I}(E_{\bullet}) \to \pi^{\flat}_p E_{\bullet},$$

and by (3.10), this map induces a map

$$b : \pi^{\flat}_p k \otimes \mathsf{I}(E_{\bullet}) \cong \pi^{\flat}_p E_{\bullet}.$$

For any integer i, the associated graded quotient $\mathsf{gr}^i_{\tau}(b)$ is the isomorphism (2.14) of Lemma 2.6, so that b is a filtered quasiisomorphism. By the direct sum decomposition (3.11), this implies that the conjugate filtration on $\pi^p_{\flat} E_{\bullet}$ splits. □

3.3. Splitting for DG algebras

Now keep the assumptions of the previous subsection, and assume given a DG algebra $A_{\bullet}$ termwise-flat over the commutative ring k. Recall that by Corollary 2.10, the corresponding complex $i^*_p A^{\natural}_{\bullet}$ is tight in the sense of Definition 2.5, so that Lemma 3.4 applies. To finish the section, we prove one corollary of this fact.

Choose resolution data $\nu = \langle P_{\bullet}, I^{\bullet} \rangle$ for the group $G = \mathbb{Z}/p\mathbb{Z}$ that are multiplicative, so that for any DG algebra $B_{\bullet}$ termwise-flat over k and equipped with a G-action, the Tate cohomology complex $\check{C}_{\bullet}(G, B_{\bullet})$ is a DG algebra over k. Consider the p-fold tensor product $A^{\otimes_k p}_{\bullet}$, and let $\mathbb{Z}/p\mathbb{Z}$ act on it by the longest permutation. Moreover, equip this tensor product with the pth rescaling of the stupid filtration.

Definition 3.5. The DG algebra $P_{\bullet}(A_{\bullet})$ is given by

$$P_{\bullet}(A_{\bullet}) = \tau^F_{\geq 0} \check{C}^{\bullet}(\mathbb{Z}/p\mathbb{Z}, \nu, A^{\otimes_k p}_{\bullet}),$$

where $A^{\otimes_k p}_{\bullet}$ is equipped with the pth rescaling of the stupid filtration.

As for the complex (3.13), we equip $P_{\bullet}(A_{\bullet})$ with the filtration $\tau^{\bullet}$ and treat it as a filtered DG algebra. Since the filtered truncation functors $\tau^F_{\geq \bullet}$ are multiplicative, $P_{\bullet}(A_{\bullet})$ is well defined, and up to a filtered quasiisomorphism, it does not depend on the choice of resolution data ν. By Proposition 2.9, we have

$$\mathsf{gr}^i_{\tau} P_{\bullet}(A_{\bullet}) \cong A^{(1)}_{\bullet}[i] \tag{3.15}$$

for any integer $i \geq 0$. In particular, we have an augmentation map

$$a : P_{\bullet}(A_{\bullet}) \to A^{(1)}_{\bullet},$$

and it is a filtered DG algebra map (where $A^{(1)}_{\bullet}$ is in filtered degree 0).

Proposition 3.6. *Assume that the prime p is odd, and that there exists a filtered DG algebra $A'_{\bullet}$ over k and a filtered DG algebra map $s : A'_{\bullet} \to A^{(1)}_{\bullet}$ such that the composition $s \circ a : A'_{\bullet} \to A^{(1)}_{\bullet}$ is a filtered quasiisomorphism. Then the spectral sequence* (2.22) *for the DG algebra $A_{\bullet}$ degenerates.*

Proof. Fix multiplicative resolution data $\langle P_{\bullet}, I^{\bullet}\rangle$ for the group $G = \mathbb{Z}/p\mathbb{Z}$, and consider the cone $\overline{P}_{\bullet}$ of the augmentation map $P_{\bullet} \to \mathbb{Z}$. By definition, both $I^{\bullet}$ and $\overline{P}_{\bullet}$ are DG algebras over k equipped with a G-action.

For every integer $n \geq 1$, the complex $\overline{P}_{\bullet}^{\otimes n}$ is concentrated in non-negative homological degrees, its degree-0 term is $\mathbb{Z}$, while all the other terms are free $\mathbb{Z}[G]$-modules. Therefore $\overline{P}_{\bullet}^{\otimes n} = \overline{P}_{\bullet}^{n}$ for some free left resolution $P_{\bullet}^{n}$ of the trivial module $\mathbb{Z}$. The complex $I^{\bullet\otimes n}$ is a right free resolution of $\mathbb{Z}$, so that $\langle P_{\bullet}^{n}, I^{\bullet\otimes n}\rangle$ also gives resolution data for G in the sense of Definition 3.1.

Now, since both $\overline{P}_{\bullet}$ and $I^{\bullet}$ are DG algebras equipped with a G-action, we have natural complexes $\overline{P}_{\bullet}^{\natural}$, $(I^{\bullet})^{\natural}$ in the category $\operatorname{Fun}(\mathsf{pt}_p \wr \Lambda, k)$. Restricting them to $\Lambda_p \subset \mathsf{pt}_p \wr \Lambda$ with respect to the embedding (2.25), we obtain complexes

$$\overline{P}_{\bullet}^{\lambda} = \lambda^* \overline{P}_{\bullet}^{\natural}, \qquad I_{\lambda}^{\bullet} = \lambda^*(I^{\bullet})^{\natural}$$

in the category $\operatorname{Fun}(\Lambda_p, k)$, and for any object $[n] \in \Lambda_p$, we have natural identifications

$$\overline{P}_{\bullet}^{\lambda}([n]) \cong \overline{P}_{\bullet}^{\otimes n}, \qquad I_{\lambda}^{\bullet}([n]) \cong I^{\bullet\otimes n}.$$

Thus we can put together resolution data $\langle P_{\bullet}^{n}, I^{\bullet\otimes n}\rangle$, $n \geq 1$ for the group G into resolution data $\langle P_{\bullet}^{\lambda}, I_{\lambda}^{\bullet}\rangle$ for the bifibration $\pi_p : \Lambda_p \to \Lambda$ such that $\overline{P^{\lambda}}_{\bullet} \cong \overline{P}_{\bullet}^{\lambda}$, and these resolution data can be then used for computing the relative Tate cohomology functor $\pi_p^{\flat}$. This gives a natural identification

$$\pi_p^{\flat} i_p^* A^{\natural} \cong \pi_{p*}\left(i_p^* A_{\bullet}^{\natural} \otimes \overline{P}_{\bullet}^{\lambda} \otimes I_{\lambda}^{\bullet}\right). \tag{3.16}$$

Now denote $B_{\bullet} = A_{\bullet}^{\otimes_k p} \otimes \overline{P}_{\bullet} \otimes I^{\bullet}$, and consider it as a G-equivariant DG algebra over k. Then by virtue of (2.26), we can rewrite (3.16) as

$$\pi_p^{\flat} i_p^* A_{\bullet}^{\natural} \cong \pi_{p*}\lambda^* B_{\bullet}^{\natural}. \tag{3.17}$$

Note that we have a natural map

$$\pi_* B_{\bullet}^{\natural} \to \pi_{p*}\lambda^* B_{\bullet}^{\natural}, \tag{3.18}$$

where $\pi : \mathsf{pt}_p \wr \Lambda \to \Lambda$ is the natural fibration. At each object $[n] \in \Lambda$, we can evaluate $\pi_* B_{\bullet}^{\natural}$ by (2.23), and then this map is just the natural embedding

$$\left(B_{\bullet}^{G}\right)^{\otimes_k n} = \left(B_{\bullet}^{\otimes_k n}\right)^{G^n} \to \left(B_{\bullet}^{\otimes_k n}\right)^{G}.$$

Moreover, by (2.24) and (3.17), the map (3.18) actually gives a natural map

$$\overline{B}_{\bullet}^{\natural} \cong \pi_* B_{\bullet}^{\natural} \to \pi_p^{\flat} i_p^* A_{\bullet}^{\natural}. \tag{3.19}$$

where $\overline{B}_{\bullet} \subset B_{\bullet}$ is the subalgebra of G-invariants.

Now equip $A_\bullet^{\otimes_k p}$ with the pth rescaling of the stupid filtration, and consider the corresponding filtrations on the algebras $B_\bullet$, $\overline{B}_\bullet$. Then since $\tau^F_{\geq 0}$ is a multiplicative functor, the natural map (3.19) induces a map

$$(\tau^F_{\geq 0}\overline{B}_\bullet)^\natural \to \tau^F_{\geq 0}\pi^\flat_p i^*_p A^\natural_\bullet = \overline{\pi}^p_\flat i^*_p A^\natural_\bullet. \tag{3.20}$$

By construction, if we compose this map with the projection

$$\overline{\pi}^\flat_p i^*_p A^\natural_\bullet \to \mathsf{gr}^0_\tau \overline{\pi}^\flat_p i^*_p A^\natural_\bullet \cong \mathsf{I}(i^*_p A^\natural_\bullet) \cong A^{(1)\natural}_\bullet,$$

then the resulting map is induced by the augmentation map

$$\tau^F_{\geq 0}\overline{B}_\bullet \to \tau^F_{[0,0]}\overline{B}_\bullet \cong A^{(1)}_\bullet.$$

It remains to spell out the notation. By definition, we actually have

$$\overline{B}_\bullet = \check{C}_\bullet(G, \nu, A^{\otimes_k n}),$$

where ν stands for our original resolution data $\langle P_\bullet, I^\bullet\rangle$. Therefore $\tau^F_{\geq 0}(\overline{B}_\bullet) \cong P_\bullet(A_\bullet)$, and the map (3.20) is a map

$$P_\bullet(A_\bullet)^\natural \to A^{(1)}_\natural.$$

If there exists a filtered DG algebra $A'_\bullet$ and a filtered DG algebra map $s : A'_\bullet \to P_\bullet(A_\bullet)$ satisfying the assumptions of the Proposition, then we also have a filtered map $A'^\natural_\bullet \to P_\bullet(A_\bullet)^\natural$, and the composition map

$$A'^\natural_\bullet \longrightarrow P_\bullet(A_\bullet) \longrightarrow \overline{\pi}^\flat_p i^*_p A^\natural_\bullet \longrightarrow A^{(1)\natural}_\bullet$$

is a filtered quasiisomorphism. Then we are done by Lemma 3.4. □

4. Characteristic 2

We now make a digression and explain how to modify the arguments of [Ka5] to obtain the conjugate spectral sequence (2.22) in characteristic 2. The problem here is Lemma 2.8: while (i) is true in any characteristic, (ii) is definitely wrong in characteristic 2, and it is currently unknown whether (iii) is true or not. However, there is the following weaker result.

Proposition 4.1. *For any DG algebra $A_\bullet$ over a perfect field k of positive characteristic p, there exists a natural isomorphism*

$$HC_\bullet(\mathsf{gr}^0_V \pi_{p\flat} i^*_p A^\natural_\bullet) \cong HH_\bullet(A^{(1)}_\bullet). \tag{4.1}$$

While weaker than (2.19), this identification still allows one to rewrite (2.20) in the form (2.22), at least for DG algebras over a perfect field. For degeneration questions, this is irrelevant; the reader who is only interested in degeneration of the spectral sequences can safely skip this section.

4.1. Trace functors

To get a better handle on the complex $\pi_{pb}i_p^* A_\bullet^\natural$, we use the formalism of trace functors of [Ka3, Section 2]. Here are the basic ingredients.

One starts by "categorifying" the construction of the object $A^\natural$ of Subsection 1.3. To every small monoidal category $\mathcal{C}$, one associates a covariant functor from Λ to the category of small categories that sends $[n] \in \Lambda$ to $\mathcal{C}^{V([n])}$, and sends a morphism $f : [n'] \to [n]$ to the product of multiplication functors $m_{f^{-1}(v)} : \mathcal{C}^{f^{-1}(v)} \to \mathcal{C}$, $v \in V([n])$ induced by the monoidal structure on $\mathcal{C}$, as in (1.15). Applying the Grothendieck construction, one obtains a category $\mathcal{C}^\natural$ and a cofibration $\rho : \mathcal{C}^\natural \to \Lambda$. This is somewhat similar to the wreath product construction $\mathcal{C} \wr \Lambda$, except that the functor is covariant, not contravariant, and the projection $\rho : \mathcal{C}^\natural \to \Lambda$ is a cofibration, not a fibration.

Explicitly, objects of $\mathcal{C}^\natural$ are pairs $\langle [n], \{c_v\}\rangle$ of an object $[n] \in \Lambda$ and a collection $\{c_v\}$ of objects in $\mathcal{C}$ numbered by vertices $v \in V([n])$. A morphism from $\langle [n'], \{c'_v\}\rangle$ to $\langle [n], \{c_v\}\rangle$ is given by a morphism $f : [n'] \to [n]$ and a collection of morphisms

$$f_v : \bigotimes_{v' \in f^{-1}(v)} c'_{v'} \to c_v, \qquad v \in V([n]).$$

A morphism is *cartesian* if all the components f_v are invertible. Note that stated in this way, the definition makes perfect sense even when the category $\mathcal{C}$ is not small.

Definition 4.2. A *trace functor* from a monoidal category $\mathcal{C}$ to some category $\mathcal{E}$ is a functor $F : \mathcal{C}^\natural \to \mathcal{E}$ that sends cartesian maps in $\mathcal{C}^\natural$ to invertible maps in $\mathcal{E}$.

Explicitly, a trace functor is given by a functor $F : \mathcal{C} \to \mathcal{E}$ and a collection of isomorphisms

$$\tau_{M,N} : F(M \otimes N) \cong F(N \otimes M), \qquad M, N \in \mathcal{C} \tag{4.2}$$

satisfying some compatibility constraints (see [Ka3, Subsection 2.1]). A trivial example of a trace functor is obtained by fixing a commutative ring k, and taking $\mathcal{C} = \mathcal{E} = k\text{-mod}$, the category of k-modules, with $\tau_{-,-}$ being the standard commutativity isomorphisms. There are also non-trivial examples. One such was considered in [Ka3] in detail. We still take $\mathcal{C} = \mathcal{E} = k\text{-mod}$, fix an integer $l \geq 1$, and let

$$F(V) = V_\sigma^{\otimes_k l}, \tag{4.3}$$

where $\sigma : V^{\otimes_k l} \to V^{\otimes_k l}$ is the order-l permutation. The maps (4.2) are given by $\tau_{M,N} = \tau'_{M,N} \circ (\sigma_M \otimes \mathsf{id})$, where $\tau'_{M,N}$ are the commutativity maps, and σ_M is the order-n permutation acting on $M^{\otimes_k l}$.

Every algebra object A in the monoidal category $\mathcal{C}$ defines a section $\alpha : \Lambda \to \mathcal{C}^\natural$ of the cofibration $\rho : \mathcal{C}^\natural \to \Lambda$, and composing this section with a trace functor F gives a natural functor

$$FA^\natural = F \circ \alpha : \Lambda \to \mathcal{E}.$$

If $\mathcal{C} = \mathcal{E} = k$-mod, what we obtain is an object $FA^\natural$ in the category $\operatorname{Fun}(\Lambda, k)$ associated to any associative unital algebra A over k. If F is the identity functor with the tautological trace functor structure, then this just the object $A^\natural$ of Subsection 1.3. In general, we obtain a version of cyclic homology twisted by the trace functor F, the main object of study in [Ka3].

Another way to express this is to say that a trace functor $F : \mathcal{C} \to k$-mod defines an object $F^\natural \in \operatorname{Fun}(\mathcal{C}^\natural, k)$, and we have $FA^\natural = \alpha^* F^\natural$. Analogously, a trace functor from $\mathcal{C}$ to the category $C_\bullet(k)$ of complexes of k-modules gives a complex $F^\natural$ in $\operatorname{Fun}(\mathcal{C}^\natural, k)$, and $FA^\natural = \alpha^* F^\natural$ is a complex in $\operatorname{Fun}(\Lambda, k)$.

Remark 4.3. Strictly speaking, when $\mathcal{C}$ is not small, $\operatorname{Fun}(\mathcal{C}^\natural, k)$ is not a well-defined category (Hom-sets might be large). A convenient solution is to only consider functors that commute with filtered colimits. Then each such functor from say k-mod to k-mod is completely determined by its restriction to the full subcategory spanned by finitely generated projective k-modules, and since this category is small, the problem does not arise. The same works for complexes (and the subcategory of perfect complexes). In our example, all large monoidal categories will be of this sort, so we will adopt this point of view. By abuse of notation, we will still use notation of the form $\operatorname{Fun}(\mathcal{C}^\natural, k)$ for the category for functors that commute with filtered colimits.

Let us now construct the cyclic power trace functor (4.3) more canonically. Fix a monoidal category $\mathcal{C}$ and an integer $l \geq 1$, and define a category $\mathcal{C}_l^\natural$ by the cartesian square

$$\begin{array}{ccc} \mathcal{C}_l^\natural & \xrightarrow{\pi_l} & \mathcal{C}^\natural \\ {\scriptstyle \rho_l}\downarrow & & \downarrow{\scriptstyle \rho} \\ \Lambda_l & \xrightarrow{\pi_l} & \Lambda. \end{array} \tag{4.4}$$

Then the functor $i_l : \Lambda_l \to \Lambda$ fits into a commutative square

$$\begin{array}{ccc} \mathcal{C}_l^\natural & \xrightarrow{i_l} & \mathcal{C}^\natural \\ {\scriptstyle \rho_l}\downarrow & & \downarrow{\scriptstyle \rho} \\ \Lambda_l & \xrightarrow{i_l} & \Lambda. \end{array} \tag{4.5}$$

Explicitly, $\mathcal{C}_l^\natural$ is the category of pairs $\langle [n], \{c_v\} \rangle$, $[n] \in \Lambda_l$, $c_v \in \mathcal{C}$, $v \in V([n])$. Then the top arrow in (4.5) is the functor sending a sequence $\{c_v\}$, $v \in V([n])$ to the same sequence repeated l times. In particular, it sends cartesian maps to cartesian maps.

We now observe that for any trace functor F from $\mathcal{C}$ to the category of k-modules, with the corresponding object $F^\natural \in \operatorname{Fun}(\mathcal{C}^\natural.k)$, the objects $\pi_{l!} i_l^* F^\natural$, $\pi_{l*} i_l^* F^\natural$ also correspond to trace functors from $\mathcal{C}$ to k-mod. If $\mathcal{C} = k$-mod and F is the tautological functor, $\pi_{l!} i_l^* F^\natural$ corresponds to the cyclic power trace functor (4.3).

4.2. Quotients of the conjugate filtration

We now fix a perfect field k of characteristic p, and we let $\mathcal{C} = C_{\bullet}(k)$ be the category of complexes of k-vector spaces. Moreover, denote by $I_{\bullet}$ the complex in $\operatorname{Fun}(\mathcal{C}^{\natural}, k)$ corresponding to the identity trace functor $C_{\bullet}(k) \to C_{\bullet}(k)$, and denote by $I_{\bullet}^{[p]}$ the complex $I_{\bullet}$ equipped with the pth rescaling of the stupid filtration.

Then the projection $\pi_p : \mathcal{C}_p^{\natural} \to \mathcal{C}^{\natural}$ of (4.4) is a bifibration with fiber pt_p, so that we can consider relative Tate cohomology functor $\pi_p^{\flat}$. By base change [Ka2, Lemma 1.7], we have $\rho^* \circ \pi_p^{\flat} \cong \pi_p^{\flat} \circ \rho^*$, so that (3.11) yields a direct sum decomposition

$$\pi_p^{\flat} k \cong \rho^* \mathbb{K}_{\bullet}(k)[1]((u)) = \bigoplus_{i \in \mathbb{Z}} \rho^* \mathbb{K}_{\bullet}(k)[2i+1] \tag{4.6}$$

in the derived category $\mathcal{D}(\mathcal{C}^{\natural}, k)$. We also have the multiplication map (3.10) and all that it entails – in particular, the isomorphisms (3.12) for complexes $E_{\bullet}$ in $\operatorname{Fun}(\mathcal{C}_p^{\natural}, k)$ that are tight in the obvious sense. We note that the pullback $i_p^* I_{\bullet}$ of the tautological complex $I_{\bullet}$ is tight.

Now fix some resolution data for π_p, so that $\pi_p^{\flat}$ is defined as a complex, and consider the complex

$$C_{\bullet} = \tau^F_{[0,1]} \pi_p^{\flat} i_p^* I_{\bullet}^{[p]}$$

in the category $\operatorname{Fun}(\mathcal{C}^{\natural}, k)$. It must correspond to some trace functor from $\mathcal{C} = C_{\bullet}(k)$ to itself. Explicitly, the trace functor sends a complex $V_{\bullet}$ to the complex

$$C_{\bullet}(V_{\bullet}) = \tau^F_{[0,1]} \check{C}^{\bullet}(\mathbb{Z}/p\mathbb{Z}, V_{\bullet}^{\otimes_k p}), \tag{4.7}$$

where we equip $V_{\bullet}^{\otimes_k p}$ with the pth rescaling of the stupid filtration. By Proposition 2.9, we have a natural sequence

$$0 \longrightarrow V_{\bullet}^{(1)}[1] \xrightarrow{\ b\ } C_{\bullet}(V_{\bullet}) \xrightarrow{\ a\ } V_{\bullet}^{(1)} \longrightarrow 0 \tag{4.8}$$

of functorial complexes of k-vector spaces that is *quasiexact* in the sense of [Ka6, Definition 1.2] – this means that $a \circ b = 0$, the map a is surjective, the map b is injective, and the complex $\operatorname{Ker} a / \operatorname{Im} b$ is acyclic. The map a corresponds to a map

$$a : C_{\bullet} \to I_{\bullet}^{(1)}$$

in the category $\operatorname{Fun}(\mathcal{C}^{\natural}, k)$, where $I_{\bullet}^{(1)}$ is the Frobenius twist of the tautological complex $I_{\bullet}$.

Moreover, consider the ring $W_2(k)$ of second Witt vectors of the field k, and let $\mathcal{C}_1 = C_{\bullet}(W_2(k))$ be the category of complexes of flat $W_2(k)$-modules. Denote by $q : \mathcal{C}_1 \to \mathcal{C}$ the quotient functor sending a complex $V_{\bullet}$ to its quotient $V_{\bullet}/p$. Note that $\mathcal{C}_1$ is a monoidal category, and q is a monoidal functor. Moreover, extend (4.7) to $W_2(k)$-modules by setting

$$C_{\bullet}(V_{\bullet}) = \tau^F_{[0,1]} \check{C}^{\bullet}(\mathbb{Z}/p\mathbb{Z}, V_{\bullet}^{\otimes_{W_2(k)} p}) \tag{4.9}$$

for any complex $V_{\bullet} \in \mathcal{C}_1$, where as in (4.7), we equip $V_{\bullet}^{\otimes_{W_2(k)} p}$ with the pth rescaling of the stupid filtration. Then we have the following somewhat surprising result.

Lemma 4.4.

(i) *For any complex* $V_\bullet \in \mathcal{C}_1$*, we have a short exact sequence of complexes*

$$0 \longrightarrow C_\bullet(V_\bullet/p) \xrightarrow{\ p\ } C_\bullet(V_\bullet) \xrightarrow{\ q\ } C_\bullet(V_\bullet/p) \longrightarrow 0,$$

where p *stands for multiplication by* p*, and* q *is the quotient map.*

(ii) *For any* $V_\bullet \in \mathcal{C}$*, let* $\overline{C}_\bullet(V_\bullet)$ *be the kernel of the map* a *of* (4.8)*, and for any* $V_\bullet \in \mathcal{C}_1$*, let* $\widetilde{C}_\bullet(V_\bullet) = C_\bullet(V_\bullet)/p\overline{C}_\bullet(V_\bullet/p)$*. Then the composition map*

$$\widetilde{C}_\bullet(V_\bullet) \xrightarrow{\ q\ } C_\bullet(V_\bullet/p) \xrightarrow{\ a\ } V_\bullet^{(1)}$$

is a quasiisomorphism, and the functor $\widetilde{C}_\bullet : \mathcal{C}_1 \to \mathcal{C}_1$ *factors through the quotient functor* $q : \mathcal{C}_1 \to \mathcal{C}$.

Proof. (i) is [Ka6, Lemma 6.9], and (ii) is [Ka6, Proposition 6.11]. □

We note that $\widetilde{C}_\bullet(-)$ is the cokernel of a map of trace functors, thus itself inherits the structure of a trace functor. Then Lemma 4.4 (ii) implies that this trace functor is actually defined on $\mathcal{C}$ – namely, we have the following.

Corollary 4.5. *Let* $\widetilde{C}_\bullet$ *be the complex in* $\operatorname{Fun}(\mathcal{C}_1^\natural, W_2(k))$ *corresponding to the trace functor* $\widetilde{C}_\bullet(-)$ *of Lemma* 4.4 (ii)*. Then there exists a complex* $W_\bullet \in \operatorname{Fun}(\mathcal{C}, W_2(k))$ *such that*

$$\widetilde{C}_\bullet \cong q^* W_\bullet,$$

where $q : \mathcal{C}_1^\natural \to \mathcal{C}^\natural$ *is induced by the monoidal quotient functor* $q : \mathcal{C}_1 \to \mathcal{C}$.

Proof. The quotient functor q is surjective on isomorphism classes of objects, so that it suffices to prove that the action of morphisms in $\mathcal{C}_1^\natural$ on $\widetilde{C}_\bullet$ canonically factors through q. For cartesian morphisms, this is clear, so it suffices to check it for morphisms in the categories $\mathcal{C}_1^V([n])$, $[n] \in \Lambda$. This immediately follows from Lemma 4.4 (ii). □

Proof of Proposition 4.1. Every k-vector space can be considered as a $W_2(k)$-module via the quotient map $W_2(k)$, so that we have a natural functor from k-vector spaces to $W_2(k)$-module. Moreover, for any small category I, we can apply this pointwise and obtain a functor

$$\xi : \operatorname{Fun}(I, k) \to \operatorname{Fun}(I, W_2(k)).$$

This functor is exact and fully faithful. It induces a functor

$$\xi : \mathcal{D}(I, k) \to \mathcal{D}(I, W_2(k)),$$

and for any complex $E_\bullet$ in $\operatorname{Fun}(I, k)$, we have a natural quasiisomorphism

$$H_\bullet(I, \xi(I_\bullet)) \cong \xi(H_\bullet(I, E_\bullet)).$$

In particular, we can take $I = \Lambda$. Then by (1.11) and (1.14), to construct an isomorphism (4.1), it suffices to construct a functorial isomorphism

$$\xi(\mathsf{gr}_V^0 \pi_{p\flat} i_p^* A_\bullet^\natural) \cong \xi(\mathbb{K}(A_\bullet^{(1)\natural}))$$

in the derived category $\mathcal{D}(\Lambda, W_2(k))$.

Consider first the universal situation. Denote

$$R_\bullet = \tau^F_{[1,2]}\pi_p^\flat i_p^* I_\bullet^{[p]}[-1],$$

and recall that we have the multiplication map (3.10) for the bifibration $\pi_p : \mathcal{C}_p^\natural \to \mathcal{C}^\natural$. This map is compatible with filtered truncations, so by (4.6), it yields a map

$$\rho^*\mathbb{K}_\bullet(k) \otimes_k C_\bullet \cong \tau_{[1,2]}\pi_p^\flat k[-1] \otimes_k \tau^F_{[0,1]}\pi_p^\flat i_p^* I_\bullet^{[p]} \to \tau_{[1,2]}\pi_p^\flat i_p^* I_\bullet^{[p]}[-1] = R_\bullet.$$

This map induces a map

$$\rho^*\mathbb{K}_\bullet \otimes \xi(C_\bullet) \to \xi(R_\bullet).$$

On the other hand, we have the complex $W_\bullet$ of Corollary 4.5 and the natural map $q : W_\bullet \to \xi(C_\bullet)$. Denote by b the composition map

$$\rho^*\mathbb{K}_\bullet \otimes W_\bullet \xrightarrow{\mathsf{id} \otimes q} \rho^*\mathbb{K}_\bullet \otimes \xi(C_\bullet) \longrightarrow \xi(R_\bullet).$$

Then b is filtered, hence compatible with the filtration $\tau^\bullet$. Moreover, since the composition $W_\bullet \to \xi(C_\bullet) \to \xi(I_\bullet^{(1)})$ is a quasiisomorphism by Lemma 4.4, the source and the target of the map b only have two non-trivial associated graded quotients gr^i_τ, for $i = 0$ and 1, and in both cases, $\mathsf{gr}^i_\tau(b)$ is one of the isomorphisms (3.12) for the tight complex $I_\bullet^{[p]}$. Therefore b is an isomorphism in the derived category $\mathcal{D}(\mathcal{C}^\natural, W_2(k))$.

Now let $\alpha : \Lambda \to \mathcal{C}^\natural$ be the section of the projection $\rho : \mathcal{C}^\natural \to \Lambda$ corresponding to the DG algebra $A_\bullet$, and consider the induced isomorphism

$$\alpha^*(b) : \mathbb{K}_\bullet(\alpha^* W_\bullet) \cong \alpha^*(\rho^*\mathbb{K}_\bullet \otimes W_\bullet) \to \alpha^*\xi(R_\bullet).$$

Then the right-hand side is exactly $\xi(\mathsf{gr}^0_V \pi_{p\flat} i_p^* A_\bullet^\natural)$, and the left-hand side is naturally isomorphic to $\mathbb{K}_\bullet(\alpha^*\xi(I_\bullet^{(1)})) \cong \mathbb{K}_\bullet(\xi(A_\bullet^{(1)\natural})) \cong \xi(\mathbb{K}_\bullet(A_\bullet^{(1)\natural}))$. $\square$

5. Degeneration

We now turn to degeneration results for the spectral sequences for cyclic homology. There are two statements: one for the conjugate spectral sequence (2.22), and one for the Hodge-to-de Rham spectral sequence (2.2).

5.1. Conjugate degeneration

Recall that for any field k, a *square-zero extension* $A'_\bullet$ of a DG algebra $A_\bullet$ over k by an $A_\bullet$-bimodule $M_\bullet$ is a filtered DG algebra $\langle A'_\bullet, \tau^\bullet\rangle$ such that $\tau^0 A'_\bullet = A'_\bullet$, $\tau^2 A'_\bullet = 0$, $\mathsf{gr}^0_\tau A'_\bullet \cong A_\bullet$, and $\mathsf{gr}^1_\tau A'_\bullet$ is quasiisomorphic to $M_\bullet$ as a bimodule over $\mathsf{gr}^0_\tau A'_\bullet \cong A_\bullet$. Recall also that up to a quasiisomorphisms, square-zero extensions are classified by elements in the *reduced Hochschild cohomology group*

$$\overline{HH}^2(A_\bullet, M_\bullet) = \operatorname{Ext}^1_{A_\bullet^o \otimes A_\bullet}(I_\bullet, M_\bullet),$$

where $I_\bullet$ is the kernel of the augmentation map $A^o_\bullet \otimes_k A_\bullet \to A_\bullet$ (see, e.g., [Ka2, Subsection 4.3] but the claim is completely standard). Reduced Hochschild cohomology groups are related to the usual ones by the long exact sequence

$$\overline{HH}^\bullet(A_\bullet, M_\bullet) \longrightarrow HH^\bullet(A_\bullet, M_\bullet) \longrightarrow M_\bullet \longrightarrow \qquad (5.1)$$

In particular, if $\overline{HH}^2(A_\bullet, M_\bullet) = 0$ for some $M_\bullet$, then every square-zero extension $A'_\bullet$ of $A_\bullet$ by $M_\bullet$ splits – there exists a DG algebra $A''_\bullet$ and a map $A''_\bullet \to A'_\bullet$ such that the composition map $A''_\bullet \to A'_\bullet \to A_\bullet$ is a quasiisomorphism.

Now fix a perfect field k of some positive characteristic $p = \mathsf{char}\, k$, and assume given a DG algebra $A_\bullet$ over k.

Theorem 5.1. *Assume that the DG algebra $A_\bullet$ over the field k satisfies the following two properties:*

(i) *There exist a DG algebra $\widetilde{A}_\bullet$ over the second Witt vectors ring $W_2(k)$ and a quasiisomorphism $\widetilde{A}_\bullet \overset{\mathsf{L}}{\otimes}_{W_2(k)} k \cong A_\bullet$.*

(ii) *The reduced Hochschild cohomology $\overline{HH}^i(A_\bullet)$ vanishes for $i \geq 2p$.*

Then the conjugate spectral sequence (2.22) *degenerates at first term, so that there exists an isomorphism $\overline{HP}_\bullet(A_\bullet) \cong HH_\bullet(A^{(1)}_\bullet)((u^{-1}))$.*

Remark 5.2. By Proposition 4.1, the spectral sequence (2.22) also exists for $p = 2$, and Theorem 5.1 holds in this case, too. Of course in this case, the condition (ii) is pretty strong.

Proof. By Proposition 3.6, it suffices to construct a filtered DG algebra $A'_\bullet$ over k and a filtered map $A'_\bullet \to P_\bullet(A_\bullet)$ such that the composition map $A'_\bullet \to A^{(1)}_\bullet$ is a filtered quasiisomorphism (the assumption "p is odd" in Proposition 3.6 is only needed to insure that the conjugate spectral sequence is well defined, and it is not used in the proof).

To define the DG algebra $P_\bullet(A_\bullet)$, we need to choose multiplicative resolution data for the group $G = \mathbb{Z}/p\mathbb{Z}$, and we are free to do it in any way we like. Note that the G-action on the DG algebra $A^{\otimes_k p}_\bullet$ extends to the action of the symmetric group Σ_p. In particular, we have an action of the normalizer $\widehat{G} = (\mathbb{Z}/p\mathbb{Z}) \rtimes (\mathbb{Z}/p\mathbb{Z})^* \subset \Sigma_p$ of $G \subset \Sigma_p$. Choose some multiplicative resolution data for $\widehat{G}$, and restrict it to $G \subset \widehat{G}$. Then the resulting DG algebra $P_\bullet(A_\bullet)$ carries a natural action of $(\mathbb{Z}/p\mathbb{Z})^* = \widehat{G}/G$. This action preserves the filtration $\tau^\bullet$, and the augmentation map

$$a : P_\bullet(A_\bullet) \to \mathsf{gr}^0_\tau P_\bullet(A_\bullet) \cong A^{(1)}_\bullet$$

is $(\mathbb{Z}/p\mathbb{Z})^*$-invariant. For $i \geq 1$, we have the isomorphisms (3.15) induced by the isomorphisms (2.13); invariantly, they can be written as

$$\mathsf{gr}^i_\tau P_\bullet(A_\bullet) \cong A^{(1)}_\bullet[i] \otimes_k \check{H}^{-i}(G, k),$$

where the group $(\mathbb{Z}/p\mathbb{Z})^*$ acts on the right-hand side via its action on the Tate comology group $\check{H}^{-i}(G, k)$.

Now, the cohomology $H^{\bullet}(G,k)$ is given by (2.10). The group $(\mathbb{Z}/p\mathbb{Z})^*$ acts trivially on the generator ε, and it acts on the generator u via its standard one-dimensional representation given by the action on $\mathbb{Z}/p\mathbb{Z} \subset k$. Therefore we have

$$\check{H}^{-i}(G,k)^{(\mathbb{Z}/p\mathbb{Z})^*} \cong k$$

if $i = 0,1 \mod 2(p-1)$, and 0 otherwise. We conclude that if we denote

$$\overline{P}_{\bullet}(A_{\bullet}) = P_{\bullet}(A_{\bullet})^{(\mathbb{Z}/p\mathbb{Z})^*} \subset P_{\bullet}(A_{\bullet})$$

and equip this DG algebra with the filtration induced by $\tau^{\bullet}$, then we have

$$\operatorname{gr}^i_\tau \overline{P}_{\bullet}(A_{\bullet}) \cong \begin{cases} A^{(1)}_{\bullet}[i], & i = 0,1 \mod 2(p-1), \\ 0, & \text{otherwise.} \end{cases} \tag{5.2}$$

If we denote by $e : \overline{P}_{\bullet}(A_{\bullet}) \to P_{\bullet}(A_{\bullet})$ the embedding map, then $\operatorname{gr}^i_\tau(e)$ is a quasiisomorphism for $i = 0,1 \mod 2p$, and $\operatorname{gr}^i_\tau(e) = 0$ otherwise.

To prove degeneration, it suffices to construct a filtered DG algebra $A'_{\bullet}$ and a filtered map $s : A'_{\bullet} \to \overline{P}_{\bullet}(A_{\bullet})$, since then we can simply compose it with the embedding e. Moreover, a filtered algebra $\langle A'_{\bullet}, \tau^{\bullet}\rangle$ is completely defined by its quotients $A^n_{\bullet} = A'_{\bullet}/\tau^{n+1}A'_{\bullet}$, $n \geq 0$, together with the quotient maps $r_n : A^{n+1}_{\bullet} \to A^n_{\bullet}$, and by assumption, $A^0_{\bullet}$ must be identified with the Frobenius twist $A^{(1)}_{\bullet}$ of the DG algebra $A_{\bullet}$. Thus if we denote

$$P^n_{\bullet} = \overline{P}_{\bullet}(A_{\bullet})/\tau^{n+1}\overline{P}_{\bullet}(A_{\bullet}), \qquad n \geq 1,$$

and denote by $p_n : P^{n+1}_{\bullet} \to P^n_{\bullet}$ the quotient maps, then it suffices to construct a collection of DG algebras $A^n_{\bullet}$ over k for all $n \geq 1$, equipped with DG algebra maps $r_n : A^{n+1}_{\bullet} \to A^n_{\bullet}$, $s_n : A^n_{\bullet} \to P^n_{\bullet}$ such that

- for every $n \geq 1$, we have $s_n \circ r_n = p_n \circ s_{n+1}$, and the composition map $a \circ e \circ s_n : A^n_{\bullet} \to A^{(1)}_{\bullet}$ is a quasiisomorphism.

We use induction on n. To start it, we take $n = 1$; we need to find a DG algebra $A^1_{\bullet}$ over k and a map $s_1 : A^1_{\bullet} \to P^1_{\bullet}$ such that $a \circ e \circ s_1 : A^1_{\bullet} \to A^{(1)}_{\bullet}$ is a quasiisomorphism. By [Ka6, Proposition 6.13], this is possible precisely because the DG algebra $A_{\bullet}$ satisfies the assumption (i) of the Theorem.

For the induction step, assume given $A^{n-1}_{\bullet}$ and $s_{n-1} : A^{n-1}_{\bullet} \to P^{n-1}_{\bullet}$, and consider the DG algebra $A''_{\bullet}$ defined by the cartesian square

$$\begin{array}{ccc} A''_{\bullet} & \longrightarrow & A^{n-1}_{\bullet} \\ \big\downarrow & & \big\downarrow {\scriptstyle s_{n-1}} \\ P^n_{\bullet} & \xrightarrow{p_{n-1}} & P^{n-1}_{\bullet}. \end{array}$$

Then up to a quasiisomorphism, $A''_{\bullet}$ is a square-zero extension of $A^{(1)}_{\bullet}$ by $\operatorname{gr}^n_\tau\overline{P}_{\bullet}(A_{\bullet})$, and finding $A^n_{\bullet}$ with the maps s_n, r_{n-1} satisfying ($\bullet$) is equivalent to finding a DG algebra $A^n_{\bullet}$ and a map $A^n_{\bullet} \to A''_{\bullet}$ such that the composition map

$$A^n_{\bullet} \longrightarrow A''_{\bullet} \longrightarrow A^{(1)}_{\bullet}$$

is a quasiisomorphism. In other words, we have to split the extension $A''_{\bullet}$. The obstruction to doing this lies in the reduced Hochschild cohomology group

$$\overline{HH}^2(A^{(1)}_{\bullet}, \mathrm{gr}^n_{\tau} \overline{P}_{\bullet}(A_{\bullet})).$$

By (5.2), this group vanishes unless $i = 0, 1 \mod 2p$, and in this case, we have

$$\overline{HH}^2(A^{(1)}_{\bullet}, \mathrm{gr}^n_{\tau} \overline{P}_{\bullet}(A_{\bullet})) \cong \overline{HH}^{2+n}(A^{(1)}_{\bullet}).$$

Since $n \geq 2$ and $n = 0, 1 \mod 2(p-1)$, we have $n \geq 2(p-1)$, and then this reduced Hochschild cohomology group vanishes by the assumption (ii). □

Remark 5.3. The condition (ii) of Theorem 5.1 is slightly unnatural: while Hochschild homology and the conjugate spectral sequence are derived Morita-invariant, reduced Hochschild cohomology groups are not (because of the third term in (5.1)). One would like to have the same statement but with $HH^{\bullet}(-)$ instead of $\overline{HH}^{\bullet}(-)$. The simplest way to obtain such a statement would be to repeat the whole argument for DG categories instead of DG algebras. In fact, [Ka5] also deals with the DG category case, so that this looks like a straightforward exercise. However, since our main interest is in degeneration in $\mathrm{char}\, 0$, we do not go into it to save space.

5.2. Hodge-to-de Rham degeneration

We now fix a field K of characteristic 0, and a DG algebra $A_{\bullet}$ over K. In this case, a Hodge-to-de Rham degeneration theorem is an immediate corollary of Theorem 2.22, and the argument is exactly the same as in [Ka2, Subsection 5.3]. We reproduce it for the sake of completeness and for the convenience of the reader.

Theorem 5.4. *Assume that the DG algebra $A_{\bullet}$ is homologically smooth and homologically proper. Then the Hodge-to-de Rham spectral sequence* (2.2) *degenerates, so that there exists an isomorphism $HP_{\bullet}(A_{\bullet}) \cong HH_{\bullet}(A_{\bullet})((u))$.*

We recall that *homologically proper* simply means that $A_{\bullet}$ is a perfect complex over K (in particular, it is homologically bounded).

Proof. By a theorem of B.Toën [T], there exists a finitely generated subring $R \subset K$ and a homologically smooth and homologically proper DG algebra $A^R_{\bullet}$ over R such that $A_{\bullet} \cong A^R_{\bullet} \overset{\mathsf{L}}{\otimes}_R K$. Since R is finitely generated, the residue field $k = R/\mathfrak{m}$ for any maximal ideal $\mathfrak{m} \subset R$ is a finite, hence perfect field of some characteristic p. Since $A^R_{\bullet}$ is homologically proper and homologically smooth, there is at most a finite number of non-trivial Hochschild homology groups $HH_{\bullet}(A^R_{\bullet})$ and reduced Hochschild cohomology groups $\overline{HH}^{\bullet}(A^R_{\bullet})$, and these groups are finitely generated R-modules. Then there exists a constant N such that $\overline{HH}^i(A^R_{\bullet}) = 0$ for $i \geq N$. Moreover, localizing R if necessary, we can further assume that $HH_i(A^R_{\bullet})$ is a projective finitely generated R-module for every i, and that for any maximal ideal $\mathfrak{m} \subset R$, $p = \mathrm{char}\, R/\mathfrak{m}$ is non-trivial in $\mathfrak{m}/\mathfrak{m}^2$ (that is, p is unramified in R), and

$2p > N$. Then for any $\mathfrak{m} \subset R$ with $k = R/\mathfrak{m}$, the DG algebra $A^k_\bullet = A^R_\bullet \overset{\mathsf{L}}{\otimes}_R k$ satisfies the assumptions of Theorem 2.22. Therefore we have an isomorphism

$$\overline{HP}_\bullet(A^k_\bullet) \cong HH_\bullet(A^k_\bullet)^{(1)}((u^{-1}))$$

of finite-dimensional graded k-vector spaces. Since $HH_\bullet(A^k_\bullet)$ is concentrated in a finite range of degrees, we can replace Laurent power series in u^{-1} with Laurent power series in u, and since $A^R_\bullet$ is homologically smooth and proper, $A^k_\bullet$ is also homologically smooth and proper. In particular, it is cohomologically bounded, so that Theorem 1.3 (ii) allows us to replace $\overline{HP}_\bullet(A^k_\bullet)$ with $HP_\bullet(A^k_\bullet)$. We thus have an isomorphism

$$HP_\bullet(A^k_\bullet) \cong HH_\bullet(A^k_\bullet)^{(1)}((u))$$

of finite-dimensional graded k-vector spaces, so that the Hodge-to-de Rham spectral sequence (2.2) for $A^k_\bullet$ degenerates for dimension reasons.

Finally, since all the Hochschild homology R-modules $HH_i(A^R_\bullet)$ are finitely generated and projective, and any differential in the Hodge-to-de Rham spectral sequence for $A^R_\bullet$ vanishes modulo any maximal ideal $\mathfrak{m} \subset R$, the differential must vanish identically. Thus the Hodge-to-de Rham spectral sequence for the DG algebra $A^R_\bullet$ degenerates, and then so does the Hodge-to-de Rham spectral sequence for $A_\bullet = A^R_\bullet \overset{\mathsf{L}}{\otimes}_R K$. □

Acknowledgement

For several reasons at once, it is a great pleasure to dedicate this paper to Maxim Kontsevich. It was him who posed the original problem. He then actively encouraged my earlier work on it, and at the same time, was consistently unhappy about the unnecessary technical complications of [Ka2]. In retrospect, it was also him who showed a better way to approach the subject, although at the time, I didn't quite understand what he meant. Another mathematician to whom this paper owes a lot is Vadim Vologodsky. In particular, he always insisted that the notion of a conjugate spectral sequence omitted in [Ka2] should be an integral part of the story. Moreover, a crucial technical idea that appears in [Ka6, Section 6] is also due to him. I am also very grateful to David Kazhdan and Vladimir Hinich for their interest in this work, and for encouraging me to try to settle the question once and for all. Finally, in the years that passed since [Ka2], I had an opportunity to discuss the subject with many people; I want to specifically mention very useful conversations with Alexandre Beilinson, Bhargav Bhatt, Boris Feigin, and Boris Tsygan.

References

[Be] A. Beilinson, *On the crystalline period map*, Camb. J. Math. **1** (2013), 1–51, or a more recent version at `arXiv:1111.3316`.

[BBD] A. Beilinson, J. Bernstein, and P. Deligne, *Faisceaux Pervers*, Astérisque **100**, Soc. Math. de France, (1983).

[Bh] B. Bhatt, *p-adic derived de Rham cohomology*, arXiv:1204.6560.

[C] A. Connes, *Cohomologie cyclique et foncteur* Ext^n, Comptes Rendues Ac. Sci. Paris Sér. A-B, **296** (1983), 953–958.

[DI] P. Deligne and L. Illusie, *Relèvements modulo* p^2 *et décomposition du complexe de de Rham*, Inv. Math. **89** (1987), 247–270.

[G] A. Grothendieck, *Catégories fibrée et descente*, SGA I, Exposé VI, SMF, (2003).

[Ka1] D. Kaledin, *Non-commutative Cartier operator and Hodge-to-de Rham degeneration*, arXiv:math/0511665.

[Ka2] D. Kaledin, *Non-commutative Hodge-to-de Rham degeneration via the method of Deligne–Illusie*, Pure Appl. Math. Q. **4** (2008), 785–875.

[Ka3] D. Kaledin, *Trace functors and localization*, in Contemp. Math. **643** (2015).

[Ka4] D. Kaledin, *Cartier isomorphism for unital associative algebras*, Proc. Steklov Inst. Math. **290** (2015), no. 1, 35–51.

[Ka5] D. Kaledin, *Co-periodic cyclic homology*, arXiv:1509.08784.

[Ka6] D. Kaledin, *Bokstein homomorphism as a universal object*, arXiv:1510.06258.

[Ko1] M. Kontsevich, private communication, (2005).

[Ko2] M. Kontsevich, *XI Solomon Lefschetz Memorial Lecture Series: Hodge structures in non-commutative geometry*, in Contemp. Math. **462** (2008).

[KS] M. Kontsevich and Y. Soibelman, *Notes on A-infinity algebras, A-infinity categories and non-commutative geometry, I*, 153–219, Lecture Notes in Phys. **757**, Springer, Berlin, (2009).

[L] J.-L. Loday, *Cyclic Homology*, second ed., Springer, (1998).

[T] B. Toën, *Rings of definition of smooth and proper dg-algebras*, arXiv:math/0611546.

D. Kaledin
Steklov Mathematics Institute
8 Gubkina st.
Moscow 119991, Russia
e-mail: kaledin@mi.ras.ru

Progress in Mathematics, Vol. 324, 131–176

Derived Varieties of Complexes and Kostant's Theorem for $\mathfrak{gl}(m|n)$

M. Kapranov and S. Pimenov

To Maxim Kontsevich for his 50th birthday

Abstract. Given a graded vector space V, the variety of complexes Com(V) consists of all differentials making V into a cochain complex. This variety was first introduced by Buchsbaum and Eisenbud and later studied by Kempf, De Concini, Strickland and many other people. It is highly singular and can be seen as a proto-typical singular moduli space in algebraic geometry. We introduce a natural derived analog of Com(V) which is a smooth derived scheme RCom(V). It can be seen as the derived scheme classifying twisted complexes. We study the cohomology of the dg-algebra of regular functions on RCom(V). It turns out that the natural action of the group GL(V) (automorphisms of V as a graded space) on the cohomology has simple spectrum. This generalizes the known properties of Com(V) and the classical theorem of Kostant on the Lie algebra cohomology of upper triangular matrices.

Mathematics Subject Classification (2010). 17B56, 14M30, 58A50.

Keywords. Kostant's theorem, varieties of complexes, twisted complexes.

0. Introduction

A. The object of study. Let $\mathbf{k}$ be a field and $V^\bullet$ be a finite-dimensional graded $\mathbf{k}$-vector space. The *Buchsbaum–Eisenbud variety of complexes* [Ke] associated to $V^\bullet$ is the scheme formed by differentials D making $(V^\bullet, D)$ into a cochain complex. In other words, it is the closed subscheme

$$\mathrm{Com}(V^\bullet) = \left\{ D = (D_i) \in \prod_i \mathrm{Hom}(V^i, V^{i+1}) \;\middle|\; D_{i+1} \circ D_i = 0,\ \forall i \right\}.$$

The group $GL(V^\bullet) = \prod_i GL(V^i)$ acts on $\mathrm{Com}(V^\bullet)$ by automorphisms. The scheme $\mathrm{Com}(V^\bullet)$ has been studied in several papers [Ke] [DS] [MS] [La] [Go] [Cu] where, in particular, the following remarkable properties have been found:

(1) $\mathrm{Com}(V^\bullet)$ is reduced (i.e., is an affine algebraic variety), and each its irreducible component is normal.
(2) If $\mathrm{char}(\mathbf{k}) = 0$, then the coordinate algebra $\mathbf{k}[\mathrm{Com}(V^\bullet)]$ has, as a $GL(V^\bullet)$-module, simple spectrum: each irreducible representation appears at most once, and one has an explicit description of the sequences $[\alpha] = (\alpha^{(1)}, \dots, \alpha^{(n)})$ of dominant weights for the $GL(V^i)$ for which the corresponding representation does appear. More generally, for any $\mathbf{k}$ the algebra $\mathbf{k}[\mathrm{Com}(V^\bullet)]$ has a distinguished basis and a straightening rule.

By definition, $\mathrm{Com}(V^\bullet)$ is cut out, in the affine space $\prod_i \mathrm{Hom}(V^i, V^{i+1})$, by a system of homogeneous quadratic equations corresponding to the matrix elements of the compositions $D_{i+1} \circ D_i$. The complexity of $\mathrm{Com}(V^\bullet)$ stems from the fact that these equations are far from being "independent": we do not have a complete intersection. The goal of the present paper is to study a natural *derived extension* of $\mathrm{Com}(V^\bullet)$ which is a *derived scheme* $R\,\mathrm{Com}(V^\bullet)$, the spectrum of a certain commutative dg-algebra $A^\bullet(V^\bullet)$. The dg-algebra $A^\bullet(V^\bullet)$ is obtained by complementing the above quadratic equations by natural relations among them, then by natural syzygies and so on, see §1.1. Alternatively, we can look at $R\,\mathrm{Com}(V^\bullet)$, in terms of the functor it represents (on the category of dg-algebras). From this point of view, points of $R\,\mathrm{Com}(V^\bullet)$ correspond to *twisted complexes* [BK][Se1], see Proposition 1.1.3. We show that some of the classical properties of $\mathrm{Com}(V^\bullet)$ are extended to $R\,\mathrm{Com}(V^\bullet)$.

There are several additional reasons for interest in $R\,\mathrm{Com}(V^\bullet)$.

B. Derived moduli spaces. First, $R\,\mathrm{Com}(V^\bullet)$ can be seen as a typical example of a *derived moduli space*. The program of derived deformation theory initiated by V. Drinfeld [Dr] and M. Kontsevich [Ko1] [Ko2], proposes that any moduli space (i.e., moduli scheme or moduli stack) $\mathcal{M}$ appearing in algebraic geometry is really a shadow of a more fundamental derived object $R\mathcal{M}$ which is a derived scheme or derived stack, in particular, (local) functions on $R\mathcal{M}$ form a sheaf of commutative dg-algebras $(\mathcal{O}^\bullet_{R\mathcal{M}}, d)$ (in the case $\mathrm{char}(\mathbf{k}) > 0$ one should work with simplicial algebras). A consequence of this point of view for classical algebraic geometry is that $\mathcal{M}$ (any classical moduli space!) carries a natural "higher structure sheaf" which is the sheaf of graded commutative algebras $\mathcal{O}^\bullet_{\mathcal{M}} = H^\bullet_d(\mathcal{O}^\bullet_{R\mathcal{M}})$ whose degree 0 component is $\mathcal{O}_{\mathcal{M}}$.

The first step of this program, i.e., a construction of the $R\mathcal{M}$ in the first place, is by now largely accomplished, both in terms of defining the dg-algebras of functions [CFK1] as well as describing the functors, represented by the $R\mathcal{M}$ on appropriate (model or ∞-) categories [Lu] [TVe] [Pr1], see [Pr2] for comparison of different approaches.

However, the geometric significance (indeed the very nature) of the sheaves $\mathcal{O}^\bullet_\mathcal{M}$ remains mysterious. Only in the simplest case of *quasi-smooth* derived schemes (called $[0,1]$-manifolds in [CFK2]) we have a meaningful interpretation, suggested already in [Ko1]. In this case $\mathcal{O}^\bullet_\mathcal{M}$ is situated in finitely many degrees (as $\mathcal{O}^\bullet_{R\mathcal{M}}$ is essentially a Koszul complex), and the class

$$[\mathcal{M}]^K_{\mathrm{virt}} = \sum_i (-1)^i \operatorname{cl}(\mathcal{O}^i_\mathcal{M}) \in K_0(\mathcal{M})$$

is the K-theoretic analog of the virtual fundamental class of $\mathcal{M}$, see [CFK2] for a precise formulation and proof.

Many of the equations defining classical moduli spaces can be reduced to "zero curvature conditions", i.e., to some forms of the equation $D^2 = 0$ defining $\mathrm{Com}(V^\bullet)$ (see [Cu] for an example of such reduction). The derived schemes $R\,\mathrm{Com}(V^\bullet)$ provide, therefore, stepping stones for access to more complicated derived moduli spaces. In fact, a closely related object, the derived stack $\mathrm{Perf}_\mathbf{k}$ (of perfect complexes of $\mathbf{k}$-vector spaces), is used by B. Toen and M. Vaquié as an inductive step for constructing many such spaces in [TVa].

From this point of view, our results are suggestive of a general pattern that should hold for higher structure sheaves of many classical moduli spaces. That is, they suggest the following:

- For moduli spaces $\mathcal{M}$ related to higher-dimensional problems (i.e., beyond the quasi-smooth situation), the sheaf of rings $\mathcal{O}^\bullet_\mathcal{M}$ is typically not finitely generated over $\mathcal{O}_\mathcal{M} = \mathcal{O}^0_\mathcal{M}$. See simple examples in §1.1.
- Nevertheless, the individual graded components $\mathcal{O}^i_\mathcal{M}$ are finitely generated (coherent) over $\mathcal{O}_\mathcal{M}$, and one should be able to make sense of the statement that the generating series

$$F(t) = \sum_i t^i \operatorname{cl}(\mathcal{O}^i_\mathcal{M}) \in K_0(\mathcal{M})[[t]] \tag{0.1}$$

  is "rational". For $\mathcal{M} = \mathrm{Com}(V^\bullet)$ being the derived variety of complexes, we study, in Chapter 2, a similar generating series with K_0 being understood as the representation ring of $GL(V^\bullet)$.

C. Kostant's theorem. The dg-algebra $A^\bullet(V^\bullet)$ can be defined as the cochain algebra of a certain graded Lie superalgebra $\mathfrak{n}^\bullet(V^\bullet)$, see §1. If we contract the $\mathbb{Z}$-grading to the $\mathbb{Z}/2$-grading, then the Lie superalgebras $\mathfrak{n}^\bullet(V^\bullet)$ for all possible $V^\bullet$ are precisely the nilpotent radicals of all possible "Levi-even" parabolic subalgebras in the Lie superalgebras $\mathfrak{gl}(m|n)$ (we refer to [Mu] for a discussion of Borel and parabolic subalgebras in this setting). For a parabolic $\mathfrak{p}$ in $\mathfrak{gl}(n)$, indeed in any reductive Lie algebra, the cohomology of the nilpotent radical $\mathfrak{n} \subset \mathfrak{p}$ is found by the classical theorem of Kostant [Ko3]. In particular, this cohomology carries the action of the Levi subgroup M and each representation appears no more than once, in complete parallel with the property (2) above.

Our main result, Theorem 1.3.8 states that the cohomology of any $\mathfrak{n}^\bullet(V^\bullet)$ with constant coefficients has a simple spectrum, as a $GL(V^\bullet)$-module. Equivalently the cohomology of the coordinate ring of the derived variety of complexes has a simple spectrum. This statement can be seen as generalizing both Kostant's theorem and the properties of the varieties of complexes listed above.

The question of generalizing Kostant's theorem to $\mathfrak{gl}(m|n)$ and other reductive Lie superalgebras has attracted some attention [BS] [CKL] [CKW] [CZ] [Cou], with the results so far being restricted to some particular classes of parabolics but treating more general coefficients. Unlike the purely even case, the problem of finding the $\mathfrak{n}$-cohomology of arbitrary finite-dimensional irreducible modules L_μ of $\mathfrak{gl}(m|n)$ is very complicated. In fact, the generating functions for $GL(V^\bullet)$-isotypical components of $H^\bullet(\mathfrak{n}, L_\mu)$ are Kazhdan–Lusztig polynomials of type A (Brundan's conjecture [Br1] proved in [CLW] with some features interpreted earlier in [Br2][CWZ]). Our result for trivial coefficients $\mu = 0$ (and arbitrary parabolics) means that the corresponding KL polynomials are monomials with coefficient 1. We do not know if this can be seen from the geometry of Schubert varieties.

D. Structure of the paper and the method of the proof. Chapter 1 defines the derived varieties of complexes, and puts them into two frameworks: the representation-theoretic framework similar to that of [Ke] [DS] and the Lie superalgebra framework, where various degrees are mixed together into one $\mathbb{Z}/2$-parity.

In Chapter 2 we study the cohomology of $A^\bullet(V^\bullet)$, the coordinate dg-algebra of $R\,\mathrm{Com}(V^\bullet)$ (in the Lie superalgebra framework) at the level of Euler characteristics, proving that the Euler characteristic of each isotypic component of the cohomology is either 0 or ± 1. This is done for several reasons.

First, the evaluation of the Euler characteristic provides a logical intermediary step for our proof of Theorem 1.3.8.

Second, the proof at the level of generating functions is easy to understand conceptually. That is, in the case of the standard Borel subgroup, the Kostant cohomology with coefficients in $\mathbf{k}$ is easily found. This is based on the Cauchy formula

$$S^p(V \otimes W) \simeq \bigoplus_{|\alpha|=p} \Sigma^\alpha(V) \otimes \Sigma^\alpha(W) \qquad (0.2)$$

(simple spectrum decomposition) and the classical Kostant theorem with coefficients. This leads to a simple formula for the generating series as a rational function. The change of the Borel subgroup amounts to re-expansion of this function in different regions which is then analyzed in terms of adding formal series for the Dirac δ-function times a function of smaller number of variables. In this way we get an explicit formula for the coefficients of the generating series (Theorem 2.3.2) which, after Theorem 1.3.8 is established, becomes a complete description of the cohomology of $\mathfrak{n}$ (i.e., of the cohomology of the dg-algebra $A^\bullet(V^\bullet)$).

Third, the simple (rational) nature of generating functions here gives a reason to expect good rationality properties of the series (0.1) in more general algebro-geometric settings.

One can view our proof of Theorem 1.3.8 as a "categorification" of the generating function argument of Chapter 2. In particular, the role of the δ-function is played by the Lie superalgebra $\mathfrak{sl}(1|1)$ which appears naturally in the context of an "odd reflection". In fact, we use some rather detailed information about representations and cohomology of $\mathfrak{sl}(1|1)$. This information is recalled in Chapter 3. Another subject that is recalled is that of *mixed complexes*, the algebraic analog of topological spaces with circle action, and especially the analog of the concept of *equivariant formality* of [GKM]. This formality concept allows us, sometimes, to identify the classes, in the representation ring, of the cohomology of two equivariant complexes without these complexes being quasi-isomorphic.

The proof of Theorem 1.3.8 is given in Chapter 4. It is organized in a way parallel to Chapter 2 so that one can see how various simple steps involving generating functions are upgraded to spectral sequences, equivariant formality and so on.

E. Self-dual complexes and other classical superalgebras. An interesting version of the variety $\mathrm{Com}(V^\bullet)$ is obtained by considering *self-dual complexes* (also known as *algebraic Poincaré complexes*, see [Ra]). For this, we assume that $V^\bullet$ is equipped with a nondegenerate symmetric or antisymmetric bilinear form β of some degree n, so V^i is identified with $(V^{n-i})^*$. We can then consider the variety $\mathrm{Com}(V^\bullet, \beta)$ formed by systems of differentials $D = (D_i)$ such that $D_i = D^*_{n-1-i}$. This variety (and its natural derived analog) is similarly related to the Kostant cohomology (i.e., cohomology of nilpotent radical of parabolics) of other classical (symplectic, orthogonal) Lie superalgebras. It is natural to ask whether our statement about simple spectrum extend to the Kostant cohomology with coefficients in $\mathbf{k}$ for other simple Lie superalgebras. For the classical case it may be approached by a method similar to ours, starting with the well-known analogs of (0.2) which decompose $S^p(S^2(V))$ and $S^p(\Lambda^2(V))$ as $GL(V)$-modules with simple spectrum.

We would like to thank J. Brundan for looking at the preliminary version of this paper and pointing out the relation of our results to the Kazhdan–Lusztig theory of [Br1, Br2, CLW], carefully explaining some details. We also thank C. Stroppel for bringing some additional references to our attention and K. Coulembier for pointing out an incorrect formulation of Proposition 3.2.2 in an earlier version. In addition, we would like to thank the referee for many useful remarks. This work was supported by the World Premier International Research Center Initiative (WPI), MEXT, Japan and (M.K.) by the Max-Planck Institute für Mathematik, Bonn, Germany. The first named author would like to thank the MPI for hospitality, support and excellent working conditions.

1. Definitions and framework

1.1. The definition of $R\,\mathrm{Com}(V^\bullet)$ and initial examples

A. Main definitions. We assume $\mathrm{char}(\mathbf{k}) = 0$ and denote by $\mathcal{C}om_{\mathbf{k}}$ the category of cochain complexes $X^\bullet$ of $\mathbf{k}$-vector spaces with its standard symmetric monoidal structure, given by the usual tensor product of complexes and the Koszul sign rule for the symmetry

$$X^\bullet \otimes Y^\bullet \to Y^\bullet \otimes X^\bullet, \quad x \otimes y \mapsto (-1)^{\deg(x)\deg(y)} y \otimes x.$$

We denote by $\mathrm{Vect}_{\mathbf{k}}^{\mathbb{Z}} \subset \mathcal{C}om_{\mathbf{k}}$ the full symmetric monoidal subcategory formed by $\mathbb{Z}$-graded $\mathbf{k}$-vector spaces with zero differential.

By a *dg-category* we mean a category $\mathcal{C}$ enriched in $(\mathcal{C}om_{\mathbf{k}}, \otimes)$, so for each two objects $a, b \in \mathcal{C}$ we have a complex $\mathrm{Hom}^\bullet(a, b)$ with composition maps satisfying the graded Leibniz rule. In particular, a dg-category with one object pt is the same as an associative unital dg-algebra $R^\bullet = \mathrm{Hom}^\bullet(\mathrm{pt}, \mathrm{pt})$ over $\mathbf{k}$. We will be particularly interested in (graded) commutative, unital dg-algebras over $\mathbf{k}$ and denote by $\mathrm{dgAlg}_{\mathbf{k}}$ the category of such dg-algebras concentrated in degrees ≤ 0.

By a *graded Lie superalgebra* we will mean a Lie algebra in $\mathrm{Vect}_{\mathbf{k}}^{\mathbb{Z}}$ (with the symmetric monoidal structure defined above).

Fix a graded $\mathbf{k}$-vector space $V^\bullet$ as in the Introduction. Consider the graded vector space

$$\mathfrak{n}^\bullet = \mathfrak{n}^\bullet(V^\bullet) = \bigoplus_{i<j} \mathrm{Hom}(V^i, V^j)[i-j]$$

(i.e., $\mathrm{Hom}(V^i, V^j)$ is put in the degree $j - i > 0$). It is a graded associative algebra (without unit) and we consider it as a graded Lie superalgebra via the supercommutator

$$[x, y] = xy - (-1)^{\deg(x)\deg(y)} yx.$$

Let

$$A^\bullet = A^\bullet(V^\bullet) = C^\bullet_{\mathrm{Lie}}(\mathfrak{n}^\bullet) \;=\; (S^\bullet(\mathfrak{n}^*[-1]), d)$$

be the Chevalley–Eilenberg cochain complex of the graded Lie superalgebra $\mathfrak{n}^\bullet$. We consider it equipped with the total grading, combining the grading induced by that on $\mathfrak{n}^\bullet$ and the cohomological grading by the degree of cochains. With this grading $A^\bullet$ a graded commutative dg-algebra situated in degrees ≤ 0, i.e., it is an object of $\mathrm{dgAlg}_{\mathbf{k}}$, and

$$A^0 \;=\; \bigotimes_i S^\bullet(V^i \otimes (V^{i+1})^*) \;=\; \mathbf{k}\Big[\prod_i \mathrm{Hom}(V^i, V^{i+1})\Big]$$

is the coordinate algebra of the ambient affine space for $\mathrm{Com}(V^\bullet)$.

Definition 1.1.1. The derived variety of complexes is the dg-scheme

$$R\,\mathrm{Com}(V^\bullet) \;=\; \mathrm{Spec}(A^\bullet(V^\bullet)).$$

The name "derived variety of complexes" for $R\,\mathrm{Com}(V^\bullet)$ is justified by the following fact.

Proposition 1.1.2. (a) *The variety of complexes is identified with the scheme of Maurer–Cartan elements of* $\mathfrak{n}^\bullet(V^\bullet)$*:*

$$\mathrm{Com}(V^\bullet) \;=\; \{D \in \mathfrak{n}^1 \mid [D,D]=0\}.$$

(b) *We have an identification of commutative algebras*

$$\mathbf{k}[\mathrm{Com}(V^\bullet)] \;\simeq\; H^0(A^\bullet) \;=\; H^0_{\mathrm{Lie}}(\mathfrak{n}^\bullet). \qquad \square$$

Additionally, the name "derived variety of complexes" is justified by the following description of the functor represented by $A^\bullet(V^\bullet)$ on the category $\mathrm{dgAlg}_{\mathbf{k}}$.

Proposition 1.1.3. *For any* $(R, d_R) \in \mathrm{dgAlg}_{\mathbf{k}}$ *the set* $\mathrm{Hom}_{\mathrm{dgAlg}_{\mathbf{k}}}(A^\bullet(V^\bullet), R)$ *is naturally identified with the set of differentials* δ *in* $R \otimes V^\bullet$ *of total degree* $+1$ *satisfying the following properties:*

(1) *The Leibniz rule*

$$\delta(rm) = d_R(r)m + (-1)^{\deg(r)} r\delta(m), \quad r \in R, m \in R \otimes V^\bullet.$$

(2) *The filtration condition:*

$$\delta(R^i \otimes V^j) \;\subset\; \bigoplus_{i' \le i} R^{i'} \otimes V^{i+j+1-i'}.$$

(c) $(d_R \otimes 1 + \delta)^2 = 0.$ $\square$

Remarks 1.1.4. (a) Doubly graded complexes with the above properties are known as *twisted complexes*, cf. [BK][Se1]. The construction of $R\,\mathrm{Com}(V^\bullet)$ can now be said to consist in passing from usual complexes to twisted complexes.

(b) The bidegree (0,1) part of a differential δ above, gives rise to R-linear maps $\widetilde{D}_i : R \otimes V^i \to R \otimes V^{i+1}$. These maps commute with d and so induce $H^\bullet_d(R)$-linear maps $D_i : H^\bullet_d(R) \otimes V^i \to H^\bullet_d(R) \otimes V^{i+1}$. Further, the other components of δ yield the following:

- Each composition $D_{i+1}D_i = 0$, and therefore one can speak about the triple (matrix) Massey products $\langle D_{i+2}, D_{i+1}, D_i\rangle : H^\bullet_d(R) \otimes V^i \to H^\bullet_d(R) \otimes V^{i+2}$.
- Each $\langle D_{i+2}, D_{i+1}, D_i\rangle$ is equal to 0 modulo indeterminacy, and so one can speak about the quadruple Massey products $\langle D_{i+3}, D_{i+2}, D_{i+1}, D_i\rangle : H^\bullet_d(R) \otimes V^i \to H^\bullet_R \otimes V^{i+3}$, and so on, so that all the consecutive Massey products $\langle D_{i+p}, \dots, D_{i+2}, D_{i+1}, D_i\rangle$ are defined and equal to 0 modulo indeterminacy, see [Ma].

The cohomology of the dg-algebra $A^\bullet(V^\bullet)$ can be seen therefore as consisting of "residual scalar Massey products": its elements represent universal elements of $H^\bullet_d(R)$ which can be constructed whenever we have a datum exhibiting the existence and vanishing of all the matrix Massey products.

We recall, see, e.g., [CFK1] [TVe] that for any derived scheme X over $\mathbf{k}$ and any $\mathbf{k}$-point $x \in X$ we have the tangent dg-space $T_x^\bullet X$ which is a $\mathbb{Z}_{\geq 0}$-graded complex of $\mathbf{k}$-vector spaces defined canonically up to quasi-isomorphism.

We denote by Vect the abelian category of $\mathbf{k}$-vector spaces, by $\mathcal{C}(\mathrm{Vect})$ the abelian category of cochain complexes over Vect and by Ho(Vect) the homotopy category of $\mathcal{C}(\mathrm{Vect})$, which is the same as the derived category of Vect.

The following property of $R\,\mathrm{Com}(V^\bullet)$ is analogous to similar properties of other derived moduli schemes [Ka] [CFK1].

Proposition 1.1.5. *Let D be a $\mathbf{k}$-point of $\mathrm{Com}(V^\bullet)$, so that $C^\bullet = (V^\bullet, D)$ is a complex. Then*

$$H^i T_D^\bullet R\,\mathrm{Com}(V^\bullet) = \begin{cases} \mathrm{Hom}_{\mathcal{C}(\mathrm{Vect})}(C^\bullet, C^\bullet[1]), & \text{if } i = 0,\\ \mathrm{Hom}_{\mathrm{Ho(Vect)}}(C^\bullet, C^\bullet[i+1]), & \text{if } i > 0. \end{cases}$$

In the case $i = 0$ this reduces to the description of the tangent spaces to the scheme $\mathrm{Com}(V^\bullet)$ given in [Cu, Prop. 17].

Proof. By definition, $T_D^\bullet R\,\mathrm{Com}(V^\bullet) = \mathrm{Der}_{A^\bullet}(A^\bullet, \mathbf{k})$ is the dg-space of derivations $A^\bullet \to \mathbf{k}$ where $\mathbf{k}$ is considered as a dg-algebra over $A^\bullet$ via evaluation at D. Looking at the space $\bigoplus_{i<j} \mathrm{Hom}_k(V^i, V^j)^*$ of free generators defining $A^\bullet$, we identify, first as a vector space and then, more precisely, as a complex:

$$\mathrm{Der}_{A^\bullet}(A^\bullet, \mathbf{k}) \;=\; \bigoplus_{i<j} \mathrm{Hom}_k(V^i, V^j) \;=\; \mathrm{Hom}^{\geq 0}(C^\bullet, C^\bullet[1]).$$

Here the RHS is the degree ≥ 0 part of the Hom-complex between $C^\bullet$ and $C^\bullet[1]$. The cohomology of this part of the Hom-complex is precisely the RHS of the identification claimed in the proposition. □

The goal of this paper is the study of the cohomology algebra $H^\bullet(A^\bullet)$, i.e., of the Lie algebra cohomology of $\mathfrak{n}^\bullet$.

B. First examples. Given a $\mathbb{Z}_{\leq 0}$-graded commutative dg-algebra $R^\bullet$, we define

$$H^{\overline{0}}(R^\bullet) \;=\; \bigoplus_i H^{2i}(R^\bullet), \quad H^{\overline{1}}(R^\bullet) \;=\; \bigoplus_i H^{2i+1}(R^\bullet).$$

Thus $H^{\overline{0}}(R^\bullet)$ is a commutative algebra in the ordinary sense, and $H^{\overline{1}}(R^\bullet)$ is an $H^{\overline{0}}(R^\bullet)$-module. We have the following natural

Question 1.1.6. Suppose $R^\bullet$ is finitely generated.

(a) Is $H^{\overline{0}}(R^\bullet)$ finitely generated as an algebra? Is $H^{\overline{1}}(R^\bullet)$ finitely generated as a module?

(b) What is the algebro-geometric nature of the scheme $\mathrm{Spec}\, H^{\overline{0}}(R^\bullet)$ and of the quasi-coherent sheaf on this scheme corresponding to $H^{\overline{1}}(R^\bullet)$?

It turns out that the answer to the first part is no, and algebras $A^\bullet(V^\bullet)$ provide easy counterexamples. We have not seen such examples in literature.

Example 1.1.7. Suppose that $V^1 = V^2 = V^4 = \mathbf{k}$ and all the other $V^i = 0$. We denote by u, v, w the generators of the 1-dimensional vector spaces $\mathrm{Hom}(V^2, V^4)$, $\mathrm{Hom}(V^1, V^2)$ and $\mathrm{Hom}(V^1, V^4)$. The algebra $\mathfrak{n}$ is then an odd version of the Heisenberg Lie algebra, spanned by u, v, w:

$$\deg(u) = 2,\ \deg(v) = 1,\ \deg(w) = 3, \quad [u, v] = w,\ [u, v] = [u, w] = 0.$$

Let x, y, z be the generators of $A^\bullet$ dual to u, v, w, so

$$A^\bullet = \mathbf{k}[y, z] \otimes \Lambda[x], \quad d(z) = xy,\ d(x) = d(y) = 0,$$
$$\deg(x) = -1,\ \deg(y) = 0,\ \deg(z) = -2.$$

We find easily that the basis of $H^\bullet(A^\bullet)$ is formed by the classes of y^a, $z^b x$ with $a, b \geq 0$, with the product given by

$$y^a y^b = y^{a+b}, \quad y^a(z^b x) = 0, \quad (z^b x)(z^{b'} x) = 0.$$

That is, $H^{\overline{0}}(A^\bullet) = \mathbf{k}[y]$ is a polynomial algebra (finitely generated) but $H^{\overline{1}}(A^\bullet)$ is the infinite-dimensional square-zero ideal spanned by the $z^b x$, so it is not finitely generated as a module. In particular, the full cohomology algebra $H^\bullet(A^\bullet)$ is not finitely generated.

Example 1.1.8. Let $V^1 = V^2 = V^3 = V^4 = \mathbf{k}$, with all the other $V^i = 0$. Thus $A^\bullet$ is generated by

$$x_{ij},\ 1 \leq i < j \leq 4,\ \deg(x_{ij}) = j - i - 1,$$

with the differential

$$d(x_{i,i+1}) = 0,$$
$$d(x_{13}) = x_{12}x_{23},\ d(x_{24}) = x_{23}x_{34},$$
$$d(x_{14}) = x_{12}x_{24} - x_{13}x_{34}.$$

Separating the even and odd parts, we have

$$A^{\overline{0}} = \mathbf{k}[x_{12}, x_{23}, x_{34}, x_{14}] \ \oplus\ \mathbf{k}[x_{12}, x_{23}, x_{34}, x_{14}]x_{13}x_{24},$$
$$A^{\overline{1}} = \mathbf{k}[x_{12}, x_{23}, x_{34}, x_{14}]x_{13} \ \oplus\ \mathbf{k}[x_{12}, x_{23}, x_{34}, x_{14}]x_{24}.$$

We write an element of $A^{\overline{0}}$ as $f + gx_{13}x_{24}$ with $f, g \in \mathbf{k}[x_{12}, x_{23}, x_{34}, x_{14}]$. Then the differential $d\colon A^{\overline{0}} \to A^{\overline{1}}$ can be written, in the functional notation, as

$$\begin{aligned} d(f + gx_{13}x_{24}) &= (x_{12}x_{24} - x_{13}x_{34})\frac{\partial f}{\partial x_{14}} + gx_{12}x_{23}x_{24} - gx_{13}x_{23}x_{34} \\ &= \left(-x_{34}\frac{\partial f}{\partial x_{14}} - x_{23}x_{34}g\right)x_{13} + \left(x_{12}\frac{\partial f}{\partial x_{14}} + x_{12}x_{23}g\right)x_{24}. \end{aligned}$$

Similarly, we write an element of $A^{\bar{1}}$ as $\phi x_{13}+\psi x_{24}$, with $\phi,\psi\in\mathbf{k}[x_{12},x_{23},x_{34},x_{14}]$, and the differential $d\colon A^{\bar{1}}\to A^{\bar{0}}$ as

$$d(\phi x_{13}+\psi x_{24}) = (\phi x_{12}x_{23}+\psi x_{23}x_{34}) - \left(x_{12}\frac{\partial\phi}{\partial x_{14}} + x_{34}\frac{\partial\psi}{\partial x_{14}}\right)x_{13}x_{24}.$$

The cohomology ring of $A^\bullet$ is concentrated in even degrees and is infinitely generated as an algebra, with generators given by the classes of cocycles x_{12}, x_{23}, x_{34} and $e_i = x_{14}^i x_{23} - i x_{14}^{i-1}x_{13}x_{24}$, for $i\geq 1$, and relations $x_{12}x_{23} = x_{23}x_{34} = x_{12}e_i = x_{34}e_i = 0$, $e_ie_j = e_{i+j}x_{23}$.

1.2. Representation-theoretic framework

Note that the algebraic group $GL(V^\bullet) = \prod_i GL(V_i)$ acts on the dg-algebra $A^\bullet$ and thus on the dg-scheme $R\,\mathrm{Com}(V^\bullet)$. Let us introduce a representation-theoretic notation in order to analyze the action of $GL(V^\bullet)$ on $H^\bullet(A^\bullet)$.

First of all, we can and will assume that the only possible non-zero components of $V^\bullet$ are $V^1,\ldots,V^n$ for some $n\geq 1$. We denote $r_i = \dim(V^i)$ and by $r=\sum r_i$ the total dimension of $V^\bullet$. We also denote

$$r_{\bar{0}} = \sum_{i \text{ even}} r_i, \quad r_{\bar{1}} = \sum_{i \text{ odd}} r_i, \quad r_{\bar{0}} + r_{\bar{1}} = r.$$

We recall that an irreducible finite-dimensional representation of $GL(V^i)$ is given by a highest weight $\alpha^{(i)} = (\alpha_1^{(i)}\geq\cdots\alpha_{ri}^{(i)})$. We denote the representation corresponding to $\alpha^{(i)}$ by $\Sigma^{\alpha^{(i)}}(V^i)$ (the Schur functor).

An irreducible representation of $GL(V^\bullet) = \prod GL(V^{(i)})$ is then given by a sequence $[\alpha] = (\alpha^{(1)},\ldots,\alpha^{(n)})$ of dominant weights for the $GL(V^i)$. We denote by

$$\Sigma^{[\alpha]}(V^\bullet) \;=\; \bigotimes_{i=1}^{n}\Sigma^{\alpha^{(i)}}(V^i)$$

the irreducible representation corresponding to $[\alpha]$. For any algebraic representation U of $GL(V^\bullet)$ we denote by $U_{[\alpha]}\subset U$ the $[\alpha]$-isotypic component of U.

Note that we can regard a sequence $[\alpha]$ of weights for $GL(V^i) = GL(r_i)$ as a single weight for $GL(r)$. More precisely, for each integral weight $\beta = (\beta_1,\ldots,\beta_r)$ of $GL(r)$ we will denote by

$$[\beta] \;=\; ((\beta_1,\ldots,\beta_{r_1}),(\beta_{r_1+1},\ldots,\beta_{r_1+r_2}),\ldots,(\beta_{r_1+\cdots+r_{n-1}+1},\ldots,\beta_{r_1+\cdots+r_n}))$$

the corresponding sequence of weights of the $GL(r_i)$. We say that β is *block-dominant*, if each component of $[\beta]$ is a dominant weight for $GL(r_i)$.

In addition, for any dominant integral weight β for $GL(r)$ we denote by $\Sigma^\beta(V)$ the corresponding irreducible representation of $GL(r)$, i.e., the result of applying the Schur functor Σ^β to V, the total space of the graded vector space $V^\bullet$.

We will be interested in the following.

Question 1.2.1. Which $\Sigma^{[\alpha]}(V^\bullet)$ appear in $H^\bullet(A^\bullet)$ and with which multiplicity? In other words, what are the multiplicities of the isotypic components $H^i(A^\bullet)_{[\alpha]}$?

1.3. Lie superalgebra framework

Every graded Lie superalgebra gives, after contracting the $\mathbb{Z}$-grading to a $\mathbb{Z}/2$-grading, a Lie superalgebra in the usual sense. In particular, for each graded vector space $V^\bullet$ one can consider $\mathfrak{n} = \mathfrak{n}(V^\bullet)$ as a Lie superalgebra. Considered for all graded vector spaces $V^\bullet$, the Lie superalgebras $\mathfrak{n}(V^\bullet)$ are precisely the nilpotent radicals of all even parabolic subalgebras in all Lie superalgebras of type $\mathfrak{gl}$. We recall that a parabolic subalgebra is called *even*, if the corresponding Levi subgroup is a purely even (non super) reductive algebraic group. In the sequel all the parabolic subalgebras will be assumed even.

Therefore Question 1.2.1 can be regarded as a super-analog of the classical question studied by B. Kostant: determine the cohomology of the nilpotent radical of a parabolic subalgebra in a semi-simple (or reductive) Lie algebra as a module over the Levi.

However the setup of parabolic subalgebras is notationally quite different from that of graded vector spaces, because it is customary to consider different parabolic subalgebras in the same Lie superalgebra $\mathfrak{g}$, something that is not natural from the point of view of §1. Let us describe this second setup, referring to [Mu] for general background on Lie supergroups and superalgebras.

Fix $m_{\bar{0}}, m_{\bar{1}} \geq 0$ and let $m = m_{\bar{0}} + m_{\bar{1}}$. Consider the Lie superalgebra $\mathfrak{g} = \mathfrak{gl}(m_{\bar{0}}|m_{\bar{1}})$. A standard realization of $\mathfrak{g}$ is as the space of m by m matrices B viewed as 2 by 2 block matrices

$$B = \begin{pmatrix} B_{\bar{0}\bar{0}} & B_{\bar{0}\bar{1}} \\ B_{\bar{1}\bar{0}} & B_{\bar{1}\bar{1}} \end{pmatrix} \in \mathrm{Mat}(m \times m, \mathbf{k}), \quad B_{\bar{i}\bar{j}} \in \mathrm{Mat}(m_{\bar{i}} \times m_{\bar{j}}, \mathbf{k}).$$

Let $\mathfrak{b} \subset \mathfrak{g}$ be the standard Borel subalgebra formed by upper triangular matrices B. *Standard parabolic subalgebras* $\mathfrak{p} \supset \mathfrak{b}$ are in bijection with decompositions

$$m_{\bar{0}} = m_1 + \cdots + m_M, \quad m_{\bar{1}} = m_{M+1} + \cdots + m_{M+N}, \quad m_i > 0.$$

Such a decomposition realizes each $B \in \mathfrak{g}$ as a block matrix $\|B_{ij}\|$, $i, j = 1, \ldots, M+N$, and

$$\mathfrak{p} = \mathfrak{p}(m_1, \ldots, m_M|m_{M+1}, \ldots, m_{M+N}) \;=\; \{B \mid B_{ij} = 0 \text{ unless } i \leq j\}.$$

In addition, let $\pi \in S_{M+N}$ be any permutation. We associate to it the π-permuted parabolic subalgebra

$$\mathfrak{p}_\pi = \mathfrak{p}_\pi(m_1, \ldots, m_M|m_{M+1}, \ldots, m_{M+N}) \;=\; \{B \mid B_{ij} = 0 \text{ unless } \pi(i) \leq \pi(j)\}.$$

We denote by

$$\mathfrak{n}_\pi = \mathfrak{n}_\pi(m_1, \ldots, m_M|m_{M+1}, \ldots, m_{M+N})$$

the nilpotent radical of $\mathfrak{p}_\pi$. We will be interested in (even) parabolic subalgebras up to conjugation (in fact, up to isomorphism). Therefore it will be enough for us to restrict to the case when π is a (M, N)-shuffle as the case of a general π is reduced to this one by conjugation, cf. [Mu], §3.3. For convenience, we recall the concept of shuffles in general.

Definition 1.3.1. Let $c_1, \ldots, c_r \in \mathbb{Z}_{>0}$ and $c = c_1 + \cdots + c_r$. A permutation $s \in S_c$ is called a $(c_1, \ldots, c_r)$-*shuffle*, if, whenever

$$c_1 + \cdots + c_\nu + 1 \leq i < j \leq c_1 + \cdots + c_{\nu+1}, \quad \nu = 0, \ldots, r-1,$$

we have $s(i) < s(j)$.

We denote the set of $(c_1, \ldots, c_r)$-shuffles by $\mathrm{Sh}(c_1, \ldots, c_r) \subset S_c$. Alternatively, $\mathrm{Sh}(c_1, \ldots, c_r)$ can be defined as the set of minimal length representatives of right cosets of S_c by $S_{c_1} \times \cdots \times S_{c_r}$.

Example 1.3.2. The class of Borel subalgebras corresponds to the case when all $m_i = 1$, so $M = m_{\bar{0}}, N = m_{\bar{1}}$. For an (M, N)-shuffle π we denote the corresponding Borel subalgebra simply by $\mathfrak{b}_\pi$ and its nilpotent radical by $\mathfrak{n}_\pi$.

Consider now the setting of §1C, that is, we have a graded vector space $V^\bullet$. Let us assume that each $V^i = \mathbf{k}^{r_i}$ is a standard coordinate space and put

$$V_{\bar{0}} = \bigoplus V^{2i} = \mathbf{k}^{r_{\bar{0}}}, \quad V_{\bar{1}} = \bigoplus V^{2i+1} = \mathbf{k}^{r_{\bar{1}}}.$$

Then we have an embedding of vector spaces

$$\mathfrak{n} = \mathfrak{n}(V^\bullet) \; = \; \bigoplus_{i<j} \mathrm{Hom}(V^i, V^j) \; \subset \; \mathfrak{gl}(r_{\bar{0}} | r_{\bar{1}}). \tag{1.3.3}$$

Note that we can consider $\mathfrak{n}$ as a Lie superalgebra (again, by reducing the $\mathbb{Z}$-grading modulo 2). Then the embedding (1.3.3) is a morphism of Lie superalgebras.

Let $m_1, \ldots, m_M$ be all the nonzero r_{2i} written in the order of increasing i. That is,

$$m_i = r_{\nu_i}, \quad \nu_i \text{ even}, \, i = 1, \ldots, M.$$

Similarly, let $m_{M+1}, \ldots, m_{M+N}$ be all the nonzero r_{2i+1} in the order of increasing i. That is,

$$m_{M+j} = r_{\nu_{M+j}}, \quad \nu_{M+j} \text{ odd}, \, j = 1, \ldots, N.$$

In particular, we have exactly $M+N$ nonzero spaces V^i. We have then an (M, N)-shuffle permutation $\pi \in S_{M+N}$ which arranges the set $\{\nu_1, \ldots, \nu_{M+N}\}$ in the increasing order. With these notations, we have

Proposition 1.3.4. *The embedding* (1.3.3) *identifies* $\mathfrak{n}(V^\bullet)$ *with the nilpotent radical*

$$\mathfrak{n}_\pi(m_1, \ldots, m_M | m_{M+1}, \ldots, m_{M+N}). \qquad \square$$

Let us also denote by

$$\mathfrak{L}_\pi = \mathfrak{L}_\pi(m_1, \ldots, m_M | m_{M+1}, \ldots, m_{M+N}) \; = \; \prod_{i=1}^{M+N} GL(m_i)$$

the Levi subgroup corresponding to $\mathfrak{p}_\pi$. As usual, $\mathfrak{L}_\pi$ acts on $\mathfrak{n}_\pi$ by conjugation, and Question 1.2.1 is equivalent to the following analog of the question studied by Kostant.

Question 1.3.5. Which irreducible representations of $\mathfrak{L}_\pi$ appear in the Lie superalgebra cohomology $H^\bullet(\mathfrak{n}_\pi, \mathbf{k})$ and with which multiplicity?

Example 1.3.6 (Kostant's theorem). Suppose $m_{\bar{1}} = 0$, so $m = m_{\bar{0}}$, and therefore each (M, N)-shuffle is trivial. This corresponds to the fact that each parabolic subalgebra in $\mathfrak{gl}(m)$ is conjugate to a standard one $\mathfrak{p}(m_0, \ldots, m_M)$. We recall the classical result of Kostant which describes the action of $\mathfrak{L}(m_1, \ldots, m_M)$ on the cohomology of $\mathfrak{n}(m_1, \ldots, m_M))$ with coefficients in any irreducible representation of $GL(m)$.

Theorem 1.3.7. *Let $\mathfrak{L} = \mathfrak{L}(m_1, \ldots, m_M) = \prod_{i=1}^M GL(m_i)$ and $\mathfrak{n} = \mathfrak{n}(m_1, \ldots, m_M)$. Then for any integral dominant weight $\beta = (\beta_1 \geq \cdots \geq \beta_m)$ we have an identification of $\mathfrak{L}$-modules*

$$H^p(\mathfrak{n}, \Sigma^\beta(\mathbf{k}^m)) \simeq \bigoplus_{w \in \mathrm{Sh}(m_1, \ldots, m_M)} \Sigma^{[w(\beta+\rho_m)-\rho_m]}(\mathbf{k}^m).$$

Here $\rho_m = (m-1, m-2, \ldots, 0)$, and $\mathrm{Sh}(m_1, m_2, \ldots, m_M)$ is the set of $(m_1, m_2, \ldots, m_M)$-shuffles inside the symmetric group S_m. For each such shuffle the weight $w(\beta + \rho_r) - \rho_r$ is block-dominant, and $\Sigma^{[w(\beta+\rho_m)-\rho_m]}(\mathbf{k}^m)$ is the corresponding irreducible representation of $\prod_{i=1}^M GL(m_i)$. □

This has the following interpretation in terms of derived varieties of complexes. Suppose the degrees of all the nonzero components of $V^\bullet$ have the same parity. Let $m_1, \ldots, m_M$ be the dimensions of all the nonzero V^i in the order of increasing i. Then the graded Lie algebra $\mathfrak{n}(V^\bullet)$ is identified with the ungraded Lie algebra $\mathfrak{n}(m_1, \ldots, m_M)$ as a Lie superalgebra. Therefore we have an isomorphism of commutative $\mathbb{Z}/2$-graded differential algebras

$$A^\bullet \simeq C^\bullet_{\mathrm{Lie}}(\mathfrak{n}(m_1, \ldots, m_M), \mathbf{k}) = \Lambda^\bullet(\mathfrak{n}^*).$$

In particular, $A^\bullet$ is finite-dimensional. Further, Kostant's theorem (case $\beta = 0$) implies that in this case the action of $GL(V^\bullet)$ on $H^\bullet(A^\bullet)$ has simple spectrum.

Our main result is a generalization of this fact.

Theorem 1.3.8. (a) *For any graded vector space $V^\bullet$ the representation of $GL(V^\bullet)$ on $H^\bullet(A^\bullet)$ has simple spectrum.*

(b) *Equivalently, for any $m_{\bar{0}}, m_{\bar{1}} \geq 0$ and for any (even) parabolic subalgebra $\mathfrak{p} \subset \mathfrak{gl}(m_{\bar{0}}|m_{\bar{1}})$ with nilpotent radical $\mathfrak{n}$ and Levi subgroup $\mathfrak{L}$, the action of $\mathfrak{L}$ on $H^\bullet(\mathfrak{n}, \mathbf{k})$ has simple spectrum.*

2. Euler characteristic analysis

2.1. Formulation of the result and the nature of the generating function

As a first step towards proving Theorem 1.3.8, in this section we prove the following Euler characteristic analog.

Theorem 2.1.1.

(a) *For any graded vector space $V^\bullet$ and any sequence $[\alpha] = (\alpha^{(1)}, \ldots, \alpha^{(n)})$ of Young diagrams, the Euler characteristic of the graded vector space*

$$\mathrm{Hom}_{GL(V^\bullet)}(\Sigma^{[\alpha]}(V^\bullet), H^\bullet(A^\bullet))$$

is equal to 0 or ±1.

(b) *Equivalently, for any (even) parabolic subalgebra $\mathfrak{p} \subset \mathfrak{gl}(m_{\bar{0}}|m_{\bar{1}})$ and for any irreducible representation $\Sigma^{[\alpha]}$ of the corresponding Levi subgroup $\mathfrak{L}$, the Euler characteristic of the graded vector space*

$$\mathrm{Hom}_{\mathfrak{L}}(\Sigma^{[\alpha]}, H^\bullet(\mathfrak{n}, \mathbf{k}))$$

is equal to 0 *or* ± 1.

We will prove the formulation (b). Let

$$\mathfrak{p} = \mathfrak{p}_\pi = \mathfrak{p}_\pi(m_1, \dots, m_M | m_{M+1}, \dots, m_{M+N}),$$

where π is an (M, N)-shuffle and $\mathfrak{n} = \mathfrak{n}_\pi$ be its nilpotent radical. For each $i = 1, \dots, M+N$ we introduce a vector of variables $s_i = (s_{i1}, \dots, s_{i,m_i})$ which are the coordinates in the maximal torus in $GL(m_i)$ and write

$$s = (s_1, \dots, s_{M+N}) \;=\; ((s_{11}, \dots, s_{1,m_1}), \dots (s_{M+N,1}, \dots, s_{M+N,m_{M+N}})) \tag{2.1.2}$$

for the corresponding element of the maximal torus of $\mathfrak{L} = \prod_{i=1}^{M+N} GL(m_i)$, which we denote by T. Also we write $i \underset{\pi}{<} j$ for $\pi(i) < \pi(j)$. We will study the generating series for the (Euler characteristic) character of T on $H^\bullet(\mathfrak{n}, \mathbf{k})$:

$$F_\pi(s) = \sum_{\lambda \in \mathbb{Z}^m} \chi(H^\bullet(\mathfrak{n}_\pi, \mathbf{k})_\lambda) s^\lambda$$

Here λ can be seen as a character of the T and for any T-module U we denote by U_λ the λ-isotypic component of U.

For any Young diagram

$$\alpha = (\alpha_1 \geq \dots \geq \alpha_m) \in \mathbb{Z}^m$$

we denote by $\sigma_\alpha(z_1, \dots, z_m)$ the Schur polynomial corresponding to α. Because $\mathfrak{n}_\pi$ and $H^\bullet(\mathfrak{n}_\pi)$ are $\mathfrak{L}$-modules, we can write

$$F(s) = \sum_{[\alpha] = (\alpha^{(1)}, \dots, \alpha^{(M+N)})} c_{[\alpha]} \sigma_{\alpha^{(1)}}(s_1) \dots \sigma_{\alpha^{(M+N)}}(s_{M+N}),$$

$$c_{[\alpha]} = \chi\big(\mathrm{Hom}_{\mathfrak{L}}(\Sigma^{[\alpha]}, H^\bullet(\mathfrak{n}, \mathbf{k}))\big),$$

with the sum being over sequences $[\alpha]$ of Young diagrams.

Proposition 2.1.3. *The series $F_\pi(s)$ is an expansion of the rational function*

$$\Phi_\pi(s) = \frac{\displaystyle\prod_{\substack{i<j \\ i,j \leq M \text{ or } i,j > M}} \prod_{p=1}^{m_i} \prod_{q=1}^{m_j} \left(1 - \frac{s_{ip}}{s_{jq}}\right)}{\displaystyle\prod_{\substack{i \underset{\pi}{<} j \\ i \leq M < j \text{ or } j \leq M < i}} \prod_{p=1}^{m_i} \prod_{q=1}^{m_j} \left(1 - \frac{s_{ip}}{s_{jq}}\right)}$$

in the region

$$U_\pi \;=\; \{s \mid \; |s_{ip}| < |s_{jq}| \text{ whenever } \pi(i) < \pi(j)\}.$$

Proof. Since the Euler characteristic of a complex is equal to the Euler characteristic of its cohomology, we have

$$F_\pi(s) = \sum_{\lambda \in \mathbb{Z}^m} \chi(C^\bullet(\mathfrak{n}_\pi, \mathbf{k})_\lambda) s^\lambda$$

Now notice that

$$C^\bullet(n_\pi, \mathbf{k}) = S^\bullet(\mathfrak{n}_\pi^*[-1]) \;=\; S^\bullet((\mathfrak{n}_\pi^{\bar{0}})^*) \otimes \Lambda^\bullet((\mathfrak{n}_\pi^{\bar{1}})^*),$$

and

$$\mathfrak{n}_\pi^{\bar{0}} = \bigoplus_{\substack{i \underset{\pi}{<} j \\ i \le M < j \text{ or } j \le M < i}} \mathrm{Hom}(\mathbf{k}^{m_i}, \mathbf{k}^{m_j}), \quad \mathfrak{n}_\pi^{\bar{1}} = \bigoplus_{\substack{i < j \\ i,j \le M \text{ or } i,j > M}} \mathrm{Hom}(\mathbf{k}^{m_i}, \mathbf{k}^{m_j}).$$

Here $\mathbf{k}^{m_i}$ is the standard representation of $GL(m_i)$ considered a direct factor of $\mathfrak{L}$. Notice further that the character of $\mathrm{Hom}(\mathbf{k}^{m_i}, \mathbf{k}^{m_j})$ is $\sum_{p=1}^{m_i} \sum_{q=1}^{m_j} s_{jq}/s_{ip}$. Our statement now follows from the classical formulas for the characters of symmetric and exterior powers (summation of the geometric series). □

Our proof of Theorem 2.1.1 will be based on the following fact which will allow us to pass between different parabolic subalgebras of the same $\mathfrak{gl}(m_{\bar{0}}, m_{\bar{1}})$. It shows that in order to obtain the generating series $F_\pi(s)$ for all choices of parabolic subalgebra $\mathfrak{p}_\pi$ we need to understand how the expansion of Φ_π changes when we go from one region to another.

Proposition 2.1.4. (a) *For any two (M, N)-shuffles π, π' the ratio of the rational functions $\Phi_\pi(s)/\Phi_{\pi'}(s)$ is a Laurent monomial in s taken with coefficient ± 1.*
(b) *More precisely, let π and π' be two (M, N)-shuffles that differ only by one transposition of neighboring elements, say i and j, $i \le M < j$, so that $\pi(j) = \pi(i) + 1$ and therefore $i \underset{\pi}{<} j$ and $j \underset{\pi'}{<} i$. Then*

$$\Phi_{\pi'}(s) = (-1)^{m_i m_j} \frac{\prod_{p=1}^{m_i} s_{ip}^{m_j}}{\prod_{q=1}^{m_j} s_{jq}^{m_i}} \Phi_\pi(s).$$

Proof. The numerators of Φ_π and $\Phi_{\pi'}$ coincide. Further, all the factors in the denominators also coincide except for the factors $(1 - \frac{s_{ip}}{s_{jq}})$ in Φ_π being replaced by the factors $(1 - \frac{s_{jq}}{s_{ip}})$ in $\Phi_{\pi'}$. Our statement now follows by applying the identity

$$\left(1 - \frac{a}{b}\right) \Big/ \left(1 - \frac{b}{a}\right) \;=\; -\frac{a}{b}.$$

□

We now prove Theorem 2.1.1 in several steps in the rest of this section.

2.2. Standard Borel

Let $\mathfrak{p} = \mathfrak{b} = \mathfrak{b}_e$ be the standard Borel subalgebra of upper triangular matrices in $\mathfrak{gl}(m_{\bar{0}}, m_{\bar{1}})$. It corresponds to the $\pi = e$ being the identity permutation. In this case $M = r_{\bar{0}}$ and $N = r_{\bar{1}}$. To simplify notation we will write s_i instead of $s_{i,1}$. We have then

$$\Phi(s) = \Phi_e(s) = \frac{\prod\limits_{1\leq i<j\leq M}\left(1-\frac{s_i}{s_j}\right)\prod\limits_{M+1\leq i<j\leq M+N}\left(1-\frac{s_i}{s_j}\right)}{\prod\limits_{i\leq M<j}\left(1-\frac{s_i}{s_j}\right)}, \tag{2.2.1}$$

and $F(s)= F_e(s)$ is the expansion of this rational function in the region $|s_i| < |s_j|$ for $i < j$. The two products in the numerators are the Vandermonde determinants:

$$\prod_{1\leq i<j\leq M}\left(1-\frac{s_i}{s_j}\right) = \sum_{w\in S_M}(-1)^{|w|}s^{w(\rho_M^*)-\rho_M^*}, \quad \rho_M^* = (0,1,\ldots,M-1), \tag{2.2.2}$$

and similarly for the second product. Recall the Cauchy formula:

$$\prod_{i=1}^{a}\prod_{j=1}^{b}(1-u_i/v_j)^{-1} = \sum_\alpha \sigma_\alpha(u)\sigma_\alpha(v^{-1}), \quad |u_i| < |v_j|, \tag{2.2.3}$$

where the sum is taken over all partitions $\alpha = (\alpha_1 \geq \cdots \geq \alpha_l \geq 0)$, $l = \min(a, b)$. Recall also the Weyl character formula:

$$\sigma_\alpha(u_1,\ldots,u_a) \;=\; \det(s^{\alpha+\rho_a})/\det(s^{\rho_a}), \quad \rho_a = (a-1,\ldots,1,0), \tag{2.2.4}$$

where

$$\det(s^\beta) \;=\; \sum_{w\in S_a}(-1)^{|w|}s^{w(\beta)}.$$

(For the identities (2.2.2), (2.2.3) and (2.2.4), see [M], Ch. I, §3.) Combining this together we obtain:

Proposition 2.2.5. *The expansion of $\Phi(s)$ in the region $\{|s_i| < |s_j|$ for $i < j\}$ has the form*

$$F(s) = \sum_\alpha \sum_{\substack{w\in S_M\\ v\in S_N}} (-1)^{|w|+|v|}(-1)^{M(M-1)/2}(s_1,\ldots,s_M)^{w(\alpha+\rho_M)-\rho_M^*}$$
$$\cdot\,(s_{M+1}^{-1},\ldots,s_{M+N}^{-1})^{v(\alpha+\rho_N)-\rho_N},$$

where α runs over partitions $(\alpha_1 \geq \cdots \geq \alpha_l \geq 0)$, $l = \min(M, N)$. □

This is an explicit Laurent series with coefficients ± 1, and Theorem 2.1.1 for the standard Borel is proved.

2.3. Arbitrary Borel in $\mathfrak{gl}(N|N)$

Let us assume that $M = N$ and π is an arbitrary (N, N)-shuffle. We prove Theorem 2.1.1 for $\mathfrak{p} = \mathfrak{b}_\pi$.

A. Explicit formula for the generating series. In fact we prove a more precise statement identifying all the coefficients of the series $F_\pi(s)$.

Recall that the roots of $\mathfrak{g}$ have the form $\alpha_{ij} = e_i - e_j \in \mathbb{Z}^{2N}$ for $i, j \in \{1, \ldots, 2N\}$. A root α_{ij} is odd, if $i \leq M < j$ or $j \leq M < j$. Further, α_{ij} is positive for $\mathfrak{b}_\pi$, if $\pi(i) < \pi(j)$. We denote

$$\delta^\pi = \sum_{\substack{i \leq M < j \\ \pi(j) < \pi(i)}} \alpha_{ji}$$

the sum of odd roots which are positive for $\mathfrak{b}_\pi$ but negative for the standard Borel.

Let S be the set of bijections $\{1, \ldots, N\} \to \{N+1, \ldots, 2N\}$. First of all, for each $\phi \in S$ we define the sign factor $(-1)^{N(\phi,\pi)}$ as follows. Take $2N$ points $a_1, \ldots, a_{2N}$ on the unit circle in $\mathbb{R}^2$ in the clockwise order but in general position otherwise. Suppose $\phi \in S$ is given. For each $i = 1, \ldots, N$ we connect $a_{\pi(i)}$ with $a_{\pi\phi(i)}$ by an arc (straight line interval) in the disk. We define $N(\phi, \pi)$ be the number of pairwise intersection points of these arcs.

Next, for each $\phi \in S$ we define the (ϕ, π)-*sector* $\mathfrak{S}(\phi, \pi) \subset \mathbb{Z}^{2N}$ to consist of linear integral combinations

$$\sum_{\substack{i \in \{1,\ldots,N\} \\ \pi(i) < \pi\phi(i)}} a_i \alpha_{i,\phi(i)} + \sum_{\substack{i \in \{1,\ldots,N\} \\ \pi(i) > \pi\phi(i)}} b_i \alpha_{i,\phi(i)}, \quad a_i \geq 0,\, b_i < 0. \tag{2.3.1}$$

Note the strict inequalities for the b_i. Geometrically, $\mathfrak{S}(\phi, \pi)$ can be viewed as the set of integer points inside a coordinate orthant of an N-dimensional linear subspace in $\mathbb{Z}^{2N}$.

For $\lambda \in \mathbb{Z}^{2N}$ define

$$I_\phi^\pi(\lambda) = \begin{cases} (-1)^{N(\phi,\pi)}, & \text{if } \lambda \in \mathfrak{S}(\phi, \pi), \\ 0, & \text{otherwise.} \end{cases}$$

Theorem 2.3.2. *Let $F_\pi(s) = \sum_{\lambda \in \mathbb{Z}^{2N}} c_\lambda^\pi s^\lambda$ be the generating series for the Euler characteristics of the λ-isotypical components of $H^\bullet(\mathfrak{n}_\pi, \mathbf{k})$. Then:*

(a) *We have*

$$c_\lambda^\pi = \sum_{\phi \in S} I_\pi^\phi(\lambda - \delta^{\pi_1} + \delta^\pi),$$

where π_1 is the shuffle sending $1, 2, \ldots, 2N$ into $1, N+1, 2, N+2, \ldots, N, 2N$.

(b) *The sum in the RHS of* (a) *is always equal to* 0 *or* ± 1.

The remainder of this section will be devoted to the proof of Theorem 2.3.2. We proceed by induction on the length of π and on N.

B. Base of induction: standard Borel. We first consider the case $\pi = e$ and arbitrary N. In this case the statement is deduced from Proposition 2.2.5 as follows. By putting $\phi = wv^{-1}$ and $\beta = v(\alpha + \rho_N)$, we rewrite the RHS of the formula in the proposition as follows:

$$(-1)^{N(N-1)/2}(s_1, \ldots, s_N)^{-\rho_N^*} \cdot (s_{n+1}, \ldots, s_{2N})^{\rho_N} \cdot \sum_{\beta \in (\mathbb{Z}_+^N)_{\neq}} \sum_{\phi \in S} \operatorname{sgn}(\phi) \prod_{i=1}^{N} \left(\frac{s_i}{s_{\phi(i)}} \right)^{\beta_i}, \tag{2.3.3}$$

where $(\mathbb{Z}_+^N)_{\neq}$ is the set of integer vectors $(\beta_1, \ldots, \beta_N)$ such that $\beta_i \geq 0$ and $\beta_i \neq \beta_j$ for $i \neq j$.

Observe that extending the summation in (2.3.3) from $(\mathbb{Z}_+^N)_{\neq}$ to $(\mathbb{Z}_+^N)$ does not change the result. Indeed, if β is such that $\beta_i = \beta_j$ for some $i \neq j$, then the monomial $\prod_{i=1}^N \left(s_i/s_{\phi(i)}\right)^{\beta_i}$ is unchanged if we compose ϕ with any permutation preserving β. Therefore the sum of such monomials with coefficients $\operatorname{sgn}(\phi)$ will vanish.

Observe further that for $\pi = e$ the number $N(\phi, e)$ is the number of "orders" of ϕ, i.e., of pairs $(i < j)$ such that $\phi(i) < \phi(j)$ and so $(-1)^{N(\phi,e)} = (-1)^{N(N-1)/2} \cdot \operatorname{sgn}(\phi)$.

We now note that $\delta^e = 0$ and the series

$$\sum_{\lambda \in \mathbb{Z}^{2N}} \sum_{\phi \in S} I_\phi^e(\lambda - \delta^{\pi_1}) s^\lambda$$

is precisely the series in (2.3.3) in which the summation over β is extended to $\mathbb{Z}_+^N$.

This establishes the base of induction.

C. Inductive step: the two expansions. We now assume that Theorem 2.3.2 is proved for a given (N, N)-shuffle π and for all $(N-1, N-1)$-shuffles τ. Suppose $|\pi'| = |\pi| + 1$ and π' and π differ only by one transposition of neighboring elements, say i and j, as in Proposition 2.1.4(b). We show how to deduce Theorem 2.3.2 for π'.

We recall (Proposition 2.1.3) that $F_\pi(s) = F_\pi^N(s)$ is the expansion of the rational function

$$\Phi_\pi(s) = \Phi_\pi^N(s) = \frac{\prod\limits_{1 \leq i < j \leq N} \left(1 - \frac{s_i}{s_j}\right) \prod\limits_{N+1 \leq i < j \leq 2N} \left(1 - \frac{s_i}{s_j}\right)}{\prod\limits_{\substack{i \leq N < j \text{ or } j \leq N < i \\ \pi(i) < \pi(j)}} \left(1 - \frac{s_i}{s_j}\right)}$$

in the region

$$U_\pi = \left\{ |s_i| < |s_j| \text{ for } \pi(i) < \pi(j) \right\}.$$

By inductive assumption, $F_\pi(s)$ is given by the formula in Theorem 2.3.2(a).

Let $G_{\pi'}^N(s)$ be the Laurent expansion of Φ_π in the region $U_{\pi'}$. Because of Proposition 2.1.4, $G_{\pi'}^N$ is a product of $F_{\pi'}^N$ and a monomial with coefficient ± 1, so we start by analyzing the coefficients of $G_{\pi'}$.

The regions U_π and $U_{\pi'}$ are separated by one hyperplane $H = \{s_i = s_j\}$ on which our function $\Phi_\pi(s)$ has a first-order pole. To compare F_π and $G_{\pi'}$ we need therefore to study how the Laurent expansion changes after crossing H, i.e., after replacing the condition $|s_i/s_j| < 1$ to $|s_i/s_j| > 1$ while keeping the magnitudes of all the other ratios intact. For convenience of the reader we recall an elementary 1-dimensional situation of such change.

D. Laurent re-expansion via the δ-function. Let $\mathbb{C}[[z, z^{-1}]]$ be the vector space of formal Laurent series $\sum_{n=-\infty}^{\infty} a_n z^n$, $a_n \in \mathbb{C}$. We will use the series

$$\delta(z) = \sum_{n=-\infty}^{\infty} z^n \in \mathbb{C}[[z, z^{-1}]] \tag{2.3.4}$$

which is the Laurent expansion of the delta function at $z = 1$. Note that $\mathbb{C}[[z, z^{-1}]]$ contains the two fields $\mathbb{C}((z))$ and $\mathbb{C}((z^{-1}))$. For a rational function $\phi(z) \in \mathbb{C}(z)$ we denote by $\phi_0 \in \mathbb{C}((z))$ and $\phi_\infty \in \mathbb{C}((z^{-1}))$ its Laurent expansions near 0 and ∞. Then

$$\left(\frac{1}{1-z}\right)_0 - \left(\frac{1}{1-z}\right)_\infty = \left(\sum_{n=0}^{\infty} z^n\right) - \left(\sum_{n=-\infty}^{-1} -z^n\right) = \delta(z). \tag{2.3.5}$$

E. Laurent re-expansion of Φ_π^N. Returning to our situation, let us view Φ_π^N as a rational function on $\mathbb{C}^N \times \mathbb{C}^N$ with coordinates $s_1, \dots, s_{2N}$ and consider the space $\mathbb{C}^{N-1} \times \mathbb{C}^{N-1}$ with coordinates $s_p, p \neq i, j$. Let $\varpi : \mathbb{C}^N \times \mathbb{C}^N \to \mathbb{C}^{N-1} \times \mathbb{C}^{N-1}$ be the projection forgetting s_i and s_j. We can use ϖ to define pullback ϖ^* on rational functions as well as on formal Laurent series in the s_p. Let $\tau \in S_{2N-2}$ be the $(N-1, N-1)$-shuffle which arranges the set

$$\pi(\{1, \dots, 2N\} - \{i, j\}) = \pi'(\{1, \dots, 2N\} - \{i, j\})$$

in the increasing order.

Lemma 2.3.6. (a) *We have*

$$F_\pi^N - G_{\pi'}^N = \left(\varpi^* F_\tau^{N-1}\right) \cdot \delta(s_i/s_j).$$

(b) *We have* $F_{\pi'} = -(s_i/s_j) G_{\pi'}$.

Proof. Consider the diagram

$$\begin{array}{ccc} H & \xrightarrow{\ \epsilon\ } & \mathbb{C}^N \times \mathbb{C}^N \\ & {\scriptstyle \eta} \searrow & \downarrow {\scriptstyle \varpi} \\ & & \mathbb{C}^{N-1} \times \mathbb{C}^{N-1}, \end{array}$$

where ϵ is the embedding of H and η is the composite projection. By inspection of (2.2.1) we note that, first of all, $\Phi_\pi^N(s) = \Psi(s)\left(1 - \frac{s_i}{s_j}\right)^{-1}$, where Ψ does not have pole along H. Further, we note that the restriction $\epsilon^*\Psi = \Psi|_H$ does not depend on s_i or s_j, more precisely,

$$\epsilon^*\Psi = \eta^*\Phi^{N-1}.$$

Part (a) of our lemma follows from this and from (2.3.5). Part (b) is a particular case of Proposition 2.1.4 (b). □

We now define $\widetilde{F}_\pi(s) = \widetilde{F}_\pi^N(s) = \sum_\lambda c_\lambda^{\pi'} s^\lambda$, where c_λ^π is given by the formula of Theorem 2.3.2(a), and similarly for all other shuffles and other values of N. By Lemma 2.3.6 it is enough to establish the following.

Proposition 2.3.7. *We have an identity of formal series*

$$\widetilde{F}_\pi^N + (s_j/s_i)\widetilde{F}_{\pi'}^N = (\varpi^*\widetilde{F}_\tau^{N-1}) \cdot \delta(s_i/s_j). \tag{2.3.8}$$

Proof. We reformulate the required identity in terms of coefficients. The coefficient of the LHS of (2.3.8) at a given s^λ is equal to

$$\sum_{\phi\in S} \left(I_\pi^\phi(\lambda - \delta^{\pi_1} + \delta^\pi) + I_{\pi'}^\phi(\lambda - \alpha_{ji} - \delta^{\pi_1} + \delta^{\pi'})\right). \tag{2.3.9}$$

The coefficient at s^λ in the RHS of (2.3.8) is zero unless λ is of the form

$$\lambda = m\alpha_{ij} + \varpi^*\mu, \text{ where } m \in \mathbb{Z},\ \mu \in \mathbb{Z}^{N-1} \times \mathbb{Z}^{N-1}. \tag{2.3.10}$$

In this case the coefficient at s^λ in the RHS is given by

$$\sum_{\psi\in S'} I_\tau^\psi(\mu - \delta^{\tau_1} + \delta^\tau). \tag{2.3.11}$$

Here τ_1 is the $(N-1, N-1)$-shuffle defined similarly to π_1.

We now write the LHS of (2.3.8) as the sum of two summands:

$$\sum_{\phi\in S} \left(I_\pi^\phi(\lambda - \delta^{\pi_1} + \delta^\pi) + I_{\pi'}^\phi(\lambda - \alpha_{ji} - \delta^{\pi_1} + \delta^{\pi'})\right)s^\lambda = \sum_{\phi(i)=j} + \sum_{\phi(i)\neq j}$$

The proposition will follow from:

Lemma 2.3.12.

(a) *We have* $\sum_{\phi(i)=j} = (\varpi^*\widetilde{F}_\tau^{N-1}) \cdot \delta(s_i/s_j)$.

(b) *We also have* $\sum_{\phi(i)\neq j} = 0$.

Proof. (a) If $\phi(i) = j$, then exactly one of the two summands in (2.3.9) is nonzero. Indeed, looking at the definitions (2.3.1) of the (ϕ, π) sector $\mathfrak{S}(\phi, \pi)$ and the (ϕ, π')-sector $\mathfrak{S}(\phi, \pi')$, we see that α_{ij} can appear only in the first summand of the condition for belonging to $\mathfrak{S}(\phi, \pi)$ and in the second summand of the condition for belonging to $\mathfrak{S}(\phi, \pi')$. Therefore it lies in exactly one of the two sectors.

Let

$$\psi : \{1, \dots, N-1\} \xrightarrow{\text{mon}} \{1, \dots, N\} - \{i\} \xrightarrow{\phi} \{1, \dots, N\} - \{j\} \xrightarrow{\text{mon}} \{1, \dots, N-1\}$$

where "mon" stands for the unique monotone bijection.

Lemma 2.3.13. *The element* $\mu \in \mathbb{Z}^{N-1} \times \mathbb{Z}^{N-1}$ *defined by* (2.3.10)*, has* $I_\tau^\psi(\mu)$ *equal to the nonzero summand in* (2.3.9).

Proof of Lemma 2.3.13. Indeed, by assumption $\pi(j) = \pi(i)+1$ therefore the points a_i and a_j on the circle will be adjacent. Further, because $\phi(i) = j$, the points a_i and a_j will be joined by an arc, denote it $[a_i, a_j]$. Because a_i and a_j are adjacent, we can assume that $[a_i, a_j]$ does not meet any other arcs $[a_k, a_{\phi(k)}]$. This means that $N(\phi, \pi) = N(\psi, \tau)$. This proves Lemma 2.3.13.

Since τ is an $(N-1, N-1)$-shuffle, part (a) of Lemma 2.3.12 follows by the inductive assumption, in virtue of the form (2.3.11) of the coefficient at s^λ in the RHS of (2.3.8). □

We now prove part (b) of Lemma 2.3.12. More precisely, we claim that for $\phi(i) \neq j$, the two summands in (2.3.9) are either both zero or are the negatives of each other. Indeed, if $\phi(i) \neq j$, then the point a_i on the circle is the endpoint of some arc $[a_p, a_i]$, and a_j is the endpoint of some $[a_q, a_j]$. Interchanging i and j (to pass from π to π') will result in $[a_p, a_i]$ being redirected to a_j and $[a_q, a_j]$ redirected to a_i. This procedure will change the total intersection number of arcs by 1 modulo 2. Lemma 2.3.12 is proved, and therefore Theorem 2.3.2 is established. □□

2.4. Arbitrary Borel in $\mathfrak{gl}(M|N)$

We want to establish the following.

Proposition 2.4.1. *Suppose that Theorem* 2.1.1 *holds for all Borel subalgebras in all* $\mathfrak{gl}(M|M)$. *Then it holds for all Borel subalgebras in all* $\mathfrak{gl}(M|N)$.

Proof. By Proposition 2.1.4, validity of Theorem 2.1.1 for all Borel subalgebras in $\mathfrak{gl}(M|N)$ is equivalent to the statement that all Laurent expansions of the function $\Phi = \Phi^{M,N}$ from (2.2.1) have all nonzero coefficients ± 1. Suppose $M < N$ (the case $M > N$ is treated similarly) and we know this statement for $\Phi^{M+1,N}$. Let us deduce it for $\Phi^{M,N}$. Denote the extra variable in $\Phi^{M+1,N}$ by s_0. Then

$$\Phi^{M+1,N}(s_0, s_1, \ldots, s_{M+N}) = \frac{\prod_{i=1}^{M} (1 - s_0/s_i)}{\prod_{j=M+1}^{M+N} (1 - s_0/s_j)} \cdot \Phi^{M,N}(s_1, \ldots, s_{M+N}).$$

Let $F^{M,N}(s)$ be the Laurent expansion of $\Phi^{M,N}$ in some region U, and $F^{M+1,N}$ be the expansion of $\Phi^{M+1,N}$ in the region given by the same inequalities as U together with $|s_0| < |s_p|$, $p = 1, \ldots, M+N$. Then

$$F^{M+1,N} = F^{M,N} \cdot \prod_{i=1}^{M} \prod_{j=M+1}^{M+N} \left(1 + \frac{s_0}{s_j} + \frac{s_0^2}{s_j^2} + \cdots - \frac{s_0}{s_i} - \frac{s_0^2}{s_i s_j} - \frac{s_0^3}{s_i s_j^2} - \cdots\right).$$

This means that $F^{M,N}$ is the sum of all the monomials in $F^{M+1,N}$ which are independent of s_0. So if $F^{M+1,N}$ has all the nonzero coefficients ± 1, then so does $F_{M,N}$. □

2.5. Arbitrary parabolic in $\mathfrak{gl}(m_{\bar{0}}|m_{\bar{1}})$

We now consider the general situation of §2.1. That is, we are given decompositions

$$m_{\bar{0}} = m_1 + \cdots + m_M, \quad m_{\bar{1}} = m_{M+1} + \cdots + m_{M+N}, \qquad m_i \in \mathbb{Z}_{>0},$$

an $(M|N)$-shuffle π and denote by $\mathfrak{p}_\pi = \mathfrak{p}_\pi(m_1, \ldots, m_M|m_{M+1}, \ldots, m_{M+N})$ the corresponding parabolic subgroup and by $\mathfrak{n}_\pi$ its nilpotent radical. We refine π to an $(m_{\bar{0}}, m_{\bar{1}})$-shuffle Π by replacing each entry $\pi(i)$, $i = 1, \ldots, M+N$, with the set of integers in the interval $[m_1+\cdots+m_{i-1}+1,\ m_1+\cdots+m_i]$ taken in the increasing order. We denote by $\mathfrak{b}_\Pi$ the Borel subalgebra in $\mathfrak{gl}(m_{\bar{0}}, m_{\bar{1}})$ corresponding to Π and by $\mathfrak{n}_\Pi$ its nilpotent radical.

Let $F_\pi(s)$ and $F_\Pi(s)$ be the generating series of the Euler characteristic of T-isotypic components of $H^\bullet(\mathfrak{n}_\pi, \mathbf{k})$ and $H^\bullet(\mathfrak{n}_\Pi, \mathbf{k})$ respectively. By Proposition 2.1.3, these series can be viewed as Laurent expansions of the two rational functions $\Phi_\pi(s)$ and $\Phi_\Pi(s)$, in the two regions U_π, U_Π respectively. Note that after identification (2.1.2), $U_\Pi = \{|s_i| < |s_j| \text{ for } \Pi(i) < \Pi(j)\}$ becomes a subset of U_π, so we can understand both F_π and F_Π as the Laurent expansions of Φ_π and Φ_Π in the same region U_Π.

Now, comparing the definitions and identifying the two sets of variables as in (2.1.2), we see the equality of rational functions

$$\Phi_\Pi(s) \;=\; \Phi_\pi(s) \cdot \prod_{i=1}^{M+N} \prod_{1\le p<q\le m_i} \left(1 - \frac{s_{ip}}{s_{iq}}\right).$$

Since the second factor in the RHS is a Laurent polynomial, this implies the equality of Laurent series

$$F_\Pi(s) \;=\; F_\pi(s) \cdot \prod_{i=1}^{M+N} \prod_{1\le p<q\le m_i} \left(1 - \frac{s_{ip}}{s_{iq}}\right).$$

We now write, as in §2.1,

$$F_\pi(s) = \sum_{[\alpha]=(\alpha^{(1)},\ldots,\alpha^{(M+N)})} c_{[\alpha]} \sigma_{\alpha^{(1)}}(s_1) \ldots \sigma_{\alpha^{(M+N)}}(s_{M+N}), \quad c_{[\alpha]} \in \mathbb{Z}.$$

Note that for each i and each Young diagram $\alpha^{(i)}$ the product

$$P_{\alpha^{(i)}}(s_i) \;=\; \sigma_{\alpha^{(i)}}(s_i) \cdot \prod_{1\le p<q\le m_i} \left(1 - \frac{s_{ip}}{s_{iq}}\right)$$

is a Laurent polynomial with all nonzero coefficients equal to ± 1 (Weyl character formula). Further, for different sequences $[\alpha] = (\alpha^{(1)}, \ldots, \alpha^{(M+N)})$ the corresponding product polynomials

$$P_\alpha(s) = P_{\alpha^{(1)}}(s_1) \ldots P_{\alpha^{(M+N)}}(s_{M+N}) \tag{2.5.1}$$

have disjoint sets of nonzero monomials. Therefore, if some $c_{[\alpha]} \notin \{0, \pm 1\}$, then some coefficient of the Laurent series $F_\Pi(s)$ will be different from 0 or ± 1. But this is impossible in virtue of the results of §2.4. Theorem 2.1.1 is now completely proved.

To be more specific, let $\rho^*_{m_i} = (0, \ldots, 0, 1, \ldots, m_i - 1, 0, \ldots, 0)$ be the weights with non-zero components only in positions from $(m_1 + \cdots + m_{i-1} + 2)$ to $(m_1 + \cdots + m_i)$, for $1 \leq i \leq M + N$. Assume also that $m_{\bar{0}} > m_{\bar{1}}$ and complete the set of variables with $s_{m+1}, \ldots, s_{2m_{\bar{0}}}$ as in the section 2.4. Then we have the following

Corollary 2.5.2. *The generating function $F_\pi(s)$ for the Euler characteristic of the cohomology of $\mathfrak{n}_\pi$ has the following shape*

$$F_\pi(s) = \sum_{\substack{\alpha \in Z^{M+N} \\ \alpha_i = 0 \text{ if } i > M+N \\ \alpha \text{ - block-dominant}}} c^\pi_{[\alpha]} \sigma_{\alpha^{(1)}}(s_1) \ldots \sigma_{\alpha^{(M+N)}}(s_{M+N}),$$

where coefficients

$$c^\pi_{[\alpha]} = \sum_{\phi \in S} I^\phi_\Pi \left(\alpha - \delta^{\Pi_1} + \delta^\Pi + \sum_{i=1}^{M+N} \rho^*_{m_i} \right).$$

3. Recollections

3.1. Mixed complexes and formality

A. Topological motivation. Recall the notion of equivariant formality following [GKM]. Let X be a CW-complex endowed with an S^1-action. We write $H^\bullet_{S^1}(X) = H^\bullet(X \times_K ES^1, \mathbf{k})$ for the S^1-equivariant cohomology of X with coefficients in $\mathbf{k}$. The fibration $X \times_K ES^1 \to BS^1$ gives rise to the Serre spectral sequence

$$E_2^{pq} = H^p(BS^1, H^q(X)) \Rightarrow H^{p+q}_{S^1}(X). \tag{3.1.1}$$

Definition 3.1.2. We say that space X with the S^1-action is *equivariantly formal* if the spectral sequence (3.1.1) degenerates at E_2.

We recall from [GKM]:

Proposition 3.1.3. (a) *X is equivariantly formal if and only if $H^\bullet_{S^1}(X)$ is a free module over $H^\bullet(BS^1) = \mathbf{k}[u]$.*

(b) *If X is equivariantly formal, then $H^\bullet(X) \simeq H^\bullet_{S^1}(X)/uH^\bullet_{S^1}(X)$.*

B. Mixed complexes. A *mixed complex* is, by definition, a graded $\mathbf{k}$-vector space $C^\bullet$ equipped with two anti-commuting differentials d, δ of degrees $(+1)$ and (-1) respectively. See [L]. Alternatively, we may think of a mixed complex $C^\bullet$ as a graded module over the exterior algebra $\Lambda[d, \delta]$ on two generators: d of degree $+1$ and δ of degree -1.

Given two mixed complexes $C^\bullet$ and $D^\bullet$, the tensor product $C^\bullet \otimes_{\mathbf{k}} D^\bullet$ of graded vector spaces has a natural structure of a mixed complex, with d and δ extended by the Leibniz rule.

A mixed complex can be seen as a linear algebra analog of a space X with an S^1-action. More precisely, let X be a smooth manifold with a smooth action of S^1. Denote by θ the vector field on X which is the infinitesimal generator of the action. Then the space of invariant differential forms

$$(\Omega^\bullet(X)^{S^1}, d = d_{\mathrm{DR}}, \delta = i_\theta)$$

equipped with the de Rham differential and the contraction with θ, is a mixed complex.

Given a mixed complex $(C^\bullet, d, \delta)$, we form a spectral sequence which is an algebraic analog of (3.1.1):

$$\left(E_2^{pq} = H_d^q(C^\bullet)^p, d_2 = \delta\right) \Rightarrow \mathrm{HC}^{p+q}(C^\bullet). \tag{3.1.4}$$

Here $\mathrm{HC}^\bullet(C^\bullet)$, known as the *cyclic cohomology* of $C^\bullet$, is the cohomology of the total complex of the *Connes double complex* $\mathcal{B}C^{\bullet\bullet}$ defined by $(\mathcal{B}C)^{pq} = C^{q-p}$ with vertical differential given by d and the horizontal one given by δ. In fact, (3.1.4) is nothing but the standard spectral sequence $('E_n)$ of $\mathcal{B}C^{\bullet\bullet}$ but re-graded by inserting zeroes at all odd levels, so that $H_d^\bullet(C^\bullet)$ which normally constitutes E_1, appears as E_2, see [L]. This motivates the following.

Definition 3.1.5. A mixed complex $(C^\bullet, d, \delta)$ is called *formal*, if the spectral sequence of the double complex $\mathcal{B}C^{\bullet\bullet}$ degenerates at E_1 or, what is the same, if the spectral sequence (3.1.4) degenerates at E_2.

Note that for the trivial 1-dimensional mixed complex $\mathbf{k}$ (in degree 0) we have

$$\mathrm{HC}^\bullet(\mathbf{k}) = \mathbf{k}[u], \quad \deg(u) = 2. \tag{3.1.6}$$

Therefore for each mixed complex $C^\bullet$ we have that $\mathrm{HC}^\bullet(C^\bullet)$ is naturally a $\mathbf{k}[u]$-module. This action of u comes from its action on Connes' double complex, which is the identification $u : \mathcal{B}C^{pq} \to \mathcal{B}C^{p+1,q+1}$. The following is an algebraic analog of Proposition 3.1.3.

Proposition 3.1.7.

(a) *A mixed complex* $(C^\bullet, d, \delta)$ *is formal if and only if* $\mathrm{HC}^\bullet(C^\bullet)$ *is a free* $\mathbf{k}[u]$-*module.*

(b) *If a mixed complex* $(C^\bullet, d, \delta)$ *is formal, then* $H_d^\bullet(C^\bullet) = \mathrm{HC}^\bullet(C^\bullet)/u\,\mathrm{HC}^\bullet(C^\bullet)$.

(c) *If* $C^\bullet$ *and* $D^\bullet$ *are formal mixed complexes, then so is* $C^\bullet \otimes_{\mathbf{k}} D^\bullet$.

Proof. If $C^\bullet$ is formal, then we have $\mathrm{HC}^\bullet(C^\bullet) = H_d^\bullet(C^\bullet) \otimes \mathbf{k}[u]$ (tensor product of graded $\mathbf{k}$-vector spaces). This shows the "only if" part of (a) as well as (b). Let us now prove the "if" part of (a). Suppose $\mathrm{HC}^\bullet(C^\bullet)$ is $\mathbf{k}[u]$-free.

Let $a \in H_d^\bullet(C^\bullet)$ be an element of the lowest degree with the property that some higher differential d_r (with the lowest r), applied to a, is not trivial. Then $d_r(a) = bu^m$ for some m, with $b \in H_d^\bullet(C^\bullet)$. This implies that bu^m equals 0 in $\mathrm{HC}^\bullet(C^\bullet)$. Note further that because of our assumption, b descends to an element in $\mathrm{HC}^\bullet(C^\bullet)$ which we denote $\bar{b}$, and we have $\bar{b}u^n = 0$. Because $\mathrm{HC}^\bullet(C^\bullet)$ is $\mathbf{k}[u]$-free, it implies that $\bar{b} = 0$. But this means that some higher differential is nontrivial on b, which is impossible since it has degree smaller than a.

Finally let us show part (c). Notice that for any mixed complexes $C^\bullet$ and $D^\bullet$ we have an identification at the level of Connes' double complexes:

$$\mathcal{B}(C^\bullet \otimes_{\mathbf{k}} D^\bullet) \;\simeq\; \mathcal{B}(C^\bullet) \otimes_{\mathbf{k}[u]} \mathcal{B}^\bullet(D^\bullet).$$

This gives the spectral sequence

$$E_2 = \mathrm{Tor}_\bullet^{\mathbf{k}[u]}(\mathrm{HC}^\bullet(C^\bullet), \mathrm{HC}^\bullet(D^\bullet)) \;\Rightarrow\; \mathrm{HC}^\bullet(C^\bullet \otimes_{\mathbf{k}} D^\bullet).$$

If $C^\bullet$ and $D^\bullet$ are formal, then the higher Tors vanish and we have an identification at the level of cyclic cohomology:

$$\mathrm{HC}(C^\bullet \otimes_{\mathbf{k}} D^\bullet) \;\simeq\; \mathrm{HC}(C^\bullet) \otimes_{\mathbf{k}[u]} \mathrm{HC}^\bullet(D^\bullet),$$

which implies that the LHS is free and so $C^\bullet \otimes_{\mathbf{k}} D^\bullet$ is formal, by (a). □

Remark 3.1.8. One can also repackage (3.1.4) into a single graded spectral sequence, i.e., a sequence of single graded complexes

$$(E_r^\bullet, \lambda_r), \;\deg(\lambda_r) = 1 - 2r, \;\lambda_r^2 = 0, \quad E_{r+1}^\bullet = H^\bullet_{\lambda_r}(E_r^\bullet), \tag{3.1.9}$$

built from the same principles as (3.1.4) but without introducing several copies of the same space. Explicitly,

$$E_0^\bullet = C^\bullet, \;\lambda_0 = d, \;E_1^\bullet = H_d^\bullet(C^\bullet), \;\lambda_1 \text{ is induced by } \delta \text{ etc.}$$

Thus, $C^\bullet$ is formal if and only if all the λ_r, $r \geq 1$, are zero. Note that if some $E_r^\bullet$ is nonzero in only finite range of degrees, then (E_r) converges, and so we have a well-defined term $E_\infty^\bullet = E_\infty^\bullet(C^\bullet)$.

C. Indecomposable mixed complexes. Define morphisms of mixed complexes to be morphisms of graded $\Lambda[d, \delta]$-modules, i.e., to be morphisms of graded vector spaces, preserving the grading and commuting with the actions of d and δ. With this understanding, mixed complexes form an abelian category.

In particular, we can speak about direct sums, indecomposables etc. It turns out that classification of mixed complexes is a tame problem of linear algebra. (See [ASS] for generalities on the tame and wild representation types.)

Proposition 3.1.10. *Let $C^\bullet$ be an indecomposable mixed complex which is a union of finite-dimensional mixed subcomplexes. Then $C^\bullet$ is isomorphic to one of the following:*

(1) *The 4-dimensional "diamond" module D (the free rank 1 module over $\Lambda[d, \delta]$):*

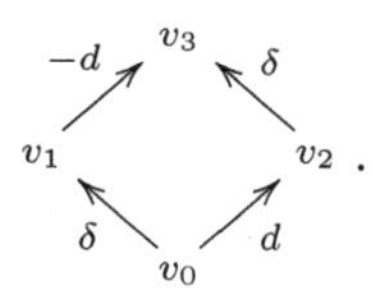

(2) *Four types of finite-dimensional "zig-zag" modules* $F_n(a,b)$, $n \geq 1$, *where a and b are the first and the last arrows respectively in the diagram*

$$v_1 \quad \cdots \xleftarrow{\delta} v_{i-1} \xrightarrow{d} v_i \xleftarrow{\delta} v_{i+1} \xrightarrow{d} \cdots \quad v_N\,,$$

and can be either d or δ. *The dimension of the module is* $N = 2n$ *if* $a = b$, *and* $N = 2n-1$ *if* $a \neq b$.

(3) *Five types of infinite-dimensional "zig-zag" modules: 2 unbounded on the left side* $I_l(d)$ *and* $I_l(\delta)$ *with the last arrow d and* δ *respectively, 2 unbounded on the right side* $I_r(d)$ *and* $I_r(\delta)$ *with the first arrow d and* δ, *and module I unbounded on both sides. We will denote by* v_0 *a generator of the vector space at the end of the "zig-zag".*

Proof. This is an easy exercise. It can also be deduced from a more general result about indecomposable modules over $\mathfrak{sl}(1|1)$ with grading liftable to a $\mathbb{Z}$-grading, see §3.2. □

We can describe formality of a mixed complex in terms of its indecomposable components.

Proposition 3.1.11. (a) *The mixed complexes* D, I, $F_n(d,d)$, $I_l(d)$ *and* $I_r(d)$ *are d-acyclic. Thus the spectral sequence* (3.1.4) *for any of these modules vanishes, and the modules are formal.*

(b) *The total d-cohomology space of each of the mixed complexes* $F_n(d,\delta)$ *and* $F_n(\delta,d)$ *is one-dimensional, generated be the class of* v_{2n} *and* v_1 *respectively. Hence all the differentials* λ_r, $r \geq 1$, *vanish and the mixed complexes are formal, with*

$$\mathrm{HC}^\bullet(F_n(d,\delta)) \simeq \mathbf{k}[u]\cdot[v_{2n}], \quad \mathrm{HC}^\bullet(F_n(\delta,d)) \simeq \mathbf{k}[u]\cdot[v_1].$$

(c) *The total d-cohomology of each of the mixed complexes* $I_l(\delta)$ *and* $I_r(\delta)$ *is one-dimensional, generated by the class of* v_0. *In particular, these mixed complexes are formal and* $\mathrm{HC}^\bullet(I_*(\delta)) \simeq \mathbf{k}[u]\cdot[v_0]$.

(d) *The total d-cohomology of the mixed complex* $F_n(\delta,\delta)$ *is two-dimensional, generated by the classes of* v_1 *and* v_{2n}. *The differential* $\lambda_i = 0$ *for* $i \neq n$, *while* $\lambda_n([v_{2n}]) = [v_1]$. *In particular,* $F_n(\delta,\delta)$ *is not formal, with* $\mathrm{HC}^\bullet(F_n(\delta,\delta)) \simeq \mathbf{k}[u]/(u^n)\cdot[v_1]$.

Proof. Follows by directly applying the definitions of the higher differentials in the spectral sequence of a double complex. More precisely, the rth differential in the spectral sequence of (d,δ) is obtained precisely by considering length $2r$ "zig-zag" patterns of elements under d and δ. □

Corollary 3.1.12. *Let* $C^\bullet$ *be mixed complex which is the union of finite-dimensional mixed subcomplexes. Then* $C^\bullet$ *is formal if and only if it does not contain indecomposable summands of type* $F_n(\delta,\delta)$. □

D. Spectral morphisms. Connes' double complex $\mathcal{B}C^{\bullet\bullet}$ of a mixed complex $C^\bullet$ can be viewed as the polynomial space $C^\bullet[u] = C^\bullet \otimes_{\mathbf{k}} \mathbf{k}[u]$ with the differentials $d = d\otimes 1$ and $u\delta = \delta \otimes u$. We will also consider the *periodized Connes complex* which is the space of Laurent polynomials $C^\bullet[u, u^{-1}]$ equipped with the differential $d+u\delta$. It is analogous to the construction of localized equivariant cohomology for spaces with S^1-action and to the construction of periodic cyclic cohomology of $\mathbf{k}$-algebras [L]. Note that the periodized Connes complex is a complex of $\mathbf{k}[u, u^{-1}]$-modules, and

$$H^\bullet_{d+u\delta} C^\bullet[u, u^{-1}] \;=\; \mathrm{HC}^\bullet(C^\bullet)[u^{-1}] := \mathrm{HC}^\bullet(C^\bullet) \otimes_{\mathbf{k}[u]} \mathbf{k}[u, u^{-1}].$$

Definition 3.1.13. Let $f : C_1^\bullet \to C_2^\bullet$ be a morphism of mixed complexes. We say that f is a *spectral morphism*, if the induced morphism of complexes of $\mathbf{k}[u, u^{-1}]$-modules

$$f_* : \big(C_1^\bullet[u, u^{-1}], d + u\delta\big) \longrightarrow \big(C_2^\bullet[u, u^{-1}], d + u\delta\big)$$

is a quasi-isomorphism.

Proposition 3.1.14. *Let C_1 and C_2 be formal mixed complexes having the total d-cohomology of finite dimension. Let $f : C_1^\bullet \to C_2^\bullet$ be a spectral morphism. Then*

$$\dim H^{\bar{0}}(C_1^\bullet) = \dim H^{\bar{0}}(C_2^\bullet), \quad \dim H^{\bar{1}}(C_1^\bullet) = \dim H^{\bar{1}}(C_2^\bullet).$$

More precisely, for each of these equalities, there are canonical filtrations in the vector spaces in the left- and right-hand sides and canonical isomorphisms between the quotients of these filtrations.

Note that the dimensions of individual cohomology spaces $H^i_d(C_1^\bullet)$ and $H^i_d(C_2^\bullet)$ may be different.

Proof of Proposition 3.1.14*:* Using Proposition 3.1.7(b), we reduce the statement to the following.

Lemma 3.1.15. *Let P, Q be free finite rank modules over $\mathbf{k}[u]$, and $\Phi\colon P \to Q$ be a morphism of modules which becomes an isomorphism after tensoring with $\mathbf{k}[u, u^{-1}]$. Then the fibers $P_0 = P/uP$ and $Q_0 = Q/uQ$ have natural filtrations such that the coefficients of the Taylor expansion of Φ near $u = 0$ define a canonical isomorphism between the associated graded of these two filtrations.*

Proof. We choose trivializations $P \simeq P_0 \otimes k[u]$ and $Q \simeq Q_0 \otimes \mathbf{k}[u]$, and then we can speak about the Taylor expansion, i.e., representation of Φ as a matrix polynomial $\Phi(u) = \sum_r \Phi_r u^r$, with $\Phi_r\colon P_0 \to Q_0$. The linear operators Φ_r do in general, depend on the choice of trivialization. However, $\Phi_0\colon P_0 \to Q_0$ is canonically defined as the map of fibers. Further, Φ_1 descends to a canonical map $\phi_1\colon \mathrm{Ker}(\Phi_0) \to \mathrm{Coker}(\Phi_0)$. Then, similarly, Φ_2 descends to a canonical map $\phi_2\colon \mathrm{Ker}(\phi_1) \to \mathrm{Coker}(\phi_1)$ and so on. These maps are canonical because "each Taylor coefficient is canonically defined modulo the previous ones". Alternatively, it is easy to verify the compatibility of the ϕ_r with changes of trivialization. Eventually, for some r, we must have $\mathrm{Ker}(\phi_r) = \mathrm{Coker}(\phi_r) = 0$ because $\Phi(u)$ becomes an isomorphism at $u \neq 0$. This means that $\phi_r\colon \operatorname{Ker}(\phi_{r-1}) \to \operatorname{Coker}(\phi_{r-1})$ is an isomorphism. Further, consider the filtrations

$$P_0 \supset \operatorname{Ker}(\Phi_0) \supset \operatorname{Ker}(\phi_1) \supset \cdots \supset \operatorname{Ker}(\phi_{r-1}),$$
$$Q_0 \supset I_{r-1} \supset I_{r-2} \supset \cdots \supset I_0, \quad I_\nu := \operatorname{Ker}(Q_0 \to \operatorname{Coker}(\phi_\nu)).$$

We see that the maps ϕ_ν define isomorphisms

$$\operatorname{Ker}(\phi_\nu)/\operatorname{Ker}(\phi_{\nu+1}) \longrightarrow I_{\nu+1}/I_\nu.$$

□

Remark 3.1.16. A classical topological situation motivating the concept of a spectral map is the localization theorem in S^1-equivariant cohomology [GKM]. Suppose X is an equivariantly formal finite CW-complex with an S^1-action, and $F = X^{S^1}$ is the fixed point locus. Then the restriction map $H^\bullet_{S^1}(X) \to H^\bullet_{S^1}(F)$ becomes an isomorphism after tensoring with $\mathbf{k}[u, u^{-1}]$ and the total dimensions of the usual cohomology spaces of X and F are equal, while the dimensions of the individual cohomology spaces may be different.

Corollary 3.1.17. *Let $C_1^\bullet, C_2^\bullet$ be formal mixed complexes equipped with an action of an algebraic torus T (commuting with the differentials), and $f\colon C_1^\bullet \to C_2^\bullet$ be a spectral morphism compatible with the action of T. Suppose that for each character χ of T the χ-isotypical components of $H_d^\bullet(C_i)$ are finite-dimensional. Then $H_d^{\bar 0}(C_1)$ is isomorphic to $H_d^{\bar 0}(C_2)$ as a T-module, and similarly for $H_d^{\bar 1}(C_1)$ and $H_d^{\bar 1}(C_2)$.* □

3.2. Representations and cohomology of $\mathfrak{sl}(1|1)$

The Lie superalgebra $\mathfrak{sl}(1|1)$ is interesting for us for two reasons. First, mixed complexes can be seen as particular representations of $\mathfrak{sl}(1|1)$. Second, we will use $\mathfrak{sl}(1|1)$ to perform "odd reflections" relating non-conjugate Borel subalgebras in $\mathfrak{gl}(m_{\bar 0}, m_{\bar 1})$.

A. Basics on $\mathfrak{sl}(1|1)$. We recall, see [FKV] for background, that the $\mathfrak{sl}(1|1)$ has the basis

$$H = \begin{pmatrix} 1 & 0 \\ 0 & 1 \end{pmatrix}, \quad Q_+ = \begin{pmatrix} 0 & 0 \\ 1 & 0 \end{pmatrix}, \quad Q_- = \begin{pmatrix} 0 & 1 \\ 0 & 0 \end{pmatrix}$$

with H being even, $Q_\pm$ being odd and with the commutation relations

$$[Q_+, Q_-] = H, \quad [H, Q_\pm] = 0.$$

In other words, $\mathfrak{sl}(1|1)$ is an odd Heisenberg algebra. Note that we have isomorphisms of associative superalgebras

$$U(\mathfrak{sl}(1|1))/(H) \simeq \Lambda[Q_+, Q_-] \simeq \Lambda[d, \delta], \tag{3.2.1}$$

where the RHS is the $\mathbb{Z}$-graded algebra whose graded modules are mixed complexes.

B. Indecomposables. We now recall the classification of indecomposable representations of $\mathfrak{sl}(1|1)$, adapting the results and notation of [Le], [Ge], [FKV] from the case of $\mathfrak{gl}(1|1)$. See also [GQS]. Call a representation of $\mathfrak{sl}(1|1)$ *liftable*, if its $\mathbb{Z}/2$-grading can be refined to a $\mathbb{Z}$-grading so that $\deg(Q_\pm) = \pm 1$. Note that a liftable representation can be extended to a representation of $\mathfrak{gl}(1|1)$, the matrix $\mathrm{diag}(1, -1)$ complementing $\mathfrak{sl}(1|1)$ to $\mathfrak{gl}(1|1)$, acting as the degree operator.

Proposition 3.2.2. *Let V be a liftable indecomposable finite-dimension representation of $\mathfrak{sl}(1|1)$. Then, up to possible change of parity, V is isomorphic to one of the following:*

(I) *The trivial 1-dimensional module* $\mathbf{k}$.

(II^+) *The 2-dimensional module* $v_1 \xrightarrow{Q_+} v_2$ *(with Q_- and H acting by 0).*

(II^-) *The 2-dimensional module* $v_1 \xleftarrow{Q_-} v_2$ *(with Q_+ and H acting by 0).*

(II_λ) *The 2-dimensional module* $v_1 \underset{Q_-}{\overset{Q_+}{\rightleftarrows}} v_2$, *with $Hv_i = \lambda v_i$, $i = 1, 2$. Here λ is an arbitrary non-zero element of* $\mathbf{k}$.

(III^0) *The 3-dimensional module* $v_1 \xleftarrow{Q_-} v_2 \xrightarrow{Q_+} v_3$, *with H acting by 0.*

(III) *The diamond module (with H acting by 0)*

$$\begin{array}{ccccc} & & v_3 & & \\ & {\scriptstyle -Q_+}\nearrow & & \nwarrow{\scriptstyle Q_-} & \\ v_1 & & & & v_2 \\ & {\scriptstyle Q_-}\nwarrow & & \nearrow{\scriptstyle Q_+} & \\ & & v_0 & & \end{array},$$

(Z) *Four types of zig-zag modules (with H acting by 0):*

$$Z^{2p} = \{\, v_1 \xleftarrow{Q_-} v_2 \xrightarrow{Q_+} \cdots \xleftarrow{Q_-} v_{2p-2} \xrightarrow{Q_+} v_{2p-1} \xleftarrow{Q_-} v_{2p} \,\},$$

$$\overline{Z}^{2p} = \{\, v_1 \xrightarrow{Q_+} v_2 \xleftarrow{Q_-} \cdots \xrightarrow{Q_+} v_{2p-2} \xleftarrow{Q_-} v_{2p-1} \xrightarrow{Q_+} v_{2p} \,\},$$

$$Z^{2p+1} = \{v_1 \xrightarrow{Q_+} v_2 \xleftarrow{Q_-} \cdots \xrightarrow{Q_+} v_{2p-2} \xleftarrow{Q_-} v_{2p-1} \xrightarrow{Q_+} v_{2p} \xleftarrow{Q_-} v_{2p+1}\},$$

$$\overline{Z}^{2p+1} = \{v_1 \xleftarrow{Q_-} v_2 \xrightarrow{Q_+} \cdots \xleftarrow{Q_-} v_{2p-2} \xrightarrow{Q_+} v_{2p-1} \xleftarrow{Q_-} v_{2p} \xrightarrow{Q_+} v_{2p+1}\}.$$

Out of these modules, the types (I)–(III) are cyclic, while the zig-zag modules are not cyclic except for the following identifications which exhaust all the isomorphisms of the modules in the list:

$$\mathrm{I} = Z^1 = \overline{Z}^1, \quad \mathrm{II}^+ = \overline{Z}^2, \quad \mathrm{II}^- = Z^2, \quad \mathrm{III}^0 = \overline{Z}^3.$$

C. Multiplication law of liftable cyclic modules. It follows from [FKV] that the tensor product of two liftable cyclic $\mathfrak{sl}(1|1)$-modules is again a direct sum of cyclic ones. More precisely, we have, adapting Prop. 2.3 of *loc. cit.* from the case of $\mathfrak{gl}(1|1)$:

Proposition 3.2.3. *Omitting the trivial identifications $I \otimes V = V$, the multiplication law of liftable cyclic $\mathfrak{sl}(1|1)$-modules is the following:*

$$\begin{aligned}
\mathrm{II}_\lambda \otimes \mathrm{II}_{\lambda'} &\simeq \begin{cases} \mathrm{II}_{\lambda+\lambda'} \oplus \mathrm{II}_{\lambda+\lambda'}, & \text{if } \lambda+\lambda' \neq 0, \\ \mathrm{III}, & \text{if } \lambda+\lambda' = 0; \end{cases} \\
\mathrm{II}_\lambda \otimes \mathrm{II}^\pm &\simeq 2 \cdot \mathrm{II}_\lambda; \\
\mathrm{II}^\pm \otimes \mathrm{II}^\pm &\simeq \mathrm{II}^\pm \oplus \mathrm{II}^\pm, \\
\mathrm{II}^+ \otimes \mathrm{II}^- &\simeq \mathrm{III}; \\
\mathrm{II}_\lambda \otimes \mathrm{III}^0 &\simeq 3 \cdot \mathrm{II}_\lambda; \\
\mathrm{II}^\pm \otimes \mathrm{III}^0 &\simeq \mathrm{II}^\pm \oplus \mathrm{III}; \\
\mathrm{III}^0 \otimes \mathrm{III}^0 &\simeq \mathrm{II}^+ \oplus \mathrm{III}^0 \oplus 2 \cdot \mathrm{I} \oplus \mathrm{II}^-; \\
\mathrm{II}_\lambda \otimes \mathrm{III} &\simeq 4 \cdot \mathrm{II}_\lambda; \\
\mathrm{II}^\pm \otimes \mathrm{III} &\simeq 2 \cdot \mathrm{III}; \\
\mathrm{III}^0 \otimes \mathrm{III} &\simeq 3 \cdot \mathrm{III}; \\
\mathrm{III} \otimes \mathrm{III} &\simeq 4 \cdot \mathrm{III}.
\end{aligned}$$

□

D. Cohomology with coefficients in cyclic modules. For any $\mathfrak{sl}(1|1)$-module V we denote by $H^i(\mathfrak{sl}(1|1), V)$ the ith cohomology of the complex of super-vector spaces

$$V \longrightarrow \mathfrak{sl}(1|1)^* \otimes V \longrightarrow \Lambda^2(\mathfrak{sl}(1|1)^*) \otimes V \longrightarrow \cdots$$

In particular it is itself a super-vector space. We denote by $x, y, t \in \mathfrak{sl}(1|1)^*$ the basis dual to Q_+, Q_-, H. According to the general rules, x, y become even generators of the Lie algebra cochain complex of $\mathfrak{sl}(1|1)$, while t becomes an odd generator.

Proposition 3.2.4. *We have the following identifications:*

(a) $H^\bullet(\mathfrak{sl}(1|1), \mathrm{I}) = \mathbf{k}[x, y]/(xy)$ *(as an algebra).*

(b) *For each* $\lambda \neq 0$ *we have* $H^\bullet(\mathfrak{sl}(1|1), \mathrm{II}_\lambda) = 0$.

(c) $H^i(\mathfrak{sl}(1|1), \mathrm{III}) = \begin{cases} \mathbf{k} \cdot [v_3] \text{ (1-dimensional space)}, & \text{if } i = 0, \\ \mathbf{k} \cdot [yv_1 + tv_3] \text{ (1-dimensional space)}, & \text{if } i = 1, \\ 0, & \text{if } i \geq 2. \end{cases}$

Proof. (a) By definition, $I = \mathbf{k}$ is the trivial 1-dimensional module, so

$$C^\bullet(\mathfrak{sl}(1|1), \mathrm{I}) = \mathbf{k}[x, y, t]$$

with x, y even, t odd and with differential $dt = xy$. So it isomorphic to the complex $\mathbf{k}[x, y] \xrightarrow{xy} \mathbf{k}[x, y]$ whence the statement.

(b) We consider $C^\bullet(\mathfrak{sl}(1|1), \mathrm{II}_\lambda)$ as a dg-module over the dg-algebra

$$C^\bullet(\mathfrak{sl}(1|1), \mathbf{k}) = \mathbf{k}[x, y, t].$$

This module is free of rank 2 if we forget the differentials. To prove that this module is acyclic, it is enough to assume $\mathbf{k}$ algebraically closed. Under this assumption, we can look at the fibers

$$F_p \ = \ C^\bullet(\mathfrak{sl}(1|1), \Pi_\lambda)) \otimes_{\mathbf{k}[x,y,t]} \mathbf{k}_p$$

of $C^\bullet(\mathfrak{sl}(1|1), \Pi_\lambda)$ at various $\mathbf{k}$-points p of Spec $H^\bullet(\mathfrak{sl}(1|1), \mathbf{k})$. Each such fiber F_p is a $\mathbb{Z}/2$-graded complex, i.e., a $\mathbb{Z}/2$-graded vector space with differential d of degree $\bar{1}$. The total dimension of F_p is twice the rank of the dg-module, i.e., is equal to 4.

In order to prove that $C^\bullet(\mathfrak{sl}(1|1), \Pi_\lambda)$ is acyclic, it is enough to prove that F_p is acyclic for each p. Now, p is a pair (x_0, y_0) with $x_0, y_0 \in \mathbf{k}$ and $x_0 y_0 = 0$. Let us consider the most degenerate situation $x_0 = y_0 = 0$ (the case when x_0 or y_0 is nonzero being similar). Assuming $p = (0,0)$, we find that F_p is identified with the $\mathbb{Z}/2$-graded complex

$$\left\{ \begin{array}{ccc} \mathbf{k}\cdot v_1 & \xrightarrow{d} & \mathbf{k}\cdot tv_1 \\ \mathbf{k}\cdot tv_2 & \xleftarrow[d]{} & \mathbf{k}\cdot v_2 \end{array} \right\}, \quad d(v_i) = \lambda t v_i, \ d(tv_i) = 0,$$

which is acyclic.

To prove (c), let U be the universal enveloping algebra of $\mathfrak{sl}(1|1)$, so that for any $\mathfrak{sl}(1|1)$-module M we have $H^\bullet(\mathfrak{sl}(1|1), M) = \operatorname{Ext}^\bullet_U(\mathbf{k}, M)$. Note that Ш is identified with $U/(H)$, see (3.2.1) and so has a 2-term resolution

$$0 \to U \xrightarrow{H} U \to \text{Ш} \to 0. \tag{3.2.5}$$

Lemma 3.2.6. *We have*

$$\dim H^i(\mathfrak{sl}(1|1), U) = \begin{cases} 1, & \text{if } i = 1, \\ 0, & \text{otherwise.} \end{cases}$$

Proof of the lemma: Consider the U-module $U[H^{-1}] = U \otimes_{k[H]} \mathbf{k}[H, H^{-1}]$. Considering $\mathbf{k}$ and $U[H^{-1}]$ as $\mathbf{k}[H]$-modules, we see that the groups $\operatorname{Ext}^i_{\mathbf{k}[H]}(\mathbf{k}, U[H^{-1}])$ vanish. Since U-homomorphisms can be identified with $\mathfrak{sl}(1|1)$-invariant subspace of $\mathbf{k}[H]$-homomorphisms, we have

$$H^\bullet(\mathfrak{sl}(1|1), U[H^{-1}]) \ = \ \operatorname{Ext}^\bullet_U(\mathbf{k}, U[H^{-1}]) = 0.$$

On the other hand, the quotient $U[H^{-1}]/U$ is identified, as an U-module, with the graded contragredient module $\operatorname{Hom}_{\mathbf{k}}(U, \mathbf{k})$, which implies that

$$\dim H^i(\mathfrak{sl}(1|1), U[H^{-1}]/U) = \begin{cases} 1, & \text{if } i = 0, \\ 0, & \text{otherwise.} \end{cases}$$

Now, the exact cohomology sequence of the short exact sequence

$$0 \to U \longrightarrow U[H^{-1}] \longrightarrow U[H^{-1}]/U \to 0$$

implies the lemma. □

We now prove part (c) of Proposition 3.2.4. The statement about H^0, i.e., about the space of $\mathfrak{sl}(1|1)$-invariants in Ш, follows at once from the definition of the diamond module in Proposition 3.2.2. Further, it is immediate to see that $yv_1 + tv_3$ is indeed a 1-cocycle of $\mathfrak{sl}(1|1)$ with coefficient in $\mathbb{III}$ and that it is not a coboundary. Using the exact cohomology sequence of the resolution (3.2.5), we conclude that there can be no other cohomology. □

4. Proof of the Simple Spectrum Theorem

This section is dedicated to the proof of the main theorem 1.3.8. The general outline of the proof is similar to the proof of the Euler characteristic analog 2.1.1. First we prove the statement for the standard Borel subalgebra, then via a sequence of reflections extend it to the case of an arbitrary Borel. And finally deduce the result about arbitrary parabolic subalgebras by relating it to the Borel case by a Hochschild–Serre spectral sequence.

4.1. Standard Borel

Let $\mathfrak{g}$ be a Lie superalgebra, and $I \subset \mathfrak{g}$ be an ideal. Then we have the Hochschild–Serre spectral sequence (cf. [Mu])

$$E_2^{pq} = H^p(\mathfrak{g}/I, H^q(I, \mathbf{k})) \Rightarrow H^{p+q}(\mathfrak{g}, \mathbf{k}). \tag{4.1.1}$$

We apply it to the case when $\mathfrak{g} = \mathfrak{n} = \mathfrak{n}_e$ is the nilpotent radical of the standard Borel subalgebra of upper triangular matrices in $\mathfrak{gl}(m_{\bar{0}}|m_{\bar{1}})$. We denote $V_{\bar{0}} = \mathbf{k}^{m_{\bar{0}}}$, $V_{\bar{1}} = \mathbf{k}^{m_{\bar{1}}}$. Further, let I, $\mathfrak{n}_1$, $\mathfrak{n}_2$ be the following subalgebras of $\mathfrak{n}$:

$$I = \left\{\begin{pmatrix} 0 & B_{\bar{0}\bar{1}} \\ 0 & 0 \end{pmatrix}\right\}, \quad \mathfrak{n}_1 = \left\{\begin{pmatrix} B_{\bar{0}\bar{0}} & 0 \\ 0 & 0 \end{pmatrix}\right\} \cap \mathfrak{n}, \quad \mathfrak{n}_2 = \left\{\begin{pmatrix} 0 & 0 \\ 0 & B_{\bar{1}\bar{1}} \end{pmatrix}\right\} \cap \mathfrak{n}.$$

Thus $\mathfrak{n}_1$ and $\mathfrak{n}_2$ are classical (non-graded) nilpotent Lie algebras, namely the nilpotent radicals of the standard Borels in $\mathfrak{gl}(m_{\bar{0}})$ and $\mathfrak{gl}(m_{\bar{1}})$. Further, I is an odd abelian ideal in $\mathfrak{n}$, identified, as an $\mathfrak{n}_1 \oplus \mathfrak{n}_2$-module, with $V_{\bar{0}} \otimes V_{\bar{1}}^*$. We may identify the quotient $\mathfrak{n}/I$ with the direct sum $\mathfrak{n}_1 \oplus \mathfrak{n}_2$. Since I is a purely odd abelian Lie superalgebra we have, applying the Cauchy formula, the following identification of $\mathfrak{n}_1 \oplus \mathfrak{n}_2$-modules:

$$H^q(I, \mathbf{k}) \simeq \Lambda^q(I^*) \simeq S^q(V_{\bar{0}}^* \otimes V_{\bar{1}}) \simeq \bigoplus_{\alpha} \Sigma^\alpha V_{\bar{0}}^* \otimes \Sigma^\alpha V_{\bar{1}}.$$

Here the sum is over all partitions α, similarly to (2.2.3) where we used the same formula at the level of characters.

Notice that $\mathfrak{n}_1$ acts only on the first factor in the last sum, and $\mathfrak{n}_2$ acts only on the second factor, so the cohomology groups in E_2 of the Hochschild–Serre spectral sequence can be written as

$$H^p(\mathfrak{g}/I, H^q(I, \mathbf{k})) = \bigoplus_{\alpha} \bigoplus_{i+j=p} H^i(\mathfrak{n}_1, \Sigma^\alpha V_{\bar{0}}^*) \otimes H^j(\mathfrak{n}_2, \Sigma^\alpha V_{\bar{1}}).$$

The classical Kostant theorem 1.3.7 tells us that the weights of $H^i(\mathfrak{n}_1, \Sigma^\alpha V_{\bar{0}}^*)$ are precisely the vertices of the permutahedron $w(\alpha + \rho_{m_{\bar{0}}}) - \rho_{m_{\bar{0}}}$, and similarly the weights of $H^j(\mathfrak{n}_1, \Sigma^\alpha V_{\bar{1}})$ are precisely the $w(\alpha + \rho_{m_{\bar{1}}}) - \rho_{m_{\bar{1}}}$. Observe that these sets are disjoint for different α. Now the differential d_r in the Hochschild–Serre spectral sequence for $r \geq 2$ goes from the row q to $q - r + 1$, but $q = |\alpha|$, so the spectral sequence degenerates at E_2. And the theorem follows.

4.2. Arbitrary Borel: outline of the proof

Suppose $M \geq N$. Let π be an (M, N)-shuffle, in this and the next sections we will establish Theorem 1.3.8 for $\mathfrak{p} = \mathfrak{b}_\pi$. We proceed by induction on the length of π. As before let π' be another shuffle that differs from π by one transposition of neighboring elements i and j, that is $\pi(j) = \pi(i) + 1$. Assume that theorem holds for $\mathfrak{b}_\pi$ and for all Borel subalgebras in $\mathfrak{gl}(M-1|N-1)$. We will show that it also holds for $\mathfrak{b}_{\pi'}$. Notice that the basis of induction, the case of $\mathfrak{gl}(M-N|0)$, follows from the classical Kostant theorem.

A. Main objects. We start by introducing the main objects that appear in the proof. Our situation is that of an "odd reflection", familiar in representation theory of simple Lie superalgebras, see [Se2], [Se3], [Mu, §3.5].

Consider the Lie superalgebra L generated by the nilpotent radicals $\mathfrak{n}_\pi$ and $\mathfrak{n}_{\pi'}$ in $\mathfrak{gl}(M|N)$. The structure of L can be understood from Figure 1. More precisely, this figure represents the conjugate subalgebra $\pi^{-1} L \pi$.

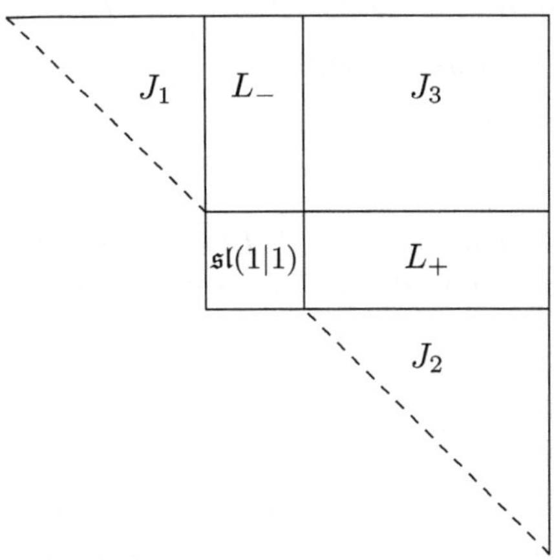

FIGURE 1. The structure of L.

The direct sum

$$J_1 \oplus J_2 \oplus J_3 \oplus L_+ \oplus L_- = \mathfrak{n}_\pi \cap \mathfrak{n}_{\pi'}$$

is an ideal in L. The center square $\mathfrak{sl}(1|1)$ is a subalgebra spanned by $e_{ij}, e_{ji}, h = [e_{ij}, e_{ji}]$. The whole Lie superalgebra L is a semidirect sum

$$L = \mathfrak{sl}(1|1) \ltimes (\mathfrak{n}_\pi \cap \mathfrak{n}_{\pi'}). \tag{4.2.1}$$

The subalgebra

$$J := J_1 \oplus J_2 \oplus J_3 = (\mathfrak{n}_\pi \cap \mathfrak{n}_{\pi'})^{\mathfrak{sl}(1|1)} \quad (4.2.2)$$

is the nilpotent radical of a Borel subalgebra in $\mathfrak{gl}(M-1|N-1)$. Note that J commutes with $\mathfrak{sl}(1|1)$, as indicated by (4.2.2).

We denote the basis of the dual space $\mathfrak{sl}(1|1)^*$ by

$$t = h^*, \ x = e_{ij}^*, \ y = e_{ji}^*,$$

so in the cochain complex $C^\bullet(\mathfrak{sl}(1|1))$ we have the relation $dt = xy$. Using the decomposition (4.2.1), we will view x, y, t as elements of the dual space L^*.

Proposition 4.2.3. *The relation $dt = xy$ holds also in $C^\bullet(L)$ as well.*

Proof. Since the quotient $L/(\mathfrak{n}_\pi \cap \mathfrak{n}_{\pi'})$ can be identified with $\mathfrak{sl}(1|1)$, we have a map of cochain complexes $C^\bullet(\mathfrak{sl}(1|1)) \to C^\bullet(L)$, so that $dt = xy$, holding in $C^\bullet(\mathfrak{sl}(1|1))$, holds in $C^\bullet(L)$ as well. □

B. Main steps of the proof. We now present the list of main results that lead to the proof. These results will be proved one by one later, unless the proof is immediate and presented right away.

Recall that $C^\bullet(L)$ is a commutative dg-algebra, and we denote by (t, dt) the (dg-)ideal in this algebra generated by t and $dt = xy$.

Proposition 4.2.4. *The natural projection*

$$q\colon C^\bullet(L) \longrightarrow C^\bullet(L)/(t, dt)$$

is a quasi-isomorphism.

Proposition 4.2.5. *The restriction map $C^\bullet(L) \to C^\bullet(\mathfrak{n}_\pi)$ induced by the embedding $\mathfrak{n}_\pi \hookrightarrow L$, annihilates the ideal (t, dt) and gives a short exact sequence of complexes*

$$0 \to C^\bullet(\mathfrak{n}_{\pi'})y \to C^\bullet(L)/(t, dt) \to C^\bullet(\mathfrak{n}_\pi) \to 0.$$

Proof. This follows immediately by inspecting the Lie superalgebras involved and their cochain complexes. □

Propositions 4.2.4 and 4.2.5 show that the Lie superalgebra L can be seen as "interpolating" between $\mathfrak{n}_\pi$ and $\mathfrak{n}_{\pi'}$, as far as the cohomology is concerned.

Recall that we are studying the *spectrum* of various cochain and cohomology spaces, i.e., their decomposition under the action of $T = \mathbb{G}_m^{M+N}$, the maximal torus in $GL(M|N)$. It will be useful to focus in particular on the weight decomposition with respect to $h \in \mathrm{Lie}(T)$. That is, for any T-module V and any $c \in \mathbb{Z}$ we denote

$$V^{h=c} := \bigoplus_{\lambda \in T^\vee,\ \lambda(h)=c} V_\lambda$$

the space of h-weight c. We write $V^h = V^{h=0}$ for the subspace of h-weight 0, i.e., for the kernel of h.

Proposition 4.2.6. *The embedding $C^\bullet(L)^h \hookrightarrow C^\bullet(L)$ is a quasi-isomorphism.*

Consider now the restriction map from the chain complex of L to that of $\mathfrak{sl}(1|1) \oplus J$, and denote by

$$\rho\colon C^\bullet(L)^h \longrightarrow C^\bullet(\mathfrak{sl}(1|1) \oplus J)^h = C^\bullet(\mathfrak{sl}(1|1) \oplus J) \tag{4.2.7}$$

the induced map on the h-weight 0 parts.

Proposition 4.2.8. *The differential $\delta = i_h$ of degree -1 given by the contraction with h, makes (together with the cochain differential d), both $C^\bullet(L)^h$ and $C^\bullet(\mathfrak{sl}(1|1) \oplus J)$ into mixed complexes, and ρ is a morphism of mixed complexes.*

Proof. In $C^\bullet(L)$, as in the cochain complex of any Lie algebra, we have the Cartan formula

$$[d, i_h] \;=\; \mathrm{Lie}_h, \tag{4.2.9}$$

where Lie_h is the operator induced by the coadjoint action of h on L^*. The proof of (4.2.9) is the same as of the classical Cartan formula for the Lie derivative on differential forms. That is, both sides of (4.2.9) are (super-)derivations of $\Lambda^\bullet(L^*)$, a free commutative superalgebra, which coincide (with the coadjoint action of h) on the space L^* of generators.

So in the h-weight 0 part we have $[d, \delta] = [d, i_h] = 0$. □

Proposition 4.2.10. *Both $C^\bullet(L)^h$ and $C^\bullet(\mathfrak{sl}(1|1) \oplus J)$ are formal mixed complexes, and ρ is a spectral morphism. Therefore (Corollary 3.1.17), we have an isomorphism of T-modules*

$$\begin{aligned} H^\bullet(L) \;=\; H^\bullet(L)^h \;&\simeq\; H^\bullet(\mathfrak{sl}(1|1) \oplus J) \\ =\; H^\bullet(\mathfrak{sl}(1|1)) \otimes_{\mathbf{k}} H^\bullet(J) \;&=\; \mathbf{k}[x, y]/(xy) \otimes_{\mathbf{k}} H^\bullet(J). \end{aligned}$$

Remark 4.2.11. The previous constructions and results can be seen as "materialization" of the argument in §2.3E, more precisely, of Lemma 2.3.6 which compares the Laurent expansions of the rational function Φ_π^N in two adjacent regions in terms of a function of a smaller number of variables and a δ-function. More precisely, the cohomology spaces $H^\bullet(\mathfrak{sl}(1|1)) = \mathbf{k}[x, y]/(xy)$ "materializes" the δ-function, since the T-weights of this cohomology form an arithmetic progression infinite on both sides. The Lie superalgebra J, being the nilpotent radical of a Borel in $\mathfrak{gl}(M-1|N-1)$, gives, as the generating series of $H^\bullet(J)$, an expansion of a rational function of smaller number of variables: the analog of the function F_τ^{N-1} from Lemma 2.3.6. The equality in part (a) of Lemma 2.3.6 is now upgraded to the short exact sequence of Proposition 4.2.5 combined with the quasi-isomorphism of Proposition 4.2.4 and the isomorphism of T-modules of Proposition 4.2.10.

We now refine Proposition 4.2.5 as follows.

Proposition 4.2.12. *The long exact sequence of cohomology associated to the short exact sequence of complexes from Proposition 4.2.5 gives:*

(a) *On the part with h-weight $c \neq 0$, isomorphisms of T-modules*

$$H^\bullet(\mathfrak{n}_\pi)^{h=c} \simeq (H^\bullet(\mathfrak{n}_{\pi'}) \cdot y)^{h=c}.$$

(b) *On the h-weight 0 part, short exact sequences of complexes*

$$0 \to H^{p-1}(\mathfrak{n}_{\pi'})^h \xrightarrow{\cdot y} H^p(L) \longrightarrow H^p(\mathfrak{n}_\pi)^h \to 0.$$

(*Note that $H^\bullet(L) = H^\bullet(L)^h$ by Proposition* 4.2.6.)

C. Completion of the proof modulo the main steps. Assuming results of n°B, the proof of Theorem 1.3.8 for $\mathfrak{p} = \mathfrak{b}_{\pi'}$ is completed as follows.

By inductive assumption, $H^\bullet(\mathfrak{n}_\pi)$ has simple spectrum. So the isomorphisms of Proposition 4.2.12(a) imply that the spectrum of the part of $H^\bullet(\mathfrak{n}_{\pi'})$ with nonzero h-weight is also simple.

Let us concentrate on the part of zero h-weight. Recall that J is the nilpotent radical of a Borel in $\mathfrak{gl}(M-1|N-1)$, so by our inductive assumption, $H^\bullet(J)$ has simple spectrum. Denote by $\alpha_x, \alpha_y \in T^\vee$ the T-weights of x and y. Then $\alpha_x = -\alpha_y$ and the subgroup $\mathbb{Z}\cdot\alpha_x \subset T^\vee$ is linearly independent with the subgroup of weights lifted from the maximal torus in $GL(M-1|N-1)$. Therefore the isomorphism of Proposition 4.2.10 implies that $H^\bullet(L)$ has simple spectrum. Now, the short exact sequences from Proposition 4.2.12(b) imply that $H^\bullet(\mathfrak{n}_{\pi'})^h$ has simple spectrum, since it is a T-submodule of $H^\bullet(L)$. □

4.3. Arbitrary Borel: end of proof

A. Proof of Proposition 4.2.4. Let $I = (t, dt) \subset C^\bullet(L)$. Here we prove that $q\colon C^\bullet(L) \to C^\bullet(L)/I$ is a quasi-isomorphism.

Consider the filtration of the graded algebra $C^\bullet(L)$ by the powers of the ideal I. The associated graded algebra $\mathrm{gr}_I C^\bullet(L)$ is isomorphic to the tensor product $C^\bullet(L)/I \otimes S^\bullet(A)$, where A is the acyclic complex $\{t \to dt\}$. All symmetric powers of A are also acyclic, hence the quotients of the filtration I^p/I^{p+1}, for $p \geq 1$ are acyclic as well. This implies that the map in question is indeed a quasi-isomorphism.

B. Proof of Proposition 4.2.6. Here we prove that the embedding $C^\bullet(L)^h \hookrightarrow C^\bullet(L)$ is a quasi-isomorphism. We start with a general reminder.

Let $\mathfrak{g}$ be a Lie superalgebra and $\mathfrak{h} \subset \mathfrak{g}$ a Lie sub(-super)algebra. Then we have the Hochschild–Serre spectral sequence

$$E_1^{pq} = H^q(\mathfrak{h}, \Lambda^p(\mathfrak{g}/\mathfrak{h})^*) \Rightarrow H^{p+q}(\mathfrak{g}, \mathbf{k}). \tag{4.3.1}$$

See [Fu] for background. If $\mathfrak{h}$ is an ideal, then the $E_{\geq 2}$-part of this spectral sequence is identified with (4.1.1).

We apply this to $\mathfrak{g} = L$ and $\mathfrak{h} = \mathfrak{sl}(1|1)$, so that $\mathfrak{g}/\mathfrak{h}$ is identified with $\mathfrak{n}_\pi \cap \mathfrak{n}_{\pi'}$. Let us look at the structure of $\Lambda^\bullet(\mathfrak{n}_\pi \cap \mathfrak{n}_{\pi'})^*$ as a representation of $\mathfrak{sl}(1|1)$. Notice that $(\mathfrak{n}_\pi \cap \mathfrak{n}_{\pi'})^*$ is a direct sum of the $\mathfrak{sl}(1|1)$-representations of types

$$\mathbf{k} = \mathrm{I},\ \mathbf{k}^{1|1} = \mathrm{II}_1,\ (\mathbf{k}^{1|1})^* = \mathrm{II}_{-1}.$$

Therefore all the representations of $\mathfrak{sl}(1|1)$ appearing in the sequel (being derived from these three by tensor operations) are liftable (possess a $\mathbb{Z}$-grading refining the $\mathbb{Z}/2$-grading). Using the multiplication table in Proposition 3.2.3 we find that each tensor power and therefore, each exterior power of $(\mathfrak{n}_\pi \cap \mathfrak{n}_{\pi'})^*$ decomposes into sum of representations of types I, II_τ ($\tau \neq 0$) and III only.

According to Proposition 3.2.4, the cohomology of $\mathfrak{sl}(1|1)$ with coefficients in II_τ, $\tau \neq 0$, vanishes. Further, the h-weight of the $\mathfrak{sl}(1|1)$-cohomology with coefficients in I and III is zero. This follows because h acts trivially on these representations as well as on $\mathfrak{sl}(1|1)$ itself (lies in the center). Therefore the spectral sequence is trivial in non-zero h-weight, which in turn implies Proposition 4.2.6.

C. Proposition 4.2.10: **formality of** $C^\bullet(\mathfrak{sl}(1|1) \oplus J)$. Here we prove the part of Proposition 4.2.10 which says that $C^\bullet(\mathfrak{sl}(1|1) \oplus J)$ is a formal mixed complex. In fact, we establish a more detailed statement to be used later. We start with a general lemma.

Lemma 4.3.2. *Let* $(C^\bullet, d, \delta)$ *be a mixed complex, which is free as a* $\Lambda[\delta]$*-module.*

a) *The canonical map* $H^p(\mathrm{Ker}\,\delta) \to \mathrm{HC}^p(C^\bullet)$ *is an isomorphism.*

b) *Let* $c \in \mathrm{Ker}\,\delta$ *be a cocycle with respect to d. Because of the freeness assumption there exists* $\tilde{c} \in C^\bullet$*, such that* $\delta(\tilde{c}) = c$*. Then the cohomology class of* $d\tilde{c}$ *is independent of the choice of* $\tilde{c}$ *and under the isomorphism of* (a) *is identified with* $u \cdot [c] \in \mathrm{HC}^\bullet(C^\bullet)$.

Proof. (a) By the assumption of the lemma the rows of the Connes bicomplex $\mathcal{BC}$ are exact, except at the first term. So the corresponding first quadrant spectral sequence has $E_1^{pq} = 0$, if $q > 0$ and $E_1^{p0} = \mathrm{Ker}(\delta\colon C^p \to C^{p-1})$, which implies the statement of the lemma.

(b) Let $\tilde{c} + \delta c'$ be another lift, then $d(\tilde{c} + \delta c') - d\tilde{c} = d\delta c'$ is a boundary in the complex $\mathrm{Ker}\,\delta$. The identification with $u \cdot [c]$ follows immediately from the definition of the Connes bicomplex. □

Proposition 4.3.3. *We have the isomorphism*

$$\mathrm{HC}^\bullet(C^\bullet(\mathfrak{sl}(1|1) \oplus J)) \simeq \mathbf{k}[x, y] \otimes_{\mathbf{k}} H^\bullet(J, \mathbf{k}),$$

such that the multiplication by u *in the cyclic cohomology corresponds to the multiplication by* xy *on the right-hand side. In particular,* $C^\bullet(\mathfrak{sl}(1|1) \oplus J)$ *is formal.*

Proof. The mixed complex $(C^\bullet(\mathfrak{sl}(1|1) \oplus J, d, i_h)$ can be represented as a tensor product of two mixed complexes (i.e., both d and δ are extended to the tensor product by the Leibniz rule):

$$\begin{aligned} C^\bullet(\mathfrak{sl}(1|1) \oplus J) &= C^\bullet(\mathfrak{sl}(1|1)) \otimes_{\mathbf{k}} C^\bullet(J) \\ &= \left(\mathbf{k}[x, y, t],\, d = xy\frac{\partial}{\partial t},\, \delta = \frac{\partial}{\partial t}\right) \underset{\mathbf{k}}{\otimes} \left(C^\bullet(J),\, d,\, \delta = 0\right). \end{aligned}$$

So the complex $\mathrm{Ker}\,\delta$ is isomorphic to $\mathbf{k}[x, y] \otimes C^\bullet(J)$ with $dx = dy = 0$, which implies the isomorphism claimed in the proposition. Since $C^\bullet(\mathfrak{sl}(1|1) \oplus J)$

is a mixed complex with algebra structure it is enough to show that $u \cdot 1 = xy$, which is clear from $\delta t = 1$ and $dt = xy$. Now, the polynomial ring $\mathbf{k}[x, y]$ is free as a $\mathbf{k}[xy]$-module, and formality follows from Proposition 3.1.7(a). □

D. $C^\bullet(L)$ as a double complex. Here we present some preparatory constructions to be used in the proof of the remaining parts of Proposition 4.2.10.

In the general situation of §B, assume, in addition, that $\mathfrak{g} = \mathfrak{h} \ltimes \mathfrak{k}$ is a semi-direct product of a subalgebra $\mathfrak{h}$ and an ideal $\mathfrak{k}$. Then the Hochschild–Serre spectral sequence (4.3.1) comes from the fact that $C^\bullet(\mathfrak{g})$ is realized as the total complex of a double complex. More precisely,

$$C^\bullet(\mathfrak{g}) = C^\bullet(\mathfrak{h}, C^\bullet(\mathfrak{k}))$$

is nothing but the cochain complex of $\mathfrak{h}$ with coefficients in the cochain complex of $\mathfrak{k}$ (on which $\mathfrak{h}$ acts). Unraveling the definition, we see that

$$C^\bullet(\mathfrak{g}) = \mathrm{Tot}(C^{\bullet\bullet}, d', d''), \text{ where } C^{pq} = C^p(\mathfrak{k}) \otimes_{\mathbf{k}} C^q(\mathfrak{h}), \quad d' = d_{\mathfrak{k}} \otimes \mathrm{Id},$$

and $d_{\mathfrak{k}}$ is the cochain differential of $\mathfrak{k}$. The differential d'' is the cochain differential of $\mathfrak{h}$ with coefficients in $\Lambda^\bullet(\mathfrak{k}^*)$ as an $\mathfrak{h}$-module. In particular, if $\mathfrak{h}$ is abelian, d'' is the standard Koszul differential corresponding to a family of (super)commuting operators.

We apply this to the case $\mathfrak{g} = L$, with the basis formed by the matrix units e_{kl} and the element h. We take $\mathfrak{h} = \mathbf{k} \cdot e_{ji}$. It is a 1-dimensional (odd) abelian Lie subalgebra. Let also $\mathfrak{k}$ be the subspace in L spanned by all basis elements other than e_{ji}. Then $\mathfrak{k}$ is an ideal and $\mathfrak{g} = \mathfrak{h} \ltimes \mathfrak{k}$. We recall that y is the dual basis element corresponding to e_{ji}, so $C^\bullet(\mathfrak{h}) = \mathbf{k}[y]$ is the usual polynomial algebra. The dg-algebra $C^\bullet(L)$ then contains $\mathbf{k}[y]$ as a subalgebra and is free as a $\mathbf{k}[y]$-module. The space of free generators is canonically identified with $C^\bullet(\mathfrak{k}) = C^\bullet(L)/(y)$. The above representation of $C^\bullet(\mathfrak{g})$ as a double complex amounts to refining the $\mathbb{Z}$-grading on $C^\bullet(L)$ to a $\mathbb{Z}^2$-grading, by putting

$$C^p(L) = \bigoplus_q C^{p-q,q}, \quad C^{p-q,q} = (C^\bullet(L)/(y))^{p-q} \cdot y^q. \tag{4.3.4}$$

Further, the Koszul differential corresponding to the action of the 1-dimensional abelian subalgebra $\mathfrak{h}$ is the operator induced by the coadjoint action of e_{ji}. We denote this operator by $\ell_y = \mathrm{Lie}_{e_{ji}}$. So we obtain the following.

Lemma 4.3.5. *With respect to the decomposition* (4.3.4), *the differential in* $C^\bullet(L)$ *splits into the sum* $d = \bar{d} + \ell_y$, *where* $\bar{d}$ *is the differential in* $C^\bullet(L)/(y)$ *and* $\ell_y = \mathrm{Lie}_{e_{ji}}$. *In other words,* $C^\bullet(L)$ *is represented as the total complex of the double complex* $(C^{\bullet\bullet}(L), \bar{d}, \ell_y)$. □

E. Proposition 4.2.10: spectrality of ρ. Here we prove that ρ is a spectral morphism. For this, we apply the double complex realization of §D to various subcomplexes in $C^\bullet(L)$. First, we apply it to the h-weight 0 part $C^\bullet(L)^h$. Second, we denote by

K the kernel of the restriction morphism ρ. We note that ρ is surjective so that we have a short exact sequence of mixed complexes

$$0 \to K \longrightarrow C^\bullet(L)^h \xrightarrow{\rho} C^\bullet(\mathfrak{sl}(1|1) \oplus J) \to 0. \tag{4.3.6}$$

The double complex construction applies to K as well. In particular, we note the following.

Corollary 4.3.7. *The complex K is a free module over $\mathbf{k}[y]$ and can be represented as the total complex of the double complex $(K^{\bullet\bullet}, \bar{d}, \ell_y)$, where*

$$K^{p-q,q} = (K^\bullet/(y))^{p-q} \cdot y^q. \qquad \square$$

Lemma 4.3.8. *The cyclic cohomology $\mathrm{HC}^\bullet(K)$ is a torsion $\mathbf{k}[u]$-module.*

Proof. Using the same argument as in Proposition 4.3.3 we see that the multiplication by u in $\mathrm{HC}^\bullet(K)$ coincides with the multiplication by xy. Therefore, in order to show that it is a torsion $\mathbf{k}[u]$-module it is enough to establish that it is a torsion $\mathbf{k}[y]$-module.

Consider the localized complex $K[y^{-1}] = K \otimes_{\mathbf{k}[y]} \mathbf{k}[y, y^{-1}]$. By Corollary 4.3.7, we may think of $K[y^{-1}]$ as the total complex of bicomplex $\tilde{K}$, with rows $\tilde{K}^{\bullet q} \simeq K/yK$, for all $q \in \mathbb{Z}$, with horizontal differential induced from K, and vertical differential given by the action of ℓ_y.

Suppose the columns of this bicomplex are exact with respect to the vertical differential. Then the associated spectral sequence trivializes at E_1, so that cohomology $H^\bullet(K[y^{-1}]) = 0$. This in turn implies vanishing of the cyclic homology of $K[y^{-1}]$, which can be seen using the spectral sequence (3.1.4). Hence $\mathrm{HC}^\bullet(K)$ is a torsion $\mathbf{k}[y]$-module. $\square$

Exactness of the columns follows from the next lemma.

Lemma 4.3.9. *The complex K/yK is free as a $\Lambda[\ell_y]$-module.*

Proof. Since $C^\bullet(\mathfrak{sl}(1|1) \oplus J)$ is a free $\mathbf{k}[y]$-module we have the short exact sequence:

$$K/yK \longrightarrow C^\bullet(L)^h/(y) \xrightarrow{\rho} C^\bullet(\mathfrak{sl}(1|1) \oplus J)/(y)\ .$$

Note that we have an isomorphism of $\Lambda[\ell_y]$-modules

$$C^\bullet(\mathfrak{sl}(1|1) \oplus J)/(y) \ \simeq\ \Lambda^\bullet(\mathfrak{sl}(1|1)^*/y) \otimes_{\mathbf{k}} \Lambda^\bullet(J^*).$$

Notice further that $\Lambda^\bullet(J^*)$ has trivial action of ℓ_y. On the other hand, $\mathfrak{sl}(1|1)^*/y$ is a free $\Lambda[\ell_y]$-module, and therefore $\Lambda^{>0}(\mathfrak{sl}(1|1)^*/y)$ is again a free $\Lambda[\ell_y]$-module. This means that $C^\bullet(\mathfrak{sl}(1|1) \oplus J)/(y)$ is the direct sum of a free $\Lambda[\ell_y]$-module and a trivial module isomorphic to $\Lambda^\bullet(J^*) = \Lambda^0(\mathfrak{sl}(1|1)^*/y) \otimes \Lambda^\bullet(J^*)$.

We will prove that the ℓ_y-trivial parts of $C^\bullet(L)^h/(y)$ and of $C^\bullet(\mathfrak{sl}(1|1) \oplus J)/(y)$ are identified by the map ρ, thus showing that $K/yK = \mathrm{Ker}(\rho)$ is free. Indeed, we have

$$C^\bullet(L)^h/(y) \ \simeq\ \Lambda^\bullet(\mathfrak{sl}(1|1)^*/y) \otimes \Lambda^\bullet(L_+^* \oplus L_-^*) \otimes \Lambda^\bullet(J^*).$$

Note that $L_+ \oplus L_-$ (and so its dual $L_+^* \oplus L_-^*$) is a free $\Lambda[\ell_y]$-module. Indeed, the action on L_- sends e_{kj} to e_{ki}, identifying the two columns constituting L_- in Figure 1. Similarly, the action on L_+ identifies the two rows constituting L_+ in the figure, sending e_{ik} to e_{jk}. So the part of $C^\bullet(L)_0/(y)$ with the trivial action of ℓ_y is isomorphic to

$$\Lambda^0(\mathfrak{sl}(1|1)^*/y) \otimes \Lambda^0(L_+^* \oplus L_-^*) \otimes \Lambda^\bullet(J^*) \;=\; \Lambda^\bullet(J^*)$$

and ρ maps it isomorphically to the ℓ_y-trivial part of $C^\bullet(\mathfrak{sl}(1|1) \oplus J)/(y)$. Therefore K/yK is a free $\Lambda[\ell_y]$-module. □

This establishes Lemma 4.3.8 as well. Therefore $\mathrm{HC}^\bullet(K)[u^{-1}]$, the periodized cyclic cohomology, vanishes. Now the fact that $\rho\colon C^\bullet(L)^h \to C^\bullet(\mathfrak{sl}(1|1) \oplus J)$ is spectral, follows from the long exact sequence for cyclic cohomology.

F. Proposition 4.2.10: formality of $C^\bullet(L)^h$. Here we finish the proof of Proposition 4.2.10 by showing that $C^\bullet(L)^h$ is a formal mixed complex. By Corollary 3.1.12 this is equivalent to the statement that $C^\bullet(L)^h$ does not contain indecomposable summands of the type $F_n(\delta,\delta)$. Now, Proposition 3.1.11 classifies possible isomorphism types, as $\mathbf{k}[u]$-modules, of the cyclic cohomology $HC^\bullet(C^\bullet)$ where $C^\bullet$ is an indecomposable mixed complex. It follows from this classification that the multiplication by u on $HC^\bullet(C^\bullet)$ has no kernel, if and only if $C^\bullet$ is not of the type $F_n(\delta,\delta)$. Therefore it suffices to prove that the multiplication by u on $\mathrm{HC}^\bullet(C^\bullet(L)^h)$ has no kernel.

Using the same argument as in the proof of Proposition 4.3.3, we find that multiplication by u is equivalent to multiplication by xy. Therefore it is enough to establish that x and y are not zero divisors in $\mathrm{HC}^\bullet(C^\bullet(L)^h)$.

First assume that we know that y is not a zero divisor. Then the formality follows from the next lemma.

Lemma 4.3.10. *Assume that multiplication by y is injective on $\mathrm{HC}^\bullet(C^\bullet(L)^h)$, then multiplication by x is injective as well.*

Proof. Consider the following commutative diagram:

$$\begin{array}{ccccc}
\mathrm{HC}^\bullet(C^\bullet(L)^h) & \xrightarrow{\gamma} & \mathrm{HC}^\bullet(C^\bullet(L)^h)[y^{-1}] & \xrightarrow{\simeq} & \mathrm{H}^\bullet(J)\otimes k[x,y,y^{-1}] \\
\downarrow{\cdot x} & & \downarrow{\cdot x} & & \downarrow{\cdot x} \\
\mathrm{HC}^\bullet(C^\bullet(L)^h) & \xrightarrow{\gamma} & \mathrm{HC}^\bullet(C^\bullet(L)^h)[y^{-1}] & \xrightarrow{\simeq} & \mathrm{H}^\bullet(J)\otimes k[x,y,y^{-1}]
\end{array}$$

The map γ is the canonical map induced by localization, and since y is not a zero divisor, γ is injective. We have seen before (see the proof of Lemma 4.3.8) that the cyclic cohomology of the localized complex $K[y^{-1}]$ vanishes, so that $\mathrm{HC}^\bullet(C^\bullet(L)_0)[y^{-1}]$ is isomorphic to $\mathrm{HC}^\bullet(C^\bullet(\mathfrak{sl}(1|1)\oplus J))[y^{-1}]$, which in turn in virtue of Proposition 4.3.3 is isomorphic to $H^\bullet(J)\otimes \mathbf{k}[x,y,y^{-1}]$.

Now, multiplication by x is clearly injective in $H^\bullet(J) \otimes \mathbf{k}[x, y, y^{-1}]$. Combining this with injectivity of γ we find that x is not a zero divisor in $\mathrm{HC}^\bullet(C^\bullet(L)^h)$. □

Denote by Z the kernel complex $\operatorname{Ker} i_h : C^\bullet(L)^h \to C^\bullet(L)^h$, which by Lemma 4.3.2 computes the cyclic cohomology of $C^\bullet(L)^h$. Let us apply the double complex construction of §D to Z and denote the resulting double complex (refining Z) by $Z^{\bullet\bullet}$. The columns of $Z^{\bullet\bullet}$ are isomorphic to $Z/(y)$.

Lemma 4.3.11. *The canonical projection $C^\bullet(L) \to C^\bullet(\mathfrak{n}_\pi)$ restricted to Z induces isomorphism $Z/(y) \simeq C^\bullet(\mathfrak{n}_\pi)^h$.*

Proof. The statement follows immediately from:

$$Z/(y) = \operatorname{Ker} i_h/(y) \;\simeq\; C^\bullet(L)^h/(t, y) \;\simeq\; C^\bullet(\mathfrak{n}_\pi)^h.$$

□

Proposition 4.3.12. *The spectral sequence attached to the double complex $Z^{\bullet\bullet}$ degenerates at E_1. In particular y is not a zero divisor in $H^\bullet(Z) \simeq \mathrm{HC}^\bullet(C^\bullet(L)^h)$.*

Proof. Notice that $E_1^{pq} = H^q(C^\bullet(\mathfrak{n}_\pi)^h) = H^q(\mathfrak{n}_\pi, \mathbf{k})^h$ for all $p \geq 0$. Let us fix weight λ with respect to the action of the torus T such that $\lambda(h) = 0$, and restrict to the λ-isotypical component. We have

$$(E_1^{p\bullet})_\lambda = H^\bullet(\mathfrak{n}_\pi, \mathbf{k})_{\lambda - p\alpha_y} y^p,$$

where α_y is the root corresponding to y.

Now, the explicit formula of Theorem 2.3.2 implies that values of the Euler characteristic of $H^\bullet(\mathfrak{n}_\pi, \mathbf{k})_{\lambda - p\alpha_y}$, are either $\{0, +1\}$ for all $p \geq 0$, or $\{0, -1\}$ for all $p \geq 0$. By inductive assumption on the $H^\bullet(\mathfrak{n}_\pi, \mathbf{k})$ this implies that the cohomology in weights $\lambda - p\alpha_y$ is concentrated either in even or odd total degree (here by total degree we understand the degree in $S^\bullet(\mathfrak{n}^*[-1])$). Notice also that the total degree of y is 2. Therefore the λ-isotypic part of the spectral sequence can not have non-zero differentials, and since λ was an arbitrary weight satisfying $\lambda(h) = 0$, the proposition follows. □

This establishes the fact that $t = xy$ is not a zero divisor in $\mathrm{HC}^\bullet(C^\bullet(L)^h)$ and therefore $C^\bullet(L)^h$ is a formal mixed complex, thus completing the proof of Proposition 4.2.10.

G. Proof of Proposition 4.2.12. We include the short exact sequence of Proposition 4.2.5 into a commutative diagram of short exact sequences

$$\begin{array}{ccccccccc}
0 & \longrightarrow & (t, y) \cdot C^\bullet(L) & \longrightarrow & C^\bullet(L) & \longrightarrow & C^\bullet(L)/(t, y) & \longrightarrow & 0 \\
 & & \big\downarrow{\scriptstyle q'} & & \big\downarrow{\scriptstyle q} & & \big\downarrow{\scriptstyle =} & & \\
0 & \longrightarrow & y \cdot C^\bullet(\mathfrak{n}_{\pi'}) & \longrightarrow & C^\bullet(L)/(t, dt) & \longrightarrow & C^\bullet(\mathfrak{n}_\pi) & \longrightarrow & 0
\end{array} \tag{4.3.13}$$

Here the vertical maps are the canonical projections, q is the quasi-isomorphism of Proposition 4.2.6, the right vertical arrow is the identity, and q', the restriction of q, is a quasi-isomorphism since the two other vertical arrows are.

Proposition 4.2.12 is a statement about the long exact sequence (LES) of cohomology associated to the bottom row in (4.3.13). Part (a) follows at once from Proposition 4.2.6: since for $c \neq 0$

$$H^p\big(C^\bullet(L)/(t,dt)\big)^{h=c} = H^p(C^\bullet(L))^{h=c} = 0, \quad p \in \mathbb{Z},$$

the coboundary maps in the long exact sequence of cohomology are the required isomorphisms.

Let us prove part (b). For this we look at the h-weight 0 part of the entire diagram (4.3.13). We need to prove that the coboundary maps in the LES associated to the bottom row are zero. Because the vertical arrows are quasi-isomorphisms, this is equivalent to showing that the coboundary maps in the LES associated to the top row are zero. By Lemma 4.3.11, the top row contains a short exact sub-sequence

$$0 \to yZ \longrightarrow Z \longrightarrow Z/yZ \to 0, \tag{4.3.14}$$

with the same rightmost term $Z/yZ = C^\bullet(\mathfrak{n}_\pi)^h$. We view this subsequence as a partial splitting of the top row of the $h=0$ part of (4.3.13) and conclude that the coboundary maps associated to that top row factor through the coboundary maps associated to (4.3.14). But Proposition 4.3.12 means that the maps $H^p(yZ) \to H^p(Z)$ are injective, and so the coboundary maps $H^{p-1}(Z/yZ) \to H^p(yZ)$ are zero. This completes the proof.

4.4. Arbitrary parabolic

Our argument will be a refinement of that in §2.5, but to visualize the situation, we switch back to the (equivalent) notation of §1.2. That is, we denote by $r_1, \dots, r_n$ the dimensions of the vector spaces $V^i = \mathbf{k}^{r_i}$ (some of which may be 0). We consider the Lie superalgebra $\mathfrak{n} = \mathfrak{n}(V^\bullet) = \bigoplus_{i<j} \operatorname{Hom}(V^i, V^j)$ formed by block upper triangular r by r matrices, with the parity of the summands induced by the parity of i and j. We will prove that the $GL(V^\bullet)$-spectrum of $H^\bullet(\mathfrak{n})$ is simple. Note that $\mathfrak{n}$ is embedded into the Lie superalgebra $\mathfrak{m}$ formed by matrices which are strictly upper triangular in the usual (not block) sense. This $\mathfrak{m}$ is the nilpotent radical of a Borel subalgebra $\mathfrak{b}$ in $\mathfrak{gl}(r_{\bar{0}}|r_{\bar{1}})$ containing $\mathfrak{n}$, and we know that the T-spectrum of $H^\bullet(\mathfrak{m})$ is simple.

Let $\mathfrak{u}_m$ denote the nilpotent radical of the standard Borel in $\mathfrak{gl}(m)$, i.e., the usual Lie algebra of strictly upper triangular matrices. In this notation, we have a semidirect sum decomposition

$$\mathfrak{m} = \mathfrak{u} \ltimes \mathfrak{n}, \quad \mathfrak{u} := \mathfrak{u}_{r_1} \oplus \cdots \oplus \mathfrak{u}_{r_n}$$

and the corresponding spectral sequence

$$E_2^{pq} = H^p(\mathfrak{u}, H^q(\mathfrak{n})) \Rightarrow H^{p+q}(\mathfrak{b}). \tag{4.4.1}$$

We identify, as modules over $GL(V^\bullet)$,

$$H^q(\mathfrak{n}) = \bigoplus_{[\alpha]} \Sigma^{[\alpha]}(V^\bullet)^{\oplus c^p_{[\alpha]}},$$

with some yet unknown multiplicities $c^p_{[\alpha]}$. Here $[\alpha] = (\alpha^{(1)}, \dots, \alpha^{(n)})$ runs over sequences of dominant weights for the $GL(V^i)$. By the classical Kostant theorem, we have an identification of T-modules

$$H^p(\mathfrak{u}, \Sigma^{[\alpha]}(V^\bullet)) \simeq \bigoplus_{\substack{w_1 \in S_{r_1}, \dots, w_n \in S_{r_n} \\ l(w_1)+\dots+l(w_n)=p}} \bigotimes_{i=1}^{n} \mathbf{k}_{w_i(\alpha^{(i)}+\rho^{(i)})-\rho^{(i)}} \tag{4.4.2}$$

Here we identify $T = \prod_{i=1}^n \mathbb{G}_m^{r_i}$ and for a character $\lambda \in (\mathbb{G}_m^{r_i})^\vee$ write $\mathbf{k}_\lambda$ for the corresponding 1-dimensional representation.

Now, it follows from the observation of §2.5, that for different sequences $[\alpha]$ the supports of the T-modules given by the RHS of (4.4.2) do not intersect: their characters are given by the product polynomials $P_\alpha(s)$ from (2.5.1). Therefore the only possibility for the differentials of the spectral sequence (4.4.1) is to act between the spaces (4.4.2) with the same α and different p. Each differential increases p by 2 or more and is T-equivariant. Now, for different p, the subsets

$$\Lambda_p = \{w = (w_1, \dots, w_n) \in S_{r_1} \times \dots \times S_{r_n} \mid \sum l(w_i) = p\}$$

are, obviously, disjoint. Therefore applying w from different Λ_p to the sequence of strictly dominant weights $\alpha^{(i)} + \rho^{(i)}$, we will never get the same weights. Therefore all the differentials in the spectral sequence are trivial. It further follows that if $c^p_{[\alpha]} > 1$ for some p, then there will be a T-isotypic component of $H^\bullet(\mathfrak{m})$ with multiplicity > 1 which is impossible. This finishes the proof.

List of Notations

References

[ASS] I. Assem, D. Simson, A. Skowronski. Elements of the Representation Theory of Associative Algebras. Vol. 1. Cambridge Univ. Press, 2006

[BK] A.I. Bondal, M.M. Kapranov. Enhanced triangulated categories. *Math. USSR Sbornik*, **70** (1991) 93–107.

[Br1] J. Brundan. Kazhdan–Lusztig polynomials and character formulae for the Lie superalgebra $\mathfrak{gl}(m|n)$. *Journal AMS*, **16** (2002), 185–231.

[Br2] J. Brundan. Dual canonical bases and Kazhdan–Lusztig polynomials. *J. Algebra*, **306** (2006), 17–46.

[BLW] J. Brundan, I. Losev, B. Webster. Tensor product categorifications and the super Kazhdan–Lusztig conjecture. arXiv:1310.0349.

[BS] J. Brundan, C. Stroppel. Highest weight categories arising from Khovanov's diagram algebra IV: the general linear supergroup. *J. Eur. Math. Soc.* **14** (2012), 373–419.

[CKL] S.-J. Cheng, J.-H. Kwon, N. Lam. A BGG-type resolution for tensor modules over general linear superalgebra. *Lett. Math. Phys.* **84** (2008), 75–87.

[CKW] S.-J. Cheng, J.-H. Kwon, W. Wang. Kostant homology formula for oscillator modules of Lie superalgebras. *Adv. in Math.* **224** (2010), 1548–1588.

[CLW] S.-J. Cheng, N. Lam, W. Wang. Brundan–Kazhdan–Lusztig conjecture for general linear Lie superalgebras. arXiv:1203:0092.

[CWZ] S.-J. Cheng, W. Wang, R.B. Zhang. Super duality and Kazhdan–Lusztig polynomials. *Trans. AMS*, **360** (2008), 5983–5924.

[CZ] S.-J. Cheng, R.B. Zhang. Analogue of Kostant's $\mathfrak{u}$-cohomology formula for the general linear superalgebra. *Int. Math. Res. Notices*, 2004, No. 1, 31–53.

[CFK1] I. Ciocan-Fontanine, M. Kapranov. Derived *Quot* schemes. *Ann. Sci. ÉNS*, **34** (2001), 403–440.

[CFK2] I. Ciocan-Fontanine, M. Kapranov. Virtual fundamental classes via dg-manifolds. *Geom. Topol.* **13** (2009), 1779–1804.

[Cou] K. Coulembier. Bernstein–Gelfand–Gelfand resolutions for basic classical Lie superalgebras. *J. Algebra* **399** (2014), 131–169.

[Cu] F. Cukierman. Varieties of complexes and foliations. *J. of Singularities,* **9** (2014), 56–67.

[DS] C. De Concini, E. Strickland. On the variety of complexes. *Adv. Math.* **41** (1981), 57–77.

[Dr] V. Drinfeld. Letter to V. Schechtman, 1986.

[FKV] J.M. Figueroa-O'Farrill, T. Kimura, A. Vaintrob. The universal Vassiliev invariant for the Lie superalgebra $\mathfrak{gl}(1|1)$. *Comm. Math. Phys.* **185** (1997), 93–127.

[Fu] D.B. Fuks. Cohomology of Infinite-Dimensional Lie algebras. Translated from the Russian by A.B. Sosinskiĭ. Consultants Bureau, New York, 1986.

[Ge] J. Germoni. Indecomposable representations of special linear superalgebras. *J. Algebra* **209** (1998), 367–401.

[GQS] G. Götz, T. Quella, V. Schomerus. Representation theory of $\mathfrak{sl}(2|1)$. *J. Algebra* **312** (2007), 829–848.

[Go] N. Gonciulea. Singular loci of varieties of complexes. II. *J. Algebra* **235** (2001), 547–558.

[GKM] M. Goresky, R. Kottwitz, R. MacPherson. Equivariant cohomology, Koszul duality and the localization theorem. *Invent. Math.* **131** (1998), 25–83.

[Ka] M. Kapranov. Injective resolutions of BG and derived moduli spaces of local systems. *J. Pure Appl. Alg.* **155** (2001), 167–179.

[Ke] G.R. Kempf. Images of homogeneous vector bundles and varieties of complexes. *Bull. Amer. Math. Soc.* **81** (1975), 900–901.

[Ko1] M. Kontsevich. Topics in algebra: deformation theory. Graduate course, Berkeley 1991.

[Ko2] M. Kontsevich. Enumeration of rational curves via torus actions, in: "The Moduli Space of Curves, edited by R. Dijkgraaf, C. Faber and G. Van der Geer, *Progress in Mathematics* **129**, pp. 335–368, Birkhäuser 1995.

[Ko3] B. Kostant. Lie Algebra Cohomology and the Generalized Borel–Weil Theorem. *Ann. Math.* **74** (1961), 329–387.

[Ku] S. Kumar. Kac–Moody Groups, their Flag Varieties and Representation Theory, *Progress in Math.* **204**, Birkhäuser, Boston, 2002.

[La] V. Lakshmibai. Singular loci of varieties of complexes. *J. Pure Appl. Algebra* **152** (2000), 217–230.

[Le] D.A. Leites. Representations of Lie superalgebras. *Theor. Math. Fizika* **52** (1982), 225–228 (in Russian).

[L] Jean-Louis Loday. Cyclic Homology. Springer-Verlag, 1998.

[Lu] J. Lurie. Derived Algebraic Geometry IV: Deformation Theory. arXiv:0709.3091.

[M] I. Macdonald. Symmetric Functions and Hall Polynomials. Oxford Univ. Press, 1979.

[Ma] J.P. May. Matric Massey products. *J. of Algebra*, **12** (1969), 533–568.

[MS] C. Musili, C.S. Seshadri. Schubert varieties and the variety of complexes in: "Arithmetic and geometry" Vol. II, 329–359, Progr. Math., **36**, Birkhäuser, Boston, MA, 1983.

[Mu] I. Musson. Lie Superalgebras and Enveloping Algebras. (Graduate Studies in Mathematics, vol. **131**). Amer. Math. Soc. 2012.

[Pr1] J.P. Pridham. Derived moduli of schemes and sheaves. *K-theory* **10** (2012), 41–85.

[Pr2] J.P. Pridham. Constructing derived moduli stacks. *Geom. Topol.* **17** (2013), 1417–1495.

[Ra] A. Ranicki. Higher-Dimensional Knot Theory. Springer-Verlag, 1998.

[Se1] P. Seidel, Fukaya categories and Picard–Lefschetz theory, European Mathematical Society, Zürich, 2008.

[Se2] V. Serganova. Kazhdan–Lusztig polynomials and character formula for the Lie superalgebra $\mathfrak{gl}(m|n)$. *Selecta Math.* (*N.S.*) **2** (1996), 607–651.

[Se3] V. Serganova. Kac–Moody superalgebras and integrability, in: *"Developments and trends in infinite-dimensional Lie theory"*, pp. 169–218, *Progr. Math.* **288**, Birkhäuser Boston, 2011.

[TVe] B. Toën, G. Vezzosi. Homotopical algebraic geometry. II. Geometric stacks and applications. *Mem. Amer. Math. Soc.* **902** (2008), 1–224.

[TVa] B. Toën, M. Vaquié. Moduli of objects in dg-categories. *Ann. Sci. ÉNS* **40** (2007), 387–444.

M. Kapranov
Kavli Institute for Physics and Mathematics of the Universe (WPI)
5-1-5 Kashiwanoha, Kashiwa-shi, Chiba, 277-8583, Japan
e-mail: `mikhail.kapranov@ipmu.jp`

S. Pimenov
Department of Mathematics
University of Nottingham
University Park
Nottingham, NG7 2RD, UK
e-mail: `slava.pimenov@gmail.com`

Progress in Mathematics, Vol. 324, 177–202

Higher Symmetry and Gapped Phases of Gauge Theories

Anton Kapustin and Ryan Thorngren

Abstract. We study topological field theory describing gapped phases of gauge theories where the gauge symmetry is partially Higgsed and partially confined. The TQFT can be formulated both in the continuum and on the lattice and generalizes Dijkgraaf–Witten theory by replacing a finite group by a finite 2-group. The basic field in this TQFT is a 2-connection on a principal 2-bundle. We classify topological actions for such theories as well as loop and surface observables. When the topological action is trivial, the TQFT is related to a Dijkgraaf–Witten theory by electric-magnetic duality, but in general it is distinct. We propose the existence of new phases of matter protected by higher symmetry.

Mathematics Subject Classification (2010). 79: Mathematical Physics.

Keywords. Symmetries, higher categories, topological phases of matter.

1. Introduction and summary

Gapped phases of matter are described at long distances by unitary Topological Quantum Field Theories. Thus it is of great interest to try to classify unitary TQFTs, at least in space-time dimensions up to four. Since gauge interactions are ubiquitous in nature, it is particularly interesting to identify TQFTs which describe gapped phases of gauge theories. In the case when the microscopic gauge group is Higgsed down to a finite group G, a complete classification (at least if no further global symmetries are postulated) has been given by Dijkgraaf and Witten [1]. Namely, topological actions in d dimensions are classified by degree-d cohomology classes for G with coefficients in $U(1)$. But DW theories do not exhaust all possibilities. For example, in 3d there are Chern–Simons theories, which are more general than 3d DW theories. In this paper we study another class of TQFTs which exist in all dimensions and are more general than DW theories. These TQFTs involve both 1-form and 2-form gauge fields, as well as 0-form and 1-form gauge symmetries. 1-form gauge fields take values in the Lie algebra of a Lie group G, while 2-form fields take values in the Lie algebra of another Lie group H. There is a homomorphism t from H to G and an action α of G on H which enable one to write down a consistent set of transformation rules. The quadruple $\mathbb{G} = (G, H, t, \alpha)$ is what is known as a 2-group (see the next section). It has been argued in [2] that such TQFTs describe massive phases of gauge theories where the microscopic gauge group is partially confined and partially Higgsed. More precisely, G is the part of the microscopic gauge group which is not Higgsed, and the image of t is the confined subgroup of G. In this paper we analyze such TQFTs along the lines of [1].

In the case when H is trivial, our TQFT reduces to a DW theory. The other extreme is when $H = G$ and t is a surjection. In this case the whole gauge group is confined, but the theory may have a nontrivial "magnetic gauge group" given by ker t. Such theories have been recently analyzed in [3, 4] (see also [5] and Section 5 of [6]). The present paper generalizes both [1] and [3, 4].

Here is a brief summary of our results. We show that TQFTs which describe phases with both Higgs effect and confinement depend on the following data: a finite (possibly non-Abelian) group Π_1, a finite abelian group Π_2, an action of Π_1 on Π_2, and a degree-3 group cohomology class for Π_1 with coefficients in Π_2. Here Π_1 is interpreted as the low-energy "electric" gauge group while Π_2 is the "magnetic" gauge group. For every such quadruple and every $d > 1$ we construct a lattice model in d dimensions generalizing the DW model. It describes a lattice gauge field with gauge group Π_1 coupled to a lattice 2-form with values in Π_2. We classify possible action for such models in dimensions 2, 3 and 4. In dimension 2 the theory is equivalent to the DW theory for Π_1. In dimension 3 the action depends on two parameters: a degree-3 cohomology class for Π_1 with coefficients in $U(1)$ and a degree-1 cohomology class for Π_1 with coefficients in $\hat{\Pi}_2 = \mathrm{Hom}(\Pi_1, U(1))$. In dimension 4 the action depends on three parameters: a degree-4 cohomology class for Π_1 with coefficients in $U(1)$, a degree-2 cohomology class for Π_1 with coefficients in $\hat{\Pi}_2$, and a quadratic function on Π_2 with values in $U(1)$. We show that in dimension 3 the lattice 2-form can be dualized to a lattice scalar with values in $\hat{\Pi}_2$. In dimension 4 the lattice 2-form can be dualized to a lattice gauge field provided only the first two terms in the action are nonzero. We also classify TQFT observables, including surface observables measuring the flux of the 2-form.

Recently DW theories found a new application: it has been argued that topological actions for a DW theory with gauge group G classify gapped phases of matter with global symmetry G and no long-range entanglement [7, 8, 9]. Such gapped phases are called symmetry-protected phases. We propose that there are more general phases which are protected not by a symmetry group, but by a symmetry 2-group. While an ultralocal internal symmetry group acts on degrees of freedom living on sites of a lattice, a ultralocal 2-symmetry acts on both site and link variables. The groups Π_1 and Π_2 describe symmetry transformations which live on sites and links, respectively. Gapped phases protected by a 2-group symmetry $\mathbb{G}$ are classified by topological actions for a TQFT based on $\mathbb{G}$. In the special case when only the magnetic gauge group Π_2 is nontrivial (the case considered in detail in [3]) it appears that the 2-group TQFT is equivalent to a special case of the Walker–Wang TQFT [10, 11].

More generally, as explained below, in d space-time dimensions one can contemplate symmetry p-groups with $p \leq d$, and accordingly one can have gapped phases with short-range entanglement protected by such higher symmetry.

A.K. would like to thank Dan Freed, Sergei Gukov, Michael Hopkins, Nathan Seiberg, Yuji Tachikawa, and Constantin Teleman for discussions. R.T. would like to thank Scott Carnahan, Curt von Keyserlingk, Evan Jenkins, Alex Rasmussen, David Roberts, and Urs Schreiber for discussions. This work was supported in part by the DOE grant DE-FG02-92ER40701 and by the National Science Foundation under Grant No. PHYS-1066293 and the hospitality of the Aspen Center for Physics.

2. 2-groups

We begin by recalling the notion of a 2-group [12] and its physical interpretation [2]. The most concise definition uses the language of higher categories. A (weak) 2-group is a weak 2-category with a single object and such that all 1-morphisms are weakly-invertible and 2-morphisms are invertible. This is analogous to the definition of a group as a category with a single object and such that all morphisms are invertible.

The most pedestrian definition goes as follows. A 2-group is a quadruple $\mathbb{G} = (G, H, t, \alpha)$ where G and H are groups, $t : H \to G$ is a group homomorphism, and $\alpha : G \to \mathrm{Aut}(H)$ is an action of G on H such that the following two identities hold:

$$t(\alpha(g)(h)) = gt(h)g^{-1}, \quad \alpha(t(h))(h') = hh'h^{-1}. \tag{1}$$

A 2-group defined in this way is also known as a crossed module. The relation between the two definitions is this: G is the set of 1-morphisms, H is the set of 2-morphisms from the identity 1-morphism to all other 1-morphisms, the map t assigns to a 2-morphism its target 1-morphism.

Let us give a few examples of 2-groups. One simple class of examples is obtained by taking G to be a connected simple Lie group, $t : H \to G$ to be a covering map, and α to be given by conjugation: $h \mapsto \widetilde{g}h\widetilde{g}^{-1}$, where $\widetilde{g} = t^{-1}g$. Note that while t^{-1} is multivalued, different choices are different only by elements of the center of H, and therefore α is well defined. A nontrivial finite example is provided by $H = G = \mathbb{Z}_4$, $t(n) = 2n$, and $\alpha(n)(m) = (-1)^n m$.

We are interested in the case when G and H are Lie groups, and the maps t and α are smooth; such a 2-group is called a Lie 2-group. The physical meaning of these data is as follows: G is the subgroup of the microscopic gauge group which is not broken by the Higgs effect, im $t \subset G$ is the part of G which is confined due to monopole condensation (this subgroup is normal, as a consequences of the identities (1), coker $t = G/\mathrm{im}\ t$ is the low-energy gauge group, ker t is the group whose elements label conserved magnetic fluxes (it is a subgroup of the center of H; we will call it the magnetic gauge group). In the above finite example, the unbroken gauge group is $\mathbb{Z}_4$, its $\mathbb{Z}_2$ subgroup is confined, so the low-energy electric gauge group is $\mathbb{Z}_2$. The magnetic gauge group is also $\mathbb{Z}_2$, and the electric gauge group acts on the magnetic fluxes by "charge conjugation".

For groups we have an obvious notion of an isomorphism; for 2-groups a better notion is that of equivalence (of 2-categories). This notion does not preserve the sets G and H, it only preserves the sets coker t and ker t. This is very reasonable from the physical point of view: representations of coker t describe electric sources visible at long distances, elements of ker t describe magnetic sources, but G and H themselves are not observable at long distances. We will see that for any 2-group $\mathbb{G}$ there is a TQFT which depends only on the equivalence class of $\mathbb{G}$.

It is thus natural to ask what sort of data describe the equivalence class of a 2-group. It turns out it is a quadruple $(\Pi_1, \Pi_2, \alpha, \beta)$, where $\Pi_1 = \text{coker } t$ is a group, $\Pi_2 = \ker t$ is an abelian group, α is an action of Π_1 on Π_2 (it descends from the action of G on H, so we denote it by the same letter), and β is an element of $H^3(B\Pi_1, \Pi_2)$. Here and below BG denotes the classifying space of principal G-bundles. If the group G is finite, it is defined up to homotopy by the condition $\pi_n(BG) = 0$ for $n > 1$ and $\pi_1(BG) = G$ [13]. By definition, there is a principal G-bundle EG with base BG such that the total space of EG is contractible. Group cohomology of G is defined in terms of EG. Namely, if G acts on an abelian group M, we have an associated flat bundle over BG with fiber M, and $H^d(BG, M)$ denotes the degree-d cohomology of BG with coefficients in this bundle.

The definition of $\alpha : \Pi_1 \to \text{Aut}(\Pi_2)$ should be clear: one first checks that the action of G on H maps ker t to itself, and then that elements of im t act trivially on ker t. This implies that coker t acts on ker t. Physically, this is also quite natural: the residual gauge group may act on the magnetic flux, like in the case of the gauge group $O(2n)$, where charge conjugation is part of the gauge group and flips the sign of the magnetic flux.

The definition of β is less obvious. Mathematically, one uses the fact [14] that elements of $H^3(B\Pi_1, \Pi_2)$ can be interpreted as equivalence classes of double extensions of Π_1 by Π_2, with a fixed action of Π_1 on Π_2. That is, these are equivalence classes of four-term exact sequences

$$1 \to \Pi_2 \to H' \to G' \to \Pi_1 \to 1,$$

where (G', H', t', α') is a crossed module with coker $t' = \Pi_1$, ker $t' = \Pi_2$, and with α' inducing our chosen action of Π_1 on Π_2. To get β we simply take the tautological sequence

$$1 \to \ker t \to H \to G \to \text{coker } t \to 1$$

We will see that the TQFT depends on $(\Pi_1, \Pi_2, \alpha, \beta)$ and nothing else. However, the continuum action is more conveniently written using the data (G, H, t, α). This is because we normally assume that Π_1 and Π_2 are finite (to ensure that the low-energy theory is gapped), but to write down a continuum action it is convenient to work with fields which take values in the Lie algebras of G and H.

The most basic example of a group arising from topology is the fundamental group of a (pointed connected) topological space. This group encodes the homotopy 1-type of a space. Similarly, there is a 2-group associated to any pointed connected topological space which encodes its homotopy 2-type. This 2-group has $G = \pi_1(X_1, x_0)$, $H = \pi_2(X, X_1, x_0)$, where X_1 is a 1-skeleton of X, and x_0 is the marked point. We will call it the fundamental 2-group of X. The homotopy equivalence class of these data is encoded by the quadruple $(\pi_1(X, x_0), \pi_2(X, x_0), \alpha, \beta)$, where α is the usual monodromy action of π_1 on π_2 and $\beta \in H^3(B\pi_1, \pi_2)$ is the so-called Postnikov invariant of X.

3. 2-gauge theory in the continuum

The path-integral of the DW theory is constructed as an integral over the moduli space of flat connections with a structure group G. Similarly, we can construct a TQFT whose path-integral is an integral over the moduli space of flat 2-connections with a structure 2-group $\mathbb{G}$. But first we need to define the notion of a flat 2-connection on a manifold X.

A flat connection for a group G can be defined as a homomorphism from $\pi_1(X)$ to G. Similarly, a flat 2-connection for a 2-group $\mathbb{G}$ can be defined as a "2-homomorphism" from the fundamental 2-group of X to $\mathbb{G}$. However, both of these definitions are not manifestly local and therefore not suitable for constructing a TQFT. A local definition of a flat 2-connection goes as follows [16, 17].

Let $\mathfrak{g}$, $\mathfrak{h}$ be the Lie algebras of G and H. Let $\bar{t} : \mathfrak{h} \to \mathfrak{g}$ be the homomorphism of Lie algebras induced by the homomorphism of Lie groups $t : H \to G$. Since $\ker t$ and $\operatorname{coker} t$ were assumed to be finite, $\bar{t}$ is an isomorphism. Let $\bar{\alpha} : G \to \mathrm{Aut}(\mathfrak{h})$ be the action of G on $\mathfrak{h}$ induced by α. Let $\{U_i\}$, $i \in I$, be a cover of X such that all overlaps are contractible. A flat 2-connection with structure 2-group $\mathbb{G}$ is defined by the following data. On each chart U_i one has a $\mathfrak{g}$-valued 1-form A_i, on each U_{ij} one has a G-valued function g_{ij} and an $\mathfrak{h}$-valued 1-form λ_{ij}, and on each U_{ijk} one has an H-valued function h_{ijk} so that the following conditions are satisfied:

- On each U_{ij} one has $A_j = g_{ij} A_i g_{ij}^{-1} + g_{ij} d g_{ij}^{-1} - \bar{t}(\lambda_{ij})$.
- On each U_{ijk} one has $g_{ik} = t(h_{ijk}) g_{jk} g_{ij}$, and

$$\begin{aligned} h_{ijk}^{-1} \lambda_{ik} h_{ijk} &= \bar{\alpha}(g_{jk})(\lambda_{ij}) + \lambda_{jk} - h_{ijk}^{-1} d h_{ijk} \\ &\quad - h_{ijk}^{-1} \bar{t}^{-1}(A_k) h_{ijk} + \bar{t}^{-1}(A_k) . \end{aligned}$$

- On each U_{ijkl} one has $h_{ijl} h_{jkl} = h_{ikl} \cdot \alpha(g_{kl})(h_{ijk})$.

We can also introduce $\mathfrak{h}$-valued 2-forms B_i on each U_i by $B_i = \bar{t}^{-1} F_{A_i}$, where $f_1 = dA + A \wedge A$. Alternatively, we can treat them as independent 2-forms, with suitable gluing conditions on double overlaps, and regard the condition $F_{A_i} = \bar{t}(B_i)$ as an equation of motion.

Just like in the case of ordinary connections, there is a notion of gauge equivalence of flat 2-connections. Let $(A_i, g_{ij}, \lambda_{ij}, h_{ijk})$ and $(A_i', g_{ij}', \lambda_{ij}', h_{ijk}')$ be a pair of 2-connections. A gauge-equivalence between them is a G-valued function g_i and an $\mathfrak{h}$-valued 1-form λ_i on each U_i, together with an H-valued function h_{ij} on every U_{ij} such that

- On each U_i one has
$$A'_i = g_i A_i g_i^{-1} + g_i d g_i^{-1} - \bar{t}(\lambda_i).$$
- On each U_{ij} one has $g'_{ij} = t(h_{ij}) g_j g_{ij} g_i^{-1}$, and
$$\begin{aligned}\lambda'_{ij} = h_{ij}\left(\alpha(g_j)(\lambda_{ij}) + \lambda_j\right) h_{ij}^{-1} - \alpha(g'_{ij})(\lambda_i) \\ + h_{ij} d h_{ij}^{-1} + h_{ij} \bar{t}^{-1}(A'_j) h_{ij}^{-1} - \bar{t}^{-1}(A'_j).\end{aligned}$$
- On each U_{ijk} one has
$$h'_{ijk} = h_{ik} \alpha(g_k)(h_{ijk}) h_{jk}^{-1} \alpha(g'_{jk})(h_{ij}^{-1}).$$

There are also 2-gauge transformations between gauge transformations, see [16, 17, 2] for details.

If 2-forms B_i are regarded as independent variables, then the action in d space-time dimensions has the form
$$S = \int \langle (f_1 - \bar{t}(B)), \wedge b \rangle + S_{\text{top}},$$
where b is a Lagrange multiple $(d-2)$-form with values in $\mathfrak{g}^*$ and S_{top} is a topological action which is invariant under gauge equivalence of flat 2-connections. If we regard B_i as dependent variables, then the action contains only S_{top}. Below we will determine the most general form of S_{top} in dimension 4 and lower.

For now, let us show that the moduli space of flat 2-connections can be reformulated purely in terms of the data $(\Pi_1, \Pi_2, \alpha, \beta)$. First, we use the 1-form gauge transformations to set $A_i = 0$ (and consequently $B_i = 0$). This ensures that the 1-forms λ_{ij} is pure gauge:
$$\lambda_{ij} = \bar{t}^{-1}(g_{ij} d g_{ij}^{-1}),$$
and simultaneously that for any two points $p, p' \in U_{ij}$ we have
$$g_{ij}^{-1}(p) g_{ij}(p') \in \operatorname{im} t.$$
Thus if we project g_{ij} to $\Pi_1 = \operatorname{coker} t$, the resulting Π_1-valued function is constant. We denote this function a_{ij}.

Next, let us choose a map $s : \Pi_1 \to G$ which is an inverse of the projection $G \to \Pi_1$. In general, we cannot choose s to be a homomorphism, but we can always choose it so that $s(x)^{-1} = s(x^{-1})$ for all $x \in \Pi_1$. By definition, $s(\mathsf{a}_{ij}) g_{ij}^{-1}$ is a function on U_{ij} which takes values in im t. Thus there exists an H-valued function h_{ij} such that $t(h_{ij}) = s(\mathsf{a}_{ij}) g_{ij}^{-1}$. Performing another gauge transformation, now with $\lambda_i = 0$, $g_i = 1$ and h_{ij} chosen as above, we bring the data defining a 2-connection to the form
$$A_i = 0, \quad \lambda_{ij} = 0, \quad g_{ij} = s(\mathsf{a}_{ij}).$$
In addition, the functions $h_{ijk} : U_{ijk} \to H$ are now constant. Thus the equivalence class of the flat 2-connection is completely determined by the constant functions

$\mathsf{a}_{ij} : U_{ij} \rightarrow \Pi_1$ and $h_{ijk} : U_{ijk} \rightarrow H$. On every U_{ijk} these functions satisfy a constraint:

$$t(h_{ijk}) = s(\mathsf{a}_{ik})s(\mathsf{a}_{ji})s(\mathsf{a}_{kj}),$$

which implies that the functions a form a Cěch 1-cocycle with values in Π_1:

$$\mathsf{a}_{ik}\mathsf{a}_{ji}\mathsf{a}_{kj} = 1.$$

Furthermore, h_{ijk} can be expressed in terms of the functions a and a constant function $\mathsf{b}_{ijk} : U_{ijk} \rightarrow \Pi_2$:

$$h_{ijk} = \mathsf{b}_{ijk}\widetilde{f}(\mathsf{a}_{ik}, \mathsf{a}_{ji}),$$

where $\widetilde{f} : \Pi_1 \times \Pi_1 \rightarrow H$ is some lift of $f : \Pi_1 \times \Pi_1 \rightarrow \text{im } t$ defined by $f(x, x') = s(x)s(x')s(x \cdot x')^{-1}$. The functions b_{ijk} form a 2-cochain with values in Π_2. This cochain is a twisted cocycle, in the sense that it satisfies

$$\begin{aligned} &\alpha(\mathsf{a}_{kl})(\mathsf{b}_{ijk}) - \mathsf{b}_{ijl} + \mathsf{b}_{ikl} - \mathsf{b}_{jkl} \\ &\quad = \widetilde{f}(\mathsf{a}_{il}, \mathsf{a}_{ji})\widetilde{f}(\mathsf{a}_{jl}, \mathsf{a}_{kj}) \left(\widetilde{f}(\mathsf{a}_{il}, \mathsf{a}_{ki})\alpha(s(\mathsf{a}_{kl}))(\widetilde{f}(\mathsf{a}_{ik}, \mathsf{a}_{ji})\right)^{-1}. \end{aligned} \tag{2}$$

Here we used additive notation for the group operation on Π_2, since Π_2 is an abelian group. Note that the right-hand side takes values in $\Pi_2 = \ker t$, because $f : \Pi_1 \times \Pi_1 \rightarrow \text{im } t$ is a 2-cocycle.

The left-hand side of this equation can be interpreted as a differential δ_{a} in the Cěch complex computing the cohomology of a local system on X. This local system has Π_2 as its fiber and is associated to the Π_1 local system defined by the 1-cocycle a via the action of Π_1 on Π_2. The right-hand side also has a nice interpretation, if we recall [13] that a 1-cocycle a with values in Π_1 can be interpreted as defining a map $\mathsf{a} : X \rightarrow B\Pi_1$, and that a 2-group (G, H, t, α) defines a class $\beta \in H^3(B\Pi_1, \Pi_2)$. Then one can show that the right-hand side is a 3-cochain representing the class $\mathsf{a}^*\beta$.

To summarize, a flat 2-connection is determined by a 1-cocycle a with values in Π_1 and a 2-cochain b with values in a Π_2 local system satisfying the twisted cocycle condition

$$\delta_{\mathsf{a}}\mathsf{b} = \mathsf{a}^*\beta.$$

The residual gauge equivalences are described by constant Π_1-valued functions f_i on each U_i and by constant Π_2-valued functions μ_{ij} on each U_{ij}. The former transformations act as follows:

$$\mathsf{a}_{ij} \mapsto f_j\mathsf{a}_{ij}f_i^{-1}, \quad \mathsf{b}_{ijk} \mapsto \alpha(f_k)(\mathsf{b}_{ijk}).$$

The latter transformations act trivially on a and shift the 2-cochain b by a coboundary:

$$\mathsf{b}_{ijk} \mapsto \mathsf{b}_{ijk} - \alpha(\mathsf{a}_{jk})(\mu_{ij}) - \mu_{jk} - \mu_{ki}.$$

4. 2-gauge theory on a lattice

The lattice formulation of the $\mathbb{G}$ gauge theory is most easily defined with respect to a triangulation of the space-time X. The simplices need to be oriented so that our formulas have the proper signs. This can be achieved by ordering the vertices. We use the convention that higher vertex labels point towards lower vertex labels since our group elements act on the left. This is demonstrated in figures below. The theory depends on neither the triangulation nor this ordering.

To get a lattice description of the $\mathbb{G}$ gauge theory we use the description of flat 2-connections in terms of the data $(\Pi_1, \Pi_2, \alpha, \beta)$ obtained in the previous section. All one has to do is to rewrite the cochains a and b and conditions on them in simplicial terms. Since this is standard, we simply state the results.

A field configuration is an assignment of an element of $A \in \Pi_1$ to each edge (1-simplex) and of an element $B \in \Pi_2$ to each triangle (2-simplex). These correspond to the cochains a and b of the previous section. They are subject to the following constraints. Given a triangle with edge variables A_0, A_1, A_2, labeled in the ordinary way according to the induced ordering on the vertices we have a flatness constraint

$$A_0 A_1^{-1} A_2 = 1, \tag{3}$$

where the labeling and orientation is demonstrated in the figure below.

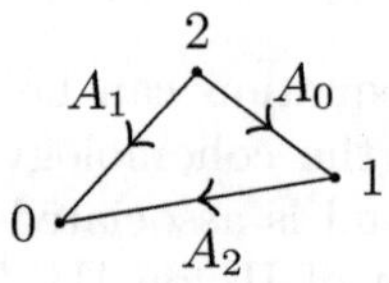

Labeling for edge variables on 2-simplices.

This ensures that the holonomy of A around any closed curve only depends on the homotopy class of that curve. Note that group elements are thought of as acting from the left.

There is also a constraint for every tetrahedron (3-simplex). For a tetrahedron with face variables $B_0, B_1, B_2, B_3 \in \Pi_2$ labeled in the ordinary way and edge variables $A_0, A_1, A_2 \in \Pi_1$ labeled as shown in the picture below, one requires:

$$\alpha(A_0)(B_0) - B_1 + B_2 - B_3 = \beta(A_0, A_1, A_2). \tag{4}$$

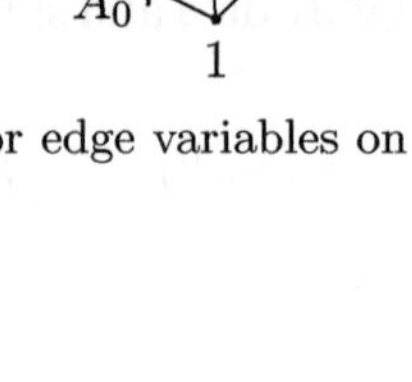

Labeling for edge variables on 3-simplices.

Using the differential d_A on twisted cochains this constraint can be written as

$$d_A B = \beta(A). \tag{5}$$

There are two kinds of gauge transformations for this theory. First, there are ordinary (0-form) gauge transformations depending on a 0-cochain f with values in Π_1. Such a transformation acts on A as follows:

$$A \mapsto A^f, \quad A_\gamma \mapsto A^f_\gamma = f_0 A_\gamma f_1^{-1}, \tag{6}$$

where A_γ is the value of A on an oriented 1-simplex γ, and f_0 and f_1 are the values of f on the two endpoints of γ. These transformations also affect the B field:

$$B_\Sigma \mapsto B^f_\Sigma = \alpha(f_0)(B_\Sigma) + \zeta_\Sigma(A, f) \tag{7}$$

on each 2-simplex Σ. Here ζ is a Π_2-valued 2-cochain satisfying

$$d_{A^f}\zeta(A, f) = \beta(A^f) - \alpha(f)(\beta(A)). \tag{8}$$

In what follows we adopt a simplified notation for the action of a 0-cochain f with values in Π_1 on a p-cochain ν with values in a local system with fiber Π_2: instead of $\alpha(f)(\nu)$ we write $f \cdot \nu$. Thus the definition of ζ takes the form

$$d_{A^f}\zeta(A, f) = \beta(A^f) - f \cdot \beta(A), \tag{9}$$

and the action of f on B takes the form

$$B \mapsto B^f = f \cdot B + \zeta(A, f). \tag{10}$$

The inhomogeneous term in the transformation (10) is necessary to preserve the twisted flatness condition (5). Indeed:

$$d_{A^f} B^f = d_{A^f}(f \cdot B + \zeta(A, f)) = d_{A^f}(f \cdot B) + \beta(A^f) - f \cdot \beta(A). \tag{11}$$

A simple calculation shows

$$d_{A^f}(f \cdot B) = f \cdot d_A B = f \cdot \beta(A),$$

hence the pair (B^f, A^f) satisfies the condition (5).

Note that the equation (9) for ζ always has a solution, since the cohomology class of $\beta(A)$ does not change when one replaces A by a gauge-equivalent one (this follows from the fact that β is closed). We can even choose ζ to satisfy the normalization condition

$$\zeta(1^g, f) = 0, \tag{12}$$

for all constant $f \in \Pi_1$ and gauge-trivial $A = 1^g$. This normalization is possible since if f is constant, the cochain β can be normalized so that $\beta(1^{gf}) = f \cdot \beta(1^g)$.

The second class of gauge transformations are 1-form gauge transformations depending on a 1-cochain a with values in Π_2. These transformations do not affect the 1-cocycle A, while B transforms as follows:

$$B_\Sigma \mapsto B^a_\Sigma = B_\Sigma + \alpha(A_2)(a_0) - a_1 + a_2, \tag{13}$$

for every 2-simplex Σ whose boundary $\partial\Sigma$ is assigned a_0, a_1, a_2. In other words, $B \mapsto B^a = B + d_A a$. These transformations preserve the twisted flatness constraint for B since $d_A^2 = 0$.

The 1-form gauge transformations have 2-gauge transformations parametrized by an assignment of an element of Π_2 to each vertex under which the 1-form gauge parameters a transform as

$$a^g_\gamma = a_\gamma + g_1 - \alpha(A_\gamma)(g_0), \tag{14}$$

where $\partial\gamma$ is assigned g_0, g_1. One can check that two 1-form gauge transformations related by such a 2-gauge transformation act identically on the fields.

1-form and 0-form gauge transformations do not commute in general. Let a be a 1-cochain with values in the Π_2 local system associated to a Π_1 local system, and f be a 0-cochain with values in Π_1. Let us denote by (a, f) as a 1-form transformation with a parameter a followed by a 0-form transformation with a parameter f. Then

$$(a, f) = (0, f) \circ (a, 1) = (f \cdot a, 1) \circ (0, f). \tag{15}$$

For future use, we record the transformations of B and A under a general transformation (a, f):

$$B^{a,f} = f \cdot (B + d_A a) + \zeta(A, f), \tag{16}$$

and

$$A^{a,f} = A^f. \tag{17}$$

When composing two gauge transformations, where the first one is (a_1, f_1), it is often convenient to write the second one as $(f_1 \cdot a_2, f_2)$ rather than (a_2, f_2). We will use this convention below.

The reader might have noticed the similarity between β and ζ on one hand, and the Chern–Simons form and its descendant on the other hand. This is not a coincidence, since the Chern–Simons form for a compact Lie group G can be regarded as a 3-cocycle on the classifying space of flat G-connections [18]. In superstring theory and supergravity, the Chern–Simons form modifies the definition of the gauge-invariant field-strength of a 2-form gauge field B, while the descendant of the Chern–Simons form describes how the B-field transforms under ordinary gauge symmetries [19]. This is clearly analogous to how β modifies the flatness constraint on B (eq. (5)) and how ζ enters the transformation law for B (eq. (7)).

It is well known that the descent procedure can be continued, so that from the 2-cochain $\zeta(A, f)$ one gets a 1-cochain $\kappa(A, f, g)$, etc. Below we will need the definition of κ only for $A = 0$, so let us describe how it is constructed. By definition,

$$\beta(A^g) - g \cdot \beta(A) = d_{A^g}\zeta(A, g),$$

where A is an arbitrary 1-cochain with values in Π_1 and g is an arbitrary 0-cochain with values in Π_1. Let us set $A = 1^f$, where f is an arbitrary 0-cochain with values in Π_1 (i.e., we let A be cohomologous to the trivial 1-cochain). Then we get

$$\beta(1^{gf}) - g \cdot \beta(1^f) = d_{1^{gf}}\zeta(1^f, g) = gf \cdot d\left((gf)^{-1} \cdot \zeta(1^f, g)\right).$$

This equation implies that the 2-cochain

$$(gf)^{-1} \cdot \zeta(1, gf) - f^{-1} \cdot \zeta(1, f) - (gf)^{-1} \cdot \zeta(1^f, g)$$

is closed for any 0-cochains f and g. Contractibility of EG implies that this cochain is exact, i.e., there exists a 1-cochain $\kappa(g,f)$ with values in Π_2 such that

$$(gf)d\kappa(g,f) = g\cdot\kappa(1,f) + \zeta(1^f,g) - \zeta(1,gf). \tag{18}$$

This 1-cochain κ is the second descendant of β.

The partition function is defined as the weighted sum over all allowed field configurations, divided by the order of the group of gauge transformations and multiplied by the order of the group of 2-gauge transformations. The weight must be gauge-invariant and topologically-invariant, in the sense that the partition function must be invariant under a subdivision of the triangulation. The most general weight function satisfying these conditions will be described in Section 6.

5. Loop and surface observables

The above construction of the 2-group TQFT works in arbitrary space-time dimension (and gives something different from the Dijkgraaf–Witten theory in dimension 3 or higher). We now discuss observables in this TQFT focusing on the 4d case.

The 2-group TQFT in 4d describes the phase with both electric gauge group Π_1 and magnetic gauge group Π_2 and therefore admits two types of loop observables and two types of surface observables. Let us begin with loop observables. There are ordinary Wilson loops for the lattice gauge field A; they are labeled by representations of Π_1. There are also disorder loop operators which correspond to 't Hooft loops in the microscopic gauge theory. In the lattice formulation, one chooses a closed path γ on the dual cell complex and modifies the twisted cocycle condition (4) on every 3-simplex T_l dual to an edge of $l\in\gamma$ as follows:

$$\alpha(A_0)(B_0) - B_1 + B_2 - B_3 = \beta(A_0,A_1,A_2) + H_l.$$

Here H_l is an element of Π_2. One can think of the elements H_l as defining a dual 1-cochain with values in a Π_2 local system on the loop γ. The above equation implies that this cochain is a cocycle, so the elements H_l for different l all lie in the same orbit of the Π_1 action on Π_2. Therefore 't Hooft loops are labeled by orbits of the Π_1-action in Π_2.

There are also order and disorder surface observables. Disorder surface observables are defined by the condition that the Π_1 gauge field has a fixed holonomy along a loop linking a homologically trivial surface Σ. In the lattice formulation, Σ should be thought of as composed of 2-cells of the dual cell complex, and the insertion of the disorder operator supported on Σ amounts to deforming the 1-cocycle condition on A for every 2-simplex dual to the 2-cells of Σ. Such surface observables are labeled by conjugacy classes in Π_1.

There are also Wilson surface observables are labeled by elements of $\hat{\Pi}_2 = \mathrm{Hom}(\Pi_2, U(1))$ which satisfy two conditions. First of all, $\eta\in\hat{\Pi}_2$ must be invariant under the action of Π_1. Second, $\eta\circ\beta\in H^3(\Pi_1,U(1))$ must vanish.

A quick way to see how these two conditions on η arise is as follows. The Wilson surface measures the flux of B though Σ. Thus it must involve a product over all 2-simplices making up Σ, with each simplex contributing a phase $\eta(\pm B)$, where the sign is determined by the mutual orientation of Σ and the 2-simplex in question. Invariance with respect to gauge transformations requires η to Π_1-invariant, giving the first condition on η. To obtain the second condition, note that the Wilson surface observable should evaluate to 1 if Σ is the boundary of a 3-simplex and no 't Hooft loop intersects this 3-simplex. Applying η to the twisted cocycle condition (4) we see that this is true if $\eta \circ \beta(A_0, A_1, A_2) = 0$ for all $A_0, A_1, A_2 \in \Pi_1$. More generally, suppose the 3-cochain $\eta \circ \beta(A_0, A_1, A_2)$ is not zero but is a coboundary of a 2-cochain on Π_1 with values in $U(1)$:

$$\eta \circ \beta(A_0, A_1, A_2) = \gamma(A_0, A_1 A_2)\gamma(A_0 A_1, A_2)^{-1}\gamma(A_1, A_2)\gamma(A_0, A_1)^{-1}.$$

Then one can modify the definition of the Wilson surface observable by multiplying the weight assigned to a 2-simplex by a factor $\gamma(A_0, A_1)$. The modified observable is trivial on an elementary 3-simplex, as required. On the other hand, if $\eta \circ \beta$ is not cohomologous to zero, no local modification of the weight can solve the problem.

Apart from the surface observables measuring the flux of B, one can also construct surface observables which are sensitive only to the 1-cocycle A. Namely, if we restrict A to a closed surface Σ, it defines a principal G-bundle on Σ with structure group Π_1, and any class in $H^2(B\Pi_1, U(1))$ gives a surface observable. Such observables are trivial if Σ is simply-connected, but are nontrivial in general.

The most general observable supported on a surface Σ involves both the B fluxes and the A-holonomies along a marked 1-skeleton of Σ. The Π_1 invariance of such operators may rely on the transformation of both A and B variables, so they are not necessarily products of Wilson lines and Wilson surfaces.

In three dimensions the analysis is very similar, so we just present the results. There are local operators (i.e., operators supported at points) labeled by orbits of Π_1 action on Π_2. These correspond to 't Hooft point operators in the microscopic gauge theory. There are two kinds of loop observables: Wilson loops labeled by representations of Π_1 and vortex loops labeled by conjugacy classes in Π_1. Finally there are Wilson surfaces defined in the same way as in four dimensions. That is, they are labeled by a Π_1-invariant element $\eta \in \hat{\Pi}_2$ which annihilates the class β, together with a class in $H^2(B\Pi_1, U(1))$.

6. Topological actions

As described in the appendix, the configuration data (A, B) are equivalent to a simplicial map $\mathcal{B}$ from the spacetime X into the classifying space $B\mathbb{G}$ of the 2-group. Gauge equivalence classes are homotopy classes of these maps. This is analogous to the fact that a flat G-connection can be viewed as a map to the classifying space BG. We therefore define an action functional for a d-dimensional theory by picking a class $\mathcal{L} \in H^d(B\mathbb{G}, \mathbb{R}/\mathbb{Z})$ and setting

$$S(\mathcal{B}) = 2\pi i \int_X \mathcal{B}^* \mathcal{L}. \tag{19}$$

This action is gauge invariant and manifestly topological.

To calculate the cohomology group $H^d(B\mathbb{G}, \mathbb{R}/\mathbb{Z})$ we need a good understanding of the classifying space $B\mathbb{G}$. This space can be taken to be any space with $\mathbb{G}$ as its homotopy type. This implies $\pi_1(B\mathbb{G}) = \Pi_1$, $\pi_2(B\mathbb{G}) = \Pi_2$, and all higher homotopy groups vanish. It also means that the action of $\pi_1(B\mathbb{G})$ on $\pi_2(B\mathbb{G})$ is given by α as well as a more complicated condition involving the Postnikov invariant β. In the appendix we describe a cell structure for this space.

It turns out $B\mathbb{G}$ is a fibration over $B\Pi_1$ with fiber $B^2\Pi_2$:

$$\begin{array}{ccc} B^2\Pi_2 & \longrightarrow & B\mathbb{G} \\ & & \downarrow \\ & & B\Pi_1 \end{array}$$

Here $B^2\Pi_2$ is a space defined up to homotopy by the condition that $\pi_2(B^2\Pi_2) = \Pi_2$ while all other homotopy groups vanish. This fibration is classified by the Postnikov class $\beta \in H^3(B\Pi_1, \Pi_2)$. We know the cohomology of the base and fibers so we can use the Serre spectral sequence to get a reasonable handle on the cohomology of $B\mathbb{G}$. Details of this spectral sequence are in the appendix.

The Serre spectral sequence tells us that for 2-dimensional theories, $\mathcal{L}$ is a sum of two terms. The first term is an element of $H^2(B^2\Pi_2, \mathbb{R}/\mathbb{Z})$. This cohomology group is isomorphic to the group $\hat{\Pi}_2$ of homomorphisms $\Pi_2 \to \mathbb{R}/\mathbb{Z}$. A character $\eta \in \hat{\Pi}_2$ gives a class on $B\mathbb{G}$ if and only if it is invariant under the action of Π_1 and $\eta(\beta) \in H^3(B\Pi_1, \mathbb{R}/\mathbb{Z})$ is zero as a cohomology class. An interpretation of these conditions was discussed in the previous section.

The second term can be any element $\omega \in H^2(B\Pi_1, \mathbb{R}/\mathbb{Z})$.

The most general 2d action is therefore

$$S(\mathcal{B}) = 2\pi i \int_X \eta(B) + 2\pi i \int_X \mathcal{A}^*\omega,$$

where here B is the Π_2-valued 2-form field and $\mathcal{A} : X \to B\Pi_1$ is the composition of the classifying map $\mathcal{B} : X \to B\mathbb{G}$ with the fibration map $B\mathbb{G} \to B\Pi_1$. Equivalently this is the map $X \to B\Pi_1$ determined by the A variables. Note that if $X = S^2$, then since $\pi_2(B\Pi_1) = 0$, the pullback $\mathcal{A}^*\omega$ always integrates to zero.

In three dimensions, we also find that $\mathcal{L}$ is a sum of two terms. The first term is an element $\omega \in H^3(B\Pi_1, \mathbb{R}/\mathbb{Z})$. The action depends on this element only up to the addition of a term of the form $\eta(\beta)$, where η is a Π_1-invariant character of Π_2.

The second term is an element $\lambda \in H^1(B\Pi_1, \hat{\Pi}_2)$. As a cochain on $B\mathbb{G}$, λ is closed only if $\langle \lambda, \cup\beta \rangle \in H^4(B\Pi_1, \mathbb{R}/\mathbb{Z})$ vanishes as a cohomology class.

The most general 3d action is therefore

$$S(\mathcal{B}) = 2\pi i \int_X \mathcal{A}^*\lambda \cup B + 2\pi i \int_X \mathcal{A}^*\omega.$$

Here the cup product between $\mathcal{A}^*\lambda$ and B is formed using the pairing between Π_2, where B takes values, and Π_2^*, where λ takes values. Note that on a simply-connected space ω does not contribute since $\mathcal{A}^*\omega$ is exact. On the other hand, although $\mathcal{A}^*\lambda$ is exact, B is not closed, so their contraction is not necessarily exact.

In four dimensions the action is a sum of three terms. The first term depends on an element of $H^4(B^2\Pi_2, \mathbb{R}/\mathbb{Z})$. This group is isomorphic to the group of quadratic functions $q : \Pi_2 \to \mathbb{R}/\mathbb{Z}$ as discussed in [3]. In order for this cocycle to extend to $B\mathbb{G}$, it needs to be invariant under the action of Π_1. This is equivalent to the quadratic function q being invariant. In order for its extension to be closed, $\langle \beta, - \rangle_q$ needs to vanish as an element of $H^3(B\Pi_1, \hat{\Pi}_2)$, where the bracket denotes the associated bilinear form for q.

The second term depends on an element $\lambda \in H^2(B\Pi_1, \hat{\Pi}_2)$. This gives a cocycle on $B\mathbb{G}$ when contracted with the B-field only if $\langle \lambda, \cup\beta \rangle$ is zero in $H^5(B\Pi_1, \mathbb{R}/\mathbb{Z})$.

The third term is an integral of a pull-back of $\omega \in H^4(B\Pi_1, \mathbb{R}/\mathbb{Z})$. The action only depends on ω up to the addition of something of the form $\langle \gamma, \cup\beta \rangle$ for some $\gamma \in H^1(B\Pi_1, \hat{\Pi}_2)$.

The most general 4d action is therefore

$$S(\mathcal{B}) = 2\pi i \int_X q_*(\mathfrak{P}B) + 2\pi i \int_X \mathcal{A}^*\lambda, \cup B + 2\pi i \int_X \mathcal{A}^*\omega,$$

where $\mathfrak{P}B$ denotes the Pontryagin square of the B-field, and q_* is the map from the universal quadratic group $\Gamma(\Pi_2)$ to $\mathbb{R}/\mathbb{Z}$ corresponding to q. This term is discussed in detail in [3]. Note that because of the Postnikov class β, it is only closed after applying q_*. For simply-connected X, as in the three-dimensional case, ω does not contribute, but λ does.

In the case of Dijkgraaf–Witten theory in dimension d, the topological action depends on a class in $H^d(BG, U(1))$, and one can give an explicit description of the corresponding cocycle as a function of d variables living in G. From the mathematical viewpoint, this explicit description arises from the standard complex computing the cohomology of BG, while from the physical viewpoint the cocycle is the weight attached to a d-simplex [1]. Similarly, one can give an explicit description of a class in $H^d(B\mathbb{G}, U(1))$ as a function of several variables, some of them living in Π_1 and some living in Π_2, satisfying a certain condition. This condition can be understood mathematically as a cocycle condition in a standard complex computing the cohomology of $B\mathbb{G}$. However, even in low degrees the formulas are quite unwieldy. For this reason we only state them for $d = 2$. In this case the cocycle is a function $\mathcal{L}$ on $\Pi_1 \times \Pi_1 \times \Pi_2$ with values in $\mathbb{R}/\mathbb{Z}$. The cocycle condition is

$$\mathcal{L}(A_0, A_1A_2; B_2) - \mathcal{L}(A_0A_1, A_2; B_1) - \mathcal{L}(A_0, A_1; B_3) + \mathcal{L}(A_1, A_2; B_0) = 0,$$

where B_0 is not independent but is expressed through the variables A_0, A_1, A_2, B_1, B_2, B_3 by means of (4). 2-cocycles for Π_1 can be identified with 2-cocycles for $B\mathbb{G}$ which do not depend on the Π_2 variable.

A 3-cocycle for a 2-group $B\mathbb{G}$ depends on three variables in Π_1 and three variables in Π_2. In general, a d-cocycle depends on d variables in Π_1 and $d(d-1)/2$ variables in Π_2. They can be thought of as labeling edge and triangles containing a given vertex of a d-simplex.

7. Duality

Since the simplicial B-field takes values in an abelian group Π_2, one could try to dualize it. In dimension d the dual variable should live on $(d-3)$-cells of the dual cell complex and take values in the Pontryagin-dual group $\hat{\Pi}_2 = \mathrm{Hom}(\Pi_2, U(1))$. But in general a nontrivial topological action for B obstructs the dualization procedure. An important special case where the dualization can be performed is when the action is either independent of B, or depends on it linearly. Let us perform the dualization procedure in various dimensions.

For $d = 2$ the 2-group TQFT is essentially equivalent to the Dijkgraaf–Witten theory with gauge group Π_1. First of all, the constraint (4) is not needed in this case. Second, the class η in the 2d action must vanish, because otherwise the partition function on S^2 vanishes after one sums over the B-fields, which contradicts the axioms of TQFT [20]. But then summation over B only produces an inessential numerical factor, and the theory is clearly equivalent to the Dikgraaf–Witten theory for Π_1 with an action given by $\omega \in H^2(B\Pi_1, \mathbb{R}/\mathbb{Z})$.

For $d = 3$ the action is always linear in B, as explained in the previous section. Therefore the dualization is always possible. We impose the constraint (4) by means of a Lagrange multiplier field C which lives on dual 0-cells and takes values in $\hat{\Pi}_2$. Thus we add to the action a term

$$2\pi i \int_X \langle C, \delta_A B - \beta(A) \rangle$$

and treat B as an unconstrained 2-cochain with values in Π_2. For simplicity, let us first assume that Π_1 acts trivially on Π_2. Then the differential δ_A becomes the usual Cěch differential δ, and summing over B produces a constraint

$$\delta C = -A^* \lambda,$$

where $\lambda \in H^1(B\Pi_1, \hat{\Pi}_2)$.

Since we assumed for now that α is trivial, the group $H^1(B\Pi_1, \hat{\Pi}_2)$ is merely the group of homomorphisms from Π_1 to $\hat{\Pi}_2$. Thus the constraint on C reads explicitly

$$C_1 - C_0 + \lambda(A_l) = 0,$$

where l is a 1-cell of the dual cell complex with source and target 0 and 1.[1] This constraint is gauge-invariant provided we assign to C a nontrivial transformation law under Π_1 gauge transformations:

$$C \mapsto C + \lambda(f),$$

where f is a 0-cochain with values in Π_1 parameterizing a gauge transformation. Thus the dual theory is a topological sigma-model with target $\hat{\Pi}_2$ coupled to a

[1]Here we neglected the fact that the field A was originally only defined on the 1-cells of the original triangulation. In more detail, one needs first to pass from the simplicial 1-cochain A to a Π_1 local system on X and then restrict to the 1-cells of the dual cell complex. The first step involves a certain arbitrariness which does not change the conclusions.

topological 3d gauge theory with gauge group Π_1. The group Π_1 acts on $\hat{\Pi}_2$ via the homomorphism λ. The action of the gauge theory is

$$S = 2\pi i \int_X A^*\omega + 2\pi i \int_X \langle C, A^*\beta\rangle.$$

Note that while the first term is the Dijkgraaf–Witten action of the 3d gauge theory, the second term is of a different nature.

The case of nontrivial α is not very different. The constraint on C now reads

$$\delta_A C + \lambda(A) = 0.$$

The transformation law for C is now more complicated:

$$C \mapsto \hat{\alpha}(f)(C) + \lambda(f),$$

where $\hat{\alpha}$ denotes the action of Π_1 on $\hat{\Pi}_2$ dual to the action of Π_1 on Π_2. The dual theory is again a gauged topological sigma-model with the same action as above.

For $d = 4$ we need to assume that the first term in the topological action vanishes (i.e., $q = 0$). Then the action is again linear in B, and B can be dualized to a 1-cochain C with values in $\hat{\Pi}_2$. Performing summation over B we find a constraint on C which reads

$$\delta_A C = \lambda(A),$$

where λ is a cocycle representing a class in $H^2(\Pi_1, \hat{\Pi}_2)$. We also have the usual constraint which says that A is a 1-cocycle with values in Π_1. One can show that these two constraints can be interpreted as the flatness constraint for a gauge field (A, C) taking values in an extension of Π_1 by $\hat{\Pi}_2$ determined by α and λ. Thus the dual theory is a topological gauge theory with this extension as the gauge group. The action is necessarily of the DW type and has the form

$$S = 2\pi i \int_X A^*\omega + 2\pi i \int_X \langle C, \cup A^*\beta\rangle.$$

Note that even if the original 2-group TQFT had a trivial action (i.e., λ and ω vanish), the dual theory has a nontrivial action which is determined by β. On the other hand, the class λ which parameterized the action of the original theory enters the dual theory only through the structure of the gauge group.

8. Phases protected by higher symmetry

8.1. SPT phases and TQFT

Recently, the Dijkgraaf–Witten TQFT was used in a novel way, as a tool to classify symmetry-protected (SPT) phases without long-range entanglement [7, 8, 9]. These are phases of matter which have three properties: (1) they are gapped; (2) they have a global symmetry group G which acts ultralocally; (3) the TQFT describing the low-energy limit is trivial. Here by an ultralocal action of a global symmetry we mean that there is a lattice realization of the phase where the symmetry transformation only mixes degrees of freedom living on a given vertex (0-cell).

It was proposed in [7] that SPT phases in space-time dimension d with a finite internal[2] symmetry group G are classified by elements of $H^d(BG, U(1))$. Let us provide an interpretation of this classification scheme in TQFT terms (see also [9] for a very similar discussion). We use the fact that a lattice system with an ultralocal internal symmetry can be canonically coupled to a flat gauge field with structure group G. Indeed, locally any flat G-connection is a pure gauge, i.e., a gauge transformation of the trivial connection. This gauge transformation is defined up to a constant symmetry transformation in G. Since the system is local, the lattice action is a sum over all vertices

$$\sum_v S_v(\phi),$$

where each term S_v depends only on the degrees of freedom in the immediate neighborhood of the vertex. Since the symmetry acts ultralocally, each S_v is separately invariant under constant symmetry transformations. We define the action of the system coupled to a flat background gauge field as a sum

$$\sum_v S_v(\phi^g),$$

where ϕ^g is a transformation of the field configuration by a local gauge transformation g describing the flat connection. Since different choices of g differ by constant symmetry transformations, this expression is well defined and gauge-covariant. Now we can integrate out the matter fields and obtain an effective action for the flat gauge field. It is necessarily topological and therefore must arise from a class in $H^d(BG, U(1))$.

It is natural to ask whether more general 2-group TQFTs we have studied here and in [2, 3] are related to new phases of matter not covered by the group cohomology classification. The role of the symmetry group G is taken by a 2-group $\mathbb{G}$. In this more general setting, a global symmetry transformation is parameterized by an element of Π_1 and a flat gauge field with gauge group Π_2. The action of $\mathbb{G}$ is assumed to be ultralocal, in the sense that the system can be canonically coupled to a flat 2-connection (A, B) with structure group $\mathbb{G}$. If the system is gapped and the ground state has short-range entanglement, one can integrate out the matter fields and obtain a topological action for (A, B) which is described by an element of $H^d(B\mathbb{G}, U(1))$.

In the case when Π_2 is trivial, this reduces to the group cohomology classification of SPT phases. The opposite extreme is when Π_1 is trivial. In this case $\mathbb{G}$ is completely determined by an abelian group Π_2, and phases protected by such a 2-symmetry are labeled by $H^d(B^2\Pi_2, U(1))$. The first nontrivial case is $d = 4$, where Π_2-protected phases are classified by quadratic functions on Π_2 with values in $U(1)$. The same data classify pre-modular braided tensor categories whose simple objects are labeled by elements of Π_2. Since pre-modular categories are also

[2]We will focus on the case of finite internal symmetry, but one can generalize it to the case when G is a compact Lie group which might involve time-reversal.

used in the construction of Walker–Wang TQFTs [10], it seems likely that 2-group TQFTs with trivial Π_1 are a special case of Walker–Wang models. Indeed, it has been conjectured in [10] (see also [11]) that in the continuum limit Walker–Wang models based on $\mathbb{Z}_n$ are described by a BF action deformed by a term $B \wedge B$. This agrees with the continuum description of the 2-group TQFT discussed in [2, 3].

A simple example of a system with an ultralocal 2-symmetry is given by a Yang–Mills theory with gauge group G where all matter fields transform trivially under the center of G. Then the symmetry 2-group has $\Pi_1 = 0$ and $\Pi_2 = Z(G)$. The system can be "minimally coupled" to a flat B-field with values in $Z(G)$. Essentially, this means that one performs the path-integral over Yang–Mills gauge fields with structure group $G/Z(G)$ and a fixed 't Hooft flux described by B. The resulting function of B must be an integral over X of a pull-back of a class in $H^d(B^2Z(G), U(1))$. In the case $d = 4$, one can interpret the elements of this group as labeling discrete theta-angles in the underlying Yang–Mills theory [21, 3].

Even more generally, if one dealing with a gapped phase in space-dimension d, symmetry transformations may live on cells of all dimensions up to d. This situation is most natural when the "matter fields" involve gauge fields of all form degrees. The symmetry structure in this case is described by a d-group, i.e., by a d-category with a single object and invertible 1-morphisms, 2-morphisms, etc. SPT phases in d dimensions with symmetry d-group $\mathbb{G}$ should be classified by degree-d cohomology of the classifying space of $\mathbb{G}$. This classifying space has homotopy groups which may be non-vanishing in degrees up to d.

8.2. Boundaries of SPT phases

In the case of an SPT phase protected by a finite symmetry group G, the nontriviality of the corresponding class in $H^d(BG, U(1))$ has interesting physical consequences for the boundary behavior of the phase [8]: the boundary cannot be gapped without either breaking symmetry or introducing long-range entanglement. The same appears to be true in the case of phases protected by a d-group symmetry. Namely, the boundary cannot be gapped by any perturbation which preserves $\mathbb{G}$ as an ultralocal symmetry and does not create long-range entanglement. Indeed, if such a perturbation existed, one could couple the system to a flat d-connection and integrate out the matter fields even in the presence of a nonempty boundary. This should produce a topological action for the flat d-connection which is is gauge-invariant on a d-manifold with a boundary. Such an action should have the form

$$S = 2\pi i \int_X \mathcal{A}^*\omega - 2\pi i \int_{\partial X} \mathcal{A}^*\psi$$

for some $\psi \in H^{d-1}(B\mathbb{G}, \mathbb{R}/\mathbb{Z})$. But this action is gauge-invariant only if $\omega = \delta\psi$, which contradicts the nontriviality of $\omega \in H^d(B\mathbb{G}, \mathbb{R}/\mathbb{Z})$. Thus if $\mathbb{G}$ is preserved on the boundary, the boundary can be gapped only at the expense of creating a topological order on the boundary. Moreover, neither the bulk action, nor the boundary TQFT action are separately gauge-invariant in this case (there is an "anomaly-inflow" from the bulk to the boundary). Thus $\mathbb{G}$ is realized anomalously on the boundary, and this anomaly is measured by ω.

Alternatively, if we insist on having no topological order on the boundary, $\mathbb{G}$ must be explicitly or spontaneously broken there, and we can describe possible patterns of symmetry breaking. Namely, $\mathbb{G}$ must be broken down to a d-subgroup $\mathbb{H}$ such that ω becomes exact when restricted to $\mathbb{H}$, $\omega|_{\mathbb{H}} = \delta\psi$. Only then is it possible to write a gauge-invariant effective action in the presence of a boundary.

8.3. Lattice realization of an SPT phase

As in [7], we can give descriptions of ungauged ground states that realize this phase in the "group cohomology" basis. Suppose that $\mathbb{G}$ is a 2-group described by a quadruple $(\Pi_1, \Pi_2, \alpha, \beta)$, as above.[3] Let M denote the spatial manifold. We will compute the ground state on M by performing the path integral in the TQFT over a spacetime with boundary M. For example, if M is a sphere, we take our spacetime B to be a ball. For these theories, spacetime does not need to be smooth, so we can always take spacetime to be CM, the cone over M.

A map from CM to $B\mathbb{G}$ is the same thing as a map from M to $B\mathbb{G}$ along with a nullhomotopy. In other words it is a gauge transformation for a $\mathbb{G}$ gauge field living just on the spatial slice M. This gives us a way of describing the map from the cone purely in terms of configurations on M and for $\mathbb{G}$ an ordinary group, and in this way we reproduce the description of ground states in [7].

Concretely, a configuration of the "matter field" Φ will be an assignment of an element of Π_1 to every vertex, an element of Π_2 to every edge, Π_3 to every face, and so on, for which a single d-simplex is pictured below with 2-group labeling, with $\phi_i \in \Pi_1$, $a_i \in \Pi_2$ (see Figure 1).

The ground state in the basis $|\Phi\rangle$ is

$$|\omega\rangle = \sum_{\Phi} \exp\left[2\pi i \int_{CM} \omega(d\Phi)\right] |\Phi\rangle. \tag{20}$$

The integral over CM indicates a sum over all simplices as above. The integrand is formed from a cocycle representative ω of the class of the action (note different representative give different states in the same phase). This d-cocycle is evaluated on each simplex using the labels given by ϕ. The notation $d\Phi$ indicates the induced map $CM \to B\mathbb{G}$. It means that we first extend Φ by the identity on all labels on the interior of CM. Then $d\Phi$ is the gauge field generated by a gauge transformation parametrized by this extension. Finally, ω is evaluated on this gauge field in the ordinary way to obtain an element of $\mathbb{R}/\mathbb{Z}$. We draw it in Figure 2 for Figure 1 above using the rules derived in Section 4.

Note that since $d\Phi$ is a gauge transformation, $\omega(d\Phi)$ is exact, and so $\int_{CM} \omega(d\Phi)$ reduces to an integral over the boundary $\partial CM = M$, ie. the spatial slice, so (20) is of a very similar form to the states considered in [7]. In particular, it is short-range entangled in the sense of [7].

[3]The case of a general d-group is algebraically more involved.

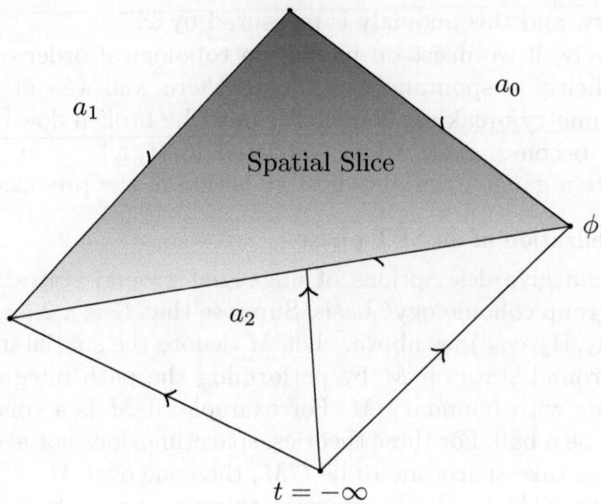

FIGURE 1. Triangulation of part of CM, with the cone point at time past infinity.

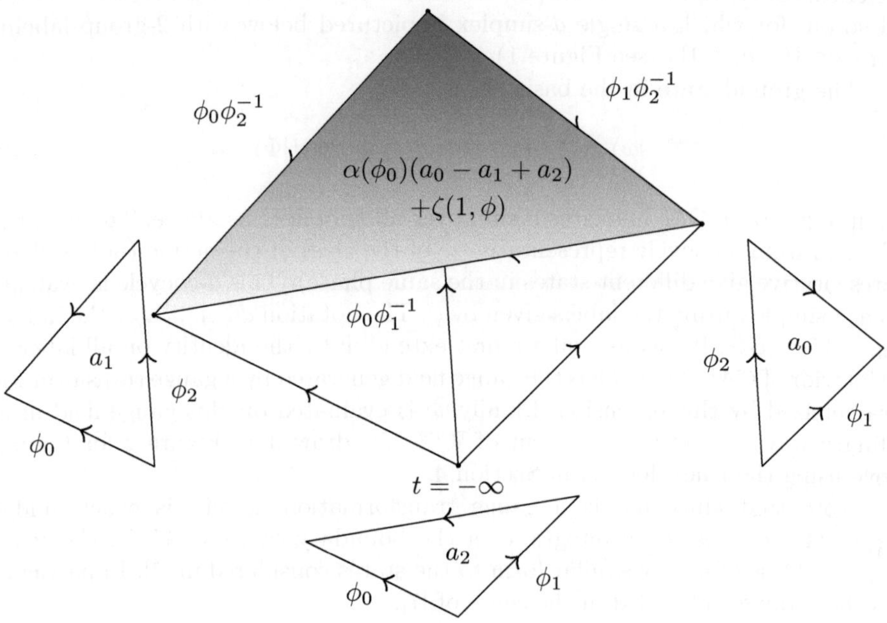

FIGURE 2

Let us derive the action of the 2-group $\mathbb{G}$ on the matter fields. When the matter field configuration $\Phi = (a, \phi)$ is trivial, a parameter of the global $\mathbb{G}$ symmetry is the same as a parameter (λ, f) of a gauge symmetry except that f is constant and $d\lambda = 0$. A global symmetry transformation (λ, f) should act on the matter fields $\Phi \mapsto \Phi'$ so that $d\Phi' = d\Phi^{\phi\cdot\lambda, f}$, where on the right-hand side we have a gauge transformation with a parameter $(\phi \cdot \lambda, f)$. This will ensure that the state $|\omega\rangle$ is invariant (see below). Let us use this property to deduce Φ'.

Let $\Phi = (a, \phi)$, then $d\Phi = (B, A)$, where

$$B = \zeta(1, \phi) + \phi \cdot da, \quad A = 1^\phi. \tag{21}$$

If we transform this field configuration by $(\phi \cdot \lambda, f)$, the 2-form part becomes

$$f \cdot \zeta(1, \phi) + (f\phi) \cdot d\lambda + \zeta(1^\phi, f) + (f\phi) \cdot da.$$

Since $d\lambda = 0$, the second term vanishes, and by (12) and since f is constant, so does the third. We are left with

$$f \cdot (\zeta(1, \phi) + \phi \cdot da).$$

Compare with (21). Meanwhile the 1-form part of $d\Phi'$ is simply $1^{f\phi}$.

One can easily check that the same 2-connection can be obtained by acting on a trivial 2-connection by a gauge transformation parametrized by $\Phi' = (a + \phi \cdot \lambda + \kappa(f, \phi), f\phi)$, where κ was defined in (18). In other words, the global symmetry transformation rules for matter fields are

$$\phi \mapsto f\phi,$$
$$a \mapsto a + \phi \cdot \lambda + \kappa(f, \phi).$$

Further, since ω is invariant under the simultaneous action of π_1 on itself and on π_2, the global symmetry transformation preserves $\omega(d\Phi)$ and therefore simply permutes the summands in $|\omega\rangle$, leaving the state invariant. Gauging the symmetry leads to the 2-group TQFT with cocycle ω.

8.4. Examples

Let us collect here a couple examples of interesting 2-groups. First is the 2-group with $\Pi_1 = (\mathbb{Z}/p)^3$, $\Pi_2 = \mathbb{Z}/p$, α trivial, and $\beta(a_1, a_2, a_3) = a_1a_2a_3$. A crossed module that realizes this 2-group is

$$\mathbb{Z}/p \to UT(p, 3) \to UT(p, 4) \to \mathbb{Z}/p^3,$$

where $UT(p, k)$ is the group of upper triangular $k \times k$ matrices over $\mathbb{F}_p$ with 1s on the diagonal (generalizing the Heisenberg group $k = 3$). The center map places the three northeast elements of a 3×3 matrix into the three northeast elements of a 4×4 matrix.

We have already computed

$$\zeta(a, f) = f_1(a_2 + df_2)(a_3 + df_3) + a_1 f_2(a_3 + df_3) + a_1a_2f_3.$$

and

$$\kappa(g, \phi) = -g_1\phi_2 d\phi_3.$$

Under a global symmetry transformation parametrized by (g, λ) we therefore have

$$\begin{aligned}\phi_1 &\mapsto \phi_1 + g_1\\ \phi_2 &\mapsto \phi_2 + g_2\\ \phi_3 &\mapsto \phi_3 + g_3\\ \alpha &\mapsto \alpha + \lambda - g_1\phi_2 d\phi_3.\end{aligned}$$

Our next example is a 2-group with $\Pi_1 = \mathbb{Z}/p$, $\Pi_2 = \mathbb{Z}/p$, α trivial, and

$$\beta(a) = \widetilde{a}\frac{\delta\widetilde{a}}{p},$$

where $\widetilde{a}$ denotes a lift of a, which is ordinarily just defined mod p, to an integer-valued cochain. Thus β is an integer defined mod p. We have used this lifting throughout the paper so far, where we might write $\beta = a\delta a/p$, but it will be important for trivializing β that we keep it explicit here. The derivation is a little bit technical, so one can feel free to skip to the global symmetry transformations.

Indeed, plugging in a gauge trivial configuration $a = \delta f$, the abridged notation makes β look identically zero since $\delta^2 = 0$. However, $\delta\widetilde{\delta f} \neq 0$. Below we illustrate for $p = 2$ how an f supported at a vertex with both an incoming and outgoing edge has this property.

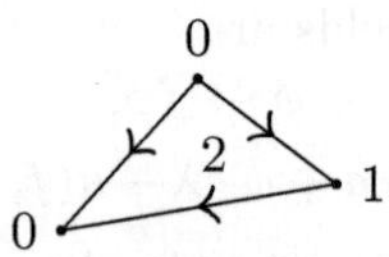

An f with $\delta\widetilde{\delta f} \neq 0$.

On the other hand, $\widetilde{\delta f} = \delta\widetilde{f} \mod p$, so since β is defined mod p we can write

$$\beta(\delta f) = \delta\widetilde{f}\frac{\delta\widetilde{\delta f}}{p} = \delta(\widetilde{f}\frac{\delta\widetilde{\delta f}}{p}).$$

Call the potential in parenthesis on the right-hand side $\zeta(0, f)$. Next we consider for constant g

$$\zeta(0, f) - \zeta(0, f + g) = -\widetilde{g}\frac{\delta\widetilde{\delta f}}{p} = -\delta(\widetilde{g}B(f)),$$

where $B(f)$ is an integral 1-cochain made by labelling each edge with a 1 if the value of $\widetilde{f}$ at the source of the edge is larger than the value at the end of the edge. Let us show this identity. When forming $\widetilde{\delta f}$, we use an integral lift with values in $[0, p-1]$. Then the edges to which we must add p to make positive are the ones that are negative. Since δf on an edge $0 \to 1$ is $f_1 - f_0$, this occurs iff $f_0 > f_1$. When computing $\delta\widetilde{\delta f}$, the values of f all cancel and we are just left with these ps, which we divide by p to get 1s. Putting this together, $\delta\widetilde{\delta f}/p = \delta B(f)$. The descendant potential we defined above for this example is thus $\kappa(g, f) = -\widetilde{g}B(f)$.

In the above description of the ground state of an SPT with this symmetry group, we have matter consisting of $\mathbb{Z}/p$ labels ϕ at vertices and $\mathbb{Z}/p$ labels α along edges. Under a global symmetry parametrized by (g, λ) we have

$$\phi \mapsto \phi + g$$
$$\alpha \mapsto \alpha + \lambda - \widetilde{g}B(\phi).$$

Gauging this symmetry produces a gauge field with gauge group given by the 2-group just described.

Appendix: Classifying space of a 2-group

This discussion follows [15]. In this section we implicitly think of $\mathbb{G}$ as a 2-category.

The classifying space of a 2-group $\mathbb{G}$ is a topological space $B\mathbb{G}$ with $\mathbb{G}$ as its homotopy type. We can construct $B\mathbb{G}$ inductively as a cell complex with 1 0-cell, 1-cells corresponding to 1-morphisms in $\mathbb{G}$, 2-cells corresponding to 2-morphisms, 3-cells corresponding to relations among the 2-morphisms, 4-cells added to kill any π_3 introduced in the previous stage, 5-cells added to kill any π_4, and so on. Note that the relations among 1-morphisms are imposed by inserting the identity 2-morphisms, and any π_2 created among these is killed by the 3-cells imposing the relations among 2-morphisms.

Consider a configuration for the $\mathbb{G}$ gauge theory. Using a section $s : \Pi_1 \to G$ to make the A variables live in G and including the B variables in H, we can interpret this configuration as a composable diagram in $\mathbb{G}$. Mapping each 0-cell of X to the unique 0-cell of $B\mathbb{G}$, 1-cells to the 1-cell of the corresponding 1-morphism (an element of G), and 2-cells to the 2-cell of the corresponding 2-morphism (this is generally an element of H but the constraint (3) implies it is actually an element of Π_2), we obtain a map from the 2-skeleton of X to $B\mathbb{G}$. The 3-cell constraint (4) implies that we can extend this map to all of X. Changing which section one uses amounts to a gauge transformation of the original configuration.

Conversely, by cellular approximation any map $X \to B\mathbb{G}$ gives us a configuration for the $\mathbb{G}$ gauge theory. Cellular homotopies are gauge transformations, so we always get gauge-equivalent configurations if we pick a different cellular approximation.

There is a map of 2-groups, which may be thought of as a functor, from $\mathbb{G}$ to the group Π_1 (considered as a 2-group with only identity 2-morphisms) given by identifying isomorphic 1-morphisms in $\mathbb{G}$. This gives a map on the 3-skeleton of $B\mathbb{G}$ to $B\Pi_1$. Since the higher cells are added to kill homotopy groups for each space, we can inductively extend this to a map $B\mathbb{G} \to B\Pi_1$. The fiber of this map over the unique 0-cell of $B\Pi_1$ is the classifying space $B^2\Pi_2$ of the group of 2-morphisms from the identity 1-morphism to itself. This space has second homotopy group Π_2 and all others vanishing.

This map is well known to be a fibration

$$\begin{array}{ccc} B^2\Pi_2 & \longrightarrow & B\mathbb{G} \\ & & \downarrow \\ & & B\Pi_1, \end{array}$$

which is classified by the Postnikov class $\beta \in H^3(B\Pi_1, \Pi_2)$. For this fibration, the E_2 page of the Serre spectral sequence is the α-equivariant cohomology

$$H^p(B\Pi_1, H^q(B^2\Pi_2, \mathbb{Z})).$$

The shape of the relevant piece is

$$\begin{array}{ccccccc} \star & & & & & & \\ 0 & 0 & 0 & & & & \\ \star & \star & \star & \star & & & \\ 0 & 0 & 0 & 0 & 0 & & \\ 0 & 0 & 0 & 0 & 0 & 0 & \\ \mathbb{Z} & 0 & \star & \star & \star & \star & \star . \end{array}$$

Note that p labels the columns and q labels the rows.

The bottom row is $H^p(B\Pi_1, \mathbb{Z})$.

The rows with $q = 1, 2, 4$ all vanish because Hurewicz' theorem implies $H_1(B^2\Pi_2, \mathbb{Z}) = H_3(B^2\Pi_2, \mathbb{Z}) = 0$. From the universal coefficient theorem it then follows $H^1(B^2\Pi_2, \mathbb{Z}) = 0$, and since all cohomology classes on $B^2\Pi_2$ in positive degree are $|\Pi_2|$-torsion, the 2nd and 4th cohomology groups also vanish.

The universal coefficient theorem also tells us that

$$H^3(B^2\Pi_2) = \mathrm{Hom}(\Pi_2, \mathbb{R}/\mathbb{Z}) = \hat{\Pi}_2,$$

so the $q = 3$ row is $H^p(B\Pi_1, \hat{\Pi}_2)$, where Π_1 acts on $\hat{\Pi}_2$ via α. For example, $H^0(B\Pi_1, \hat{\Pi}_2)$ is the subgroup of Π_1-invariant characters in $\hat{\Pi}_2$.

It is also known that $H^5(B^2\Pi_2, \mathbb{Z}) = H^4(B^2\Pi_2, \mathbb{R}/\mathbb{Z})$ is the group of quadratic functions $q : \Pi_2 \to \mathbb{R}/\mathbb{Z}$ [22]. The isomorphism is discussed in detail in [3]. The group in the $(0, 5)$ spot in the top left is then the subgroup of Π_1-invariant quadratic forms.

The first possibly non-zero differential is on the E_3 page:

$$H^0(B\Pi_1, H^5(B^2\Pi_2, \mathbb{Z})) \to H^3(B\Pi_1, \hat{\Pi}_2).$$

We find it difficult to prove, but we believe that this map sends an element of the left-hand side, which is a Π_1-invariant quadratic form $q : \Pi_2 \to \mathbb{R}/\mathbb{Z}$ to $\langle \beta, - \rangle_q$, where the bracket denotes the bilinear pairing $\langle x, y \rangle_q = q(x+y) - q(x) - q(y)$.

The next possibly non-zero differentials are on the E_4 page:

$$H^j(B\Pi_1, \hat{\Pi}_2) \to H^{j+4}(B\Pi_1, \mathbb{Z}) \simeq H^{j+3}(B\Pi_1, \mathbb{R}/\mathbb{Z}).$$

This map is contraction with β.

The last relevant possibly non-zero differential is on the E_6 page:

$$H^0(B\Pi_1, H^5(B^2\Pi_2, \mathbb{Z})) \to H^6(B\Pi_1, \mathbb{Z}).$$

We believe this differential is actually zero.

This is enough to give the description of the topological actions we give in Section 6.

References

[1] R. Dijkgraaf and E. Witten, "Topological Gauge Theories and Group Cohomology," Commun. Math. Phys. **129**, 393 (1990).

[2] S. Gukov and A. Kapustin, "Topological Quantum Field Theory, Nonlocal Operators, and Gapped Phases of Gauge Theories," arXiv:1307.4793 [hep-th].

[3] A. Kapustin and R. Thorngren, "Topological Field Theory on a Lattice, Discrete Theta-Angles and Confinement," arXiv:1308.2926 [hep-th].

[4] R. Thorngren, "Electric-Magnetic Duality of Topological Gauge Theories from Compactification," arXiv:1309.1489 [hep-th].

[5] F. Quinn, "Lectures on axiomatic topological quantum field theory," in: *Geometry and Quantum Field Theory, Park City, UT, 1991,* IAS/Park City Math. Ser. **1**, 323–453, AMS, 1995.

[6] D.S. Freed, M.J. Hopkins, J. Lurie and C. Teleman, "Topological Quantum Field Theories from Compact Lie Groups," arXiv:0905.0731 [math.AT].

[7] X. Chen, Z.-C. Gu, Z.-X. Liu and X.-G. Wen, "Symmetry protected topological orders and the cohomology class of their symmetry group," Phys. Rev. B **87**, 155114 (2013) [arXiv:1106.4772 [cond-mat.str-el]].

[8] M. Levin and Z.-C. Gu, "Braiding statistics approach to symmetry-protected topological phases," Phys. Rev. B **86**, 115109 (2012) [arXiv:1202.3120 [cond-mat.str-el]].

[9] L.-Y. Hung and X.-G. Wen, "Quantized topological terms in weakly coupled gauge theories and their connection to symmetry protected topological phases," arXiv:1211.2767 [cond-mat.str-el].

[10] K. Walker and Z. Wang, "(3+1)-TQFTs and Topological Insulators," arXiv:1104.2632 [cond-mat.str-el].

[11] C.W. von Keyserlingk, F.J. Burnell and S.H. Simon, "Three-dimensional topological lattice models with surface anyons," Phys. Rev. B **87**, 045107 (2013) [arXiv:1208.5128 [cond-mat.str-el]].

[12] J.C. Baez, A.D. Lauda, "Higher-Dimensional Algebra V: 2-groups," arXiv:math/0307200 [math.QA].

[13] A. Hatcher, "Algebraic Topology," Cambridge University Press, 2001.

[14] K. Brown, "Cohomology of groups", Springer, 1982.

[15] J. Baez and D. Stevenson, "The Classifying Space of a Topological 2-Group", [http://arxiv.org/abs/0801.3843 [math.AT]]

[16] J. Baez and U. Schreiber, "Higher gauge theory: 2-connections on 2-bundles," hep-th/0412325.

[17] U. Schreiber and K. Waldorf, "Connections on non-abelian gerbes and their holonomy," arXiv:0808.1923 [hep-th].

[18] J. Cheeger and J. Simons, "Differential characters and geometric invariants," in: *Geometry and Topology* (*College Park,* 1983/84), Springer, 1985.

[19] J. Polchinski, "String theory. Vol. 2: Superstring theory and beyond," Cambridge, UK: Univ. Pr. (1998) 531 pp.

[20] E. Witten, "Quantum Field Theory and the Jones Polynomial," Commun. Math. Phys. **121**, 351 (1989).

[21] O. Aharony, N. Seiberg and Y. Tachikawa, "Reading between the lines of four-dimensional gauge theories," arXiv:1305.0318 [hep-th].

[22] S. Eilenberg, S. MacLane, "On the groups $H(\Pi, n)$. II. Methods of computation," Ann. Math. (2) **60** (1954) 49.

Anton Kapustin
California Institute of Technology
Division of Physics, Mathematics and Astronomy
1200 E California Blvd.
Pasadena, CA 91125, USA

Ryan Thorngren
University of California
Math. Department
Evans Hall, Berkeley CA, 94720, USA

Progress in Mathematics, Vol. 324, 203–260

Constructing Buildings and Harmonic Maps

Ludmil Katzarkov, Alexander Noll, Pranav Pandit
and Carlos Simpson

Happy Birthday Maxim!

Abstract. In a continuation of our previous work [21], we outline a theory which should lead to the construction of a universal pre-building and versal building with a ϕ-harmonic map from a Riemann surface, in the case of two-dimensional buildings for the group SL_3. This will provide a generalization of the space of leaves of the foliation defined by a quadratic differential in the classical theory for SL_2. Our conjectural construction would determine the exponents for SL_3 WKB problems, and it can be put into practice on examples.

Mathematics Subject Classification (2010). Primary: 14H30; Secondary: 34E20, 51E24, 30F15, 30F30.

Keywords. Building, Grothendieck topology, polyhedral complex, spectral curve, spectral network, BPS state, WKB exponent.

1. Introduction

Let X be a Riemann surface, compact for now. The moduli spaces of representations, vector bundles with connection, and semistable Higgs bundles denoted M_B, M_{DR} and M_H respectively, are isomorphic as spaces, and we denote the common underlying space just by M. This space has three different algebraic structures, and the Betti and de Rham ones share the same complex manifold.

These algebraic varieties are noncompact, indeed M_B is affine, but they have compactifications. First, $\overline{M}_{DR}$ and $\overline{M}_H$ are natural orbifold compactifications, and when M itself is smooth, the orbifold compactifications are smooth.

The Betti moduli space, otherwise usually known as the character variety, admits many natural compactifications. Indeed the mapping class group acts on M_B but it doesn't stabilize any one of them. Gross, Hacking, Keel and Kontsevich [13] have recently studied more closely the problem of compactifying M_B and show that there are indeed some optimal choices with good properties.

Morgan–Shalen [26] for SL_2, and recently Parreau [27] for groups of higher rank, construct a compactification of M where the points at ∞ are actions of the fundamental group on $\mathbb{R}$-buildings. This compactification is related to the inverse limit of the algebraic compactifications of M_B, and we shall call it $\overline{M}^P_B$.

Let $\overline{M}^{TSC}$ be the Tychonoff–Stone–Čech compactification. It is universal, so it maps to the other ones. The points at infinity may be identified with non-principal ultrafilters[1] ω on M. We may therefore consider the maps

$$\begin{array}{ccccc} \overline{M}_H & \leftarrow & \overline{M}^{TSC} & \rightarrow & \overline{M}^P_B. \\ & & \downarrow & & \\ & & \overline{M}_{DR} & & \end{array}$$

Given a non-principal ultrafilter ω, consider its image ω_{DR} in $\overline{M}_{DR}$ (resp. ω_H in $\overline{M}_H$) and its image ω^P_B in $\overline{M}^P_B$.

Our basic question is to understand the relationship between ω_{DR} (resp. ω_H) and ω^P_B. In what follows we concentrate on ω_{DR} but there are also conjectures for ω_H as mentioned in [21], with recent results by Collier and Li [6] and Mochizuki [25].

The divisors at infinity in $\overline{M}_{DR}$ and $\overline{M}_H$ are both the same. They are

$$D = M^*_H/\mathbb{C}^*$$

where $M^*_H := M_H - \text{Nilp}$ is the complement of the nilpotent cone. Recall that $\text{Nilp} = f^{-1}(0)$ where $f : M_H \to \mathbb{A}^N$ is the Hitchin fibration.

Hence ω_{DR} (and similarly ω_H) may be identified as an equivalence class of points in M^*_H modulo the action of $\mathbb{C}^*$. We may therefore write

$$\omega_{DR} = (E, \varphi)$$

as a semistable Higgs bundle, such that the Higgs field φ is not nilpotent, and this identification holds up to nonzero complex scaling of φ.

It turns out that the essential features of the correspondence with $\overline{M}^P_B$ depend not on φ up to complex scaling, but up to real scaling. We therefore introduce the *real blowing up* $\widetilde{M}_{DR}$ of $\overline{M}_{DR}$ along the divisor D. It is still a compactification of M_{DR}. The boundary at infinity is $\widetilde{D}$ which is an S^1-bundle over D, with

$$\widetilde{D} = M^*_H/\mathbb{R}^{+,*}.$$

Let $\widetilde{\omega}_{DR}$ denote the image of ω in $\widetilde{D} \subset \widetilde{M}_{DR}$. We may again write

$$\widetilde{\omega}_{DR} = (E, \varphi)$$

but this time φ is defined up to positive real scaling.

[1]Here, by an ultrafilter on a normal (T_4) topological space, we mean a maximal filter consisting of closed subsets, as were considered by Wallman [34].

We denote by ϕ the *spectrum* of φ. It consists of a multivalued tuple of holomorphic differential forms on X. This is equivalent to saying that it is a point in the Hitchin base $\mathbb{A}^N$, or again equivalently, a spectral curve $\Sigma \subset T^*X$. For us, it will be most useful to think of writing $\phi = (\phi_1, \ldots, \phi_r)$ locally over X, where ϕ_i are holomorphic differential forms. There will generally be some singularities at points over which $\Sigma \to X$ is ramified, and as we move around in the complement of these singularities the order of the ϕ_i may change. Denote by X^* the complement of the set of singularities.

Gaiotto–Moore–Neitzke [11] define the *spectral network* associated to ϕ. This is one of the main players in our story. We refer to [11, 12] for many pictures, and to our paper [21] for some specific pictures related to the BNR example. A brief overview will be given in Section 3.1 below.

These structures constitute our understanding of $\widetilde{\omega}_{DR}$.

On the other side, the point ω_B^P may be identified with an action of $\pi_1(X)$ on an $\mathbb{R}$-building Cone_ω. The theory of Gromov–Schoen and Korevaar–Schoen allows us to choose an equivariant harmonic mapping $h : \widetilde{X} \to \mathrm{Cone}_\omega$. This is most often uniquely determined by the action. The metric on the building, and hence the differential of the harmonic mapping, are well defined up to positive real scaling.

Here, we use the terminology "harmonic map" to an $\mathbb{R}$-building to mean a continuous map such that the domain, minus a singular set of real codimension 2, admits an open covering where each open set maps into a single apartment, and these local maps are harmonic mappings to Euclidean space. The differential of a harmonic map is the real part of a multivalued differential form.

In [21] we used the groupoid version of Parreau's theory [27] to construct the harmonic mapping h, and the classical local WKB approximation to show that its differential is ϕ.

Theorem 1.1 ([21]). *The differential of the harmonic mapping h is the real part of the multivalued differential ϕ considered above:*

$$dh = \Re\phi.$$

We note that Collier and Li have proven the corresponding result for the Hitchin WKB problem in some cases [6], and more recently Mochizuki does it in general [25].

This theorem suggests the following question.

Question 1.2. To what extent does ϕ determine h and hence ω_B^P?

In what follows we shall assume that we have passed to the universal cover of the Riemann surface so $\widetilde{X} = X$. Furthermore, since our considerations now, about how to integrate the differential ϕ into a harmonic map h, concern mainly bounded regions of X, we may envision other examples. Such might originally have come from noncompact Riemann surfaces and differential equations with irregular singularities. The BNR example in [21] was of that form.

A *ϕ-harmonic map* will mean a harmonic map h from X to a building B, such that its differential is $dh = \Re\phi$.

The goal of this paper is to sketch a theory which should lead to the answer to Question 1.2, at least for the group SL_3 and under certain genericity hypotheses about the absence of "BPS states".

1.1. Contents

We start by reviewing briefly the theory of harmonic maps to trees, viewed as buildings for SL_2, with the universal map given by the leaf space of a foliation defined by a quadratic differential. The remainder of the paper is devoted to generalizing this classical theory to higher rank buildings and in particular, as discussed in Section 3, to two-dimensional buildings for SL_3.

In order to go towards an algebraic viewpoint, we indicate in Section 4 how one can view a building as a presheaf on a certain Grothendieck site of *enclosures*. This allows one to formulate a small-object argument completing a pre-building to a building.

In Section 5 we describe the initial construction associated to a spectral curve ϕ. This is the building-like object, admitting a ϕ-harmonic map from X, obtained by glueing together small local pieces.

The initial construction will have points of positive curvature, and the main work is in Sections 6 and 8 where we describe a sequence of modifications to the initial construction designed to remove the positive curvature points. One of the main difficulties is that a point where only four 60° sectors meet, admits in principle two different foldings. Information coming from the original harmonic map serves to determine which of the two foldings should be used. In order to keep track of this information, we introduce the notion of "scaffolding". The result of Sections 6 and 8 is a sequence of moves leading to a sequence of two-manifold constructions.

Our main conjecture is that when the spectral network [11] of ϕ doesn't have any BPS states, this sequence is well defined and converges locally to a two-manifold construction with only nonpositive curvature. The universal pre-building is obtained by putting back some pieces that were trimmed off during the procedure.

In Section 7, coming as an interlude between the two main sections, we present an extended example. We show how to treat the BNR example which we had already considered rather extensively in [21], but from the point of view of the general process being described here. It was through consideration of this example that we arrived at our process, and we hope that it will guide the reader to understanding how things work.

In Section 9 we present just a few pictures showing what can happen in a typical slightly more complicated example. This example and others will be considered more extensively elsewhere.

In Section 10 we consider some consequences of our still conjectural process, and indicate various directions for further study.

1.2. Convention

This paper is intended to sketch a picture rather than provide complete proofs. All stated theorems, propositions and lemmas are actually "quasi-theorems": plausible and tractable statements for which we have in mind a potential method of proof.

Statements which should be considered as open problems needing considerably more work to prove, are labeled as conjectures.

2. Trees and the leaf space

Let us first consider the case of representations into SL_2. Exact WKB analysis and its relationship with the geometry of quadratic differentials has been considered extensively by Iwaki and Nakanishi [17]. See also Dumas [7].

In the Morgan–Shalen–Parreau theory, the limiting building Cone_ω is then an $\mathbb{R}$-tree. In saying $\mathbb{R}$-*tree* we include the data of a distance function which is the standard one on any apartment. The choice of metric is well defined up to a global scalar on Cone_ω.

The spectral curve $\Sigma \to X$ is a 2-sheeted covering defined by a quadratic differential $q \in H^0(X, K_X^{\otimes 2})$. The multivalued differential form ϕ is just the set of two square roots: $\phi_i = \pm\sqrt{q}$. The singularities are the zeros of q, and we shall usually assume that these are simple zeros. It corresponds to saying that Σ is smooth which, for a 2-sheeted cover, implies having simple branch points.

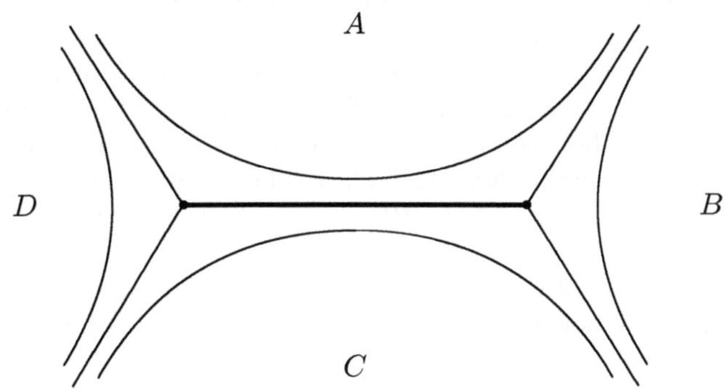

FIGURE 1. A foliation with BPS state

The differential form $\Re\phi$ is well defined up to a change of sign. Therefore, it defines a single real direction at every nonsingular point of X, which we call the *foliation direction.* These directions are the tangent directions to the leaves of a real foliation, which we call the *foliation defined by* ϕ or equivalently the *foliation defined by the quadratic differential* q.

Suppose T is an $\mathbb{R}$-tree and $h : \widetilde{X} \to T$ is a ϕ-harmonic map. Then, the closed leaves of the foliation defined by ϕ map to single points in T. This is clear from the differential condition at smooth points. By continuity it also extends across the singularities: the closed leaves are defined to be the smallest closed subsets which are invariant by the foliation. In pictures, a leaf entering a three-fold singular point therefore generates two branches of the leaf going out in the other two directions, and all three of these branches have to map to the same point in T.

We assume that the *space of leaves of the foliation* is well defined as an $\mathbb{R}$-tree. Denote it by T_ϕ. The points of T_ϕ are by definition just the closed leaves of the foliation.

In order to consider the universal property, we say that a *folding map* $u : \mathbb{R} \to T'$ is a map such that there is a locally finite decomposition of $\mathbb{R}$ into segments, such that on each segment u is an isometric embedding. A map $u : T \to T'$ between $\mathbb{R}$-trees is a *folding map* if its restriction to each apartment is a folding map. Note that an isometric embedding is a particularly nice kind of folding map which doesn't actually fold anything.

Corollary 2.1. *The projection map $h_\phi : X \to T_\phi$ is a ϕ-harmonic map, universal among ϕ-harmonic maps to $\mathbb{R}$-trees. That says that if $h : X \to T$ is any other ϕ-harmonic map, there exists a unique folding map $T_\phi \to T$ making the diagram*

$$\begin{array}{ccc} X & \to & T_\phi \\ & \searrow & \downarrow \\ & & T \end{array}$$

commute.

See [7]. In general, we have to admit the possibility that $T_\phi \to T$ be a folding map rather than an isometric embedding. Let us look at an example which illustrates this. Suppose that our quadratic differential or equivalently ϕ, has two singular points which are on the same leaf of the foliation. See Figure 1.

The leaf segment between the two singularities is emphasized, it is called a "saddle connection" or *BPS-state* [11]. These play a key role in the wallcrossing story. Here, it is the existence of the BPS state which leads to nonrigidity of the harmonic map.

The space of leaves of the foliation is a tree T_ϕ with four segments. In Figure 2 these segments are labeled in the same way as the corresponding regions in the previous picture of the foliation.

Suppose we have a ϕ-harmonic map h to another tree T. The universal property gives a factorizing map $u : T_\phi \to T$.

The map h is supposed to send small open sets in X^* to single apartments in T. The small open sets in X provide some small segments in the tree T_ϕ, and the segments in T_ϕ which are images of these neighborhoods, have to map by isometric embeddings (i.e., not be folded) under u.

Little neighborhoods along the segments joining the regions for example A to B, B to C, C to D and D to A, as well as a neighborhood along the BPS state, are shown in Figure 3.

The four outer neighborhoods project to four corner segments as shown in Figure 4. Therefore u cannot fold these segments.

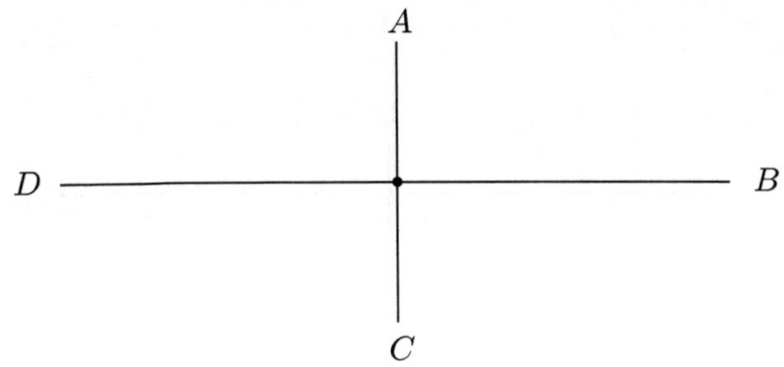

FIGURE 2. The quotient tree

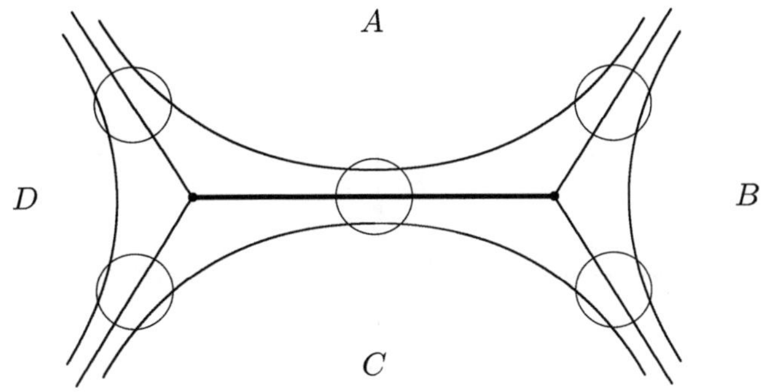

FIGURE 3. Some little neighborhoods

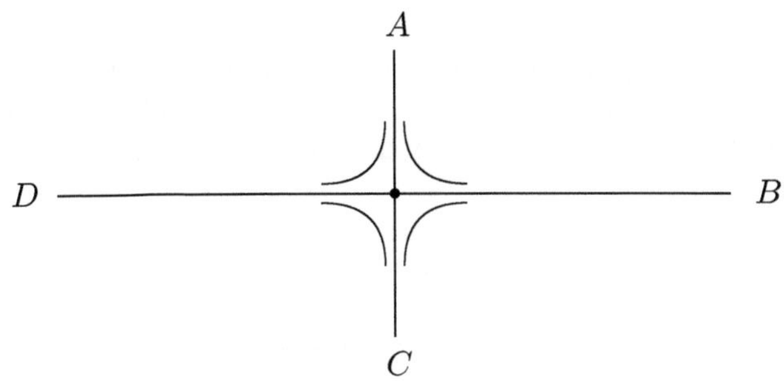

FIGURE 4. Non-folded segments

The little neighborhood along the central leaf projects to the segment shown in Figure 5 joining the upper edge A of the tree and the lower edge C. It follows that u is not allowed to fold the two segments A and C together.

However, u could very well fold the segments B and D together since they are not constrained by any neighborhoods in X. Therefore, we can have a harmonic[2]

[2]R. Wentworth pointed out to us that our composed map to the tree in Figure 6, while locally harmonic, will not however be globally energy-minimizing.

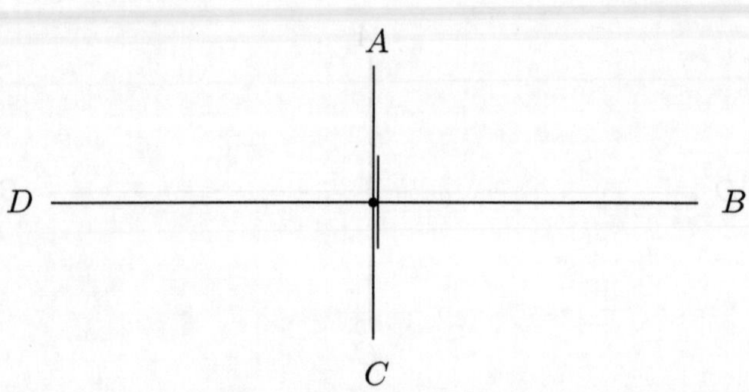

FIGURE 5. Non-folded segment from central neighborhood

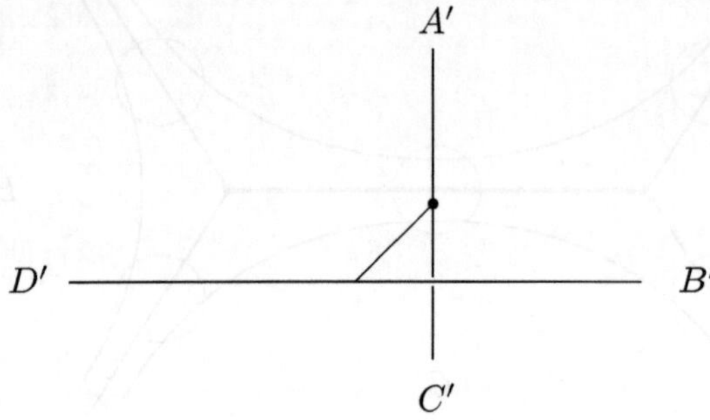

FIGURE 6. A folded tree

map to the tree T shown in Figure 6. The central dot goes to the central dot, the segments A and C go to A' and C', and the segments B and D go to the paths from the central dot out to B' and D', which are folded together along some short segment.

We see in this example that the projection to the leaf space of the foliation is not rigid, and this non-rigidity looks closely related to the presence of the BPS state.

On the contrary, if there are no BPS states, then h_ϕ is rigid:

Theorem 2.2. *Suppose that the spectral curve is smooth, i.e., the quadratic differential has simple zeros. If there are no BPS states, then any ϕ-harmonic map $h : X \to T$ factors as $h = u \circ h_\phi$ through a unique map $u : T_\phi \to T$ which is an isometric embedding. Thus, T is just T_ϕ plus some other edges not touched by the image of h.*

In the situation of the theorem, any two nonsingular points of X are joined by a strictly noncritical path, that is to say a path transverse to the foliation. The distance between the two points in T_ϕ is the length of the path using the transverse measure defined by $\Re\phi$. This noncritical path also has to go to a noncritical path in T, so the distance in T is the same as in T_ϕ.

If there is a BPS state, then some distances might not be well determined. For example the distance between points in regions B an D may change with the family of maps pictured in Figure 6.

Non-uniqueness of the distance is something that has to be expected from the point of view of the Voros resurgent expression for the transport function [33]. The resurgent expression may be viewed, roughly speaking, as a combination of contributions of different exponential orders $e^{\lambda_i t}$. See [30] for a more precise discussion of how this works. When there is a BPS state, then two exponents have the same real part, $\Re\lambda_i = \Re\lambda_j$. If these are the leading terms, then we obtain a function which becomes oscillatory along the positive real direction in t. A simple example would be $2\cos(t) = e^{it} + e^{-it}$. If the transport function $T_{PQ}(t)$ is oscillatory, then the asymptotic exponent

$$\nu^{\omega}_{PQ} = \lim_{\omega} \frac{1}{t} \log \|T_{PQ}(t)\|$$

could very well depend on the choice of ultrafilter ω used to define the limit. Thus, when there is a BPS state we can expect that the distance function defined by the harmonic map could depend on the choice of ultrafilter. In particular, it wouldn't be uniquely determined by the spectral differential ϕ.

These considerations, fitting perfectly well with the illustration of Figure 6, motivate the hypothesis about BPS states in Theorem 2.2.

3. Spectral networks and two-dimensional buildings

We feel that there should be a similar picture for higher-dimensional buildings. The basic philosophy and motivations were described in [21].

Our idea at the current stage of this project is to concentrate on mappings to 2-dimensional buildings. These buildings are asymptotic cones for symmetric spaces $SL_3/SU(3)$, and the mappings are limiting points in the sense of Parreau's theory [27] [21]. The spectral curves for such mappings are triple coverings $\Sigma \to X$.

3.1. Spectral networks and BPS states

For SL_2 and trees, a key role was played by the foliation corresponding to the quadratic differential. In the higher rank case, the notion of *spectral network* introduced by Gaiotto, Moore and Neitzke is the appropriate generalization.

It would go out of our scope to give an extensive introduction to this theory; the reader is referred to the original references of Gaiotto–Moore–Neitzke [11, 12, 10]. We will just briefly recall some of the main features as related to the situation of spectral curves for the group SL_3.

For an SL_3 spectral curve, over a general point in X there are three sheets of the spectral curve corresponding to three differentials ϕ_i, ϕ_j, ϕ_k for indices $i, j, k \in \{1, 2, 3\}$. One should note that if we go around a branch point, two of the indices are transposed. This subset of two indices is associated to the branch point, and the branch points together with their labelings are the starting points of the spectral network.

The foliations we are interested in are those defined by $\Re\phi_{ij}$ where $\phi_{ij} := \phi_i - \phi_j$. At a general point there are three of these (an interesting locus that will come into play later is the *caustic* where the three foliations are tangent). At a branch point, there is a distinguished foliation corresponding to $\pm\Re\phi_{ij}$ where i, j are the indices transposed by the branching. The singular leaf of this foliation consists of three rays exiting the branch point. These foliation lines are the start of the spectral network.

The spectral network lines are foliation lines labeled by pairs of indices. An order of the labeling is associated to a direction determined by looking at the imaginary part of ϕ_{ij}.

In the higher rank case starting with SL_3, Gaiotto–Moore–Neitzke observed the new phenomenon of *collisions*: when spectral network lines labeled ij and jk meet they can produce a new spectral network line labeled ik, following the third foliation.

Gaiotto–Moore–Neitzke introduce the terminology "BPS state" for certain configurations of spectral network lines. It generalizes to higher ranks the notion of "saddle connection" for the foliation associated to a quadratic differential, and indeed we have already employed the terminology "BPS state" for those above.

We refer the reader to the papers [11, 12, 10] as well as a recent growing body of literature for the relationship between this notion and physics. For our purposes we are interested in the geometrical combinatorial structure as follows.

Definition 3.1. A *BPS state* is a finite graphical subset of the spectral network, including all three rays at any collision point, all of whose endpoints lie at branch points.

An example with three branch points and a single collision is illustrated in Figure 7.

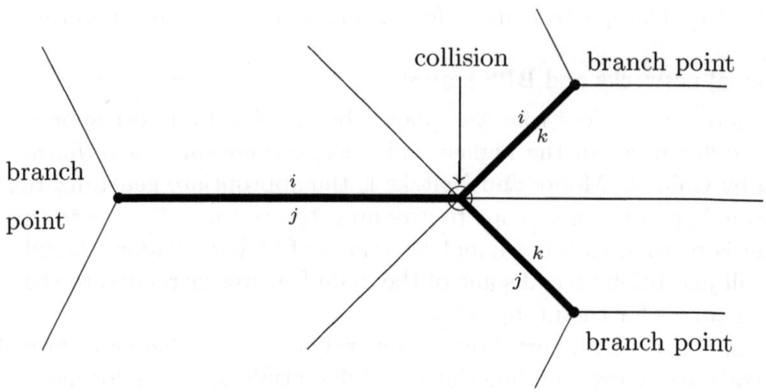

FIGURE 7. A BPS state for SL_3

3.2. Buildings for SL_3

For the SL_3 case, our goal is to sketch the outlines of a theory which should lead to a generalization of Theorem 2.2. Bringing into play the spectral network, this theory will say that for generic ϕ in a chamber where there are no BPS states, there should be a versal map to a building, such that the resulting distance function is uniquely determined by ϕ and preserved under ϕ-harmonic maps.

The SL_3 situation is in many ways an intermediate case. In the SL_2 case, the mapping h_ϕ to the tree T_ϕ was *surjective*. For SL_r with $r \geq 4$, the corresponding buildings have dimension ≥ 3, so the map from X has no hope of being surjective. We will present some speculations about that higher rank situation at the end of the paper.

In the SL_3 case, the dimension of the building is two, which is the same as the dimension of X. Therefore, we can expect that X will surject onto a subset which at least has nonempty interior. So, it presents some similarity with the case of trees, and this simplifies the geometrical aspects. We are able to develop a fairly precise although still conjectural picture.

Two-dimensional Euclidean space is the standard apartment for SL_3, furnished with a web of three foliations acted upon by the Weyl group S_3 just like the three spectral network foliations $\Re\phi_{ij}$.

The image of X will be a quotient of X, glueing together points over certain portions in a way that respects the web of foliations. This was seen in our BNR example of [21] which shall be recalled in detail in Section 7 below. Our goal is to construct a map

$$h_\phi : X \to B_\phi$$

which should play the same role as the projection to the tree of leaves in the SL_2 case.

The construction has several steps. The main part will be the construction of a map to a *pre-building*

$$h_\phi^{\mathrm{pre}} : X \to B_\phi^{\mathrm{pre}}$$

such that the building B_ϕ can then be obtained from B_ϕ^{pre} by adding on sectors not touched by the image of X. Already in the BNR example of [21] there were infinitely many additional sectors to be added here. They seem to be somewhat less related to the geometry of the situation.

The pre-building will itself be a quotient of an *initial construction*. The initial construction is obtained by glueing together small pieces. Recall that one of the main characteristics of a building is its nonpositive curvature property. The initial construction will, however, have some positively curved points: those are points where the total surrounding angle is 240° rather than 360°. As we shall see below, it leads to a process of successive pasting together of parts of the construction. We conjecture that after a locally finite number of steps this process should stop and give a pre-building.

4. Constructions as presheaves on enclosures

In order to get started, we need a precise way to manipulate the building-like objects involved in the construction. The idea for passing from B_ϕ^{pre} to B_ϕ will be to apply the small object argument. Also, the construction of B_ϕ^{pre} itself will involve successively imposing a bigger and bigger relation on the initial construction. So, it appears that we are working with algebraic rather than topological or metric objects. This makes it desirable to have an algebraic framework.

We propose to consider a *Grothendieck site* $\mathbb{E}$ of the basic building blocks, called "enclosures". Then "constructions" will be sheaves of sets on the site $\mathbb{E}$, satisfying basic local presentability and separability properties.

Intuitively, a construction is a space obtained by glueing together the basic pieces such as shown in Figure 8.

Proofs are not yet given, however we hope that they will be reasonably straightforward. The general theory is described for buildings of any dimension.

Let A be the affine space on which our buildings will be modeled. A *root half-space* is a half-space bounded by a root hyperplane. An *enclosure* is a bounded closed subset defined by the intersection of finitely many root half-spaces. For buildings corresponding to SL_3, the affine space is[3] $A = \mathbb{R}^2$ and some examples of enclosures are shown in Figure 8.

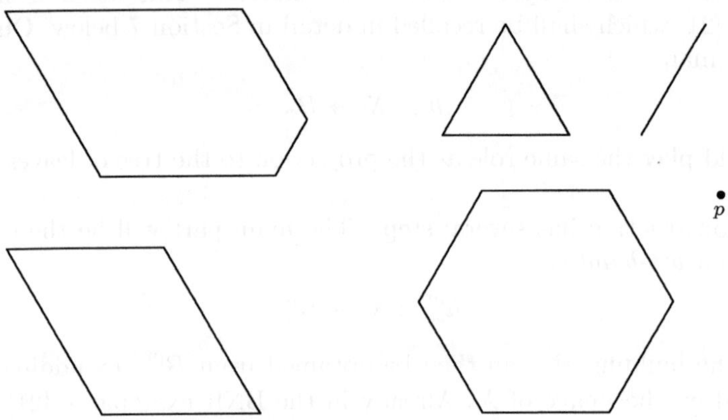

FIGURE 8. Some enclosures

An *affine map* of enclosures $E \to E'$ is a map which is the restriction of an automorphism of A given by an affine Weyl group element. Let $\mathbb{E}$ be the category of enclosures and affine maps between them. There is an object "point" denoted p consisting of a single point in A.

[3]More precisely A is the space of triples (x_1, x_2, x_3) with $\sum x_i = 0$ and the root half-spaces are defined by $x_i - x_j \geq c$.

We define a Grothendieck topology on $\mathbb{E}$ as follows: a covering of E is a finite collection of affine maps of enclosures $E_i \to E$ such that E is the union of their images.

The category $\mathbb{E}$ admits fiber products, but not products and in particular there is no terminal object.

Proposition 4.1. *The coverings define a Grothendieck topology on* $\mathbb{E}$.

If E is an enclosure, we denote by $\widetilde{E}$ the sheaf associated to the presheaf represented by E. It is different from the presheaf: if E' is another enclosure, then the sections of $\widetilde{E}$ over E', which is to say the maps $E' \to \widetilde{E}$ or equivalently the maps $\widetilde{E'} \to \widetilde{E}$, are the *folding maps* from E' to E. These are the continuous maps which are piecewise affine for a decomposition of E' into finitely many pieces which are themselves enclosures.

We can give a normalized form for coverings. Suppose $H_1, \ldots, H_k$ is a sequence of parallel Weyl hyperplanes in order. Then we obtain a sequence of *strips* $S_0, \ldots, S_{2k}$ covering A. The strip S_{2i} is the closed subset consisting of everything between and including H_i and H_{i+1}, with S_0 and S_{2k} being the two outer half-planes (we assume $k \geq 2$ so there is no question about the ordering of these). The strip S_{2i+1} is just H_{i-1} itself. Suppose we are given a collection of such sequences of strips $S^1_\cdot, \ldots, S^a_\cdot$ for various directions of the Weyl hyperplanes. Then for $J = (j_1, \ldots, j_a)$ with $0 \leq j_i \leq 2k_i$ we may consider the enclosure

$$U_J := E \cap S^1_{j_1} \cap \cdots \cap S^a_{j_a}.$$

These cover E.

Lemma 4.2. *Suppose* $\{V_k\}$ *is a covering of* E. *Then it may be refined to a* standard covering, *that is to say a covering of the form* $\{U_J\}$ *constructed above.*

We remark that in a standard covering $\{U_J\}$, the intersections of elements are again elements, since intersections of strips are included as strips too (that was why we included the H_i themselves). We may now give a more explicit description of the folding maps.

Corollary 4.3. *Suppose* E *and* F *are enclosures. Any folding map, that is to say a map to the associated sheaf* $E \to \widetilde{F}$, *is given by taking a standard covering* $\{U_J\}$ *and assigning for each* $J \subset \{H_i^+, H_i^-\}$ *an affine map* $a_J : U_J \to F$, *subject to the condition that if* $U_K \subset U_J$ *then* $a_J|_{U_K} = a_K$. *Two folding maps are the same if and only if they are the same pointwise, which is equivalent to saying that they are the same on a common refinement of the two covers; a common refinement may be obtained by taking the unions of the sequences of hyperplanes.*

Corollary 4.4. *A folding map between enclosures is finite-to-one.*

We now consider a sheaf F on $\mathbb{E}$. We say that it is *finitely generated* if there is a finite collection of maps $E_i \to F$ from enclosures, such that the map of presheaves

$$\coprod E_i \to F$$

is surjective in the sheaf-theoretical sense, i.e., it induces a surjection of associated sheaves.

We say that F is *finitely related* if, for any two maps from enclosures $E, E' \to F$, the fiber product $E \times_F E'$ is finitely generated. We say that F is *finitely presented* if it is finitely related and finitely generated.

Lemma 4.5. *If E is an enclosure, then any subsheaf $F \subset \widetilde{E}$ is finitely related. More generally if $\{E_i\}_{i=1,\ldots,n}$ is a nonempty collection of enclosures, then any subsheaf*

$$F \subset \prod_i \widetilde{E}_i$$

is finitely related.

Proof. We need to consider the fiber product $G \times_F G'$ for two maps from enclosures $G, G' \to F$. These maps correspond to sequences $(\zeta_1, \ldots, \zeta_n)$ with $\zeta_i : G \to \widetilde{E_i}$, resp. $(\zeta'_1, \ldots, \zeta'_n)$ with $\zeta'_i : G' \to \widetilde{E_i}$.

Suppose $G = \bigcup G_J$ (resp. $G' = \bigcup G'_{J'}$) are (finite) coverings by enclosures. Suppose we can prove that $G_J \times_F G'_{J'}$ are finitely generated. One can then conclude that $G \times_F G'$ is finitely generated.

Apply this to a common refinement of the coverings needed to define the folding maps ζ_i, in the above standard form. We conclude that it suffices to consider the case where ζ_i are affine maps.

Now in this case (and no longer using the notation for the coverings), the fiber product is expressed as

$$G \times_F G' = \{(x, x') \text{ s.t. } \zeta_i(x) = \zeta'_i(x') \text{ for } i = 1, \ldots, n\}.$$

This expression is somewhat heuristic as we are really talking about sheaves but it serves to indicate the proof.

Since ζ_1 is an isomorphism we may assume that it is the identity and same for ζ'_1. Therefore, the first condition says that $x' = x$ and with this normalization, we may write

$$G \times_F G' = \{x \in A \text{ s.t. } x \in G \cap G', \;\; \zeta_i(x) = \zeta'_i(x) \text{ for } i = 2, \ldots, n\}.$$

The conditions $\zeta_i(x) = \zeta'_i(x)$ define Weyl hyperplanes or are always true, so this represents $G \times_F G'$ as an enclosure. This completes the proof. □

Note that $E \times E'$ will never be finitely generated as soon as $\dim(A) \geq 1$. Therefore, the empty direct product which is to say the terminal object $*$ in sheaves on enclosures, is not finitely related. The above proof used $n \geq 1$ in an essential way.

One should not confuse the terminal object $*$ with the enclosure p when $p \in A$ is a point. There are no maps $E \to \tilde{p}$ for E different from a point or the empty set.

A *construction* is a finitely related sheaf on the site of enclosures.

Theorem 4.6. *The category of constructions is closed under finite colimits, and fiber products.*

We can define a topological space underlying a construction. If F is a construction, let $F(p)$ denote the set of points, that is to say the set of maps $p \to F$ where $p \in A$ is a point (recall from above that this is different from the terminal sheaf $*$). Give $F(p)$ a topology as follows: for any enclosure E, $E(p)$ (which is equal to $\widetilde{E}(p)$) has a topology as a subset of the affine space A. Then we say that a subset $U \subset F(p)$ is open if its pullback to $E(p)$ is open, for any enclosure E and any map $E \to F$.

Conjecture 4.7. *If F is a construction then the topological space $F(p)$ Hausdorff; furthermore it is a CW-complex.*

It might be necessary to add additional hypotheses on F in order to insure that $F(p)$ is a CW-complex.

4.1. Spherical theory

We would like to consider the local structure of a construction at a point. For this we need a spherical version of the above theory. It seems like we probably don't need to consider "enclosures" in the spherical building but only "sectors". A *sector* is a minimal closed chamber of a given dimension in the spherical complex associated to A. The set of sectors is partially ordered by inclusion and the spherical Weyl group acts on it. It is a finite topological space, in particular it has a structure of site (the only coverings of a sector must include that sector itself). Let $\mathbf{S}$ be the category of sectors.

A *spherical construction* is a presheaf or equivalently sheaf on $\mathbf{S}$.

We have a map from $\mathbf{S}$ to the filters on A located at any given point.

Suppose F is a sheaf on $\mathbb{E}$ and $x \in F(p)$. We would like to associate the spherical construction F_x, defined as follows: if $\sigma \in \mathbf{S}$ then σ corresponds to a filter of enclosures, that is to say a filtered category of enclosures E with p on the boundary of E and whose local corner at p looks like σ. Call this category $\langle\sigma\rangle$. For $E \in \langle\sigma\rangle$, consider the set $F(E)_x$ consisting of maps $E \to F$ such that p maps to x. Then we set

$$F_x(\sigma) := \lim_{\to, E \in \langle\sigma\rangle} F(E)_x.$$

An element of $F_x(\sigma)$ therefore consists of a germ of map $E \to F$ sending p to x, such that the corner of E at p is σ. These germs are up to equivalence that if two maps agree on a smaller enclosure also containing p and having σ as corner, then the two germs are said to be equivalent.

4.2. $\mathbb{R}$-trees

When the group is SL_2, the standard apartment is just $\mathbb{R}$ and the enclosures are closed bounded segments. The category of constructions gives a good point of view for the theory of $\mathbb{R}$-trees. For example, if q is a quadratic differential defining a spectral multivalued differential $\phi_i = \pm\sqrt{q}$ on a compact Riemann surface X, then the tree T_φ of leaves of the foliation $\Re\phi$ on $\widetilde{X}$ may be seen as a sheaf on $\mathbb{E}_{SL_2}$ as follows: for a segment E let $T_\phi^{\mathrm{pre}}(E)$ be the space of differentiable maps $E \to \widetilde{X}$

which are transverse to the foliation such that the pullback of $\Re\phi$ is the standard differential dx on $E \subset \mathbb{R}$. These maps are taken modulo the relation that two maps are the same if they map points of E to the same leaves of the foliation. Then T_ϕ is the associated sheaf. Thus $T_\phi(E)$ is the space of maps from E to the leaf space, which are represented on finitely many segments covering E by differentiable maps into $\widetilde{X}$.

This point of view on the leaf space is similar to the quotient construction in diffeology [16]. The objects in our category of constructions could be viewed as some kinds of diffeological spaces with affine structure.

4.3. The SL_3 case

We now specialize to the case of buildings for the group SL_3. The affine space is $A = \mathbb{R}^2$. The spherical Weyl group is the symmetric group acting through its irreducible 2-dimensional representation. Some enclosures were pictured in Figure 8. There are three directions of reflection hyperplanes. These divide the vector space at the origin into six 2-dimensional sectors, acted upon transitively by the Weyl group. On the other hand, there are two orbits for the 1-dimensional sectors, the even vertices of the hexagon and the odd vertices. Therefore our category of sectors $\mathbf{S}$ is equivalent to the following category $\mathbf{S}'$: there are an object η, corresponding to the 2-dimensional sectors, and two objects ν^+ and ν^- corresponding to the 1-dimensional sectors. We choose one of these denoted ν^+ which we say has positive orientation. The morphisms are

$$\nu^- \to \eta \leftarrow \nu^+.$$

A spherical construction H is a sheaf on $\mathbf{S}'$. This consists of three sets $H(\eta)$, $H(\nu^+)$ and $H(\nu^-)$ with morphisms

$$H(\nu^-) \leftarrow H(\eta) \to H(\nu^+).$$

Such a structure may be viewed as a graph whose edges are $H(\eta)$ with vertices grouped into the positive ones $H(\nu^+)$ and negative ones $H(\nu^-)$. Each edge joins a positive vertex to a negative vertex. A spherical construction is equivalent to such a graph.

If we have a construction F for this Weyl group and if $x \in F(p)$ is a point then we obtain a spherical construction F_x which is a graph as above.

Following the simple characterization which was given for example by Abramenko and Brown [1], we say that a spherical construction is a *spherical building* if any two vertices are at distance ≤ 3, every pair of vertices is contained in a hexagon, and if there are no loops of length ≤ 4 (the length of a loop has to be even because of the parity property of edges).

A spherical construction is a *spherical pre-building* if it is connected, if every node is contained in at least one edge, and if it has no loops of length ≤ 4. The construction of [21] gives a way of going from a spherical pre-building to a spherical building.

In a spherical pre-building, say that two nodes are *opposite* if they have opposite parity and are at distance 3. If it is spherical building this means that they are opposite nodes of any hexagon containing them.

A *segment* is a 1-dimensional enclosure $S \subset A$. We note that a segment has a natural orientation. At any point $x \in S(p)$ in the interior, the spherical building S_x has two elements, $S_{x,+}$ and $S_{x,-}$, not joined by any edge; they are of opposite parity and the positive direction in $S(p)$ is defined to be the direction going towards $S_{x,+}$. If x is an endpoint then S_x has one element; the two endpoints of S are oriented positively or negatively respectively.

Let us denote by $S^{t,+}$ the segment of length t based at the origin, such that the parity of the single element of $S_0^{t,+}$ is positive. We assume that these segments are all in the same line so that $S^{t,+} \subset S^{t',+}$ when $t < t'$. Let $S^{t,-}$ denote the segment with the opposite orientation at the origin.

Remark 4.8. Suppose F is a construction. If $\varphi : S^{t,+} \to F$ is a morphism then for $x \in S^{t,+}$, the image of the positive (resp. negative) element of $S_x^{t,+}$ under φ is denoted $\varphi_{x,+}$ (resp. $\varphi_{x,-}$) and it is a positive (resp. negative) node in the spherical building $F_{\varphi(x)}$. Same for $S^{t,+-}$.

The first of our conditions on a construction concern the spherical constructions at each point. We say that F satisfies:

SPB-loc: if for any $x \in F(p)$ the spherical construction F_x is a spherical pre-building;

SB-loc: if for any $x \in F(p)$ the spherical construction F_x is a spherical building.

Definition 4.9. Suppose E is an enclosure, and F a construction satisfying condition SPB-loc. We say that a map $f : E \to F$ is *immersive at a point* $x \in E(p)$, if the map of spherical constructions $E_x \to F_{f(x)}$ preserves distances. The map is *immersive* if it is immersive at all points of $E(p)$.

Here, the distances in E_x are calculated by considering $E_x \subset A_x$. In particular, if E is a segment then we consider the distance between the two elements of E_x to be 3.

Thus a map from a segment $\varphi : S \to F$ is immersive at an interior point $x \in S$ if $\varphi_{x,+}$ and $\varphi_{x,-}$ are opposite in the spherical pre-building $F_{\varphi(x)}$. We also use the terminology *straight* for an immersive segment, and say that a map from a segment is *angular* otherwise.

Intuitively, for enclosures of dimension 2 a map $f : E \to F$ is immersive if and only if $f(p) : E(p) \to F(p)$ is locally injective.

Lemma 4.10. *Suppose that F satisfies SPB-loc. A map from an enclosure $\varphi : E \to F$ is immersive at all but at most finitely many angular points. Hence φ is immersive if and only if the set of angular points is empty.*

We now continue with some extension conditions for a construction F (let us reiterate that we are working here in the SL_3 situation).

Ex-Seg: let t denote the other endpoint of $S^{t,+}$. Assuming SPB-loc, for any immersive map $\varphi : S^{t,+} \to F$ and any element $\nu \in F_{\varphi(t)}$ opposite to φ_t^-, and for any $t' > t$ we ask that there exist an extension of φ to an immersive map $\varphi' : [0, t'] \to F$ such that $(\varphi')_t^+ = \nu$. Similarly for segments $S^{t,-}$ in the opposite orientation.

We next get to our main extension statement, for obtuse angles. Some notation will be needed first.

Let $P^{a,b,+}$ be parallelograms centered at the origin with side lengths a and b, and positive orientation of the two edges at the origin (resp. $P^{a,b,-}$ with negative orientation). We may assume that the first edge is the segment $S^{a,+}$, and denote the second edge by $\omega S^{b,+}$ (it is obtained by rotating by 120 degrees).

Similarly let T^a denote the triangle with edge length a. We consider both segments $S^{a,+}$ and $> S^{a,-}$ to be edges of T^a starting from the origin.

Ex-Obt: Assuming SPB-loc, suppose we are given maps $\varphi : S^{a,+} \to F$ and $\psi : \omega S^{b,+} \to F$ such that $\varphi(0) = \psi(0) = x$. Suppose that the distance between the elements $\varphi_{0,+}$ and $\psi_{0,+}$ in the spherical building F_x is ≥ 2 (i.e., they are distinct). Suppose that the maps φ and ψ are immersive. Then there exists an immersive map $\zeta : P^{a,b,+} \to F$ coinciding with the given maps φ and ψ on the edges. We also ask the same condition with the other orientation, for $P^{a,b,-}$.

Ex-Side: given a segment $\varphi : S \to F$ of length a with $x = f(0)$ one of the endpoints, and an edge in the spherical building $\nu \in F_x$, φ extends to an immersed triangle $T^a \to F$ such that ν is not in the image of φ_0.

Lemma 4.11. *Suppose F is a construction satisfying SPB-loc, Ex-Obt and Ex-Side. Then F satisfies SB-loc.*

Proof. Suppose $x \in F(p)$. By hypothesis F_x is a spherical pre-building. We would like to show it is a spherical building. From Ex-Obt, we get that any two nodes of the same parity in F_x have distance ≤ 2. By hypothesis SPB-loc, the F_x are connected graphs and all nodes are contained in edges. It follows that any two nodes have distance ≤ 3. The condition Ex-Side implies that any node in any of the spherical pre-buildings F_x, is contained in at least two distinct edges. It now follows that any two nodes are contained in a hexagon. Thus F_x is a spherical building. □

Lemma 4.12. *Suppose F is a construction satisfying SB-loc, Ex-Obt and Ex-Seg. Then it satisfies Ex-Side. Furthermore, the sector of the triangle can be specified at either of the endpoints of the segment.*

Theorem 4.13. *Suppose F is a construction satisfying SB-loc, Ex-Seg, and Ex-Obt, such that $F(p)$ is connected. Suppose $x, y \in F(p)$. Then there exists an immersive map from an enclosure $f : E \to F$ such that x, y are in the image of $f(p)$.*

Proof. (Sketch) Choose a path from x to y, that is to say a continuous map from $[0, 1]$ to the topological space $F(p)$. We assume a topological result saying that the path may be covered by finitely many intervals which map into single enclosures. From this, we may assume that the path has the following form: it goes through a series of triangles which are immersive maps $\varphi_i : T^a \to F$, such that φ_i and φ_{i+1} share a common edge. Notice here that we may assume that all the triangles have the same edge length. We have x in the image of φ_1 and y in the image of φ_k.

We obtain in this way a strip of triangles. We next note that there are moves using Ex-Obt which allow us to assume that there are not consecutive turns in the same direction along an edge.

From this condition, we may then extend using Ex-Obt and Ex-Side (which follows from Ex-Seg) to turn this strip of triangles, into a single immersive map from an enclosure. □

We let A denote the sheaf represented by the affine space of the same notation. More precisely, for an enclosure E a map $E \to A$ consists of a factorization $E \to E' \to A$ where $E' \subset A$ is an enclosure in its standard position, and $E \to E'$ is a folding map (that is, a map to $\widetilde{E'}$). Thus A is a direct limit of its enclosures. In particular we know what it means to have an immersive map from A to somewhere, the map should be immersive on each of the enclosures in A. If F is a construction, an *apartment* is an immersive map $A \to F$.

Proposition 4.14. *Suppose F is a construction satisfying SB-loc, Ex-Obt and Ex-Seg. Then any immersive map from an enclosure $E \to F$ extends to an apartment $A \to F$.*

Corollary 4.15. *If F is a construction satisfying SB-loc, Ex-Obt and Ex-Seg, such that $F(p)$ is connected, then any two points $x, y \in F(p)$ are contained in a common apartment.*

Proof. Combine the previous proposition with Theorem 4.13. □

Proposition 4.16. *Suppose F is a construction satisfying SB-loc. Then an immersive map $E \to F$ from an enclosure, is injective. The same is true for an apartment $A \to F$.*

Definition 4.17. Say that F is a *building* if it is a construction with $F(p)$ connected and simply connected, satisfying SB-loc, Ex-Obt and Ex-Seg.

Conjecture 4.18. *If F is a building, then the topological space $F(p)$ has a natural structure of $\mathbb{R}$-building modeled on the affine space A.*

4.4. Pre-buildings

Definition 4.19. A *pre-building* is a construction F which satisfies SPB-loc such that $F(p)$ is simply connected.

Suppose F is a pre-building. Recall that a segment $s : S \to F$ is straight if it is immersive, that is for any $y \in S(p)$ which is not an endpoint, the two directions of S_y map to two vertices of $F_{s(y)}$ which are separated by three edges.

Definition 4.20. A map $V^{a,b} \to F$ to a pre-building is *standard* if the two edges are straight and their two directions at the origin are not the same.

We would now like to investigate the process of adding a parallelogram $P^{a,b}$ to a pre-building F along a standard inclusion $V^{a,b} \to F$ to get a new pre-building. The problem is that a similar parallelogram, or a part of it, might already be there. For example if we add $P^{a,b}$ to itself by taking a pushout along $V^{a,b}$ then we generate a fourfold point at the origin.

Thus the need for a process which we call *folding in*, of joining up the new $P^{a,b}$ with any part of it that might already have been there. This will be used in the small-object argument below as well as in our general modification procedure later.

Suppose F is a pre-building and $d : V^{a,b} \to F$ is a standard inclusion. Define $R \subset P^{a,b}$ to be the union of all of the $P^{c,d} \subset P^{a,b}$ based at the origin, such that there exists a map on the bottom making the following diagram commute:

$$\begin{array}{ccc} V^{c,d} & \hookrightarrow & V^{a,b} \\ \downarrow & & \downarrow \\ P^{c,d} & \to & F. \end{array}$$

One can show that the map, if it exists, is unique. We obtain a subsheaf $R \subset P^{a,b}$ and there is a map $R \to F$ combining all of the previously mentioned maps. The *folding-in* of $P^{a,b}$ along d is the pushout

$$G := F \cup^R P^{a,b}.$$

The relation R is designed so that G again satisfies SPB-loc. Since R was defined as a very general union of things, there remains open the somewhat subtle technical issue, which we haven't treated, of showing that G is a construction.

The folding-in G accepts a map $F \cup^{V^{a,b}} P^{a,b} \to G$ and we have the following universal property:

Lemma 4.21. *The folding-in G is a pre-building. Suppose B is a pre-building and $f : F \to B$ is a map such that the image of d is again a standard inclusion into B (as will be the case for example if f is non-folding). Suppose that $V^{a,b} \to B$ completes to a map $P^{a,b} \to B$. Then these factor through a unique map $G \to B$.*

We also conjecture a versality property for maps to buildings: if $F \to B$ is any map to a building then there exists an extension to $G \to B$, however this extension might not be unique. Because of the non-uniqueness on the interior pieces making up R, this conjecture requires a study of convergence issues and the statement might need to be modified.

4.5. The small-object argument

Free constructions of buildings have been done by Ronan [29] and Berenstein and Kapovich [4].

The small-object argument is a way of ensuring the extension properties by free addition of cells, based on the following properties of our category of constructions.

Lemma 4.22. *Suppose* $F_0 \to F_1 \to \cdots \to F_i \to \cdots$ *is a sequence of sheaves. Then the presheaf colimit* $\operatorname{colim} F_i$ *is a sheaf, and if* E *is an enclosure, then any map* $E \to \operatorname{colim} F_i$ *factors through a map* $E \to F_i$ *for some* i. *If the* F_i *are constructions and the maps are injective, then* $\operatorname{colim} F_i$ *is a construction. The local spherical constructions of the colimit are colimits of the local spherical constructions of the pieces.*

Proof. Let $\operatorname{colim} F_i$ denote the presheaf colimit. By definition, if E is an enclosure then $(\operatorname{colim} F_i)(E) = \operatorname{colim}(F_i(E))$ saying that any map from E to the colimit factors through one of the F_i. Similarly, two maps are equal if and only if they are equal as maps into one of the possibly later F_i. It follows that any finite glueing diagram in the colimit comes from a glueing diagram in one of the F_i, so if the F_i are sheaves then the colimit is a sheaf. Suppose now that the F_i are constructions and the maps are injective. Given two maps $E, E' \to \operatorname{colim} F_i$, assuming that they both factor through F_i then their fiber product over $\operatorname{colim} F_i$ is the same as their fiber product over F_i. Therefore, assuming F_i are finitely related it follows that the colimit is finitely related. The statement about local spherical constructions is obtained by applying the definition of germ. □

The property of maps from enclosures can be described as saying that enclosures are *ω-small* objects. It follows that finitely presented objects are also ω-small. The lemma allows us to make constructions by iterating and taking the colimit.

In our setup, the conditions Ex-Seg, Ex-Obt and Ex-Side are extension conditions of the following form: we have a certain finite arrangement R consisting of some points or segments, and a map $R \to E$ to an enclosure putting R on the boundary of E. Then the conditions state that for any immersive map $R \to F$ there exists an extension to an immersive map $E \to F$.

These conditions may be ensured using the small object argument. Given F, define $F \to \mathrm{Ex}(F)$ to be the construction obtained by combining pushouts along all the inclusions $R \to E$, using the folding-in process described in the previous subsection for Ex – Obt. Let $\mathrm{Ex}^{i+1}(F) := \mathrm{Ex}(\mathrm{Ex}^i(F))$ be the iterates and consider $\mathrm{Ex}^\omega(F)$ to be the colimit of these iterates over the first infinite ordinal ω (the natural numbers).

Theorem 4.23 (Small object argument). *If* F *is a pre-building, and if we form* $F' := \mathrm{Ex}^\omega(F)$ *then* F' *is a building.*

Proof. (Sketch) We first note that F' is a construction, and $F'(p)$ is simply connected. Also F' again satisfies SBP-loc. By ω-smallness of the finitely presented objects entering into the extension properties, F' satisfies Ex-Seg, Ex-Obt and Ex-Side. Indeed, in order to see an extension property with respect to $R \to E$, suppose given $R \to F'$; since R is ω-small, it factors through one of the $\mathrm{Ex}^i(F)$, and by construction of Ex we get an extension to $E \to \mathrm{Ex}^{i+1}(F)$. For Ex-Obt, the folding-in process only adds parallelograms along standard inclusions, but this turns out to be good enough to get it in general. Hence, F' satisfies SB-loc, so it is a building. In particular any two points of $F'(p)$ are contained in a common apartment. □

Conjecture 4.18 then says that $F'(p)$ has a natural structure of $\mathbb{R}$-building.

Conjecture 4.24. *Suppose F is a pre-building, and let F' be the building obtained by the small-object argument. Then it is versal: for any other map to a building $F \to B$ there exists a factorization through $F' \to B$.*

The difficulty is that we needed to use the folding-in construction to construct F', and the versality property for the folding-in construction is not easy to see because of the possibly infinite nature of the relations R coupled with non-uniqueness of extensions to $P^{a,b}$ for non-standard maps $V^{a,b} \to B$. The small-object construction may need to be modified in order to get to this versality conjecture.

5. An initial construction

The next step is to relate constructions to ϕ-harmonic maps. Consider a Riemann surface X together with a spectral curve, defining a multivalued differential ϕ. Recall that locally on X^*, ϕ may be written as $(\phi_1, \ldots, \phi_n)$ with ϕ_i holomorphic, and $\sum \phi_i = 0$. In the present paper we consider spectral curves for SL_3 so $\phi = (\phi_1, \phi_2, \phi_3)$. The *foliations* are defined by $\Re\phi_{ij} = 0$. In our SL_3 case there are three foliation lines going through each point of X^*.

For any point $x \in X^*$ there exists a neighborhood U_x, with the property [21] that U_x has to map to a single apartment under any ϕ-harmonic map. Thus locally, a ϕ-harmonic map to a building factors through the map

$$h^x : U_x \to A$$

to the standard apartment given by integrating the forms $\Re\phi_i$ with basepoint x. The foliation lines defined by $\Re\phi_{ij} = 0$ are just the preimages in U_x of the reflection hyperplanes of A. The preimage of a small enclosure near the origin in A will therefore be a domain in U_x whose boundary is composed of foliation lines for the three foliations. We may shrink U_x so that its closure is itself such a domain.

5.1. The Ω_{pq} argument

Recall briefly the proof of [21] showing existence of such a neighborhood U_x. A path $\gamma : [0,1] \to X$ is *noncritical* if the differentials $\gamma^*(\Re\phi_i)$ remain in the same order all along the path. For any ϕ-harmonic map to a building $h : X \to B$, the image of a noncritical path has to be contained in a single apartment. One shows this using the property of a building that says two opposite sectors based at any point are contained in a single apartment.

Suppose $p, q \in X$ are joined by some noncritical path. Let Ω_{pq} denote the subset swept out by all of these paths, in other words it is the set of all points $y \in X$ such that there exists a noncritical path γ with $\gamma(0) = p$, $\gamma(1) = q$ and $\gamma(s) = y$ for some $s \in [0,1]$. We claim that Ω_{pq} maps into a single apartment. Suppose $y, y' \in \Omega_{pq}$ and let γ and γ' denote the corresponding paths. We have apartments $A, A' \subset B$ such that $h \circ \gamma$ goes into A and $h \circ \gamma'$ goes into A'. Now $A \cap A'$ is a Finsler-convex subset in either of the two apartments, and it contains p and q. In particular it contains the parallelogram with opposite endpoints p and q. It follows that it contains the images of the two paths, so $h(y)$ and $h(y')$ are both in A. Letting y' vary we get that $h(\Omega_{pq}) \subset A$.

The ϕ_i admit a uniform determination over Ω_{pq}, and the map $\Omega_{pq} \to A$ is determined just by integrating the real one-forms $\Re\phi_i$.

Now if $x \in X^*$, we may find nearby points $p, q \in X$ such that x is in the interior of Ω_{pq}. This is easy to see away from the caustic lines. The caustic lines are transverse to all of the differentials so if x lies on a caustic one can choose p and q on the caustic itself, as will show up in our pictures later. This still works at an intersection of caustics too.

Our neighborhood U_x is now chosen to be any neighborhood of x contained in some Ω_{pq}. This construction is uniform, independent of the harmonic map h.

5.2. Caustics

A point $x \in X^*$ is *on a caustic* if the three foliation lines are tangent. This is equivalent to saying that the three points $\phi_i(x) \in T^*X_x$ are aligned, i.e., they are on a single real segment. The *caustics* are the curves of points in X satisfying this condition. We include also the branch points in the caustics.

There is a single caustic coming out of each ordinary branch point.

The caustics play a fundamental role in the geometry of harmonic maps to buildings, specially in the SL_3 case.

Let $C \subset X$ denote the union of the caustic curves. Then, for any $x \in X - C$ the map $h^x : U_x \to A$ is etale at x. Hence, if $h : X \to B$ is a ϕ-harmonic map to a building, h is etale onto the local apartments outside of C. In other words, the local integrals of any two of the differentials provide local coordinate systems on $X - C$. On the other hand, h folds X along C.

We may also view this as determining a flat Riemannian metric on $X - C$, pulled back from the standard Weyl-invariant metric by the local maps h^x to the standard apartment obtained by integrating $\Re\phi_i$. The metric has a distributional curvature concentrated along C. From the pictures it seems that the curvature is everywhere negative, and from our process we shall see that the total amount of curvature along a single caustic joining two branch points gives an excess angle of $120°$.

5.3. Non-caustic points

If $x \in X - C$ is a non-caustic point, then we may assume that the neighborhood U_x maps isomorphically to the interior of an enclosure in the standard apartment. The enclosure E_x could be chosen as a standard hexagon or perhaps a standard parallelogram for example.

We have chosen U_x such that for any harmonic ϕ-map to a building $h : X \to B$, the map $U_x \to B$ factors through an apartment $A \subset B$ via the map

$h^x : U_x \to A$ given by integrating the $\Re\phi_i$. The map h^x factors through an affine (non-folding) map $E_x \to A$. Altogether we get a factorization

$$U_x \to E_x \to B \tag{1}$$

and the map $E_x \to B$ doesn't fold along any edges passing through x.

When x is a branch point or on a caustic, we can still get a local factorization of the form (1) through a construction E_x, as will be discussed next.

5.4. Smooth points of caustics

Suppose $x \in C$ is a point in the smooth locus of a caustic curve. Then the local integration map folds every neighborhood of x in two along C. The image of U_x by h^x is folded along C. By intersecting with a smaller enclosure containing the image of x, we may assume that U_x has the property that there is an enclosure E_x with $h^x : U_x \to E_x^\circ$ being a proper 2 to 1 covering folded along C. We may also assume that E_x is the convex hull of the closed $h^x(\overline{U}_x)$.

For the generic situation, there are two other types of points that need to be considered: the intersections of caustics, and the branch points.

5.5. Branch points

For the branch points, locally two of the differentials say ϕ_1 and ϕ_2 come together, and the third one could be considered as independent. Therefore, a ϕ-harmonic map looks locally like the projection to the tree of leaves of a quadratic differential, crossed with a real segment. It means that we are forced to consider a singular construction rather than an enclosure. This singular construction still denoted E_x may be taken for example as the union of three half-hexagons joined along their diameters.

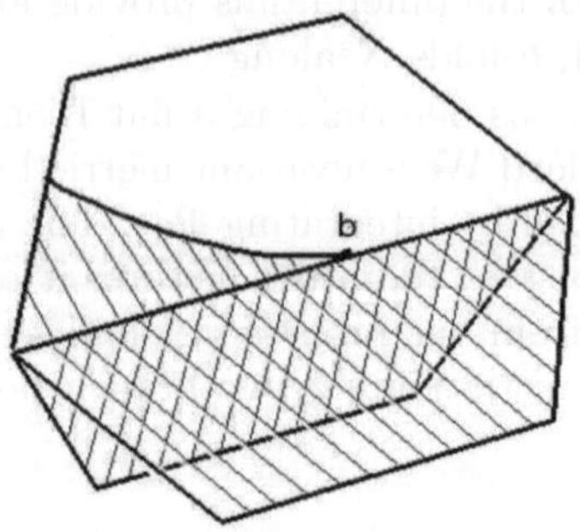

FIGURE 9. Three half-hexagons

If $x = b$ is a branch point then we may choose the neighborhood U_x together with a map $h^x : \overline{U}_x \to E_x$ such that E_x is the convex hull of the image. The image of $\overline{U}_x$, shaded in above, looks locally like the one that we saw in the BNR example [21].

5.6. Crossing of caustics

The case where x is a crossing point of two caustics is new, not appearing in the BNR example. We have therefore looked fairly closely at this situation in one of the next basic examples. A spectral network with the two crossing caustics will be shown later as Figure 23 of Section 9.

The Ω_{pq} argument works also here, so we have a neighborhood U_x of x which has to map to a single apartment in any ϕ-harmonic map to a building. The two caustics divide U_x into four sectors. The map $h^x : U_x \to A$ is 1 to 1 in the interior of two of the opposing sectors; it is 3 to 1 in the interior of the other two opposing sectors; it is 2 to 1 along the caustics and 1 to 1 at x. This is emphasized by including the images of two circles in the picture shown in Figure 10.

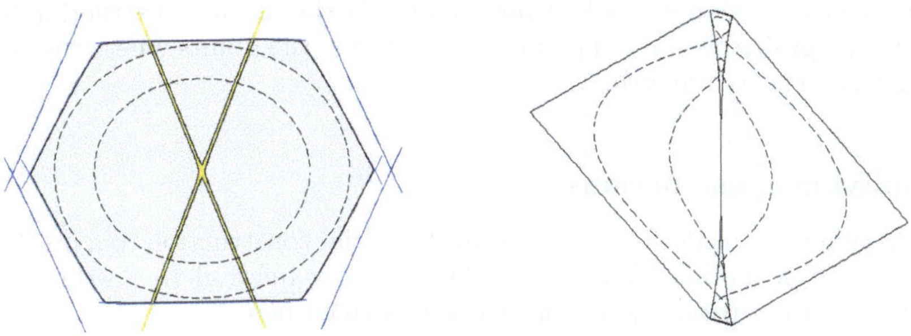

FIGURE 10. A neighborhood of the crossing

We choose $\overline{U}_x$ as a hexagonal shaped region shown in Figure 10 on the left. The image E_x in a standard apartment is a hexagon-shaped enclosure, shown on the right in Figure 10. As said above, the map is $3:1$ over the thin middle regions between the two caustics. The map $h_x : \overline{U}_x \to E_x$ is proper and $\overline{U}_x$ is the inverse image of E_x.

5.7. The initial construction

We now put together the above neighborhoods to get a good covering of X.

Theorem 5.1. *There exists a finite covering of X by open sets U_i, and constructions E_i as considered above (either an enclosure or the union of three half-hexagons), together with maps $h_i : \overline{U}_i \to E_i$ such that for any harmonic ϕ-map to a building $h : X \to B$, there is an isometric embedding $E_i \subset B$ such that h factors through h_i and indeed $\overline{U}_i$ is a connected component of $h^{-1}(E_i)$.*

For the intersections $U_{ij} := U_i \cap U_j$, we obtain constructions E_{ij} with inclusions to both E_i and E_j, such that E_{ij} is the convex hull of $h_i(U_{ij})$ in U_i (resp. the convex hull of $h_j(U_{ij})$ in U_j).

Define

$$Z := \frac{\bigcup_i E_i}{\sim}$$

where the relation $\sim$ is obtained by identifying the $E_{ij} \subset E_i$ with $E_{ij} \subset E_j$.

Theorem 5.2. *This defines a construction Z. We have a ϕ-harmonic map $h_Z : X \to Z$ and it has the following universal property: for any ϕ-harmonic map to a building $h : X \to B$, there is a factorization $h = h' \circ h_Z$ for a unique map of constructions $h' : Z \to B$.*

The construction Z is called our *initial construction.* It is not a pre-building because it will not, in general, have the required nonpositive curvature property. For example the initial construction described in Section 7 will have fourfold points a_1 and a_3.

Lemma 5.3. *We can nonetheless insure that the local spherical constructions of Z don't have cycles of length 2.*

In the next section we look at how to modify the initial construction in order to remove the positively curved points, that is to say the points where the spherical building has a cycle of length 4.

6. Modifying constructions

In this section we consider how to go from the initial construction to a pre-building by a sequence of modification steps. The reader is referred to Section 7 for an illustration of the various operations to be described here.

These operations are very similar in spirit to foldings and trimmings in the theory of Stallings graphs [31, 19, 28], and the two-manifold construction that comes out at the end should be considered as a "core".

The important role of the parallelogram was brought out in the "diamond moves" of Fontaine, Kamnitzer and Kuperberg [9, §5].

6.1. Scaffolding

In order to keep track of what kind of folding happens, we first look at some extra information that can be attached to a construction. In this subsection we remain as usual in the SL_3 case.

Let F be a construction. An *edge germ* of F is defined to be a quadruple (x, v, a, b) where $x \in F(p)$ and v is a vertex in the spherical construction F_x, and a and b are edges in F_x sharing v as a common endpoint.

There is a change of dimension when passing from the spherical building to the construction itself, so the vertex v corresponds to a germ of 1-dimensional segment based at x, and the edges a and b correspond to germs of 2-dimensional sectors based at x that are separated by the segment.

Suppose $f : F \to G$ is a map of constructions. If (x, v, a, b) is an edge germ of F, then the images $f(a)$ and $f(b)$ are edges in $G_{f(x)}$ sharing the vertex $f(v)$. We say that f *folds* along (x, v, a, b) if $f(a)$ and $f(b)$ coincide. We say that f *opens* along (x, v, a, b) if it doesn't fold.

Let $\mathbf{EG}(F)$ denote the set of edge germs of F. A *scaffolding* of a construction F is a pair $\sigma = (\sigma^o, \sigma^f)$ such that σ^o and σ^f are disjoint subsets of $\mathbf{EG}(F)$. The first set σ^o is said to be the set of edge germs which are marked "open", and the second set σ^f is said to be the set of edge germs which are marked "fold".

If $f : F \to G$ is a map of constructions, and σ is a scaffolding of F, we say that f is *compatible with* σ if f folds along the edge germs in σ^f and opens along the edge germs in σ^o.

Implicit in this terminology is that G was provided with a fully open scaffolding (such as will usually be the case for a pre-building). More generally a map between constructions both provided with scaffoldings is compatible if it maps the open edge germs in F to open ones in G, and for edge germs marked "fold" in F it either folds them or else maps them into edge germs marked "fold" for G.

Definition 6.1. A scaffolding is *full* if $\sigma^o \cup \sigma^f = \mathbf{EG}(F)$. A scaffolding is *coherent* if there exists a building B and a map $h : F \to B$ compatible with σ.

It should be possible to replace the definition of coherence with an explicit list of required properties, but we don't do that here. Recall that we will be working under the assumption of existence of some ϕ-harmonic map so the above definition is adequate for our purposes.

One of the main properties following from coherence is a propagation property. A neighborhood of a hexagonal point cannot be folded in an arbitrary way. One may list the possibilities, the main one being just folding in two; note however that there is another interesting case of three fold lines alternating with open lines. Certain cases are ruled out and we may conclude the following property:

Lemma 6.2. *If σ is coherent, and x is a hexagonal point, then if two adjacent edge germs at x are in σ^o it follows that the two opposite edge germs are also in σ^o.*

This will mainly be used to propagate the open or fold edge germs along edges which are "straight" in the following sense. This definition coincides with Definition 4.9 for a pre-building provided with the fully open scaffolding.

Definition 6.3. Suppose S is a segment and $\varphi \in F(S)$. Suppose σ is a scaffolding for F. We say that φ is *straight* if, at any point x in the interior of φ (that is to say $x \in F(p)$ is the image of a point $a \in S(p)$ which is not an endpoint), the forward and backward directions along φ in the spherical construction F_x are separated by three edges a, b, c such that the edge germs ab and bc are in σ^o. Here the edge germ denoted ab corresponds to the vertex separating the edges a and b in the spherical construction.

Corollary 6.4. *Suppose φ is a straight edge and σ a coherent scaffolding. Then the marking of both forward and backward edge germs of φ is the same all along φ.*

Straight edges are mapped to straight edges:

Lemma 6.5. *Suppose F is a construction with scaffolding σ and $\varphi \in F(S)$ is an edge. If $h : F \to B$ is any map to a building compatible with σ, and if φ is straight with respect to σ, then the image of φ is contained in a single apartment of B and is a straight line segment in that or any other apartment.*

Scaffolding is used to keep track of the additional information which comes from a ϕ-harmonic map, namely that small open neighborhoods in X^* map without folding to apartments in a building. For trees, this was illustrated in Figures 3, 4 and 5.

Proposition 6.6. *The initial construction Z of Theorem 5.2 is provided with a coherent full scaffolding σ_Z, such that if $x \in X - C$ is a non-caustic point, then all edge germs in the hexagon in $Z_{h_Z(x)}$ image of the local spherical construction of E_x at the origin, are in σ_Z^o. If B is a building, there is a one-to-one correspondence between harmonic ϕ-maps $X \to B$ and construction maps $Z \to B$ compatible with σ_Z.*

Proof. (Sketch) All edge germs based at any point $x \in X - C$ are in σ^o by the remark at the end of Section 5.3. In the regions near caustics, intersections of caustics or branch points, the local constructions E_x map into any building without folding, so all of their edge germs are in σ^o. The remaining edges are always straight segments attached to fourfold points, and at these fourfold points two of the edges are already marked "open" so the other two must be marked "fold". This is illustrated in the BNR example in Section 7. The scaffolding is coherent because we are assuming that there exists at least one ϕ-harmonic map such as can be obtained from Parreau's theory [27] [21]. □

6.2. Cutting out

We now describe an important operation. Suppose (F, σ) is a construction with coherent scaffolding. Suppose $P^{a,b}$ is a parallelogram (with either orientation), and denote by $V^{a,b}$ the union of two segments based at the origin on the boundary of $P^{a,b}$. Suppose that F may be written as a pushout

$$F = G \cup^{V^{a,b}} P^{a,b}$$

along a map $g : V^{a,b} \to G$. Let σ_G be the induced scaffolding of G. Suppose that the two segments making up $V^{a,b}$, of lengths a and b respectively, are straight in G with respect to σ_G.

Theorem 6.7. *In the above situation, if B is a building, then any map $h_G : G \to B$ compatible with σ_G extends uniquely to a map $h : F \to B$ compatible with σ. Furthermore, under this map the image of the parallelogram $P^{a,b}$ is isometrically embedded in B and doesn't fold any edge germs of $\mathbf{EG}(P^{a,b})$.*

In the above situation, we refer to G as being obtained from F by *cutting out* the image of the parallelogram $P^{a,b}$. Notice that all of the edge germs in $\mathbf{EG}(P^{a,b}) \subset \mathbf{EG}(F)$ are either marked "open", or not marked. We may extend the scaffolding σ to one σ_1 with $\sigma_1^f = \sigma^f$ and $\sigma_1^o = \sigma^o \cup \mathbf{EG}(P^{a,b})$, and the theorem implies that any map from F to a building compatible with σ must also be compatible with σ_1. If σ is full then of course coherence implies that $\sigma_1 = \sigma$.

Note also that if σ is full then σ_G is full.

We say that (F, σ) is *trimmed* if it is not possible to cut out any parallelograms in the above way.

Theorem 6.8. *The initial construction Z of Theorem* 5.2, *provided with the scaffolding σ_Z of Proposition* 6.6, *may be trimmed. The result is a two-manifold construction Z_0 again provided with a coherent full scaffolding σ_0. If B is a building, there is a one-to-one correspondence between:*

- *Harmonic ϕ-maps $X \to B$;*
- *Maps $Z \to B$ compatible with σ_Z; and*
- *Maps $Z_0 \to B$ compatible with σ_0.*

6.3. Pasting together

The construction Z_0 now obtained will have, in general, many fourfold points of positive curvature. These cannot appear in the image of a map $Z_0 \to B$, so some folding must occur. The direction of the folds is determined by the scaffolding, and indeed that was the reason for introducing the notion of scaffolding: without it, there is not *a priori* any preferred way of determining the direction to fold a fourfold point. However, once the direction is specified, we may proceed to glue together some further pieces of the construction according to the required folding. This is the "pasting together" process.

Put

$$W^{a,b} := P^{a,b} \cup^{V^{a,b}} P^{a,b}.$$

We have a projection $\pi : W^{a,b} \to P^{a,b}$ given by the identity on each of the two pieces.

Here we may use either of the two possible orientations, that doesn't need to be specified but of course it should be the same for both parallelograms.

The local spherical construction of $W^{a,b}$ at the origin $0 \in W^{a,b}$ is a graph with a single cycle of length 4, in other words the origin is a fourfold point. In particular, if $W^{a,b} \to B$ is any map to a building, two of the four edge germs at 0 must be folded. There is a choice here: at least two opposite edge germs must be folded, but the other two could either be opened or also folded. We are interested in the case when the originally given parallelograms are not folded. Let v_1, v_2 be the edge germs at the origin in $W^{a,b}$ which are in the middle of the spherical constructions of the two pieces $P^{a,b}$ (the spherical construction of $P^{a,b}$ at the origin is a graph with two adjacent edges and three vertices, the middle edge germ corresponds to the middle vertex).

Proposition 6.9. *Suppose we are given a coherent scaffolding σ_W of $W^{a,b}$. We assume that v_1 and v_2 are in σ_W^o, and that the two edges comprising $V^{a,b}$ are straight. Then any map $h_W : W^{a,b} \to B$ to a building, compatible with σ_W, factors through the projection*

$$W^{a,b} \stackrel{\pi}{\to} P^{a,b} \stackrel{h_P}{\to} B$$

via a map h_P which is an isometric embedding of the parallelogram into a single apartment of B.

Let $\sigma_{W,\max}$ be the full scaffolding defined by taking all edge germs of each $P^{a,b}$ to be open, and letting the edge germs along the edges of $V^{a,b}$ be folded. A corollary of the proposition is that any coherent scaffolding of $W^{a,b}$ satisfying the hypotheses has to be contained in $\sigma_{W,\max}$.

Now suppose F is a construction with coherent scaffolding σ. Suppose we have an inclusion

$$W^{a,b} \hookrightarrow F$$

and suppose that the induced scaffolding σ_W of $W^{a,b}$ satisfies the hypotheses of the proposition, namely the angle between v_1 and v_2 is open and the edges of $V^{a,b}$ are straight. Define the quotient

$$\widetilde{F} := F \cup^{W^{a,b}} P^{a,b}$$

which amounts to *pasting together* the two parallelograms which make up $W^{a,b}$.

Corollary 6.10. *Under the above hypotheses, any map to a building $h : F \to B$ compatible with σ factors through a map*

$$\widetilde{F} \to B.$$

This operation may be combined with the cutting-out operation. Let $\overline{V}^{a,b}$ denote the opposite copy of $V^{a,b}$ inside $P^{a,b}$. We may write

$$\partial W^{a,b} = \overline{V}^{a,b} \cup^{p \sqcup p} \overline{V}^{a,b},$$

it is a rectangular one-dimensional construction formed from two edges of length a and two edges of length b (again as usual, we should chose one of the two possible orientations for everything). We may define a projection

$$\partial W^{a,b} \to V^{a,b}$$

as the identity on the two pieces. The combination of pasting together and cutting out may be summarized in the following theorem.

Theorem 6.11. *Suppose (F, σ) is a construction with coherent scaffolding. Suppose that F can be written*

$$F = G \cup^{\partial W^{a,b}} W^{a,b}.$$

Suppose the scaffolding σ_W induced by σ on $W^{a,b}$ satisfies the hypotheses of Proposition 6.9. Define

$$\widehat{G} := G \cup^{\partial W^{a,b}} V^{a,b}.$$

We may write

$$\widetilde{F} = \widehat{G} \cup^{V^{a,b}} P^{a,b}.$$

Furthermore, σ induces scaffoldings σ_G on G and $\sigma_{\widehat{G}}$ on $\widehat{G}$. Let $\sigma_{\widetilde{F}}$ be the scaffolding of $\widetilde{F}$ obtained from $\sigma_{\widehat{G}}$ by declaring all edge germs of $P^{a,b}$ to be open.

Assume that $V^{a,b} \subset \widehat{G}$ is standard with respect to the scaffolding $\sigma_{\widehat{G}}$. Then there is a one-to-one correspondence between the sets

$$\left\{ \textit{maps } h' : \widehat{G} \to B \textit{ to a building compatible with } \sigma_{\widehat{G}} \right\}$$

and

$$\{ \textit{maps } h : F \to B \textit{ to a building compatible with } \sigma, \}$$

both being the same as

$$\left\{ \text{maps } h : \widetilde{F} \to B \text{ to a building compatible with } \sigma_{\widetilde{F}} \right\}$$

via the natural maps

$$\widehat{G} \to \widetilde{F} \leftarrow F.$$

The scaffolding $\sigma_{\widehat{G}}$ is never full, because it doesn't say what to do along the new edges which have been introduced by glueing together the appropriate edges in $\partial W^{a,b}$. This is the main problem for iterating the construction, and it eventually leads to the need to look for BPS states.

6.4. Two-manifold constructions

A *two-manifold construction* is a construction F such that the topological space $F(p)$ is a 2-dimensional manifold.

Lemma 6.12. *A construction F is a two-manifold if and only if its local spherical constructions F_x are polygons.*

The polygons have an even number of edges because of the orientations of enclosures.

Suppose F is a two-manifold construction. A point $x \in F(p)$ is a *flat point* if F_x is a hexagon; it is *positively curved point* if F_x has two or four sides, and it is a *negatively curved point* if F_x has eight or more sides. Usually we will deal with just three kinds, the rectangular points, the flat or hexagonal points, and the octagonal points.

We have the following important observation which shows that the two-manifold property is preserved by the combination of pasting together and cutting out.

Lemma 6.13. *In the situation of Theorem* 6.11*, suppose F is a two-manifold construction. Then $\widehat{G}$ is also a two-manifold construction.*

Suppose F is a two-manifold construction with no twofold points. Suppose $x \in F(p)$ is a rectangular point (i.e., F_x has four edges), and suppose σ is a full coherent scaffolding such that two of the edge germs at x are in σ^o. Then there exists a copy of $W^{a,b} \subset F$ sending the origin to x, and by fullness and coherence of σ we may assume that we are in the situation of Theorem 6.11, meaning that the two edges of $V^{a,b}$ are straight and the angle between them is open. We may assume that it is a maximal such copy. The process of Theorem 6.11 yields a new two-manifold construction $\widehat{G}$.

6.5. The reduction process

We may now schematize the reduction process obtained by iterating the above operation.

We are going to define a sequence of two-manifold constructions F_i with full coherent scaffoldings σ_i. The first one F_0 will be obtained by trimming some initial construction. We assume that there are never any twofold points. We will be happy and the sequence will stop if we reach a construction F_i which has no fourfold points, so it is a pre-building.

Suppose we have constructed the (F_i, σ_i) for $i \leq k$. We suppose that F_k has no twofold points, and that at every rectangular point, two of the edges are marked "open". Choose a rectangular point $x \in F_k$ and let $W^{a,b} \subset F_k$ be a maximal copy with the origin corresponding to x, such that the edges of $V^{a,b}$ are straight. Applying the pasting together and cutting out process of Theorem 6.11, we obtain a new two-manifold construction $F_{k+1} := \widehat{G}$.

Hypothesis 6.14. We assume $\widehat{G}$ has no twofold points, and that there is a unique way to extend the scaffolding $\sigma_{\widehat{G}}$ to a full coherent scaffolding σ_{k+1} on F_{k+1} such that there are two open edges at any new rectangular points.

Corollary 6.15. *If this hypothesis is satisfied at each stage, then the reduction process may be defined.*

One of the main ideas which we would like to suggest is that this hypothesis will be true if there are no BPS states. How the reduction process works, and the relationship with BPS states, will be discussed further in Section 8 below.

6.6. Recovering the pre-building

Suppose we start with a construction F and scaffolding σ, then trim it to obtain a two-manifold construction F_0 without twofold points. Suppose then that Hypothesis 6.14 holds at each stage so we can define the reduction process, and suppose that the process stops after a finite time with some F_k which has no positively curved points. Then we would like to say that this allows us to recover a pre-building with a map from F. We explain how this works in the present subsection.

The first stage is to undo the initial trimming. Put $G_0 := F$. Then apply Theorem 6.7 successively to obtain a sequence of constructions G_i, with G_{i-1} playing the role denoted F in Theorem 6.7. Hence, G_{i-1} is a pushout of G_i with a standard parallelogram. If this process stops at a trimmed construction $F_0 = G_k$ then we may successively do the pushouts to get $G_{k-1}, G_{k-2}, \ldots, G_0 = F$. Thus F is obtained from F_0 by a sequence of pushouts with standard parallelograms.

Next, starting with F_0 we do a sequence of pasting and cutting operations to yield two-manifold constructions F_i. Here, F_i is obtained from F_{i-1} by applying Theorem 6.11. In particular, we have a construction $\widetilde{F}_{i-1}$ which is on the one hand the result of a pasting operation applied to F_{i-1}, but on the other hand is a pushout of F_i along a standard parallelogram.

Define inductively a sequence of constructions H_i, starting with $H_0 := F$, as follows. At each stage, we will have $F_i \subset H_i$, and H_i is obtained from F_i by a sequence of pushouts along standard parallelograms (to be precise this means pushouts along the inclusion $V^{a,b} \hookrightarrow P^{a,b}$). This holds for H_0.

Suppose we know H_{i-1}. Then $\widetilde{F}_{i-1}$ is a pasting of F_{i-1} as in Corollary 6.10. Do the same pasting to H_{i-1} instead of F_{i-1}, in other words

$$H_i := H_{i-1} \cup^{F_{i-1}} \widetilde{F}_{i-1}.$$

By the inductive hypothesis, H_{i-1} is a pushout of F_{i-1} by a series of standard parallelograms, thus it follows that H_i is a pushout of $\widetilde{F}_{i-1}$ by the same series. On the other hand, $\widetilde{F}_{i-1}$ was a pushout of F_i by a standard parallelogram. Therefore we obtain the inductive hypothesis that H_i is a pushout of F_i along a series of standard parallelograms.

We have a sequence of maps $H_{i-1} \to H_i$. These are quotient maps. Composing them we obtain a series of maps $F \to H_i$. If F was the initial construction for (X, ϕ) then we would get in this way a ϕ-harmonic map $X \to H_i$.

Under the hypothesis that the process leading to a sequence of F_i stops at some finite k, we obtain a construction H_k. If F_k has no fourfold points, we would like to say that H_k is the pre-building; however, in putting back in the pushouts with standard parallelograms, we might be introducing new fourfold points. Therefore, as was done previously for the small object argument, we should apply the folding-in process of Lemma 4.21. Precisely, having $F_k \subset H_k$ obtained by a sequence of pushouts by standard parallelograms, apply the folding-in process at each stage to transform the pushout into a good pushout as described in Lemma 4.21 preserving the pre-building property. The resulting construction is a pre-building $\widetilde{H}$ accepting a map $F \to \widetilde{H}$. It is universal for maps to pre-buildings.

Theorem 6.16. *Suppose that there are no BPS states and Conjectures* 8.4 *and* 8.5 *hold. Then we obtain a pre-building* $B_\phi^{\mathrm{pre}} := \widetilde{H}$ *and a* ϕ*-harmonic map*

$$h_\phi : X \to B_\phi^{\mathrm{pre}}$$

which is universal for ϕ*-harmonic maps to pre-buildings containing extensions to the enclosures* E_x *used for the initial construction. The small object argument of Theorem* 4.23 *yields a versal* ϕ*-harmonic map to a building.*

The map to the building is only versal. Indeed, at a stage of the construction of B_ϕ^{pre} where we add a pushout with a standard parallelogram, if we are given a map $V^{a,b} \to B$ which is not itself standard (say for example the two edges coincide) then the extension to a map $P^{a,b} \to B$ is not unique.

7. The BNR example, revisited

In order to illustrate the procedure described above, we show how it works in the BNR example from [21]. It was by considering this example in a new way that we came upon the above procedure.

Recall the picture from [21, Figure 3] of the spectral network containing two singular points b_1, b_2. At each singularity, there are two spectral network lines which meet lines from the opposite singularity, and the resulting four segments delimit a diamond-shaped zone in the middle of the picture. There is a single caustic C joining b_1 to b_2 across the middle of this region.

Recall that we had found that a ϕ-harmonic map to a building would fold together this middle region along the caustic, identifying the two collision points.

The resulting pre-building is conical with a single singularity at the origin. Here, eight sectors form a two-manifold, with the collision being a negatively curved singular point of total angle 480°. There are two additional sectors, forming a link across the middle of the octagon. The image of the middle region goes into these sectors but is not surjective due to the folding along the caustic, see [21, Figure 4].

We now illustrate how this end picture comes about using the process described above. The first step is to create the initial construction. For this, let us cover the caustic by some regions as illustrated in Figure 11.

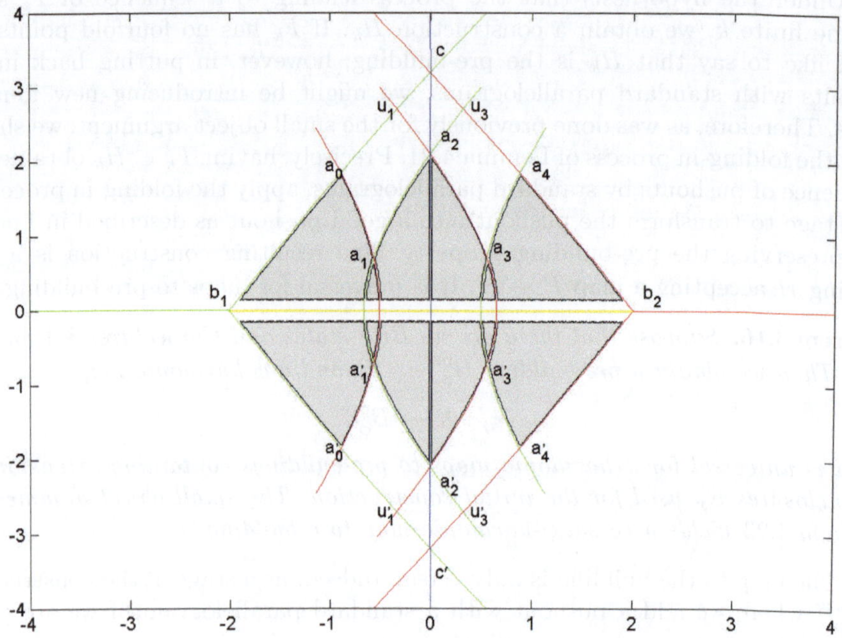

FIGURE 11. Regions covering the caustic

The regions which cover the caustic are the ones which are bounded by the two foliation lines going from a_0 to a'_0, resp. a_2 to a'_2, resp. a_4 to a'_4. The first and last ones include the singularities b_1 and b_2.

Each of these regions is folded under any ϕ-harmonic map. They are in fact Ω_{pq} regions (except at the two endpoints). The image of the regions from Figure 11 under the integration map to an apartment, are shown in Figure 12.

The enclosures corresponding to these folded pieces are sketched into the picture completing the shaded regions to full parallelograms.

The initial construction Z consists of glueing together these enclosures covering the caustics, and then adding enclosures around the remaining points which are not for the moment glued together. The illustration of Figure 12 may be viewed as a picture of Z, where the non-shaded regions of the curve correspond to two

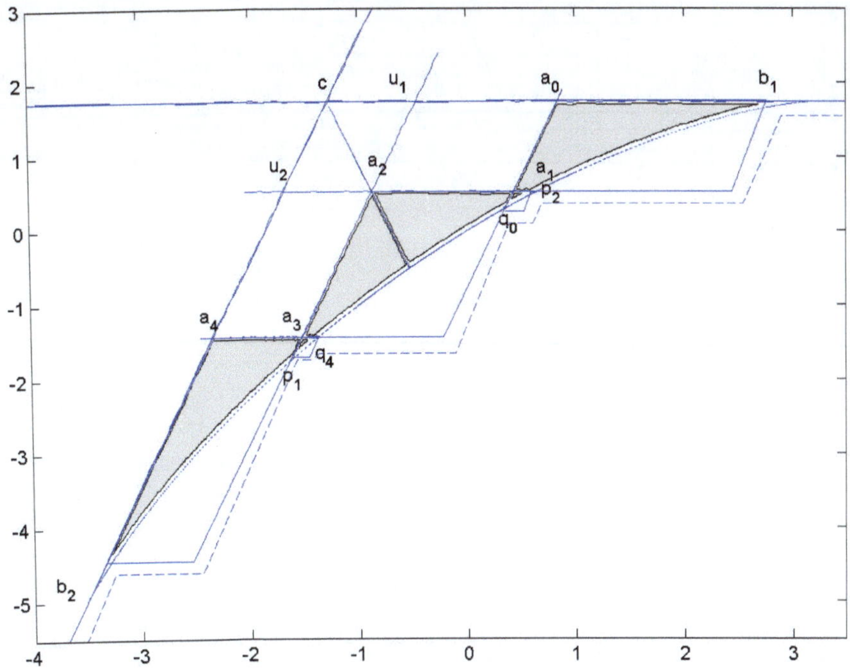

FIGURE 12. Images of the regions

different sheets, an upper and a lower one corresponding to the upper and lower regions in Figure 11. The completions of parallelograms just below the caustic are part of the initial construction Z (these pieces should be considered as shaded having only one sheet) but they are not images of points in X. The zone below the dashed line is empty, that is to say it doesn't correspond to any points in Z, in Figure 12 and similarly for the subsequent ones.

Now Z may be trimmed by taking out the regions containing folded pieces, leaving Z_0 as shown in Figure 13. Again there are two sheets, which are joined together along the edge from b_1 to b_2 (and also along the rays pointing outward from b_1 and b_2). In this case the edge where the two sheets are joined corresponds to the dotted line, and as said above, the zone below the dotted line is empty.

In Z_0 we have five points a_0, a_1, a_2, a_3, a_4 which are not hexagonal. The points a_0, a_2, a_4 are eightfold whereas a_1, a_3 are fourfold points. The segments b_1a_0 and a_4b_2 are marked "open", due to the trivalent edges in the constructions which are placed at the singularities. Indeed, on the outer side of b_1 one gets an open edge, because we are in the image of a neighborhood in x as was illustrated in Figures 3 and 4. This open edge then propagates into the segment b_1a_0 by Lemma 6.4, because in Z_0 all the points along this edge are hexagonal, including b_1, and up

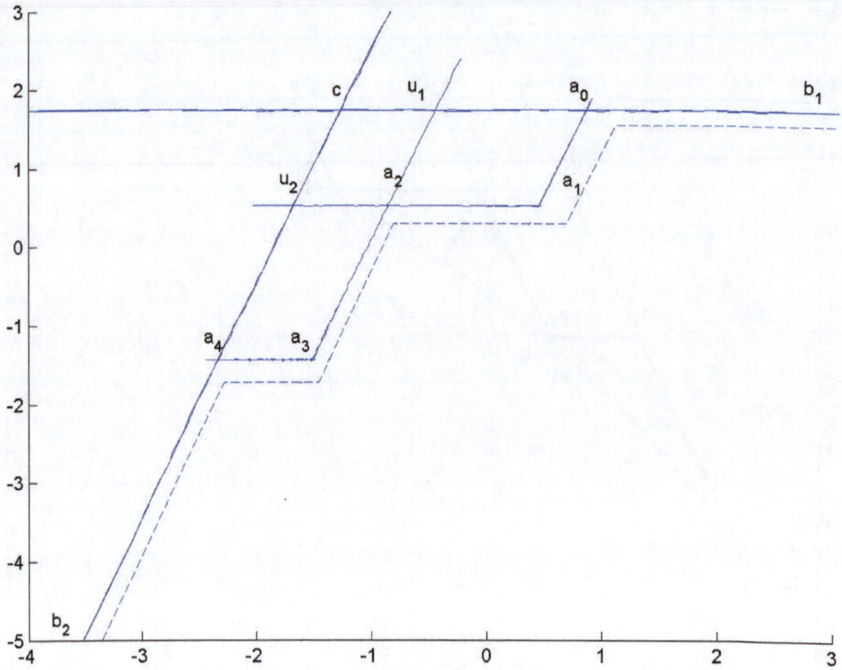

FIGURE 13. After trimming

to (but not including) a_0. The segment is straight because on either side we are in the image of neighborhoods from X.

We may now show that the four segments a_0a_1, a_1a_2, a_2a_3 and a_3a_4 are marked "fold". Consider for example the fourfold point a_1. Let u_1 and u_1' be the points which complete the two parallelograms spanned by a_0, a_1, a_2. The four sectors around a_1 are $a_1a_0u_1$, $a_1a_0u_1'$, $a_1a_2u_1$ and $a_1a_2u_1'$. Suppose we are given a map $Z_0 \to B$ to a building, coming from a ϕ-harmonic map $h : X \to B$. The two sectors $a_1a_0u_1$ and $a_1a_2u_1$ come from two adjacent sectors at a point in X (one of the lifts of the point a_1). Since the map h doesn't fold anything in $X - C$, we conclude that there is no folding along the edge a_1u_1. Similarly for a_1u_1'. But since there are no cycles of length ≤ 4 in the local spherical buildings of B, the four sectors have to be collapsed somehow. Therefore, our map must fold along the segments a_1a_0 and a_1a_2.

The same argument at a_3 shows that the map must fold along a_3a_2 and a_3a_4. Therefore the four segments a_0a_1, a_1a_2, a_2a_3 and a_3a_4 are marked "fold".

All other edges of Z_0 correspond to edges in X, so we conclude that all edges except for these four are marked "open".

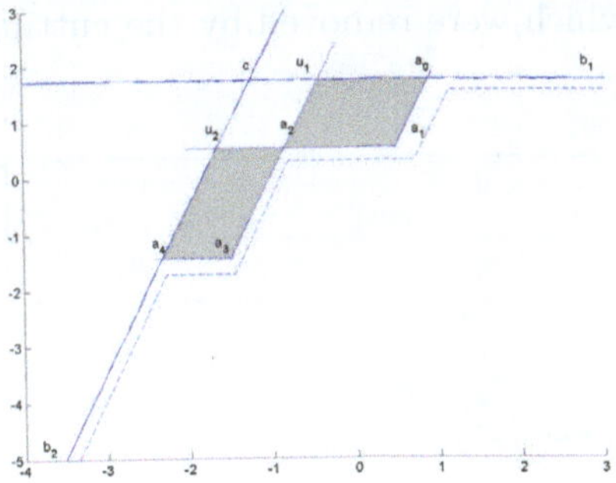

FIGURE 14

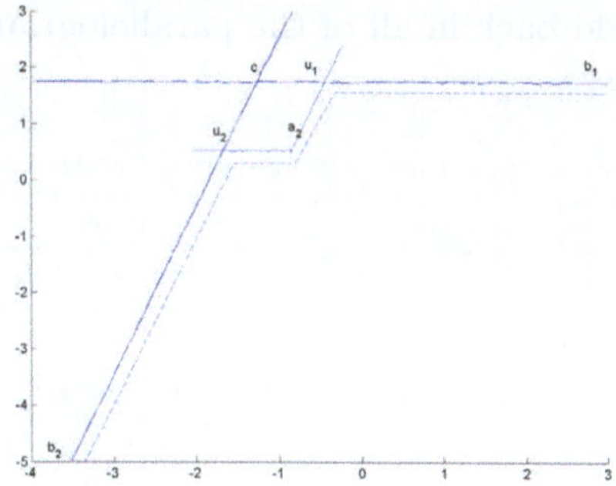

FIGURE 15

We may now look at how the pasting-together process works. The first step is to paste together the two parallelograms $a_0a_1a_2u_1$ and $a_0a_1a_2u_1'$. Similarly we paste together $a_2a_3a_4u_3$ and $a_2a_3a_4u_3'$. The result is shown in Figure 14.

Then we can cut out the parallelograms obtained from the above pasting. This gives the picture shown in Figure 15 which is again a two-manifold construction Z_1. As before there are two sheets joined along the edge.

Now there is one remaining fourfold point at a_2 (notice that what was previously an eightfold point has now become a fourfold point). Denote by u_1 and u_3 the images in Z_1 of the pairs of points (u_1, u_1') and (u_3, u_3') respectively. By the same reasoning as before, the segments a_2u_1 and a_2u_3 are marked "fold" whereas all the other edges are marked "open".

The next and last step of the pasting-together process is to paste together the two parallelograms $u_1a_2u_3c$ and $u_1a_2u_3c'$ where c and c' are the two collision points from X. This gives Figure 16.

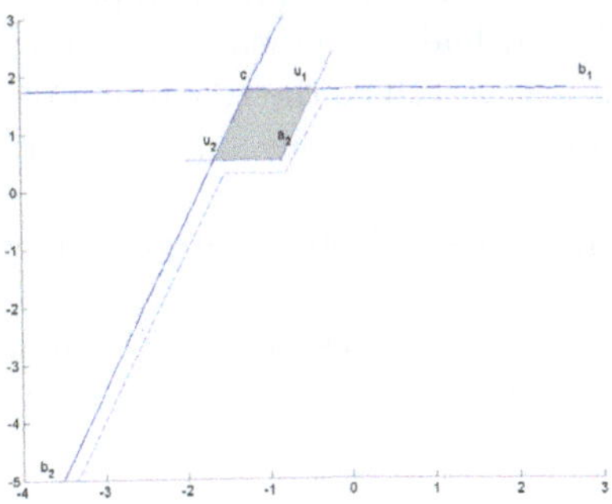

FIGURE 16

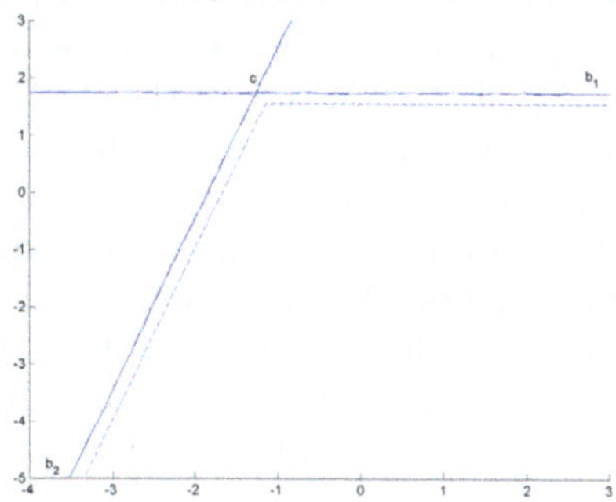

FIGURE 17

Cutting out the parallelogram obtained from the pasting-together, gives the two-manifold construction Z_2 shown in Figure 17.

It has only a single eightfold point at the image of the two collision points which are now identified. There are no fourfold points. Therefore we may back up and add back in all of the parallelograms which were removed by the cutting-out processes.

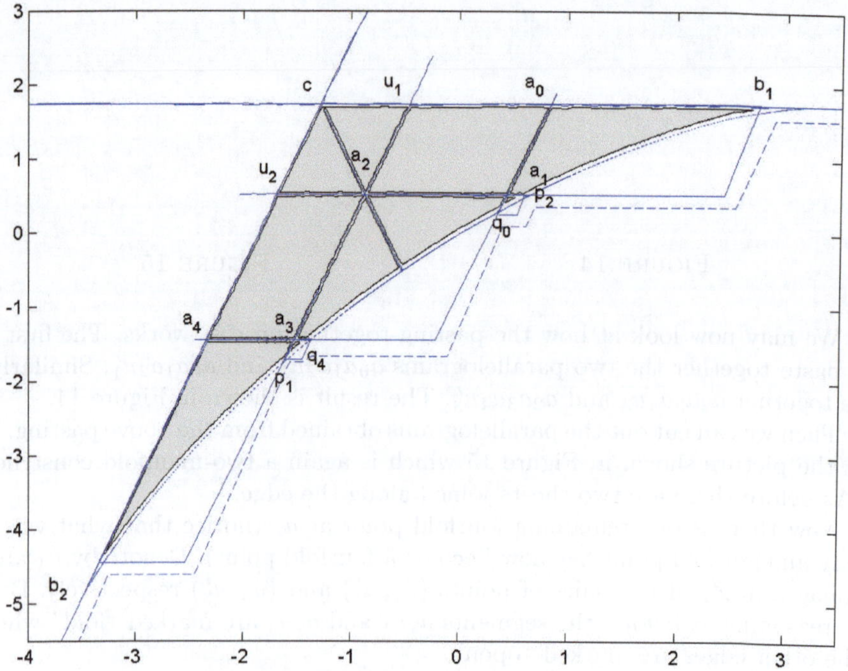

FIGURE 18. The pre-building

This gives the picture shown in Figure 18 which is the universal pre-building accepting a map from X. Here as before, the unshaded regions above and to the left correspond to two sheets; the shaded regions as well as the remaining pieces of the parallelograms just next to the dotted line, are single sheets, and everything below the dotted line is empty.

When we turn this into a building using the small object argument, we get back the universal building constructed in [21].

This was a first illustration of how the reduction process introduced in Section 6.5 leads to a universal pre-building. In the next section we discuss some further aspects of the process in the general case.

8. The process in general

In general we have the following setup: starting with (X, ϕ) we first make the initial construction Z, then trim it to get a two-manifold construction Z_0. Then we go through a sequence of steps of pasting-together then cutting-out described in Section 6.5. This yields a sequence of two-manifold constructions Z_i (these were denoted F_i in Section 6 but we change to the notation Z_i here in order to think of them as modifications of the original surface Z). The construction Z_0 has a coherent full scaffolding by Lemma 6.6. In order to proceed with the construction at each step, we would like to know that after each operation the new Z_i may still be given a uniquely defined full coherent scaffolding, which is Hypothesis 6.14.

If this hypothesis holds at each stage, then we can continue the operation and hope that it converges locally at least. In this section we explain some ideas for how to understand more precisely the sequence of constructions, and how to see Hypothesis 6.14 saying that full scaffoldings will be determined at each step, if there are no BPS states.

8.1. Properties of scaffoldings

Let us first axiomatize some properties of our two-manifold constructions and their scaffoldings.

Our two-manifold constructions Z_i are provided with scaffoldings in which almost all of the edges are marked "open" or o. Those vertices in the local spherical buildings which are marked "fold" or f, are parts of straight edges (that is, edges such that at each point, at least one side of the edge has both directions transverse to that edge, marked o). The marking is the same along the straight edge.

Fourfold points – We require that, at any fourfold point there should be two opposite edges marked o, and two opposite edges marked f:

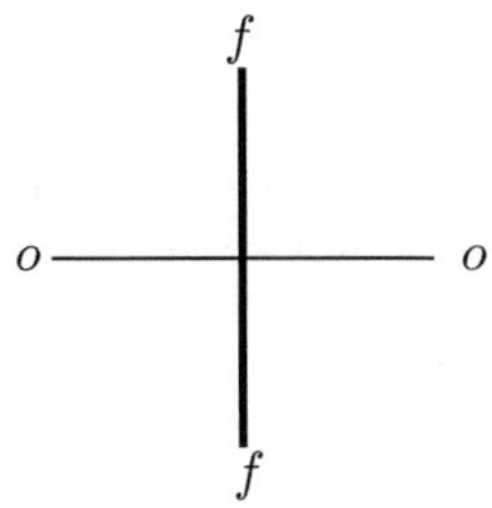

The configuration with all four edges folded is admissible from the point of view of coherence, but throughout our procedure we conjecture that it should not occur.

Hexagonal points – At a hexagonal point, the possibilities are as follows. First, either all edges are open or two opposite edges are folded.

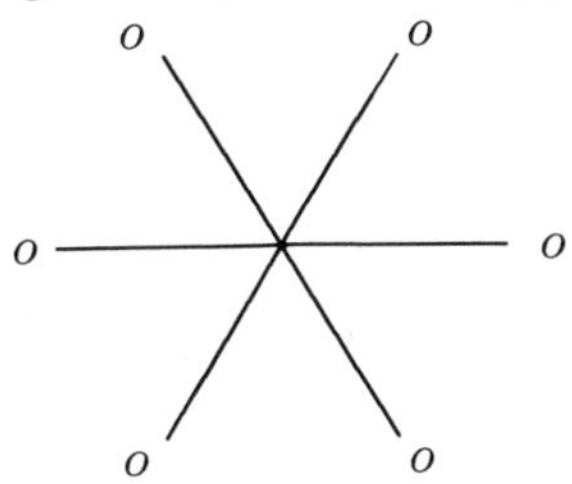

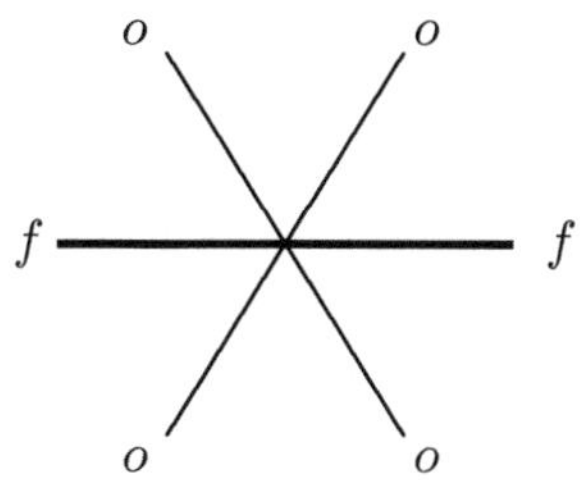

It is also possible to have three o and three f alternating; this configuration should actually be considered as a superposition of a $4v$ point plus an $8v$ point starting a new postcaustic curve. Furthermore, one additional edge can also be folded.

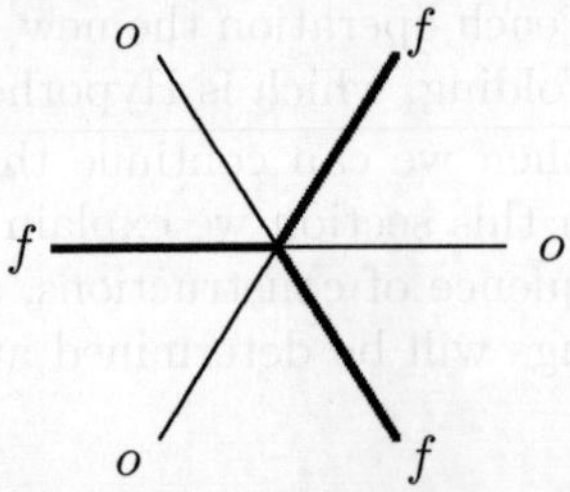

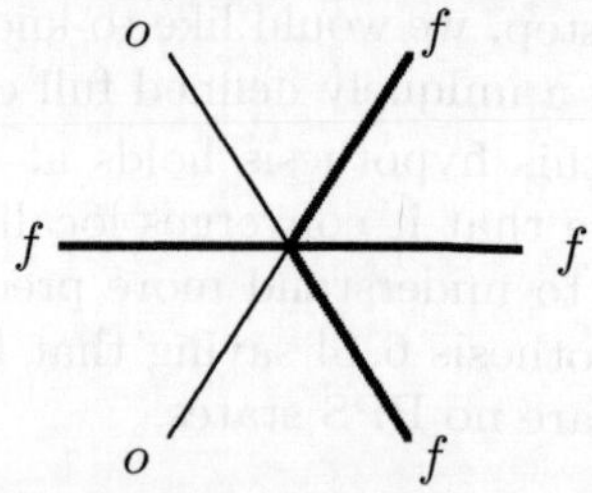

Finally, a hexagonal point can have all edges folded. Again, we conjecture that this doesn't appear in our procedure.

Eightfold points – At an eightfold point, there are the following main possibilities. First, a BNR point with all edges open. Then, a point with a single fold line, as happens at the end of a postcaustic curve, initially coming from a singularity.

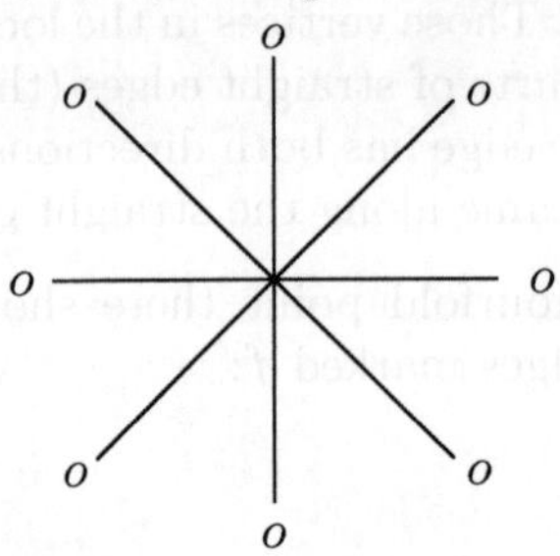

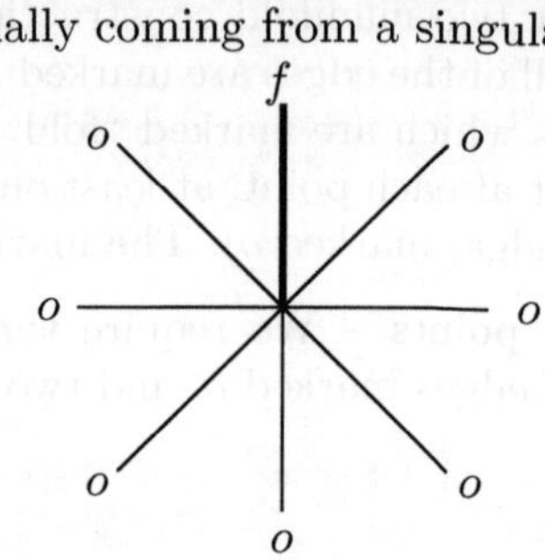

There are two ways of having a post-caustic pass through the eightfold point, either two opposite fold lines, as is most standard; or two adjacent fold lines.

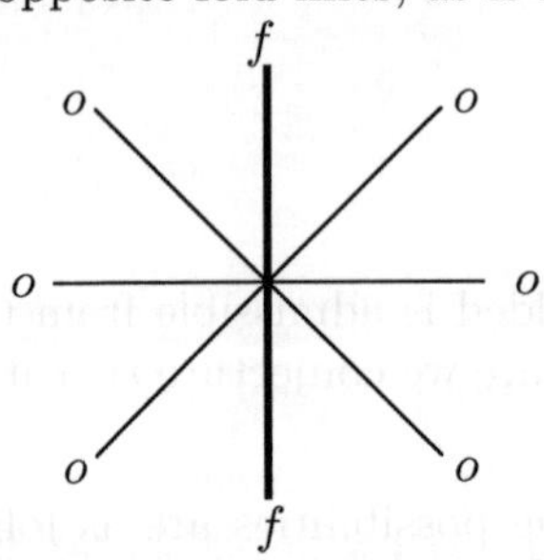

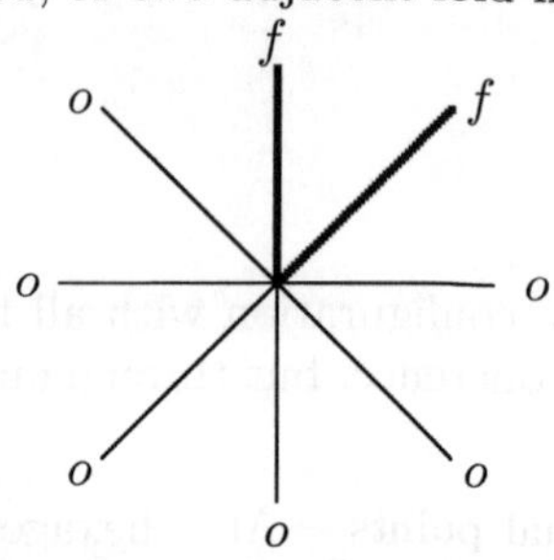

Notice that when the two adjacent fold lines get folded up, the image is a hexagon, with one sector covered three times by passing over and back.

This picture with two adjacent fold lines comes up, in particular, in the initial construction when we have a crossing of two caustic lines. It can also arise sometime later in the procedure.

Other eightfold points – several other configurations are also admissible from the point of view of coherence, although we feel that they probably will not occur in the procedure. It is left to the reader to enumerate these.

This finishes the description of the expected properties of open and fold markings of our scaffolding. The set of possible local pictures listed above covers the situations which we meet along the various steps of the process. A scaffolding of a two-manifold construction will be called *admissible* if its local pictures are in the above list (completed as per the preceding paragraph).

Proposition 8.1. *If Z_i is a two-manifold construction with a full coherent scaffolding, then the scaffolding is admissible.*

Corollary 8.2. *In the two-manifold construction obtained by trimming the initial construction, and then in the two-manifold constructions which are met along the way of our process (assuming that there are no BPS states and Hypothesis* 6.14 *holds), the full coherent scaffolding is always admissible.*

8.2. Post-caustics

It is a consequence of the above list of local pictures allowed for an admissible scaffolding, that the collection of points which have an edge marked "fold" in the scaffolding is organized into a collection of piecewise linear curves marked as folded edges. Furthermore, it follows that the endpoints of the curves are $8v$ points (or $6v$ triple points), and along a curve the successive segments have endpoints which alternate as $8v, 4v, 8v, \ldots, 4v, 8v, 4v, 8v$. All points outside of these are hexagonal, i.e., flat.

Call these curves "post-caustic" curves, because they arise in the initial construction as approximations to the caustics.

We may think of Z_i as a locally flat surface but with singularities: positive curvature at the $4v$ points (total angle 240°) and negative curvature (total angle 480°) at the $8v$ points. The total curvature of a post-caustic curve is equivalent to one $8v$ vertex.

8.3. Determining new markings

The next step is to start trying to remove the fourfold points by using the pasting-together construction of Section 6.3. Suppose we have gotten to a two-manifold construction Z_i.

At a fourfold point we choose a standard pair of parallelograms to be glued together. It may be seen from the Ω_{pq} argument that the two parallelograms must go to the same place in any harmonic map to a building. Here, by a standard parallelogram we mean one in which all of the interior edges, including interior edges along the sides, are labeled as open. Thus in a map to a building the paths which look noncritical, must actually map to noncritical paths. This applies in particular to the two outer edges which are common to both parallelograms. They form a noncritical path joining the two endpoints of the parallelograms p and q, and the Ω_{pq} argument works as usual. Both parallelograms are swept out by noncritical paths so the two images in any map to a building must be identified.

Let Y_{i+1} denote the result of pasting together the two parallelograms. The next two-manifold construction Z_{i+1} is then obtained by cutting out the resulting parallelogram from Y_{i+1}.

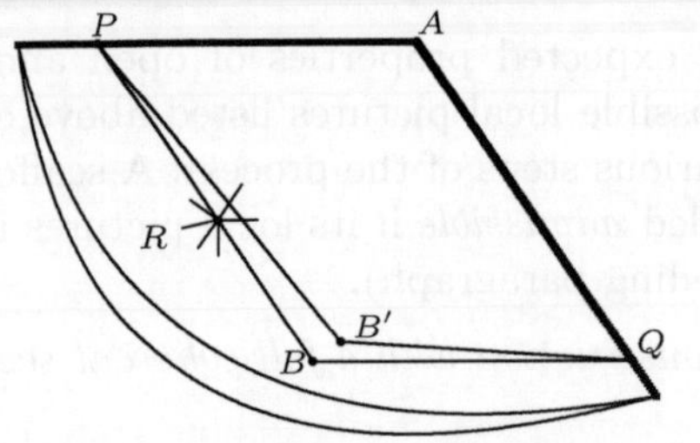

FIGURE 19

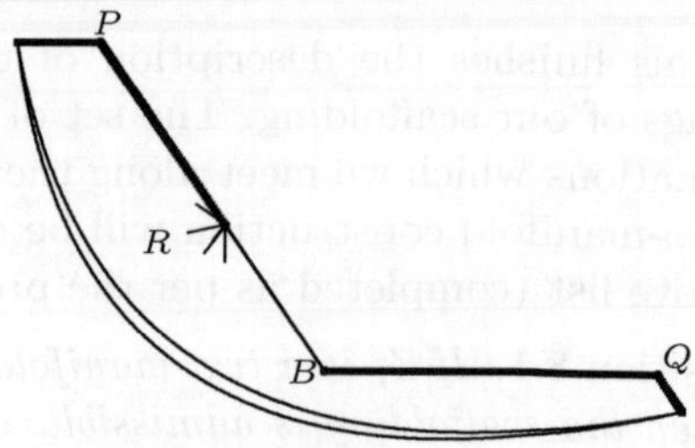

FIGURE 20

Denote by the "common edges" of the parallelograms the two edges which are common and which start from the fourfold point; the "glued edges" are the other two edges of each parallelogram.

We assume that at least one of the two edge lengths of the parallelogram is maximal. It means that a non-hexagonal point, or possibly a hexagonal point with nontrivial marking, was encountered along some glued edge of at least one of the parallelograms.

In the new two-manifold construction Z_{i+1}, the only edges which don't benefit from the previous scaffolding are the two glued edges. Therefore, the main problem is how to determine the new marking along these edges. We illustrate this in the next section.

8.4. A sample step

Let us consider an example of pasting with a singularity on the edge. We draw this in Figure 19. The two parallelograms to be pasted together are $PAQB$ and $PAQB'$. They share two common edges, PA and QA. These edges come together at a fourfold point A, and the other two edges at A are marked "open", so the edges PA and QA are marked "fold"; hence they have been drawn as thick lines. The interior edges of both parallelograms are all marked "open", in particular the four sides of each one are straight. Thus, the two parallelograms must be mapped to the same place under any harmonic map to a building, and we will paste them together and cut them out. Pasting together identifies the points B and B', we denote the image just by B too. After cutting out the parallelogram we get a new two-manifold shown in Figure 20

The eightfold point R lies on the edge PB of the first parallelogram, this is constraining the width of the parallelogram to be maximal. Let R' denote the corresponding point on the other parallelogram; we are assuming that R' is a regular hexagonal point. Now R and R' are also identified in the new two-manifold and we let their image be called R too.

In the picture we have drawn, we are assuming that P and Q are hexagonal points and the folded edges PA and QA extend beyond the points P and Q respectively. In the new two-manifold, P and Q are fourfold points, from which it follows that the edges QB and PR are folded.

Main question: what is the marking of the edge RB?

Let us look in the spherical construction at the point R. At R' we have a hexagon with all directions marked "open", whereas at the original point R we are assuming there is an eightfold point. The two directions which are interior to the parallelogram are marked "open" by the hypothesis that the parallelogram is standard. We may picture the original point R as follows, with the edges going towards P and B marked p and b respectively.

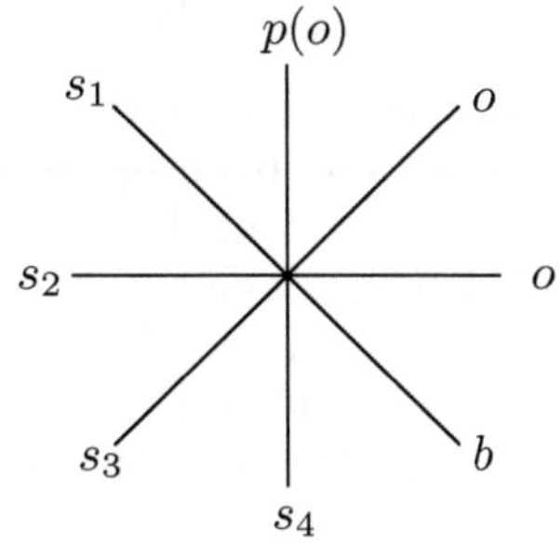

Notice that the edge p should be marked "open" otherwise we would have a non-admissible configuration at P. There are several different possibilities for the labels s_i as well as that of b, as will be discussed shortly.

At the new point R, the three interior sectors have been replaced by the three sectors which were exterior to the parallelogram at the point R'. The two directions inside here are again marked "open" because of our hypothesis that R' was a flat hexagonal point. We are left with a diagram of the following form:

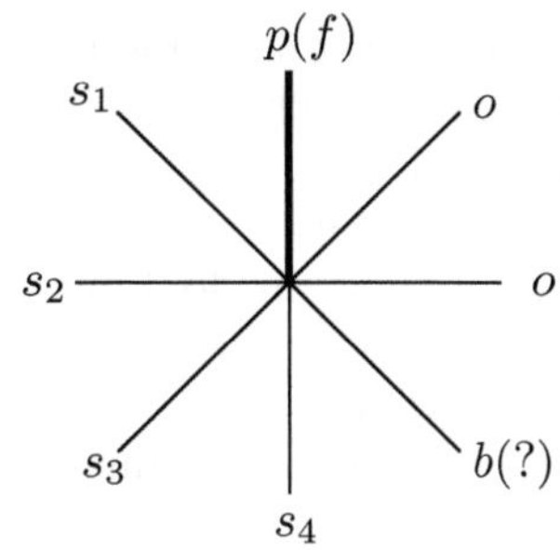

The direction p is now marked "fold" as discussed previously.

Lemma 8.3. *In the above situation, the markings of s_2, s_3 and s_4 determine the marking of b.*

Proof. Since p is marked "fold", it has to be folded leading to a hexagonal point. The edge germs to the left and right of p are joined together into a single one marked s below, and we don't know how that one will be further folded. The hexagon may be pictured as:

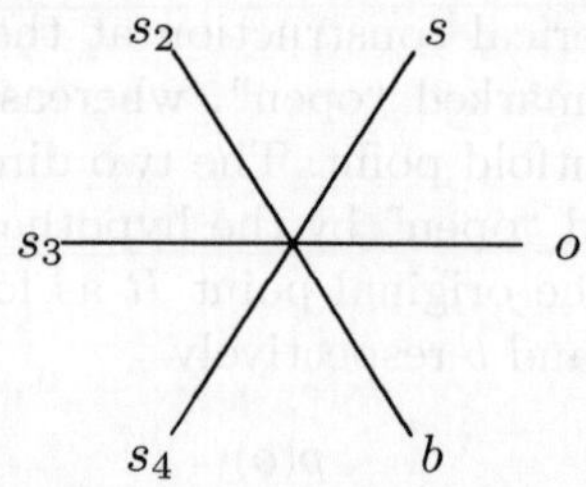

If s_2, s_3 and s_4 are open, then it follows that the full hexagon must be open. If either s_2 or s_4 is folded, then from the given open edge we cannot have a triple fold, so the opposite edge must also be folded. If s_2 is folded then it follows that b must be folded. If s_4 is folded then s must be folded. The hexagon then folds to a germ of half-apartment; and one can see b would be folded if and only if s_3 is folded. On the other hand, if s_3 is folded, then since the edge opposite to it is open, it follows that s and b must be folded. This completes the determination of the marking of b. □

Of course this lemma treats just one of the various possible situations which can arise. We expect that similar considerations should hold in the other cases.

In our example, since the edge RB is straight, the marking of b propagates along the full edge and we conclude that the marking of RB is determined; for our example, that completes the determination of the full scaffolding on the new two-manifold.

By coherence one sees that there were only a few possibilities for the folded edges at R, and the reader may find it interesting to study what happens to the post-caustic curves in these cases. Suppose for example that the marking of b determined by the lemma is "open". The possibilities for R are: a BNR point with no folding; a single folded edge s_1; two opposite folded edges s_1 and b; or two adjacent folded edges s_4 and b. In the latter two cases, the edge BR was folded in the original construction, and this fold must continue past B. In the new construction, the edge BR is no longer folded but the part of the post-caustic after B will still be present. Therefore, at the point B in the new construction we obtain two adjacent folded edges, this continuation together with the new fold along BQ. Also in these two cases, at the new point R we have two folded edges, p and either s_1 (adjacent) or s_4 (opposite). The new post-caustic coming from the point P thus attaches onto the piece of post-caustic that used to come from one side of B, whereas the old post-caustic on the other side of B is now attached to the piece going through Q.

8.5. The indeterminate case

In a less generic situation, the new marking might not be well determined. Consider a picture similar to the previous one, but where the post-caustic bounding the parallelogram ends at an eightfold point P as shown in Figure 21.

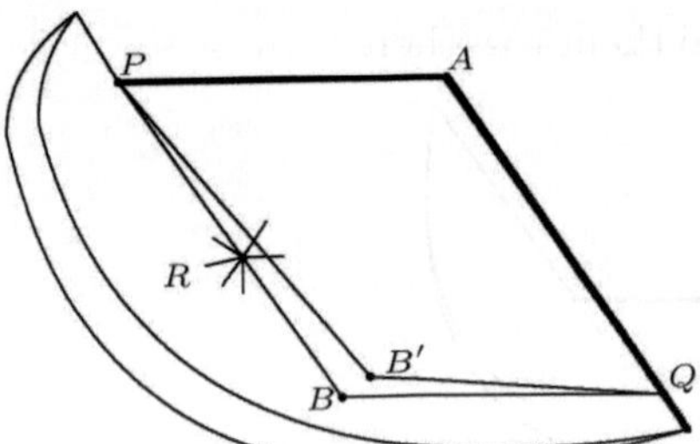

FIGURE 21

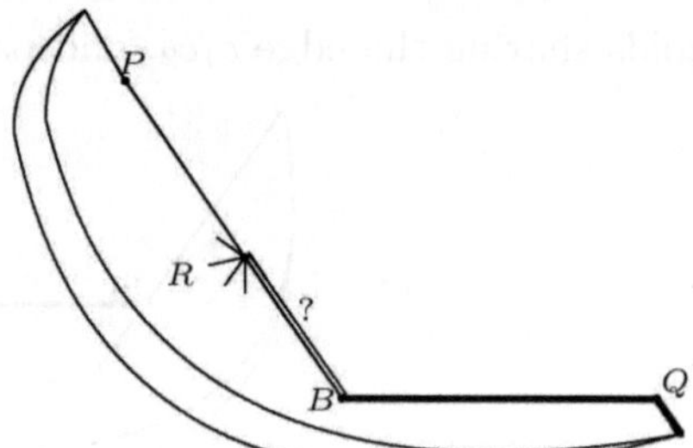

FIGURE 22

Now when we paste and cut out, the picture is in Figure 22, noting specially that the edge PR is now open. The argument used to conclude that RB should be open, is no longer valid and we don't know how to mark RB (however that might be determined for some configurations of the markings of b and the s_i).

It corresponds to a BPS state: the eightfold point P lies on a spectral network curve going in the upward direction, whereas in the cases where the markings of s_i don't tell us what to do for b, the point R is also on a spectral network curve. These two spectral network curves meet head-on along the segment PR and we get a BPS state.

One can also see that the resulting pre-building can admit folding maps to other buildings. Here two eightfold points are connected by an edge (which is not marked "open"). This is the first and most basic example where can be a nontrivial choice of maps to buildings, some of which fold and hence do not preserve distances.

Such a two-manifold construction, which we denote now by C, is analogous to the tree with a BPS state of Figure 2.

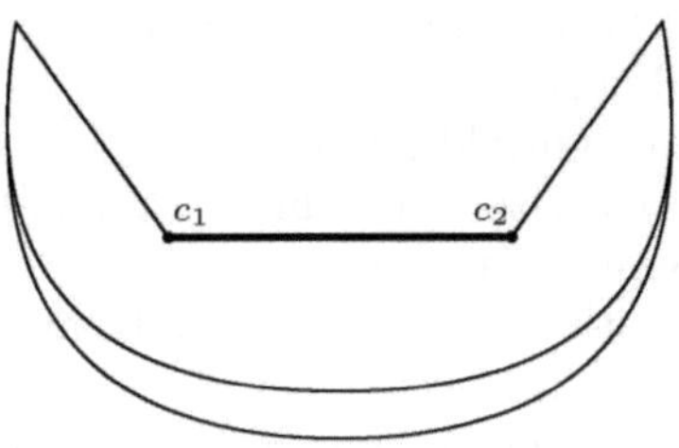

In this picture there are two sheets below the three edges, and no sheets above. The two eightfold points c_1 and c_2 are joined by an edge which we have emphasized; it corresponds to a BPS state.

This construction C is itself a pre-building, so it can be completed to a building $C \subset B$ containing it isometrically as a subset. However, it also admits a map to another construction, folding along the edge c_1c_2. To see this, choose two trapezoids sharing the edge c_1c_2, and paste them together.

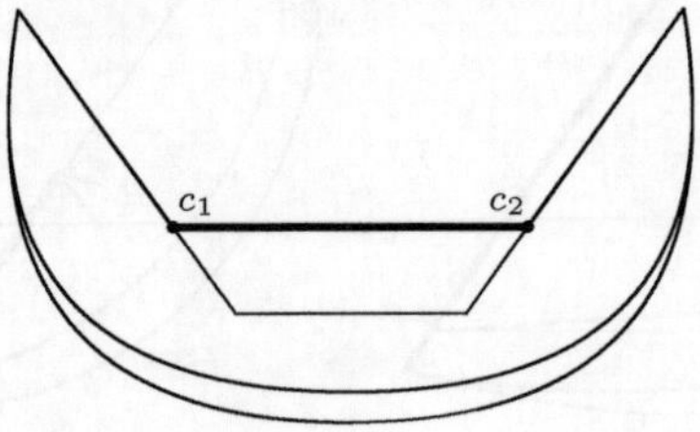

View this as a single sheet over the trapezoid and two sheets below. The resulting construction C' is again a pre-building so it can be completed to a building $C' \subset B'$. Composing with the projection $C \to C'$ therefore gives a map to a building $h' : C \to B'$. Now h' folds along the segment c_1c_2, and it is not an isometric embedding.

This family of maps h' is the analogue for SL_3 of the family of folding maps of trees pictured in Figure 6.

8.6. Keeping track of BPS states

Our main conjecture is that if there are no BPS states, then an indeterminate case doesn't occur. In order to set up the possibility to consider BPS states, we will need to pull along the spectral network (SN) lines in the process of our sequence of Z_i.

Let us consider the SN lines on the Z_i as being additional markings, which should conjecturally be subject to the following conditions.

- Some points will be marked as singularities.
- Each singularity will have a single SN line coming out of it.
- Each eightfold point which is at the end of a postcaustic, that is to say, having a single fold line, will have a single SN coming out of it. It is either opposite to the fold line, or adjacent to it (this is similar to the picture for the cases of two fold lines at an eightfold point).
- Each eightfold point with zero fold lines, that is to say a BNR point, will have two SN lines coming out of it.
- The eightfold points with two fold lines don't have SN lines.
- The SN lines should not intersect.
- The SN lines, in the middle, go through regular hexagonal points and they are straight.

The steps of our process consist of pasting together two parallelograms and cutting them out. The two edges which are common to the two parallelograms are fold lines, meeting in a fourfold point.

We choose parallelograms which have singular points somewhere on the boundary.

Each move will either decrease the total length of the postcaustics, i.e., the fold lines, or diminish the number of fourfold points. In the case of integer edge-lengths this would prove the finiteness conjecture. In general we need some other argument.

One of the basic moves, when applied to the first fourfold point along a postcaustic, will glue together some new open edges. In the case that the SN curve was opposite the fold line, the new glued edges are added to the SN curve, this way we build out the SN curve. In the adjacent case, on the contrary, the SN curve gets eaten up.

One thing that can happen here is that as the SN curve gets eaten up, we can end up passing the singularity. Then it changes to the other "opposite" case and we start spinning out the SN curve in the opposite direction. This is why all the SN directions at a singular point can actually have the opportunity to contribute.

The main problem is to analyze the transformations of this picture which can occur when the various types of singularities occur on the boundary of our parallelogram.

The markings o or f of the new edges glued in after we cut out the two parallelograms, are mostly determined; however, because of the singularity there is one segment where it isn't determined. This was illustrated in the previous subsection.

At least in a first stage, we will probably also want to assume some genericity of the positions of the singularities so that we don't reach several singularities on the boundary at once.

The basic idea is to say that if we reach a position where the new markings o or f of the indeterminate edge are not well determined by the configuration, then it must be that we had two SN curves which join up, and these form a BPS state. This phenomenon was illustrated in the example of Figures 21 and 22.

It is possible to envision finishing the process with no more fourfold points, but still having a segment marked "fold" joining two eightfold points. In this case, the pre-building is not rigid and there might be several harmonic mappings inducing different distance functions. We are conjecturing that this can only occur if there is a BPS state, indeed the fold line between eightfold points should correspond to a piece of the BPS state.

Understanding all of the possible cases currently looks complicated, but we hope that it can be done. We formulate the resulting statement as a conjecture.

Conjecture 8.4. *Suppose that (X, ϕ) has no BPS states. Then the sequence of two-manifold constructions Z_i is well defined until we get to one which has no more fourfold points, or sixfold points with triple fold lines. Furthermore, this process stops in finite time, at least locally on X. Finally, at the end of the process there are no more "fold" markings in the scaffolding.*

Associated with this is a finiteness conjecture.

Conjecture 8.5. *On bounded regions the process stabilizes in finitely many steps.*

These conjectures say that the series of two-manifold constructions will stabilize to a nonpositively curved one which can be called the *core*. It plays a role analogous to Stallings' core graphs [31, 19, 28]. Starting from the core we can then put back everything that had been trimmed off, following the discussion in Section 6.6, to get a pre-building.

Corollary 8.6. *If there are no BPS states, then on a bounded region we obtain a well-defined pre-building B^{pre}_{ϕ}. It has a ϕ-harmonic map*

$$h_{\phi} : X \to B^{\mathrm{pre}}_{\phi}$$

such that if $h : X \to B$ is any ϕ-harmonic map to a building, then there exists a unique factorization through a non-folding map $f : B^{\mathrm{pre}}_{\phi} \to B$.

8.7. Distances

One of the main statements which we would like to obtain through this theory is that the factorization from the pre-building is an isometric embedding. Recall that the *Finsler distance* is defined on an apartment A with coordinates x_1, x_2, x_3 subject to $\sum x_i = 0$, by

$$d((x_1, x_2, x_3), (x'_1, x'_2, x'_3)) := \max\{x'_i - x_i\}.$$

It is not symmetric, and indeed the two numbers $d(x, x')$ and $d(x', x)$ serve to define the *vector distance* which is the series of numbers $x'_i - x_i$ arranged in nonincreasing order.

Conjecture 8.7. *In the situation of Corollary* 8.6, *given a ϕ-harmonic map $h : X \to B$ to a building, the factorization $f : B^{\mathrm{pre}}_{\phi} \to B$ is an isometric embedding for the Finsler distance.*

To approach this statement, we should define a notion of *noncritical path* in a construction, with respect to a scaffolding.

Suppose F is a construction provided with scaffolding, and suppose S is a straight segment joining its two endpoints x and y. Let s_x and s_y denote the vertices corresponding to the directions of S in F_x and F_y respectively. We say that F is *Finsler concave* along S if there exist vertices u and v in F_x and F_y respectively, such that u is at distance 4 from s_x and v is at distance 4 from s_y. Include also the possibility of a "segment of length zero" at a point x with two vertices u, v of F_x at distance ≥ 5. Say that F is *Finsler nonconcave* if there is no segment or point at which it is Finsler concave.

The idea to show Conjecture 8.7 would be to do the following steps.

- The Finsler (or vector) distance between points in a (pre)building is calculated by adding up the distances along a noncritical path.
- If (F, σ) is a construction with scaffolding and $F \to B$ is a compatible map to a building, then the image of a noncritical path in F is a noncritical path in B with the same Finsler length.
- In a connected construction with fully open scaffolding, which is Finsler nonconcave, any two points are joined by a noncritical path.

- When there are no BPS states and assuming the previous conjectures, the pre-building B_ϕ^{pre} is Finsler nonconcave.

It should be possible to check these properties for any given example.

8.8. Isoperimetric considerations

We close this section by mentioning an important aspect. Under the pasting-together process, we glue together two parallelograms which are originally joined along two edges. The two parallelograms are supposed to be disjoint except for sharing these two edges. We need to choose a maximal such situation. In particular, when increasing the size of the parallelogram up to its maximal value (determined by when we meet some kinds of singularities along the boundary) we would like to make sure that we don't suddenly hit two points which are already previously identified. This is insured using an isoperimetric argument. Namely, suppose we have pasted together the parallelograms just up to a point right before the extra joined points. Cut out this pasted parallelogram, leaving as usual a two-manifold construction. The eightfold point from the vertex of the pasted parallelogram, is very near to a pair of points on the two sheets which are already identified. This gives a very short loop in our two-manifold construction (by very short here, it means arbitrarily short depending on how close we got to the already-joined points).

By hypothesis, our Riemann surface X is simply connected (if we started with an original problem on a non-simply connected surface the first step would have been to pass to the universal cover). It follows that the successive two-manifold constructions Z_i are also simply connected. So, we are now at a simply connected two-manifold construction which has a very short loop. This loop cuts the surface into two pieces, at least one of which is a disk. We claim that this is impossible. To see that, notice that since we are going arbitrarily close to the joined-together point, there are only two possible ways for a post-caustic curve of fold lines to pass from outside the disk to inside the disk: either through the eightfold vertex of the pasted parallelogram, or through the joined-together point. From this maximum of two post-caustics passing through the boundary of the disk, and using the fact that along any post-caustic the singular points alternate eightfold and fourfold points starting with eightfold points at the endpoints, we count that the difference

$$\text{(number of fourfold points)} - \text{(number of eightfold points)}$$

is at most 1. It follows that the total angle deficit due to the curvature inside the disk is at most 120° of positive curvature (whereas an arbitrary amount of negative curvature). Therefore the isoperimetric inequality for this disk cannot be worse than that of a single fourfold point, where a small boundary implies that the disk is small itself. We get a contradiction to the arbitrarily small nature of our loop. This completes the proof that any two parallelograms to be pasted together through our process, are joined solely on the two edges coming out of the fourfold point.

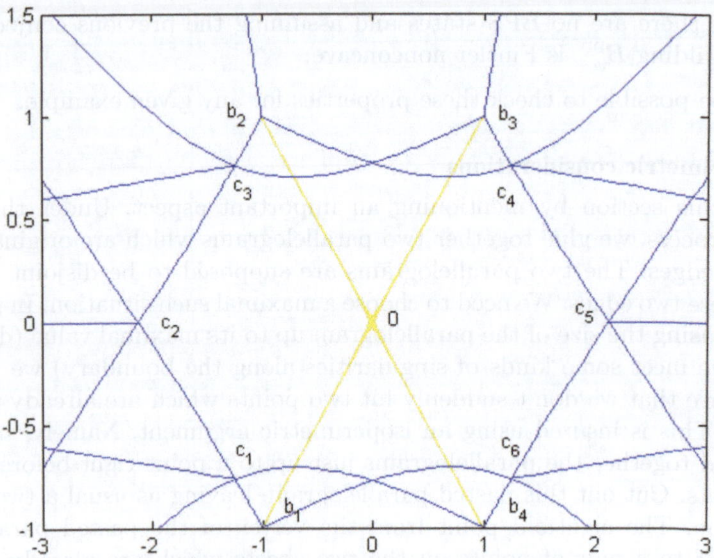

FIGURE 23. A spectral network

9. The A2 example

We include a few pictures concerning the next class of examples after BNR. It illustrates the phenomenon of crossing of caustics, and there are angular rotations leading to a BPS state. Because of space considerations we don't have room for a full analysis here, that will be done elsewhere.

The spectral network for this example is shown in Figure 23. Note that there are two caustics intersecting in the middle.

The associated two-manifold construction is, in this case, the same as the pre-building. It will look as shown in Figure 24.

There are two eightfold points c and c'. The zig-zag boundary lines are shown to emphasize that the picture represents three sheets over the triangular regions and a single sheet elsewhere.

The image of the map from X to the pre-building is shown in Figure 25. At the top is again the spectral network, reprising Figure 23, with some horizontal dashed lines introduced for visualization. Below it is the image, in the pre-building that was pictured before. One can follow the horizontal dashed lines that loop around in order to understand what is happening.

The collision points c_1 and c_6 go to the point labeled c in the prebuilding; the collision points c_3 and c_4 go to the point labeled c'. The collision points c_2 and c_5 don't go to singular points in the pre-building. Note how the regions containing points c_2 and c_5 are folded over to opposite sides as compared to the domain picture.

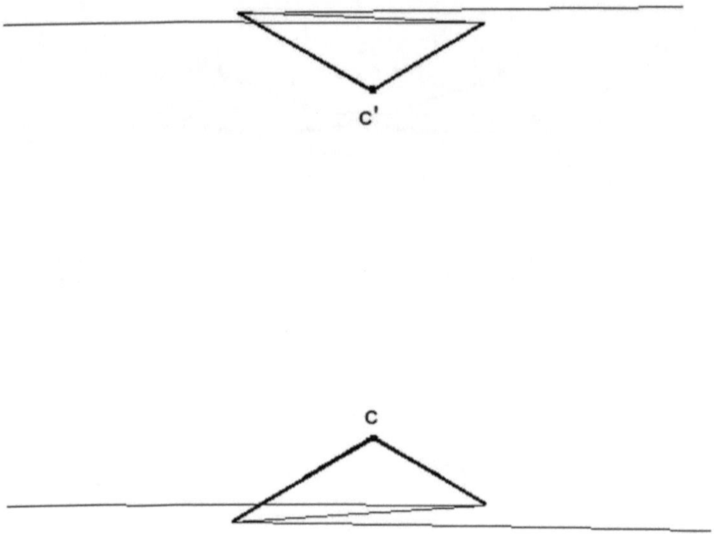

FIGURE 24. The pre-building

As the spectral differential ϕ changes, for example just by being multiplied by an angular constant $e^{i\theta}$ or some more complicated deformation, the relative positions of the two singular points c and c' will change. The BPS states occur when the line segment cc' goes in the direction of one of the reflection hyperplanes. We are then in the situation that was considered above in Subsection 8.5.

10. Consequences and further considerations

In this section we look at some consequences and other further considerations.

10.1. The universal pre-building

As pointed out in Corollary 8.6, the process conjectured above leads to the construction of a map to a pre-building

$$h_\phi : X \to B_\phi^{\mathrm{pre}}.$$

In the absence of BPS states, it will have a strong universal property: if $h : X \to B$ is any ϕ-harmonic map to a building, then there is a unique factorization $B_\phi^{\mathrm{pre}} \to B$ which is an isometric embedding.

Conjectures 8.4 and 8.5, when there are no BPS states, imply that we can construct the pre-building without necessarily following through the whole procedure. Indeed, if we can illustrate some pre-building B' with a ϕ-harmonic map $X \to B'$, such that some choice of the initial construction surjects onto B', then $B' = B_\phi^{\mathrm{pre}}$. For example, this shows that the picture given in Figure 24 of Section 9 does indeed show the pre-building for the A2 example.

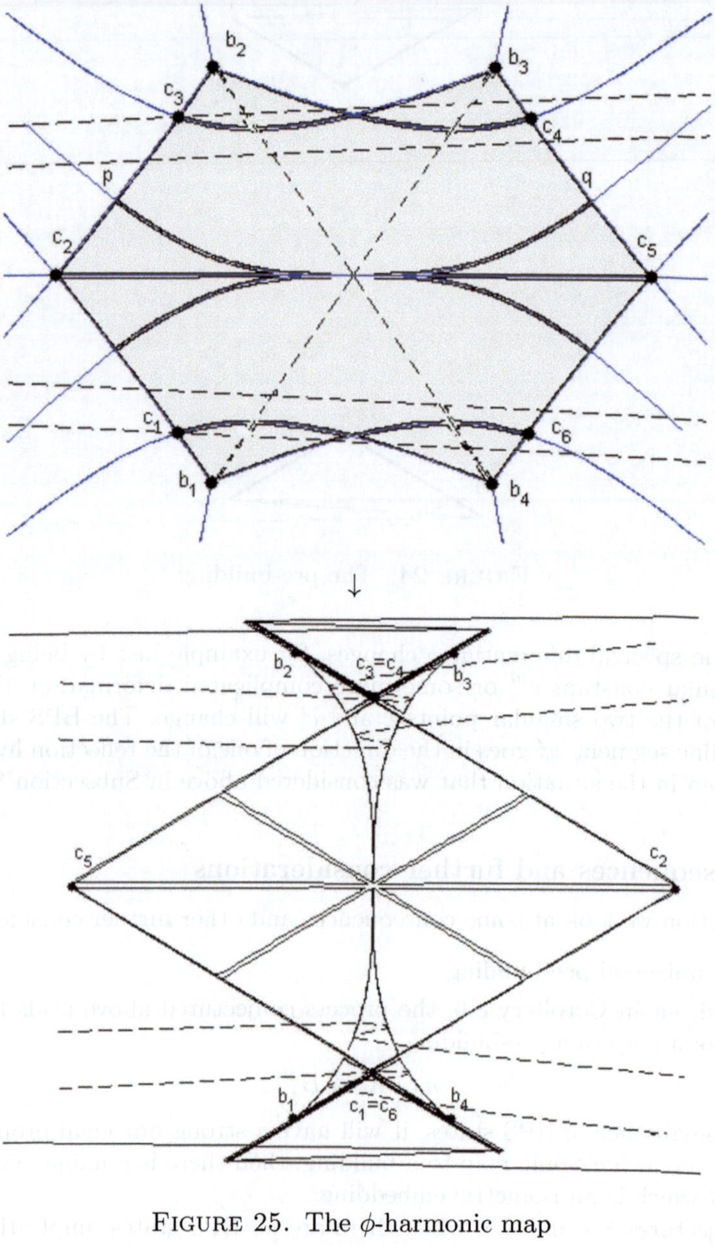

FIGURE 25. The ϕ-harmonic map

10.2. Determination of the WKB exponents

Our conjectures up through Conjecture 8.7 imply that the WKB exponents are uniquely determined, as we state in the following corollary.

Corollary 10.1. *For any WKB problem with spectral curve corresponding to* ϕ*, the ultrafilter exponent* $\nu_\omega^{WKB}(P,Q)$ *for the transport from a point* P *to a point* Q*, must be equal to the Finsler distance in* $B_\phi^{\rm pre}$ *from* $h_\phi(P)$ *to* $h_\phi(Q)$*. In particular, it doesn't depend on the choice of ultrafilter, which implies that the limit used to define the WKB exponent exists and its value is calculated as the Finsler distance in* $B_\phi^{\rm pre}$.

Even before obtaining general proofs of all of the conjectures which go into this statement, it should be possible to verify the required things when we are given a particular example of a spectral differential ϕ, thus obtaining a calculation of the WKB exponents.

One expects, from the resurgence picture, that when there is a BPS state the ultrafilter exponent should usually depend on the choice of ultrafilter.

Conjecture 10.2. *The WKB exponents determined by our process, when there are no BPS states, coincide with the exponents constructed by the spectral network procedure of Gaiotto–Moore–Neitzke.*

10.3. Relation with the work of Aoki, Kawai, Takei et al

The pictures we consider here are starting to look a lot like the pictures which occur in the work of Aoki, Kawai, and Takei [2] [3]. It will be interesting to establish a precise comparison.

10.4. Stability conditions

In the works of Bridgeland and Smith [5] and Haiden, Katzarkov and Kontsevich [14], a picture is developed whereby each pair (X,ϕ) consisting of a compact Riemann surface and a twofold spectral covering, corresponds to a Bridgeland stability condition on a Fukaya-type triangulated category. The space of (X,ϕ) is divided up into chambers whose boundaries are walls determined by the existence of BPS states. A stability condition corresponding to (X,ϕ) should really be thought of as associated to the circle of $(X,e^{i\theta}\phi)$, and the values of θ where BPS states arise are the angles of the central charge for stable objects.

This chamber and wall structure is expected to be closely related to the wallcrossing phenomenon introduced by Kontsevich and Soibelman [22]. Making this relationship precise, and extending the construction of stability conditions to higher rank spectral curves, are some of the long-term motivations for the present project.

In the SL_2 case these subjects have been treated recently by Iwaki and Nakanishi [17], and one of the goals in the higher rank case would be to extend the various different aspects of their theory.

What we obtain from our construction for SL_3 is that inside each chamber, that is to say inside a connected region where there are no BPS states, our pre-building $B_\phi^{\rm pre}$ is a combinatorial geometric object associated[4] to (X,ϕ). Over a

[4]The pre-building is essentially uniquely defined according to our conjectures. The only ambiguity comes from the choice of size of neighborhoods used to make the initial construction. Those affect

given chamber the pre-building will vary in a smooth way. As we cross a wall determined by a BPS state, the pre-building will transform in some kind of "mutation". The transformations of spectral networks are pictured in [11].

We hope that the study of the geometric form and structure of these mutations can provide some insight into the structure of the categorical stability conditions.

Kontsevich has outlined a far reaching program that associates a "Fukaya-type category with coefficients" to a "perverse sheaf of categories" on a symplectic manifold X. A major part of this program is the construction of stability structures on the Fukaya category with coefficients in a sheaf of categories starting from the data of stability structures on the stalks of this sheaf. A symplectic fibration $\pi : Y \to X$ conjecturally gives rise to such a perverse sheaf of categories $\mathfrak{S}$, whose stalks are, roughly speaking, the Fukaya categories of the fibers of π. In this situation, the Fukaya category with coefficients in the sheaf of categories $\mathfrak{S}$ on X is supposed to be equivalent to a Fukaya type category associated to Y. Recently, Kapranov and Schechtman [20] have given a precise definition of the notion of a "perverse sheaf of categories", which they call a *perverse Schober*, in the case when X is a Riemann surface.

There is a CY-manifold Y of complex dimension 3 associated to a spectral cover Σ of a Riemann surface X, namely, the conic bundle over the cotangent bundle $T^\vee_X$ whose discriminant locus is Σ. The natural morphism $\pi : Y \to X$ is a symplectic fibration, and therefore should define a perverse Schober $\mathfrak{S}$ on X. We formulate the following:

Conjecture 10.3. *The Schober $\mathfrak{S}$ on X associated with the conic bundle, together with a stability structure on it, can be recovered from a family $\{h_\theta\}_{\theta\in\mathbb{R}}$ of versal harmonic $e^{i\theta}\phi$-maps, where ϕ is the multivalued differential corresponding to the given spectral cover Σ.*

10.5. Higher ranks

In this paper we have restricted to looking at buildings for SL_3. These two-dimensional objects remain geometrically close to the points of the original Riemann surface, and this has been used in fundamental ways in our discussion. So it remains a big open problem how to go to a general theory for buildings of arbitrary rank, say for the groups SL_r. The image of X becomes a very thin subset. Maxim Kontsevich suggested on several occasions that we should think of trying to "thicken" X to something of the right dimension.

We feel that this will indeed be the way that the theory should play out. The initial construction will give a neighborhood of X of the correct dimension $r-1$. The $r-2$-dimensional boundary of this neighborhood looks something like a bundle of $r-4$-spheres over the 2-dimensional X. Probably, the analogue of our process described above will be some kind of discrete evolution of the boundary as we

the size, which is inessential, of the additional "stegosaurus dorsal plates" attached to a caustic, illustrated for example in the white areas just above the dotted line in Figure 18.

successively enlarge the region of the building that is being constructed. The two-manifold constructions which show up in our process for SL_3 should be thought of as the boundaries of what are, for us, "0-dimensional disk bundles over X".

10.6. Buildings as enriched categories

In [23], Lawvere made the fundamental observation that metric spaces can be viewed as categories enriched over the poset $([0,\infty],\geq)$ equipped with the (symmetric) monoidal structure given by addition of real numbers. Furthermore, various metric constructions, such as Cauchy completion, have natural categorical interpretations. Motivated by these observations, Flagg [8] has shown that much of the basic theory of metric spaces carries over to categories enriched over an arbitrary value distributive quantale $\mathcal{Q}$. The distance between two points in these "generalized metric spaces" is an object of the quantale $\mathcal{Q}$ instead of a real number.

Tits' definition of a (discrete) building as a chamber system with a Weyl group-valued "distance function" (see, e.g., [1]) strongly suggests that the theory of buildings should naturally be situated within the theory of $\mathcal{Q}$-enriched categories for some suitable quantale $\mathcal{Q}$ that depends on the Weyl group. This idea has been put forth by Dolan and Trimble, and is discussed in [32].

In a forthcoming work, we hope to systematically develop a theory of $\mathbb{R}$-buildings as categories enriched over a suitable variant of the category of enclosures considered in this paper. We feel that, combined with the theory of buildings as sheaves on the site of enclosures discussed earlier, such a theory will provide a powerful framework in which to formulate and study the types of constructions we have looked at in this paper in connection with the problem of constructing versal buildings.

Acknowledgement

The ideas presented here are part of our ongoing attempt to understand the geometric picture relating points in the Hitchin base and stability conditions. So this work is initiated and motivated by the work of Maxim Kontsevich and Yan Soibelman, as exemplified by their many lectures and the discussions we have had with them. It is a great pleasure to dedicate this paper to Maxim on the occasion of his 50th birthday.

The reader will note that many other mathematicians are also contributing to this fast-developing theory and we would also like to thank them for all of their input. We hope to present here a small contribution of our own.

On a somewhat more specific level, the idea of looking at harmonic maps to buildings, generalizing the leaf space tree of a foliation, came up during some fairly wide discussions about this developing theory in which Fabian Haiden also participated and made important comments, so we would like to thank Fabian.

We would like to thank many other people including Brian Collier, Georgios Daskalopoulos, Mikhail Kapranov, François Labourie, Ian Le, Chikako Mese, Richard Schoen, Richard Wentworth, and the members of the Geometric Langlands seminar at the University of Chicago, for interesting discussions contributing to this paper. We thank Doron Puder for an interesting talk at the IAS on the theory of Stallings graphs.

The authors would like to thank the University of Miami for hospitality and support during the completion of this work. The fourth named author would in addition like to thank the Fund For Mathematics at the Institute for Advanced Study in Princeton for support. The first named author was supported by the Simons Foundation as a Simons Fellow.

The authors were supported by an ERC grant, as well as the following grants: NSF DMS 0600800, NSF DMS 0652633 FRG, NSF DMS 0854977, NSF DMS 0901330, NSF DMS 150908, FWF P 24572 N25, FWF P20778, FWF P 27784-N25, ANR grant 933R03/13ANR002SRAR (Tofigrou), DMS-1265230 "Wall Crossings in Geometry and Physics", DMS-1201475 "Spectra, Gaps, Degenerations and Cycles", ERC Gemis, a Simons research grant, Simons collaborative Grant – HMS, OISE-1242272 PASI "On Wall Crossings, Stability Hodge Structures & TQFT", and Advanced Grant "Arithmetic and physics of Higgs moduli spaces" No. 320593 of the European Research Council. LK was partially supported by the Laboratory of Mirror Symmetry NRU HSE, RF government grant, ag. 14.641.31.000.

We would all like to thank the IHES for hosting Maxim's birthday conference, where we did a significant part of the work presented here.

References

[1] P. Abramenko, K. Brown. *Buildings*, GRADUATE TEXTS IN MATH. **248**, Springer (2008).

[2] T. Aoki, T. Kawai, Y. Takei. New turning points in the exact WKB analysis for higher-order ordinary differential equations. *Analyse algébrique des perturbations singulières, I* (*Luminy* 1991) TRAVAUX EN COURS **47**, Hermann, Paris (1994), 69–84.

[3] T. Aoki, T. Kawai, Y. Takei. On the exact steepest descent method: a new method for the description of Stokes curves. *J. Math. Phys.* **42** (2001), 3691–3713.

[4] A. Berenstein, M. Kapovich. Affine buildings for dihedral groups. *Geom. Dedicata* **156** (2012), 171–207.

[5] T. Bridgeland, I. Smith. Quadratic differentials as stability conditions. *Publ. Math. I.H.E.S.* (DOI) 10.1007/s10240-014-0066-5 (2014).

[6] B. Collier, Q. Li. Asymptotics of certain families of Higgs bundles in the Hitchin component. Preprint arXiv:1405.1106 (2014).

[7] D. Dumas. Holonomy limits of complex projective structures. Preprint arXiv: 1105.5102 (2011).

[8] R. C. Flagg. Quantales and continuity spaces. *Algebra Universalis* **37** (1997), no. 3, 257–276.

[9] B. Fontaine, J. Kamnitzer, G. Kuperberg. Buildings, spiders, and geometric Satake. *Compos. Math.* **149** (2013), 1871–1912.

[10] D. Gaiotto, G. Moore, A. Neitzke. Wall-crossing, Hitchin systems and the WKB approximation. *Adv. Math.* **234** (2013), 239–403.

[11] D. Gaiotto, G. Moore, A. Neitzke. Spectral networks. *Ann. Henri Poincaré* **14** (2013), 1643–1731.

[12] D. Gaiotto, G. Moore, A. Neitzke. Spectral networks and snakes. *Ann. Henri Poincaré* **15** (2014), 61–141.

[13] M. Gross, P. Hacking, S. Keel, M. Kontsevich. Canonical bases for cluster algebras. Preprint arXiv:1411.1394 (2014).

[14] F. Haiden, L. Katzarkov, M. Kontsevich. Stability in Fukaya categories of surfaces. Preprint arXiv:1409.8611 (2014).

[15] F. Herrlich, G. Schmithüsen. Dessins denfants and origami curves. *Handbook of Teichmüller theory, II* IRMA LECT. MATH. THEOR. PHYS. **13**, EMS, Zürich (2009), 767–809.

[16] P. Iglesias-Zemmour. *Diffeology.* MATHEMATICAL SURVEYS AND MONOGRAPHS **185**, AMS, Providence (2013).

[17] K. Iwaki, T. Nakanishi. Exact WKB analysis and cluster algebras. Preprint arXiv:1401.7094 (2014).

[18] K. Iwaki, T. Nakanishi. Exact WKB analysis and cluster algebras II: simple poles, orbifold points and generalized cluster algebras. Preprint arXiv:1409.4641 (2014).

[19] I. Kapovich, A. Myasnikov. Stallings foldings and subgroups of free groups. *J. Algebra* **248** (2002), 608–668.

[20] M. Kapranov, V. Schechtman. Perverse Schobers. Preprint arXiv: 1411.2772 (2014).

[21] L. Katzarkov, A. Noll, P. Pandit, C. Simpson. Harmonic maps to buildings and singular perturbation theory. *Commun. Math. Phys.* (DOI) 10.1007/s00220-014-2276-6 (2015).

[22] M. Kontsevich, Y. Soibelman. Wall-crossing structures in Donaldson–Thomas invariants, integrable systems and Mirror Symmetry. Preprint arXiv:1303.3253 (2013).

[23] F.W. Lawvere. Metric spaces, generalized logic, and closed categories. *Rend. Sem. Mat. Fis. Milano* **43** (1973), 135–166.

[24] R. Mazzeo, J. Swoboda, H. Weiss, F. Witt. Ends of the moduli space of Higgs bundles. Preprint arXiv:1405.5765 (2014).

[25] T. Mochizuki. Asymptotic behaviour of certain families of harmonic bundles on Riemann surfaces. Preprint arXiv:1508.05997 (2015).

[26] J. Morgan, P. Shalen. Valuations, trees, and degenerations of hyperbolic structures. I, *Ann. of Math.* **120** (1984), 401–476.

[27] A. Parreau. Compactification d'espaces de représentations de groupes de type fini. *Math. Z.* **272** (2012), 51–86.

[28] O. Parzanchevski, D. Puder. Stallings graphs, algebraic extensions and primitive elements. Preprint arXiv:1210.6574 (2012).

[29] M. Ronan. A construction of buildings with no rank 3 residues of spherical type. *Buildings and the geometry of diagrams* (*Como,* 1984), LECTURE NOTES IN MATH. **1181**, Springer (1986), 242–248.

[30] C. Simpson. *Asymptotic behavior of monodromy.* LECTURE NOTE IN MATH. **1502**, Springer (1991).

[31] J. Stallings. Topology of finite graphs. *Invent. Math.* **71** (1983), 551–565.

[32] T. Trimble. Buildings for category theorists. *ncatlab*, available online at `http://ncatlab.org/toddtrimble/published/Buildings+for+category+theorists`.

[33] A. Voros. The return of the quartic oscillator: the complex WKB method. *Ann. Henri Poincaré* Sect. A (N.S.) **39** (1983), 211–338.

[34] H. Wallman. Lattices and topological spaces. *Ann. of Math.* (2) **39** (1938), 112–126.

Ludmil Katzarkov and Pranav Pandit
Fakultät für Mathematik
Universität Wien
Oskar-Morgenstern-Platz 1
A-1090 Wien Austria
e-mail: lkatzarkov@gmail.com
pranav.pandit@univie.ac.at

Alexander Noll
Mühlegasse 5
CH-6340 Baar, Switzerland
e-mail: alexander.noll.a@gmail.com

Carlos Simpson
Laboratoire J.A. Dieudonné
Université Nice Sophia Antipolis
Parc Valrose
F-06108 Nice Cedex 02, France
e-mail: Carlos.Simpson@unice.fr

Progress in Mathematics, Vol. 324, 261–293

Cohomological Hall Algebras, Semicanonical Bases and Donaldson–Thomas Invariants for 2-dimensional Calabi–Yau Categories (with an Appendix by Ben Davison)

Jie Ren and Yan Soibelman

To Maxim Kontsevich on his 50th birthday

Abstract. We discuss semicanonical bases from the point of view of Cohomological Hall algebras via the "dimensional reduction" from 3-dimensional Calabi–Yau categories to 2-dimensional ones. Also, we discuss the notion of motivic Donaldson–Thomas invariants (as defined by M. Kontsevich and Y. Soibelman) in the framework of 2-dimensional Calabi–Yau categories. In particular we propose a conjecture which allows one to define Kac polynomials for a 2-dimensional Calabi–Yau category (this is a theorem of S. Mozgovoy in the case of preprojective algebras).

Mathematics Subject Classification (2010). 16G20, 14F42, 81R99.

Keywords. Cohomological Hall algebras, motivic Donaldson–Thomas invariants, 2-dimensional Calabi–Yau categories, semicanonical bases.

1. Introduction

Representations of a quiver Q give rise in general to the category of cohomological dimension one. In this sense quivers are analogous to curves.

In a similar vein representations of the preprojective algebra Π_Q give rise in general to the category of cohomological dimension two. Since preprojective algebras are obtained as simplectic reductions, we can say that they are analogous to Higgs bundles on a curve. Likewise the deformed preprojective algebras Π^λ are analogous to algebraic vector bundles with connections on a curve. The interplay between algebraic and geometric sides of this dictionary is a subject of many papers (here are few random samples: [2], [7], [17]).

The above categories can be "upgraded" to 3-dimensional Calabi–Yau categories. On the algebraic side it can be achieved by constructing a "triple" quiver $\widehat{Q}$ endowed with a cubic potential W, see, e.g., [16] or Section 2 below. On the geometric side one takes the total space of an appropriate rank two vector bundle on a curve or the total space of the anticanonical bundle on a surface. From the point of view of, say, gauge theory, the lower-dimensional categories can be thought of as "dimensional reductions" of the corresponding 3-dimensional Calabi–Yau categories.

The framework of 3-dimensional Calabi–Yau categories is appropriate for the theory of motivic Donaldson–Thomas invariants (see [12], [13]). The above-mentioned dimensional reduction gives rise to the corresponding theory in a lower dimension. It is natural to ask about the meaning of the objects arising as a result of such dimensional reduction. Our paper illustrates this philosophy in two examples.

In our first example we discuss semicanonical bases (see [14], [15]) from the point of view of Cohomological Hall algebras (see [13]). In our second example we formulate a conjecture about an analog of the Kac polynomial of a 2-dimensional Calabi–Yau category. The conjecture is motivated by [16] in which the motive of the stack of indecomposable representations of a quiver (Kac polynomial) was expressed in terms of the motives of stacks of representations of the corresponding preprojective algebra and the Donaldson–Thomas invariants of the corresponding $3CY$ category.

It is known after Lusztig (see [14], [15]) that semicanonical bases can be interpreted in terms of top-dimensional components of the stack of nilpotent representations of the preprojective algebra Π_Q associated with the quiver Q. On the other hand, the critical loci of $Tr(W)$ contains the stack of representations of Π_Q. This gives an idea that the semicanonical basis can be derived from the pair $(\widehat{Q}, W)$.

Indeed, the pair $(\widehat{Q}, W)$ gives rise to the corresponding Cohomological Hall algebra (COHA for short), see [13]. It is natural to transport the associative product from the COHA to the vector space generated by the above-mentioned top-dimensional components (in a more invariant way one can speak about the Galois-invariant part of COHA, cf. [13]). We will show that in this way we obtain an associative algebra endowed with a basis which enjoys the properties of the semi-canonical one.

From the point of view of the above dictionary between algebra and geometry, the preprojective algebra Π_Q should be replaced in general by a 2-dimensional Calabi–Yau category ($2CY$ category for short), while the pair $(\widehat{Q}, W)$ should be replaced by a 3-dimensional Calabi–Yau category ($3CY$ category for short). The "dimensional reduction" of this $3CY$ categories to a 2-dimensional category (as well as the relation between the corresponding cohomology groups of stacks of objects) was described in [13], Section 4.8. One can expect a similar story in the general categorical framework.

In particular we expect that the notion of semicanonical basis has intrinsic categorical meaning for $2CY$ categories. A class of such categories is proposed in Section 3.2. The categories from our class are parametrized by quivers.

One can also hope that the reduction from $3CY$ categories to 2-dimensional categories will help to understand the relation between motivic Donaldson–Thomas theory (see [12], [13]) and some invariants of $2CY$ categories, e.g., Kac polynomials.[1]

It is not surprising that some notions introduced for Kac–Moody algebras can be interpreted in terms of $2CY$ categories, since Kac–Moody algebras arise naturally in the framework of $2CY$ categories generated by spherical collections à la [13]. Spherical objects generate a t-structure of a $2CY$ category. Its heart corresponds to a Borel subalgebra, and the spherical objects themselves correspond to simple roots. In a similar vein Schur objects correspond to positive roots, etc. Change of the t-structure corresponds to the change of the cone of positive roots (and hence to the change of Borel subalgebra in the corresponding Kac–Moody algebra). Reflections at spherical objects correspond to the action of the Weyl group. Space of stability functions (i.e., central charges) on a fixed t-structure corresponds to the Cartan subalgebra. It contains the lattice which is dual to the K-theory classes of spherical objects. The Euler bilinear form on the K-group of the category (it is symmetric due to the $2CY$ condition) corresponds to the Killing form on the Cartan subalgebra. The moduli stack of objects of the heart of the t-structure is symplectic, and it contains a Lagrangian substack (in fact subvariety).[2]

Notice that every $2CY$ category gives rise to a $3CY$ category with the trivial Euler form. Geometrically, the underlying Calabi–Yau 3-fold is the product $S \times \mathbb{A}^1$, where S is a Calabi–Yau surface. Upgrade from Π_Q to $(\widehat{Q}, W)$ illustrates the algebraic side of the dictionary (in general the $3CY$ category should have a t-structure with the heart containing pairs (E, f), where E is an object of the heart of a fixed t-structure of our $2CY$ category and $f \in \mathrm{Hom}(E, E)$).

Motivic Donaldson–Thomas invariants of such "upgraded" $3CY$ categories do not change inside of a connected component of the space of stability conditions. As a result, the DT-invariants are in fact invariants of the t-structure of the underlying $2CY$ category.

Another interesting question is the relation of the representation theory of COHA as described in [19] to the one of the algebras acting on the cohomologies of the moduli spaces of instantons on $\mathbb{CP}^2$ (AGT conjecture, etc.). In the Appendix written by Ben Davison the relation between COHA and cohomological Hall algebra of the commuting variety introduced in [18] is explained (in fact the relation takes place for a general quiver). On the other hand, representations of

[1]The relationship between motivic DT-invariants and Kac polynomials was studied in many papers, including [6], [16], [8].

[2]This Lagrangian subvariety should be compared with Lusztig's Lagrangian nilpotent cone, whose irreducible components correspond to positive roots. In the spirit of the analogy with Higgs bundles it is similar to the Laumon nilpotent cone.

W-algebras used in the loc. cit. for the proof of the AGT conjecture are derived from the cohomological Hall algebra of the commuting variety. We hope that the appearance of COHA in this story is not accidental and plan to study that topic in the future.

Finally, we remark that there are $2CY$ categories which have purely geometric origin (e.g., the category of coherent sheaves on a $K3$ surface). The above questions about semicanonical basis or Kac polynomial make sense in the geometric case as well.

2. COHA and semicanonical basis

We start by recalling the definition of the product on the critical COHA proposed in [13]. For the convenience of the reader we will closely follow the very detailed exposition from [5] which contains proofs of several statements sketched in [13] as well as several useful improvements of the loc. cit.

2.1. Reminder on the critical cohomology

Let Y be a complex manifold, and $f : Y \to \mathbb{C}$ a holomorphic function. We define the vanishing cycles functor φ_f as follows: $\varphi_f \mathcal{F}[-1] := (R\Gamma_{\{Re(f)\leq 0\}}\mathcal{F})_{f^{-1}(0)}$. This is a nonstandard definition of this functor, which is equivalent to the usual one in the complex case. From now on we will abbreviate $R\mathscr{F}$ to $\mathscr{F}$ for any functor $\mathscr{F}$. There is an isomorphism $\mathbb{Q}_Y \otimes \mathbb{T}^{\dim Y} \to D\mathbb{Q}_Y$, where $\mathbb{T}$ is the Tate motive, and D is the Verdier duality. The above isomorphism induces an isomorphism

$$\varphi_f \mathbb{Q}_Y \otimes \mathbb{T}^{\dim Y} \to \varphi_f D\mathbb{Q}_Y. \tag{1}$$

If $g : Y' \to Y$ is a map between manifolds, then the natural transformation of functors $\Gamma_{\{\mathrm{Re}(f)\leq 0\}} \to g_*\Gamma_{\{\mathrm{Re}(fg)\leq 0\}} g^*$ induces a natural transformation

$$\varphi_f \to g_*\varphi_{fg} g^*. \tag{2}$$

Assume that g is an affine fibration, then by [5, Cor. 2.3] there is a natural equivalence

$$\varphi_f g_! g_* \to g_! \varphi_{fg} g_*. \tag{3}$$

Definition 2.1. For any submanifold $Y^{sp} \subset Y$, the critical cohomology with compact support $H_c^{\bullet,\mathrm{crit}}(Y^{sp}, f)$ is defined as the cohomology of the following object in $\mathscr{D}^b(\mathbf{MMHS})$ (**MMHS** denotes the category of monodromic mixed Hodge structures):

$$(\mathbb{C}^* \to \mathbb{A}^1)_!(Y^{sp} \times \mathbb{C}^* \to \mathbb{C}^*)_!(Y^{sp} \times \mathbb{C}^* \to Y \times \mathbb{C}^*)^* \varphi_{\frac{f}{u}} \mathbb{Q}_{Y\times\mathbb{C}^*},$$

where u is the coordinate on $\mathbb{C}^*$.

Let $Y = X \times \mathbb{A}^n$ be the total space of the trivial vector bundle, endowed with the $\mathbb{C}^*$-action that acts trivially on X and with weight one on $\mathbb{A}^n$. Let $f : Y \to \mathbb{A}^1$ be a $\mathbb{C}^*$-equivariant holomorphic function, where $\mathbb{C}^*$ acts with weight one on $\mathbb{A}^1$. Then $f = \sum_{k=1}^{k=n} f_k x_k$, where $\{x_k, k = 1, \ldots, n\}$ is a linear coordinate system on

$\mathbb{A}^n$, and f_k are functions on X. Let $Z \subset X$ be the reduced scheme which is the vanishing locus of all functions f_k. Then Z is independent of the choice of x_k. Let $\pi : Y \to X$ be the natural projection, and $i : Z \to X$ be the closed inclusion.

Theorem 2.2 (see [5, Cor. A.6]). *There is a natural isomorphism of functors in* $\mathscr{D}^b(\mathbf{MHM}(X))$*:*

$$\pi_! \varphi_f \pi^* \xrightarrow{\sim} \pi_! \pi^* i_* i^*.$$

In particular,

$$H_c^{\bullet,\mathrm{crit}}(Y, f) \simeq H_c^\bullet(Z \times \mathbb{A}^n, \mathbb{Q}) \simeq H_c^\bullet(Z, \mathbb{Q}) \otimes \mathbb{T}^n.$$

Here $\mathbf{MHM}(X)$ *denotes the category of mixed Hodge modules on* X.

If $Y_i = X_i \times \mathbb{A}^{n_i}$ with equivariant functions f_i satisfy the above conditions for $i = 1, 2$, then we have

Theorem 2.3 (see [5, Prop. A.5]). *The following diagram of isomorphisms commutes:*

$$\begin{array}{ccc} H_c^{\bullet,\mathrm{crit}}(Y_1 \times Y_2, f_1 \boxplus f_2) & \xrightarrow{TS} & H_c^{\bullet,\mathrm{crit}}(Y_1, f_1) \otimes H_c^{\bullet,\mathrm{crit}}(Y_2, f_2) \\ \downarrow & & \downarrow \\ H_c^\bullet(Z_1 \times Z_2 \times \mathbb{A}^{n_1+n_2}, \mathbb{Q}) & \xrightarrow{Ku} & H_c^\bullet(Z_1 \times \mathbb{A}^{n_1}, \mathbb{Q}) \otimes H_c^\bullet(Z_2 \times \mathbb{A}^{n_2}, \mathbb{Q}) \end{array}$$

Here TS *denotes the Thom–Sebastiani isomorphism, and* Ku *the Künneth isomorphism (see loc. cit.).*

Corollary 2.4 (see [5, Cor. A.7]). *Let* $X^{sp} \subset X$ *be a subvariety of* X *and* $Y^{sp} = X^{sp} \times \mathbb{A}^n$, $Z^{sp} = Z \cap X^{sp}$. *There is a natural isomorphism in* **MMHS**

$$H_c^{\bullet,\mathrm{crit}}(Y^{sp}, f) \simeq H_c^\bullet(Z^{sp} \times \mathbb{A}^n, \mathbb{Q}).$$

The above statements also hold in equivariant case. Let us recall that framework. Assume that Y is a G-equivariant vector bundle over X, where G is an algebraic group embedded in $GL(n, \mathbb{C})$, and $f : Y \to \mathbb{A}^1$ is G-invariant. Let $fr(n, N)$ be the space of n-tuples of linearly independent vectors in $\mathbb{C}^N$ for $N \geqslant n$, and $\overline{(Y, G)}_N := Y \times_G fr(n, N)$. We denote the induced function by $f_N : \overline{(Y, G)}_N \to \mathbb{A}^1$. For a G-invariant closed subset $Y' \subset Y$, we define the equivariant cohomology with compact support by $H_{c,G}^{\bullet,\mathrm{crit}}(Y', f) := \lim_{N\to\infty} H_c^{\bullet,\mathrm{crit}}(Y'_N, f_N) \otimes \mathbb{T}^{-\dim(fr(n,N))}$, where $Y'_N \subset \overline{(Y, G)}_N$ is the subspace of points projected to Y'.

Theorem 2.5 (see [5, Cor. A.8]). *Let* $Y^{sp} = X^{sp} \times \mathbb{A}^n$ *be the total space of a sub* G*-bundle. Then there is an isomorphism in* **MMHS**

$$H_{c,G}^{\bullet,\mathrm{crit}}(Y^{sp}, f) \simeq H_{c,G}^\bullet(Z^{sp} \times \mathbb{A}^n, \mathbb{Q}).$$

Moreover, the following diagram of isomorphisms commutes:

$$\begin{array}{ccc} H^{\bullet,\mathrm{crit}}_{c,G}(Y_1^{sp}\times Y_2^{sp}, f_1\boxplus f_2) & \xrightarrow{TS} & H^{\bullet,\mathrm{crit}}_{c,G}(Y_1^{sp}, f_1)\otimes H^{\bullet,\mathrm{crit}}_{c,G}(Y_2^{sp}, f_2) \\ \downarrow & & \downarrow \\ H^{\bullet}_{c,G}(Z_1^{sp}\times Z_2^{sp}\times \mathbb{A}^{n_1+n_2}, \mathbb{Q}) & \xrightarrow{Ku} & H^{\bullet}_{c,G}(Z_1^{sp}\times \mathbb{A}^{n_1}, \mathbb{Q})\otimes H^{\bullet}_{c,G}(Z_2^{sp}\times \mathbb{A}^{n_2}, \mathbb{Q}) \end{array}$$

2.2. COHA and preprojective algebras

Let Q be a quiver with the set of vertices I and the set of arrows Ω. We denote by $a_{ij}\in\mathbb{Z}_{\geqslant 0}$ the number of arrows from i to j for $i,j\in I$.

One constructs the double quiver $\overline{Q}$, the preprojective algebra Π_Q, and the triple quiver with potential $(\widehat{Q}, W)$ as follows. For any arrow $a : i\to j\in\Omega$, we add an inverse arrow $a^* : j\to i$ to Q to get $\overline{Q}$, then $\Pi_Q = \mathbb{C}\overline{Q}/\sum_{a\in\Omega}[a,a^*]$. Adding loops $l_i : i\to i$ at each vertex $i\in I$ to $\overline{Q}$ gives us the "triple" quiver $\widehat{Q}$. It is endowed with cubic potential $W=\sum_{a\in\Omega}[a,a^*]l$, where $l=\sum_{i\in I} l_i$. For any dimension vector $\gamma=(\gamma^i)_{i\in I}\in\mathbb{Z}^I_{\geqslant 0}$ we have the following algebraic varieties:

a) the space $\mathbf{M}_{\overline{Q},\gamma}$ of representations of the double quiver $\overline{Q}$ in coordinate spaces $(\mathbb{C}^{\gamma^i})_{i\in I}$;
b) the similar space of representations $\mathbf{M}_{\Pi_Q,\gamma}$ of Π_Q;
c) the similar space of representations $\mathbf{M}_{\widehat{Q},\gamma}$ of $\widehat{Q}$.

All these spaces of representations are endowed with the action by conjugation of the complex algebraic group $\mathbf{G}_\gamma=\prod_{i\in I} GL(\gamma^i,\mathbb{C})$.

Let

$$\chi_Q(\gamma_1,\gamma_2)=\chi(\mathrm{Ext}^\bullet(x_1,x_2)) = -\sum_{i,j\in I} a_{ij}\gamma_1^i\gamma_2^j + \sum_{i\in I}\gamma_1^i\gamma_2^i$$

be the Euler form on the K_0 group of the category of finite-dimensional representations of Q, where x_1 and x_2 are arbitrary representations of Q of dimension vectors γ_1 and γ_2 respectively.

In the context of the previous section, let $X=\mathbf{M}_{\overline{Q},\gamma}$, $Y=\mathbf{M}_{\widehat{Q},\gamma}=\mathbf{M}_{\overline{Q},\gamma}\times\mathbb{A}^{\gamma\cdot\gamma}$ (dot denotes the inner product), and

$$f = Tr(W)_\gamma = \sum_{i\in I, k=1,\dots,(\gamma^i)^2} f_{ik}x_{ik},$$

where f_{ik} are functions on $\mathbf{M}_{\overline{Q},\gamma}$, and $\{x_{ik}\}$ is a linear coordinate system on $\mathbb{A}^{\gamma\cdot\gamma}$. Then $Z=\mathbf{M}_{\Pi_Q,\gamma}$. Denote by $\mathbf{M}_{\Pi_Q,\gamma_1,\gamma_2}$ the space of representations of $\overline{Q}$ in coordinate spaces of dimension $\gamma_1+\gamma_2$ such that the standard coordinate subspaces of dimension γ_1 form a subrepresentation, and the restriction of $\rho\in\mathbf{M}_{\Pi_Q,\gamma_1,\gamma_2}$ on the block-diagonal part is an element in $\mathbf{M}_{\Pi_Q,\gamma_1}\times\mathbf{M}_{\Pi_Q,\gamma_2}$. The group $\mathbf{G}_{\gamma_1,\gamma_2}\subset\mathbf{G}_\gamma$ consisting of transformations preserving subspaces $(\mathbb{C}^{\gamma_1^i}\subset\mathbb{C}^{\gamma^i})_{i\in I}$ acts on $\mathbf{M}_{\Pi_Q,\gamma_1,\gamma_2}$. Suppose that we are given a collection of $\mathbf{G}_\gamma$-invariant closed subsets

$\mathbf{M}^{sp}_{\overline{Q},\gamma} \subset \mathbf{M}_{\overline{Q},\gamma}$ satisfying the following condition: for any short exact sequence $0 \to E_1 \to E \to E_2 \to 0$ of representations of $\overline{Q}$ with dimension vectors $\gamma_1, \gamma := \gamma_1 + \gamma_2, \gamma_2$ respectively, $E \in \mathbf{M}^{sp}_{\overline{Q},\gamma}$ if and only if $E_1 \in \mathbf{M}^{sp}_{\overline{Q},\gamma_1}$, and $E_2 \in \mathbf{M}^{sp}_{\overline{Q},\gamma_2}$. Then $\mathbf{M}^{sp}_{\widehat{Q},\gamma} = \mathbf{M}^{sp}_{\overline{Q},\gamma} \times \mathbb{A}^{\gamma\cdot\gamma}$.

The COHA of $(\widehat{Q}, W)$ (see [13]) induces the coproduct on the vector space $\bigoplus_{\gamma\in\mathbb{Z}^I_{\geqslant 0}} H^\bullet_{c,\mathbf{G}_\gamma}(\mathbf{M}^{sp}_{\Pi_Q,\gamma}, \mathbb{Q})$ as follows:

• $H^\bullet_{c,\mathbf{G}_\gamma}(\mathbf{M}^{sp}_{\Pi_Q,\gamma}, \mathbb{Q}) \to H^\bullet_{c,\mathbf{G}_{\gamma_1,\gamma_2}}(\mathbf{M}^{sp}_{\Pi_Q,\gamma}, \mathbb{Q})$, which is the pullback associated to the closed embedding of groups $\mathbf{G}_{\gamma_1,\gamma_2} \to \mathbf{G}_\gamma$ with proper quotient. The projections $pr_{\gamma_1,\gamma_2,N} : \overline{(\mathbf{M}_{\widehat{Q},\gamma}, \mathbf{G}_{\gamma_1,\gamma_2})}_N \to \overline{(\mathbf{M}_{\widehat{Q},\gamma}, \mathbf{G}_\gamma)}_N$ induce natural transformations of functors $\varphi_{\gamma,N} \to (pr_{\gamma_1,\gamma_2,N})_! \varphi_{\gamma,\gamma_1,\gamma_2,N} (pr_{\gamma_1,\gamma_2,N})^*$ by (2) and properness of $pr_{\gamma_1,\gamma_2,N}$, thus give us

$$\begin{aligned}&(\pi_{\gamma,N})_! \varphi_{\gamma/u,N} (\pi_{\gamma,N})^*[-1]\\ &\quad\to (\pi_{\gamma,N})_! (pr_{\gamma_1,\gamma_2,N})_! \varphi_{(\gamma,\gamma_1,\gamma_2)/u,N} (pr_{\gamma_1,\gamma_2,N})^* (\pi_{\gamma,N})^*[-1].\end{aligned}$$

Here $\varphi_{\gamma,N} = \varphi_{Tr(W)_{\gamma,N}}$ is the vanishing cycles functor of the function $tr(W)_{\gamma,N}$ on $\overline{(\mathbf{M}_{\widehat{Q},\gamma}, \mathbf{G}_\gamma)}_N$, and $\varphi_{\gamma,\gamma_1,\gamma_2,N}$ corresponds to $Tr(W)_{\gamma,\gamma_1,\gamma_2,N}$ on $\overline{(\mathbf{M}_{\widehat{Q},\gamma}, \mathbf{G}_{\gamma_1,\gamma_2})}_N$. (Note that in subscript of $\varphi_{\gamma,\gamma_1,\gamma_2,N}$, γ indicates the dimension vector of $\mathbf{M}_{\widehat{Q},\gamma}$, and γ_1, γ_2 indicate those of $\mathbf{G}_{\gamma_1,\gamma_2}$. We will use similar notations in the subsequent steps.)

Since the following diagram commutes:

$$\begin{array}{ccc} \overline{(\mathbf{M}_{\widehat{Q},\gamma}, \mathbf{G}_{\gamma_1,\gamma_2})}_N \times \mathbb{C}^* & \xrightarrow{\quad pr_{\gamma_1,\gamma_2,N}\quad} & \overline{(\mathbf{M}_{\widehat{Q},\gamma}, \mathbf{G}_\gamma)}_N \times \mathbb{C}^* \\ \Big\downarrow{\scriptstyle \pi_{\gamma,\gamma_1,\gamma_2,N}} & & \Big\downarrow{\scriptstyle \pi_{\gamma,N}} \\ \overline{(\mathbf{M}_{\overline{Q},\gamma}, \mathbf{G}_{\gamma_1,\gamma_2})}_N \times \mathbb{C}^* & \xrightarrow{\quad pr_{\overline{Q},\gamma_1,\gamma_2,N}\quad} & \overline{(\mathbf{M}_{\overline{Q},\gamma}, \mathbf{G}_\gamma)}_N \times \mathbb{C}^* \end{array}$$

we have

$$\begin{aligned}&(\pi_{\gamma,N})_! (pr_{\gamma_1,\gamma_2,N})_! \varphi_{(\gamma,\gamma_1,\gamma_2)/u,N} (pr_{\gamma_1,\gamma_2,N})^* (\pi_{\gamma,N})^*[-1]\\ &\quad\simeq (pr_{\overline{Q},\gamma_1,\gamma_2,N})_! (\pi_{\gamma,\gamma_1,\gamma_2,N})_! \varphi_{(\gamma,\gamma_1,\gamma_2)/u,N} (\pi_{\gamma,\gamma_1,\gamma_2,N})^* (pr_{\overline{Q},\gamma_1,\gamma_2,N})^*[-1].\end{aligned}$$

By Theorem 2.2, we have two isomorphisms:

$$(\pi_{\gamma,N})_! \varphi_{\gamma/u,N} (\pi_{\gamma,N})^*[-1] \simeq (\pi_{\gamma,N})_! (\pi_{\gamma,N})^* (i_{\gamma,N})_* (i_{\gamma,N})^*$$

and

$$\begin{aligned}&(pr_{\overline{Q},\gamma_1,\gamma_2,N})_! (\pi_{\gamma,\gamma_1,\gamma_2,N})_! \varphi_{(\gamma,\gamma_1,\gamma_2)/u,N} (\pi_{\gamma,\gamma_1,\gamma_2,N})^* (pr_{\overline{Q},\gamma_1,\gamma_2,N})^*[-1]\\ &\quad\simeq (pr_{\overline{Q},\gamma_1,\gamma_2,N})_! (\pi_{\gamma,\gamma_1,\gamma_2,N})_! (\pi_{\gamma,\gamma_1,\gamma_2,N})^* (i_{\gamma,\gamma_1,\gamma_2,N})_* (i_{\gamma,\gamma_1,\gamma_2,N})^* (pr_{\overline{Q},\gamma_1,\gamma_2,N})^*.\end{aligned}$$

Here $i_{\gamma,N}$ and $i_{\gamma,\gamma_1,\gamma_2,N}$ are inclusions, and the subscripts have the same meaning as the vanishing cycles functors above.

Pulling back to $\mathbf{M}^{sp}_{\overline{Q},\gamma,N} \times \mathbb{C}^*$ gives us the commutative diagram

$$\begin{array}{ccc}
H^{\bullet,\mathrm{crit}}_{c,\mathbf{G}_\gamma}(\mathbf{M}^{sp}_{\widehat{Q},\gamma}, W_\gamma) & \longrightarrow & H^{\bullet,\mathrm{crit}}_{c,\mathbf{G}_{\gamma_1,\gamma_2}}(\mathbf{M}^{sp}_{\widehat{Q},\gamma}, W_\gamma) \\
\big\downarrow \wr & & \big\downarrow \wr \\
H^{\bullet}_{c,\mathbf{G}_\gamma}(\mathbf{M}^{sp}_{\Pi_Q,\gamma}, \mathbb{Q}) \otimes \mathbb{T}^{\gamma\cdot\gamma} & \longrightarrow & H^{\bullet}_{c,\mathbf{G}_{\gamma_1,\gamma_2}}(\mathbf{M}^{sp}_{\Pi_Q,\gamma}, \mathbb{Q}) \otimes \mathbb{T}^{\gamma\cdot\gamma}
\end{array}$$

- $H^{\bullet}_{c,\mathbf{G}_{\gamma_1,\gamma_2}}(\mathbf{M}^{sp}_{\Pi_Q,\gamma}, \mathbb{Q}) \to H^{\bullet}_{c,\mathbf{G}_{\gamma_1,\gamma_2}}(\widetilde{\mathbf{M}}^{sp}_{\Pi_Q,\gamma_1,\gamma_2}, \mathbb{Q}) \otimes \mathbb{T}^{-\gamma_1\cdot\gamma_2}$, where $\mathbf{M}^{sp}_{\Pi_Q,\gamma_1,\gamma_2} = \mathbf{M}^{sp}_{\Pi_Q,\gamma} \cap \mathbf{M}_{\Pi_Q,\gamma_1,\gamma_2}$, and $\widetilde{\mathbf{M}}^{sp}_{\Pi_Q,\gamma_1,\gamma_2} \subset \mathbf{M}_{\Pi_Q,\gamma_1,\gamma_2}$ is the pullback of $\mathbf{M}^{sp}_{\Pi_Q,\gamma_1} \times \mathbf{M}^{sp}_{\Pi_Q,\gamma_2}$ under the projection $\mathbf{M}_{\Pi_Q,\gamma_1,\gamma_2} \to \mathbf{M}_{\Pi_Q,\gamma_1} \times \mathbf{M}_{\Pi_Q,\gamma_2}$. This is the pullback associated to the closed embedding $\mathbf{M}_{\Pi_Q,\gamma_1,\gamma_2} \to \mathbf{M}_{\Pi_Q,\gamma}$. The inclusions $j_{\gamma_1,\gamma_2,N}$: $\overline{(\mathbf{M}_{\widehat{Q},\gamma_1,\gamma_2}, \mathbf{G}_{\gamma_1,\gamma_2})}_N \to \overline{(\mathbf{M}_{\widehat{Q},\gamma}, \mathbf{G}_{\gamma_1,\gamma_2})}_N$ induce natural transformations of functors $\varphi_{\gamma,\gamma_1,\gamma_2,N} \to (j_{\gamma_1,\gamma_2,N})_*\varphi_{\gamma_1,\gamma_2,N}(j_{\gamma_1,\gamma_2,N})^*$ by (2). So we have

$$\begin{aligned}
&(\pi_{\gamma,\gamma_1,\gamma_2,N})_!\varphi_{(\gamma,\gamma_1,\gamma_2)/u,N}(\pi_{\gamma,\gamma_1,\gamma_2,N})^*[-1] \\
&\quad \to (\pi_{\gamma,\gamma_1,\gamma_2,N})_!(j_{\gamma_1,\gamma_2,N})_*\varphi_{(\gamma_1,\gamma_2)/u,N}(j_{\gamma_1,\gamma_2,N})^*(\pi_{\gamma,\gamma_1,\gamma_2,N})^*[-1].
\end{aligned}$$

By the commutative diagram

$$\begin{array}{ccc}
\overline{(\mathbf{M}_{\widehat{Q},\gamma_1,\gamma_2}, \mathbf{G}_{\gamma_1,\gamma_2})}_N \times \mathbb{C}^* & \xrightarrow{\;j_{\gamma_1,\gamma_2,N}\;} & \overline{(\mathbf{M}_{\widehat{Q},\gamma}, \mathbf{G}_{\gamma_1,\gamma_2})}_N \times \mathbb{C}^* \\
\big\downarrow \pi_{\gamma_1,\gamma_2,N} & & \big\downarrow \pi_{\gamma,\gamma_1,\gamma_2,N} \\
\overline{(\mathbf{M}_{\overline{Q},\gamma_1,\gamma_2}, \mathbf{G}_{\gamma_1,\gamma_2})}_N \times \mathbb{C}^* & \xrightarrow{\;j_{\overline{Q},\gamma_1,\gamma_2,N}\;} & \overline{(\mathbf{M}_{\overline{Q},\gamma}, \mathbf{G}_{\gamma_1,\gamma_2})}_N \times \mathbb{C}^*
\end{array}$$

we have

$$\begin{aligned}
&(\pi_{\gamma,\gamma_1,\gamma_2,N})_!(j_{\gamma_1,\gamma_2,N})_*\varphi_{(\gamma_1,\gamma_2)/u,N}(j_{\gamma_1,\gamma_2,N})^*(\pi_{\gamma,\gamma_1,\gamma_2,N})^*[-1] \\
&\quad \simeq (j_{\overline{Q},\gamma_1,\gamma_2,N})_*(\pi_{\gamma_1,\gamma_2,N})_!\varphi_{(\gamma_1,\gamma_2)/u,N}(\pi_{\gamma_1,\gamma_2,N})^*(j_{\overline{Q},\gamma_1,\gamma_2,N})^*[-1].
\end{aligned}$$

Then the isomorphisms

$$\begin{aligned}
&(\pi_{\gamma,\gamma_1,\gamma_2,N})_!\varphi_{(\gamma,\gamma_1,\gamma_2)/u,N}(\pi_{\gamma,\gamma_1,\gamma_2,N})^*[-1] \\
&\quad \simeq (\pi_{\gamma,\gamma_1,\gamma_2,N})_!(\pi_{\gamma,\gamma_1,\gamma_2,N})^*(i_{\gamma,\gamma_1,\gamma_2,N})_*(i_{\gamma,\gamma_1,\gamma_2,N})^*
\end{aligned}$$

and

$$\begin{aligned}
&(j_{\overline{Q},\gamma_1,\gamma_2,N})_*(\pi_{\gamma_1,\gamma_2,N})_!\varphi_{(\gamma_1,\gamma_2)/u,N}(\pi_{\gamma_1,\gamma_2,N})^*(j_{\overline{Q},\gamma_1,\gamma_2,N})^*[-1] \\
&\quad \simeq (j_{\overline{Q},\gamma_1,\gamma_2,N})_*(\pi_{\gamma_1,\gamma_2,N})_!(\pi_{\gamma_1,\gamma_2,N})^*(i_{\gamma_1,\gamma_2,N})_*(i_{\gamma_1,\gamma_2,N})^*(j_{\overline{Q},\gamma_1,\gamma_2,N})^*
\end{aligned}$$

obtained from the theorem give us the commutative diagram by pulling back to $\mathbf{M}^{sp}_{\overline{Q},\gamma,\gamma_1,\gamma_2,N} \times \mathbb{C}^*$:

$$\begin{array}{ccc}
 & H^{\bullet,\mathrm{crit}}_{c,\mathbf{G}_{\gamma_1,\gamma_2}}(\mathbf{M}^{sp}_{\widehat{Q},\gamma_1,\gamma_2}, W_\gamma) \xrightarrow{\ \sim\ } H^{\bullet,\mathrm{crit}}_{c,\mathbf{G}_{\gamma_1,\gamma_2}}(\widetilde{\mathbf{M}}^{sp}_{\widehat{Q},\gamma_1,\gamma_2}, W_\gamma) & \\
 \nearrow & & \searrow \\
H^{\bullet,\mathrm{crit}}_{c,\mathbf{G}_{\gamma_1,\gamma_2}}(\mathbf{M}^{sp}_{\widehat{Q},\gamma}, W_\gamma) & & H^{\bullet,\mathrm{crit}}_{c,\mathbf{G}_{\gamma_1,\gamma_2}}(\widetilde{\mathbf{M}}^{sp}_{\widehat{Q},\gamma_1,\gamma_2}, W_{\gamma_1,\gamma_2}) \\
\downarrow \wr & & \downarrow \wr \\
H^{\bullet}_{c,\mathbf{G}_{\gamma_1,\gamma_2}}(\mathbf{M}^{sp}_{\Pi_Q,\gamma}, \mathbb{Q}) \otimes \mathbb{T}^{\gamma\cdot\gamma} & \longrightarrow & H^{\bullet}_{c,\mathbf{G}_{\gamma_1,\gamma_2}}(\widetilde{\mathbf{M}}^{sp}_{\Pi_Q,\gamma_1,\gamma_2}, \mathbb{Q}) \otimes \mathbb{T}^{l_1}
\end{array}$$

where $l_1 = \gamma\cdot\gamma - \gamma_1\cdot\gamma_2$.

- $H^{\bullet}_{c,\mathbf{G}_{\gamma_1,\gamma_2}}(\widetilde{\mathbf{M}}^{sp}_{\Pi_Q,\gamma_1,\gamma_2}, \mathbb{Q}) \xrightarrow{\sim} H^{\bullet}_{c,\mathbf{G}_{\gamma_1}\times\mathbf{G}_{\gamma_2}}(\widetilde{\mathbf{M}}^{sp}_{\Pi_Q,\gamma_1,\gamma_2}, \mathbb{Q}) \otimes \mathbb{T}^{-\gamma_1\cdot\gamma_2}$. The affine fibrations $q_{\gamma_1,\gamma_2,N} : \overline{(\mathbf{M}_{\widehat{Q},\gamma_1,\gamma_2}, \mathbf{G}_{\gamma_1}\times\mathbf{G}_{\gamma_2})}_N \to \overline{(\mathbf{M}_{\widehat{Q},\gamma_1,\gamma_2}, \mathbf{G}_{\gamma_1,\gamma_2})}_N$ induce isomorphisms

$$\varphi_{(\gamma_1,\gamma_2)/u,N}(\mathbb{Q}_{\overline{(\mathbf{M}_{\widehat{Q},\gamma_1,\gamma_2},\mathbf{G}_{\gamma_1,\gamma_2})}_N\times\mathbb{C}^*} \xrightarrow{\sim} (q_{\gamma_1,\gamma_2,N})_*\mathbb{Q}_{\overline{(\mathbf{M}_{\widehat{Q},\gamma_1,\gamma_2},\mathbf{G}_{\gamma_1}\times\mathbf{G}_{\gamma_2})}_N\times\mathbb{C}^*}).$$

By applying the Verdier duality we get

$$\varphi_{(\gamma_1,\gamma_2)/u,N}((q_{\gamma_1,\gamma_2,N})_! D\mathbb{Q}_{\overline{(\mathbf{M}_{\widehat{Q},\gamma_1,\gamma_2},\mathbf{G}_{\gamma_1}\times\mathbf{G}_{\gamma_2})}_N\times\mathbb{C}^*} \xrightarrow{\sim} D\mathbb{Q}_{\overline{(\mathbf{M}_{\widehat{Q},\gamma_1,\gamma_2},\mathbf{G}_{\gamma_1,\gamma_2})}_N\times\mathbb{C}^*}).$$

Then

$$\varphi_{(\gamma_1,\gamma_2)/u,N}((q_{\gamma_1,\gamma_2,N})_! \mathbb{Q}_{\overline{(\mathbf{M}_{\widehat{Q},\gamma_1,\gamma_2},\mathbf{G}_{\gamma_1}\times\mathbf{G}_{\gamma_2})}_N\times\mathbb{C}^*} \xrightarrow{\sim} \mathbb{Q}_{\overline{(\mathbf{M}_{\widehat{Q},\gamma_1,\gamma_2},\mathbf{G}_{\gamma_1,\gamma_2})}_N\times\mathbb{C}^*}\otimes\mathbb{T}^{\gamma_1\cdot\gamma_2})$$

by (1), and

$$\begin{aligned}
&(q_{\gamma_1,\gamma_2,N})_!\varphi_{(\gamma_1,\gamma_2,\gamma_1\times\gamma_2)/u,N}\mathbb{Q}_{\overline{(\mathbf{M}_{\widehat{Q},\gamma_1,\gamma_2},\mathbf{G}_{\gamma_1}\times\mathbf{G}_{\gamma_2})}_N\times\mathbb{C}^*} \\
&\quad\simeq (q_{\gamma_1,\gamma_2,N})_!\varphi_{(\gamma_1,\gamma_2,\gamma_1\times\gamma_2)/u,N}(q_{\gamma_1,\gamma_2,N})^*\mathbb{Q}_{\overline{(\mathbf{M}_{\widehat{Q},\gamma_1,\gamma_2},\mathbf{G}_{\gamma_1,\gamma_2})}_N\times\mathbb{C}^*} \\
&\quad\xrightarrow{\sim} \varphi_{(\gamma_1,\gamma_2)/u,N}(\mathbb{Q}_{\overline{(\mathbf{M}_{\widehat{Q},\gamma_1,\gamma_2},\mathbf{G}_{\gamma_1,\gamma_2})}_N\times\mathbb{C}^*}\otimes\mathbb{T}^{\gamma_1\cdot\gamma_2})
\end{aligned}$$

by (3). Then we have isomorphisms

$$\begin{aligned}
&(\pi_{\gamma_1,\gamma_2,N})_!\varphi_{(\gamma_1,\gamma_2)/u,N}(\pi_{\gamma_1,\gamma_2,N})^*(\mathbb{Q}_{\overline{(\mathbf{M}_{\overline{Q},\gamma_1,\gamma_2},\mathbf{G}_{\gamma_1,\gamma_2})}_N\times\mathbb{C}^*}\otimes\mathbb{T}^{\gamma_1\cdot\gamma_2}) \\
&\quad\to (\pi_{\gamma_1,\gamma_2,N})_!(q_{\gamma_1,\gamma_2,N})_!\varphi_{(\gamma_1,\gamma_2,\gamma_1\times\gamma_2)/u,N}(q_{\gamma_1,\gamma_2,N})^*(\pi_{\gamma_1,\gamma_2,N})^* \\
&\qquad\times\mathbb{Q}_{\overline{(\mathbf{M}_{\overline{Q},\gamma_1,\gamma_2},\mathbf{G}_{\gamma_1,\gamma_2})}_N\times\mathbb{C}^*}.
\end{aligned}$$

The commutative diagram

$$\begin{array}{ccc}
\overline{(\mathbf{M}_{\widehat{Q},\gamma_1,\gamma_2},\mathbf{G}_{\gamma_1}\times\mathbf{G}_{\gamma_2})}_N\times\mathbb{C}^* & \xrightarrow{q_{\gamma_1,\gamma_2,N}} & \overline{(\mathbf{M}_{\widehat{Q},\gamma_1,\gamma_2},\mathbf{G}_{\gamma_1,\gamma_2})}_N\times\mathbb{C}^* \\
\downarrow{\scriptstyle \pi_{\gamma_1,\gamma_2,\gamma_1\times\gamma_2,N}} & & \downarrow{\scriptstyle \pi_{\gamma_1,\gamma_2,N}} \\
\overline{(\mathbf{M}_{\overline{Q},\gamma_1,\gamma_2},\mathbf{G}_{\gamma_1}\times\mathbf{G}_{\gamma_2})}_N\times\mathbb{C}^* & \xrightarrow{q_{\overline{Q},\gamma_1,\gamma_2,N}} & \overline{(\mathbf{M}_{\overline{Q},\gamma_1,\gamma_2},\mathbf{G}_{\gamma_1,\gamma_2})}_N\times\mathbb{C}^*
\end{array}$$

gives us isomorphisms

$$\begin{aligned}
&(\pi_{\gamma_1,\gamma_2,N})_!(q_{\gamma_1,\gamma_2,N})_!\varphi_{(\gamma_1,\gamma_2,\gamma_1\times\gamma_2)/u,N}(q_{\gamma_1,\gamma_2,N})^*(\pi_{\gamma_1,\gamma_2,N})^*[-1]\\
&\quad\simeq (q_{\overline{Q},\gamma_1,\gamma_2,N})_!(\pi_{\gamma_1,\gamma_2,\gamma_1\times\gamma_2,N})_!\varphi_{(\gamma_1,\gamma_2,\gamma_1\times\gamma_2)/u,N}(\pi_{\gamma_1,\gamma_2,\gamma_1\times\gamma_2,N})^*\\
&\qquad\times (q_{\overline{Q},\gamma_1,\gamma_2,N})^*[-1].
\end{aligned}$$

Theorem 2.2 implies isomorphisms

$$(\pi_{\gamma_1,\gamma_2,N})_!\varphi_{(\gamma_1,\gamma_2)/u,N}(\pi_{\gamma_1,\gamma_2,N})^*[-1]\simeq(\pi_{\gamma_1,\gamma_2,N})_!(\pi_{\gamma_1,\gamma_2,N})^*(i_{\gamma_1,\gamma_2,N})_*(i_{\gamma_1,\gamma_2,N})^*$$

and

$$\begin{aligned}
&(q_{\overline{Q},\gamma_1,\gamma_2,N})_!(\pi_{\gamma_1,\gamma_2,\gamma_1\times\gamma_2,N})_!\varphi_{(\gamma_1,\gamma_2,\gamma_1\times\gamma_2)/u,N}(\pi_{\gamma_1,\gamma_2,\gamma_1\times\gamma_2,N})^*(q_{\overline{Q},\gamma_1,\gamma_2,N})^*[-1]\\
&\quad\simeq (q_{\overline{Q},\gamma_1,\gamma_2,N})_!(\pi_{\gamma_1,\gamma_2,\gamma_1\times\gamma_2,N})_!(\pi_{\gamma_1,\gamma_2,\gamma_1\times\gamma_2,N})^*(i_{\gamma_1,\gamma_2,\gamma_1\times\gamma_2,N})_*\\
&\qquad\times (i_{\gamma_1,\gamma_2,\gamma_1\times\gamma_2,N})^*(q_{\overline{Q},\gamma_1,\gamma_2,N})^*.
\end{aligned}$$

Pulling back to $\mathbf{M}^{sp}_{\overline{Q},\gamma_1,\gamma_2,N}\times\mathbb{C}^*$ gives us the commutative diagram

$$\begin{array}{ccc}
H^{\bullet,\mathrm{crit}}_{c,\mathbf{G}_{\gamma_1,\gamma_2}}(\widetilde{\mathbf{M}}^{sp}_{\widehat{Q},\gamma_1,\gamma_2},W_{\gamma_1,\gamma_2}) & \xrightarrow{\sim} & H^{\bullet,\mathrm{crit}}_{c,\mathbf{G}_{\gamma_1}\times\mathbf{G}_{\gamma_2}}(\widetilde{\mathbf{M}}^{sp}_{\widehat{Q},\gamma_1,\gamma_2},W_{\gamma_1,\gamma_2})\otimes\mathbb{T}^{-\gamma_1\cdot\gamma_2} \\
\downarrow\wr & & \downarrow\wr \\
H^{\bullet}_{c,\mathbf{G}_{\gamma_1,\gamma_2}}(\widetilde{\mathbf{M}}^{sp}_{\Pi_Q,\gamma_1,\gamma_2},\mathbb{Q})\otimes\mathbb{T}^{\gamma\cdot\gamma-\gamma_1\cdot\gamma_2} & \xrightarrow{\sim} & H^{\bullet}_{c,\mathbf{G}_{\gamma_1}\times\mathbf{G}_{\gamma_2}}(\widetilde{\mathbf{M}}^{sp}_{\Pi_Q,\gamma_1,\gamma_2},\mathbb{Q})\otimes\mathbb{T}^{\gamma\cdot\gamma-2\gamma_1\cdot\gamma_2}
\end{array}$$

• $H^{\bullet}_{c,\mathbf{G}_{\gamma_1}\times\mathbf{G}_{\gamma_2}}(\widetilde{\mathbf{M}}^{sp}_{\Pi_Q,\gamma_1,\gamma_2},\mathbb{Q})\xrightarrow{\sim}H^{\bullet}_{c,\mathbf{G}_{\gamma_1}\times\mathbf{G}_{\gamma_2}}(\mathbf{M}^{sp}_{\Pi_Q,\gamma_1}\times\mathbf{M}^{sp}_{\Pi_Q,\gamma_2},\mathbb{Q})\otimes\mathbb{T}^{-\chi_{\widetilde{Q}}(\gamma_2,\gamma_1)}$.
Similar as the previous step, the affine fibrations $p_{\gamma_1,\gamma_2,N}$:

$$\overline{(\mathbf{M}_{\widehat{Q},\gamma_1,\gamma_2},\mathbf{G}_{\gamma_1}\times\mathbf{G}_{\gamma_2})}_N\to\overline{(\mathbf{M}_{\widehat{Q},\gamma_1}\times\mathbf{M}_{\widehat{Q},\gamma_2},\mathbf{G}_{\gamma_1}\times\mathbf{G}_{\gamma_2})}_N$$

induce isomorphisms

$$\begin{aligned}
&(p_{\gamma_1,\gamma_2,N})_!\varphi_{\gamma_1,\gamma_2,\gamma_1\times\gamma_2,N}(p_{\gamma_1,\gamma_2,N})^*\mathbb{Q}_{\overline{(\mathbf{M}_{\widehat{Q},\gamma_1}\times\mathbf{M}_{\widehat{Q},\gamma_2},\mathbf{G}_{\gamma_1}\times\mathbf{G}_{\gamma_2})}_N\times\mathbb{C}^*}\\
&\quad\xrightarrow{\sim}\varphi_{\gamma_1\boxplus\gamma_2,N}(\mathbb{Q}_{\overline{(\mathbf{M}_{\widehat{Q},\gamma_1}\times\mathbf{M}_{\widehat{Q},\gamma_2},\mathbf{G}_{\gamma_1}\times\mathbf{G}_{\gamma_2})}_N\times\mathbb{C}^*}\otimes\mathbb{T}^l),
\end{aligned}$$

where

$$l=\sum_{a:i\to j\in\widetilde{Q}_1}\gamma_1^j\gamma_2^i.$$

Then we have isomorphisms

$$(\pi_{\gamma_1\times\gamma_2,N})_!(p_{\gamma_1,\gamma_2,N})_!\varphi_{(\gamma_1,\gamma_2,\gamma_1\times\gamma_2)/u,N}(p_{\gamma_1,\gamma_2,N})^*(\pi_{\gamma_1\times\gamma_2,N})^* \times \mathbb{Q}_{\overline{(\mathbf{M}_{\overline{Q},\gamma_1}\times\mathbf{M}_{\overline{Q},\gamma_2},\mathbf{G}_{\gamma_1}\times\mathbf{G}_{\gamma_2})}_N\times\mathbb{C}^*} \xrightarrow{\sim} (\pi_{\gamma_1\times\gamma_2,N})_!\varphi_{(\gamma_1\boxplus\gamma_2)/u,N}(\pi_{\gamma_1\times\gamma_2,N})^*(\mathbb{Q}_{\overline{(\mathbf{M}_{\overline{Q},\gamma_1}\times\mathbf{M}_{\overline{Q},\gamma_2},\mathbf{G}_{\gamma_1}\times\mathbf{G}_{\gamma_2})}_N\times\mathbb{C}^*}\otimes\mathbb{T}^l).$$

The commutative diagram

$$\begin{array}{ccc}
\overline{(\mathbf{M}_{\widehat{Q},\gamma_1,\gamma_2},\mathbf{G}_{\gamma_1}\times\mathbf{G}_{\gamma_2})}_N\times\mathbb{C}^* & \xrightarrow{p_{\gamma_1,\gamma_2,N}} & \overline{(\mathbf{M}_{\widehat{Q},\gamma_1}\times\mathbf{M}_{\widehat{Q},\gamma_2},\mathbf{G}_{\gamma_1}\times\mathbf{G}_{\gamma_2})}_N\times\mathbb{C}^* \\
\Big\downarrow{\scriptstyle \pi_{\gamma_1,\gamma_2,\gamma_1\times\gamma_2,N}} & & \Big\downarrow{\scriptstyle \pi_{\gamma_1\times\gamma_2,N}} \\
\overline{(\mathbf{M}_{\overline{Q},\gamma_1,\gamma_2},\mathbf{G}_{\gamma_1}\times\mathbf{G}_{\gamma_2})}_N\times\mathbb{C}^* & \xrightarrow{p_{\overline{Q},\gamma_1,\gamma_2,N}} & \overline{(\mathbf{M}_{\overline{Q},\gamma_1}\times\mathbf{M}_{\overline{Q},\gamma_2},\mathbf{G}_{\gamma_1}\times\mathbf{G}_{\gamma_2})}_N\times\mathbb{C}^*
\end{array}$$

implies isomorphisms

$$\begin{aligned}&(\pi_{\gamma_1\times\gamma_2,N})_!(p_{\gamma_1,\gamma_2,N})_!\varphi_{(\gamma_1,\gamma_2,\gamma_1\times\gamma_2)/u,N}(p_{\gamma_1,\gamma_2,N})^*(\pi_{\gamma_1\times\gamma_2,N})^*[-1]\\ &\simeq (p_{\overline{Q},\gamma_1,\gamma_2,N})_!(\pi_{\gamma_1,\gamma_2,\gamma_1\times\gamma_2,N})_!\varphi_{(\gamma_1,\gamma_2,\gamma_1\times\gamma_2)/u,N}(\pi_{\gamma_1,\gamma_2,\gamma_1\times\gamma_2,N})^*\\ &\quad\times(p_{\overline{Q},\gamma_1,\gamma_2,N})^*[-1].\end{aligned}$$

By Theorem 2.2, we have

$$\begin{aligned}&(\pi_{\gamma_1\times\gamma_2,N})_!\varphi_{(\gamma_1\boxplus\gamma_2)/u,N}(\pi_{\gamma_1\times\gamma_2,N})^*[-1]\\ &\simeq(\pi_{\gamma_1\times\gamma_2,N})_!(\pi_{\gamma_1\times\gamma_2,N})^*(i_{\gamma_1\times\gamma_2,N})_*(i_{\gamma_1\times\gamma_2,N})^*\end{aligned}$$

and

$$\begin{aligned}&(p_{\overline{Q},\gamma_1,\gamma_2,N})_!(\pi_{\gamma_1,\gamma_2,\gamma_1\times\gamma_2,N})_!\varphi_{(\gamma_1,\gamma_2,\gamma_1\times\gamma_2)/u,N}(\pi_{\gamma_1,\gamma_2,\gamma_1\times\gamma_2,N})^*(p_{\overline{Q},\gamma_1,\gamma_2,N})^*[-1]\\ &\simeq(p_{\overline{Q},\gamma_1,\gamma_2,N})_!(\pi_{\gamma_1,\gamma_2,\gamma_1\times\gamma_2,N})_!(\pi_{\gamma_1,\gamma_2,\gamma_1\times\gamma_2,N})^*(i_{\gamma_1,\gamma_2,\gamma_1\times\gamma_2,N})_*\\ &\quad\times(i_{\gamma_1,\gamma_2,\gamma_1\times\gamma_2,N})^*(p_{\overline{Q},\gamma_1,\gamma_2,N})^*.\end{aligned}$$

By pulling back to $\overline{(\mathbf{M}^{sp}_{\overline{Q},\gamma_1}\times\mathbf{M}^{sp}_{\overline{Q},\gamma_2},\mathbf{G}_{\gamma_1}\times\mathbf{G}_{\gamma_2})}_N\times\mathbb{C}^*$, we have

$$\begin{array}{ccc}
H^{\bullet,\mathrm{crit}}_{c,\mathbf{G}_{\gamma_1}\times\mathbf{G}_{\gamma_2}}(\widetilde{\mathbf{M}}^{sp}_{\widehat{Q},\gamma_1,\gamma_2},W_{\gamma_1,\gamma_2}) & \xrightarrow{\sim} & H^{\bullet,\mathrm{crit}}_{c,\mathbf{G}_{\gamma_1}\times\mathbf{G}_{\gamma_2}}(\mathbf{M}^{sp}_{\widehat{Q},\gamma_1}\times\mathbf{M}^{sp}_{\widehat{Q},\gamma_2},W_{\gamma_1}\boxplus W_{\gamma_2})\otimes\mathbb{T}^l \\
\Big\downarrow\wr & & \Big\downarrow\wr \\
H^{\bullet}_{c,\mathbf{G}_{\gamma_1}\times\mathbf{G}_{\gamma_2}}(\widetilde{\mathbf{M}}^{sp}_{\Pi_Q,\gamma_1,\gamma_2},\mathbb{Q})\otimes\mathbb{T}^{l_1} & \xrightarrow{\sim} & H^{\bullet}_{c,\mathbf{G}_{\gamma_1}\times\mathbf{G}_{\gamma_2}}(\mathbf{M}^{sp}_{\Pi_Q,\gamma_1}\times\mathbf{M}^{sp}_{\Pi_Q,\gamma_2},\mathbb{Q})\otimes\mathbb{T}^{l_2}
\end{array}$$

where $l_1=\gamma\cdot\gamma-\gamma_1\cdot\gamma_2$, and $l_2=\gamma_1\cdot\gamma_1+\gamma_2\cdot\gamma_2+l$.

- $H^{\bullet}_{c,\mathbf{G}_{\gamma_1}\times\mathbf{G}_{\gamma_2}}(\mathbf{M}^{sp}_{\Pi_Q,\gamma_1}\times\mathbf{M}^{sp}_{\Pi_Q,\gamma_2},\mathbb{Q})\xrightarrow{\sim}H^{\bullet}_{c,\mathbf{G}_{\gamma_1}}(\mathbf{M}^{sp}_{\Pi_Q,\gamma_1},\mathbb{Q})\otimes H^{\bullet}_{c,\mathbf{G}_{\gamma_2}}(\mathbf{M}^{sp}_{\Pi_Q,\gamma_2},\mathbb{Q})$. This is the Künneth isomorphism compatible with the Thom–Sebastiani isomorphism by Theorem 2.3.

The above computations can be summarized for convenience of the reader in the form of the following statement.

Proposition 2.6. *The coproduct making the vector space* $\bigoplus_{\gamma\in\mathbb{Z}^I_{\geqslant 0}} H^\bullet_{c,\mathbf{G}}(\mathbf{M}^{sp}_{\Pi_Q,\gamma},\mathbb{Q})$ *into a coalgebra is given by the composition of the maps*

$$\begin{aligned}
&H^\bullet_{c,\mathbf{G}_\gamma}(\mathbf{M}^{sp}_{\Pi_Q,\gamma},\mathbb{Q}) \to H^\bullet_{c,\mathbf{G}_{\gamma_1,\gamma_2}}(\mathbf{M}^{sp}_{\Pi_Q,\gamma},\mathbb{Q})\\
&\quad\longrightarrow H^\bullet_{c,\mathbf{G}_{\gamma_1,\gamma_2}}(\widetilde{\mathbf{M}}^{sp}_{\Pi_Q,\gamma_1,\gamma_2},\mathbb{Q})\otimes\mathbb{T}^{-\gamma_1\cdot\gamma_2}\\
&\quad\xrightarrow{\sim} H^\bullet_{c,\mathbf{G}_{\gamma_1}\times\mathbf{G}_{\gamma_2}}(\widetilde{\mathbf{M}}^{sp}_{\Pi_Q,\gamma_1,\gamma_2},\mathbb{Q})\otimes\mathbb{T}^{-2\gamma_1\cdot\gamma_2}\\
&\quad\xrightarrow{\sim} H^\bullet_{c,\mathbf{G}_{\gamma_1}\times\mathbf{G}_{\gamma_2}}(\mathbf{M}^{sp}_{\Pi_Q,\gamma_1}\times\mathbf{M}^{sp}_{\Pi_Q,\gamma_2},\mathbb{Q})\otimes\mathbb{T}^{-\chi_Q(\gamma_1,\gamma_2)-\chi_Q(\gamma_2,\gamma_1)}\\
&\quad\xrightarrow{\sim} H^\bullet_{c,\mathbf{G}_{\gamma_1}}(\mathbf{M}^{sp}_{\Pi_Q,\gamma_1},\mathbb{Q})\otimes H^\bullet_{c,\mathbf{G}_{\gamma_2}}(\mathbf{M}^{sp}_{\Pi_Q,\gamma_2},\mathbb{Q})\otimes\mathbb{T}^{-\chi_Q(\gamma_1,\gamma_2)-\chi_Q(\gamma_2,\gamma_1)}.
\end{aligned}$$

Now let

$$\mathcal{H}_\gamma := H^\bullet_{c,\mathbf{G}_\gamma}(\mathbf{M}^{sp}_{\Pi_Q,\gamma},\mathbb{Q})^\vee\otimes\mathbb{T}^{-\chi_Q(\gamma,\gamma)},\quad\text{and}\quad \mathcal{H}=\bigoplus_{\gamma\in\mathbb{Z}^I_{\geqslant 0}}\mathcal{H}_\gamma.$$

Then the above coproduct makes $\mathcal{H}$ an associative algebra with product

$$\begin{aligned}
&\mathcal{H}_{\gamma_1}\otimes\mathcal{H}_{\gamma_2}\\
&\quad= H^\bullet_{c,\mathbf{G}_{\gamma_1}}(\mathbf{M}^{sp}_{\Pi_Q,\gamma_1},\mathbb{Q})^\vee\otimes\mathbb{T}^{-\chi_Q(\gamma_1,\gamma_1)}\otimes H^\bullet_{c,\mathbf{G}_{\gamma_2}}(\mathbf{M}^{sp}_{\Pi_Q,\gamma_2},\mathbb{Q})^\vee\otimes\mathbb{T}^{-\chi_Q(\gamma_2,\gamma_2)}\\
&\quad= H^\bullet_{c,\mathbf{G}_{\gamma_1}}(\mathbf{M}^{sp}_{\Pi_Q,\gamma_1},\mathbb{Q})^\vee\otimes H^\bullet_{c,\mathbf{G}_{\gamma_2}}(\mathbf{M}^{sp}_{\Pi_Q,\gamma_2},\mathbb{Q})^\vee\otimes\mathbb{T}^{-\chi_Q(\gamma_1,\gamma_1)-\chi_Q(\gamma_2,\gamma_2)}\\
&\quad\to H^\bullet_{c,\mathbf{G}_{\gamma_1+\gamma_2}}(\mathbf{M}^{sp}_{\Pi_Q,\gamma_1+\gamma_2},\mathbb{Q})^\vee\otimes\mathbb{T}^{-\chi_Q(\gamma_1,\gamma_2)-\chi_Q(\gamma_2,\gamma_1)}\otimes\mathbb{T}^{-\chi_Q(\gamma_1,\gamma_1)-\chi_Q(\gamma_2,\gamma_2)}\\
&\quad= H^\bullet_{c,\mathbf{G}_{\gamma_1+\gamma_2}}(\mathbf{M}^{sp}_{\Pi_Q,\gamma_1+\gamma_2},\mathbb{Q})^\vee\otimes\mathbb{T}^{-\chi_Q(\gamma_1+\gamma_2,\gamma_1+\gamma_2)}=\mathcal{H}_{\gamma_1+\gamma_2}.
\end{aligned}$$

Definition 2.7. The associative algebra $\mathcal{H}$ is called the *Cohomological Hall algebra* of the preprojective algebra Π_Q associated with the quiver Q.

Remark 2.8. In the framework of equivariant K-theory a similar notion was introduced in [20].

Corollary 2.9. *This product preserves the modified cohomological degree, thus the zero degree part*

$$\mathcal{H}^0=\bigoplus_{\gamma\in\mathbb{Z}^I_{\geqslant 0}}\mathcal{H}^0_\gamma=\bigoplus_{\gamma\in\mathbb{Z}^I_{\geqslant 0}} H^{-2\chi_Q(\gamma,\gamma)}_{c,\mathbf{G}_\gamma}(\mathbf{M}^{sp}_{\Pi_Q,\gamma},\mathbb{Q})^\vee\otimes\mathbb{T}^{-\chi_Q(\gamma,\gamma)}$$

is a subalgebra of $\mathcal{H}$.

We can reformulate the definition of COHA of Π_Q using language of stacks. The natural morphism of stacks $\mathbf{M}_{\Pi_Q,\gamma_1,\gamma_2}/\mathbf{G}_{\gamma_1,\gamma_2}\to\mathbf{M}_{\Pi_Q,\gamma}/\mathbf{G}_\gamma$ is proper, hence it induces the pushforward map on $\mathcal{H}$. Composting it with the pullback by the morphism $\mathbf{M}_{\Pi_Q,\gamma_1,\gamma_2}/\mathbf{G}_{\gamma_1,\gamma_2}\to\mathbf{M}_{\Pi_Q,\gamma_1}/\mathbf{G}_{\gamma_1}\times\mathbf{M}_{\Pi_Q,\gamma_2}/\mathbf{G}_{\gamma_2}$, we obtain the product.

2.3. Lusztig's seminilpotent Lagrangian subvariety

In this subsection we work in the framework close to the one from [1].

Let Q be a quiver (possibly with loops) with vertices I and arrows Ω, and denote by Ω_i the set of loops at $i \in I$. We call i imaginary if the number of loops $\omega_i = |\Omega_i| \geqslant 1$, and real if $\omega_i = 0$. Let I^{im} be the set of imaginary vertices and Ire real vertices.

Definition 2.10. A representation $x \in \mathbf{M}_{\overline{Q},\gamma}$ is *seminilpotent* if there is an I-graded filtration $W = (W_0 = V_\gamma \supset \cdots \supset W_r = \{0\})$ of the representation space $V_\gamma = (V_i)_{i\in I}$, such that $x_{a^*}(W_\bullet) \subseteq W_{\bullet+1}$, and $x_a(W_\bullet) \subseteq W_\bullet$ for $a \in \Omega$.

Remark 2.11. Our definition of seminilpotency is slightly different from that in [1]. We put nilpotent condition on the dual arrows a^* rather than a. But main results of [1] hold in our situation as well.

We denote by $\mathbf{M}^{sp}_{\overline{Q},\gamma}$ the space of seminilpotent representations of dimension γ. Then by [1, Th. 1.15], the space of seminilpotent representations of Π_Q of dimension γ, $\mathbf{M}^{sp}_{\Pi_Q,\gamma} \subset \mathbf{M}^{sp}_{\overline{Q},\gamma}$, is a Lagrangian subvariety of $\mathbf{M}_{\overline{Q},\gamma}$.

Let

$$\mathbf{M}^{sp}_{\Pi_Q,\gamma,i,l} = \left\{ x \in \mathbf{M}^{sp}_{\Pi_Q,\gamma} | \mathrm{codim}\left(\bigoplus_{j\neq i, a:j\to i \mathrm{in} \overline{Q}} \mathrm{Im} x_a \right) = l \right\}.$$

Then $\mathbf{M}^{sp}_{\Pi_Q,\gamma} = \bigcup_{i\in I, l\geqslant 1} \mathbf{M}^{sp}_{\Pi_Q,\gamma,i,l}$ by the seminilpotency condition. There is a one-to-one correspondence of the sets of irreducible components (see [1, Prop.1.14])

$$\mathrm{Irr}(\mathbf{M}^{sp}_{\Pi_Q,\gamma,i,l}) \xrightarrow{\sim} \mathrm{Irr}(\mathbf{M}^{sp}_{\Pi_Q,\gamma-le_i,i,0}) \times \mathrm{Irr}(\mathbf{M}^{sp}_{\Pi_Q,le_i}), \tag{4}$$

where $e_i = (\delta_{ij})_{j\in I}$. For any vertex i, we have $\mathrm{Irr}(\mathbf{M}^{sp}_{\Pi_Q,\gamma}) = \bigsqcup_{l\geqslant 0} \mathrm{Irr}(\mathbf{M}^{sp}_{\Pi_Q,\gamma,i,l})$. If $i \in I^{re}$ then $\mathrm{Irr}(\mathbf{M}^{sp}_{\Pi_Q,le_i})$ consists of only one element, namely the zero representation. We denote by $Z_{i,l}$ the only element in $\mathrm{Irr}(\mathbf{M}^{sp}_{\Pi_Q,le_i})$. If $i \in I^{im}$, then there are two cases. If the number of loops $\omega_i = 1$, then $\mathrm{Irr}(\mathbf{M}^{sp}_{\Pi_Q,le_i})$ is parametrized by $\mathfrak{C}_{i,l} = \{c = (c_k)\}$, the set of partitions of l (i.e., $\sum_k c_k = l$, $c_k > 0, \forall k$, and $c_{k+1} \geqslant c_k$). If $\omega_i > 1$, then it is parametrized by the set of compositions also denoted by $\mathfrak{C}_{i,l}$ (i.e., $\sum_k c_k = l$, $c_k > 0, \forall k$).

We put $|c| = \sum_k c_k$ for $c \in \mathfrak{C}_{i,l}$, and denote by $Z_{i,c} \in \mathrm{Irr}(\mathbf{M}^{sp}_{\Pi_Q,le_i})$ the irreducible component corresponding to c. Let $Z \in \mathrm{Irr}(\mathbf{M}^{sp}_{\Pi_Q,\gamma})$, then there exists $i \in I$ and $l \geqslant 1$ such that $Z \bigcap \mathbf{M}^{sp}_{\Pi_Q,\gamma,i,l}$ is dense in Z. We denote by $\varepsilon_i(Z)$ the corresponding partition or composition if $i \in I^{im}$, and $\varepsilon_i(Z) = l$ if $i \in I^{re}$, via the one-to-one correspondence (4).

Now let $\mathscr{M}_\gamma$ be the $\mathbb{Q}$-vector space of constructible functions $f : \mathbf{M}^{sp}_{\Pi_Q,\gamma} \to \mathbb{Q}$ which are constant on any $\mathbf{G}_\gamma$-orbit, and $\mathscr{M} = \bigoplus_\gamma \mathscr{M}_\gamma$. Then one can define a product $*$ on $\mathscr{M}$ in the way which is analogous to the definition of Lusztig for nilpotent case in [14, Section 12].

More precisely, let us denote by $\mathbf{M}^{sp}_{\Pi_Q,V}$ the space of seminilpotent representations of Π_Q with I-graded vector space V, and $\mathscr{M}_V$ the $\mathbb{Q}$-vector space of constructible functions $f : \mathbf{M}^{sp}_{\Pi_Q,V} \to \mathbb{Q}$ constant on any $\mathbf{G}_\gamma$-orbit. Let V_1, V_2 and V be I-graded vector spaces of dimensions γ_1, γ_2 and $\gamma = \gamma_1 + \gamma_2$ respectively, and $f_i \in \mathscr{M}_{V_i}, i = 1, 2$. Then $f_1 * f_2 \in \mathscr{M}_V$ is defined using the diagram

$$\mathbf{M}^{sp}_{\Pi_Q,V_1} \times \mathbf{M}^{sp}_{\Pi_Q,V_2} \xleftarrow{p_1} \mathbf{F}' \xrightarrow{p_2} \mathbf{F}'' \xrightarrow{p_3} \mathbf{M}^{sp}_{\Pi_Q,V}$$

where the notations are as follows: F'' is the variety of pairs (x, U) with $x \in \mathbf{M}^{sp}_{\Pi_Q,V}$ and U an x-stable I-graded subspace of V with dimension γ_2; F' is the variety of quadruples (x, U, R'', R') where $(x, U) \in F''$, $R'' : V_2 \xrightarrow{\sim} U$ and $R' : V_1 \xrightarrow{\sim} V/U$. The maps $p_1(x, U, R'', R') = (x_1, x_2)$ where $xR' = R'x_1$ and $xR'' = R''x_2$, $p_2(x, U, R'', R') = (x, U)$, and $p_3(x, U) = x$. Note that p_2 is a $\mathbf{G}_{V_1} \times \mathbf{G}_{V_2}$-principal bundle and p_3 is proper. Let $f(x_1, x_2) = f_1(x_1)f_2(x_2)$, then there is a unique function $f_3 \in \mathscr{M}_{F''}$ such that $p_1^* f = p_2^* f_3$. Finally, define $f_1 * f_2 = (p_3)_!(f_3)$. By identifying the vector spaces $\mathscr{M}_V$ for various V with $\mathscr{M}_\gamma$ in a coherent way $(\dim(V) = \gamma)$, we define the product $*$ on $\mathscr{M}$, making it an associative $\mathbb{Q}$-algebra.

One can also reformulate this product using the diagram of stacks

$$\mathbf{M}_{\Pi_Q,\gamma_2}/\mathbf{G}_{\gamma_2} \times \mathbf{M}_{\Pi_Q,\gamma_1}/\mathbf{G}_{\gamma_1} \leftarrow \mathbf{M}_{\Pi_Q,\gamma_2,\gamma_1}/\mathbf{G}_{\gamma_2,\gamma_1} \to \mathbf{M}_{\Pi_Q,\gamma}/\mathbf{G}_\gamma.$$

We denote by $1_{i,c}$ (resp. $1_{i,l}$) the characteristic function of $Z_{i,c}$ (resp. $Z_{i,l}$), and $\mathscr{M}_0 \subseteq \mathscr{M}$ the subalgebra generated by $1_{i,(l)}$ and $1_{i,1}$. For any $Z \in \mathrm{Irr}(\mathbf{M}^{sp}_{\Pi_Q,\gamma})$ and $f \in \mathscr{M}_\gamma$, let $\rho_Z(f) = c$ if $Z \bigcap f^{-1}(c)$ is open dense in Z.

Theorem 2.12 (see [1, Prop. 1.18]). *For any $Z \in \mathrm{Irr}(\mathbf{M}^{sp}_{\Pi_Q,\gamma})$ there exists $f_Z \in \mathscr{M}_{0,\gamma} = \mathscr{M}_0 \cap \mathscr{M}_\gamma$ such that $\rho_Z(f_Z) = 1$, and $\rho_{Z'}(f_Z) = 0$ for $Z' \neq Z$.*

2.4. Generalized quantum group

We recall some definitions and facts about generalized quantum group introduced in [1].

Let $(\bullet, \bullet)$ be the symmetric Euler form on $\mathbb{Z}^I$ defined by $(i, j) = 2\delta_{ij} - a_{ij} - a_{ji}$, and $(\iota, j) = l(i, j)$ if $\iota = (i, l) \in I_\infty = (I^{re} \times \{1\}) \bigcup (I^{im} \times \mathbb{N}_{\geqslant 1})$ and $j \in I$.

Definition 2.13. Let F be the $\mathbb{Q}(v)$-algebra generated by $(E_\iota)_{\iota \in I_\infty}$, $\mathbb{N}^I$-graded by $|E_\iota| = li$ for $\iota = (i, l)$. If $A \subseteq \mathbb{N}^I$, then let $\mathrm{F}[A] = \{E \in \mathrm{F} \mid |E| \in A\}$.

For any $\gamma = (\gamma^i)_{i \in I} \in \mathbb{Z}^I$, let $\mathrm{ht}(\gamma) = \sum_i \gamma^i$ be its height, and $v_\gamma = \prod_i v_i^{\gamma^i}$, where $v_i = v^{(i,i)/2}$. We endow F with a coproduct $\delta(E_{i,l}) = \sum_{l_1+l_2=l} v_i^{l_1 l_2} E_{i,l_1} E_{i,l_2}$, where $E_{i,0} = 1$. Then for any family $(v_\iota)_{\iota \in I_\infty} \subseteq \mathbb{Q}(v)$, there is a bilinear form $\{\bullet, \bullet\}$ on F such that

- $\{E, E'\} = 0$ if $|E| \neq |E'|$,
- $\{E_\iota, E_\iota\} = v_\iota, \forall \iota \in I_\infty$,
- $\{EE', E''\} = \{E \otimes E', \delta(E'')\}, \forall E, E', E'' \in \mathrm{F}$.

It turns out that $\sum_{l_1+l_2=-(\iota,j)+1} (-1)^{l_1} \frac{E_{j,1}^{l_1}}{l_1!} E_\iota \frac{E_{j,1}^{l_2}}{l_2!}$ is in the radical of $\{\bullet, \bullet\}$.

Definition 2.14. Let $\widetilde{\mathcal{U}}^+$ be the quotient of F by the ideal generated by the above element and the commutators $[E_{i,l}, E_{i,k}]$ for $\omega_i = 1$. Then $\{\bullet, \bullet\}$ is well defined on $\widetilde{\mathcal{U}}^+$. Let $\mathcal{U}^+$ be the quotient of $\widetilde{\mathcal{U}}^+$ by the radical of $\{\bullet, \bullet\}$.

Theorem 2.15 (see [1, Th. 3.34]). *There is an isomorphism of algebras*

$$\begin{aligned}\phi:\ & \mathcal{U}^+_{v=1} \to \mathscr{M}_0,\\ & E_{i,(l)} \mapsto 1_{i,(l)}, \quad i \in I^{im},\\ & E_{i,1} \mapsto 1_{i,1}, \quad i \in I^{re}.\end{aligned}$$

Definition 2.16. The semicanonical basis of $\mathcal{U}^+_{v=1}$ is $\phi^{-1}(\{f_Z | Z \in \mathrm{Irr}(\mathbf{M}^{sp}_{\Pi_Q})\})$.

2.5. Semicanonical basis of $\mathcal{H}^0$

We have already seen that for an appropriate subspace $\mathbf{M}^{sp}_{\overline{Q},\gamma} \subset \mathbf{M}_{\overline{Q},\gamma}$, the degree 0 part $\mathcal{H}^0$ of COHA is a subalgebra. In particular, we can take $\mathbf{M}^{sp}_{\overline{Q},\gamma}$ to be the space of seminilpotent representations of $\overline{Q}$. Then $\mathbf{M}^{sp}_{\Pi_Q,\gamma}$ is the space of seminilpotent representations in $\mathbf{M}_{\Pi_Q,\gamma}$, and $\dim(\mathbf{M}^{sp}_{\Pi_Q,\gamma}/\mathbf{G}_\gamma) = -\chi_Q(\gamma,\gamma)$, so the classes of irreducible components $\{[Z] | Z \in \mathrm{Irr}(\mathbf{M}^{sp}_{\Pi_Q,\gamma})\}$ lie in $\mathcal{H}^0$. In fact, these classes form a basis of $\mathcal{H}^0$ by the following theorem.

Theorem 2.17. *Let X be a scheme with top-dimensional irreducible components $\{C^k\}$, and a connected algebraic group G acts on it. Then $H^{2\mathrm{top}}_{c,G}(X)$ has a basis one-to-one corresponding to $\{C^k\}$, where top is the dimension of the stack X/G.*

Proof. Choose an embedding $G \hookrightarrow GL(n,\mathbb{C})$. Let $fr(n,N)$ be the space of n-tuples of linearly independent vectors in $\mathbb{C}^N$ for $N \geqslant n$. Then $X \times fr(n,N)$ has irreducible components $\{C^k \times fr(n,N)\}$, thus $X \times_G fr(n,N) = (X \times fr(n,N))/G$ has irreducible components $\{\overline{C^k}\}$ one to one corresponding to $\{C^k\}$ since G is irreducible. Then the Borel–Moore homology $H^{BM}_{2\bullet}(X \times_G fr(n,N))$ has a basis $\{[\overline{C^k}]\}$, where $\bullet = \dim(X) + \dim(fr(n,N)) - \dim G$, implying that $H^{2\bullet}_c(X \times_G fr(n,N))^\vee = H^{BM}_{2\bullet}(X \times_G fr(n,N))$ has basis one to one corresponding to $\{C^k\}$ (For details of Borel–Moore homology, see [4, Section 2.6].) Then $H^{2top}_{c,G}(X) = \lim_{N\to\infty} H^{2\bullet}_c(X \times_G fr(n,N)) \otimes \mathbb{T}^{-\dim fr(n,N)}$ has basis one-to-one corresponding to $\{C^k\}$, where top $= \bullet - \dim(fr(n,N)) = \dim(X/G)$. □

Definition 2.18. We call the basis defined above the *semicanonical basis* of the subalgebra $\mathcal{H}^0$.

Given an element $\mathcal{F}$ in $\mathscr{D}^b(X)$ with constructible cohomology, and $x \in X$, the function $\chi(\mathcal{F})(x) = \chi(\mathcal{F}_x) = \sum_i (-1)^i \dim(H^i(\mathcal{F}_x))$ is constructible. Moreover, the standard operations (pullback, pushforward, etc.) in $\mathscr{D}^b(X)$ and the corresponding operations on constructible functions are compatible.

Recall the family of constructible functions $\{f_Z | Z \in \mathrm{Irr}(\mathbf{M}^{sp}_{\Pi_Q})\}$. Then $U_Z = f_Z^{-1}(1)$ is constructible. Let $f_{Z,N}$ be the characteristic function of $\overline{(U_Z, \mathbf{G}_\gamma)_N}$, and $\mathbb{Q}_{Z,N}$ be the constant sheaf on $\overline{(U_Z, \mathbf{G}_\gamma)_N}$. Since the operations on constructible functions and constructible sheaves agree, there is an isomorphism of algebras $\Psi : \mathcal{H}^0 \to \mathscr{M}_0^{op}, [Z] \mapsto f_Z$. It is obtained by taking the dual of compactly supported cohomology and passing to the limit.

Furthermore, notice that $\mathcal{H}^0 \simeq (\mathcal{U}_{v=1}^+)^{op}$, and that Lusztig's product $*$ is opposite to the product of COHA (see the end of Section 2.3).

The semicanonical basis of $\mathcal{H}^0$ is compatible with a certain filtration. More precisely, we have the following result.

Theorem 2.19. *Fix $d = (d_i) \in \mathbb{Z}^I_{\geqslant 0}$. Then the subspace spanned by*

$$\{[Z] | \exists i, s.t. |\varepsilon_i(Z)| \geqslant d_i\}$$

coincides with $\sum_{i \in I, |c| = d_i} \mathcal{H}^0[Z_{i,c}]$, *where* $Z_{i,c} \in \mathrm{Irr}(\mathbf{M}^{sp}_{\Pi_Q, le_i})$ *is the irreducible component corresponding to c (defined in Section 2.3), and $c = l$ if $i \in I^{re}$.*

Proof. By definitions, $\sum_{i \in I, |c| = d_i} \mathcal{H}^0[Z_{i,c}]$ is contained in the subspace spanned by $\{[Z] | \exists i, s.t. |\varepsilon_i(Z)| \geqslant d_i\}$. To prove the reverse inclusion it suffices to show that for any $i \in I$, $\gamma \in \mathbb{Z}^I_{\geqslant 0}$, and $[Z] \in \mathcal{H}^0$ such that $Z \in \mathrm{Irr}(\mathbf{M}^{sp}_{\Pi_Q, \gamma})$ and $|\varepsilon_i(Z)| = l$, we have $[Z] \in \sum_{|c| = l} \mathcal{H}^0[Z_{i,c}]$. We use descending induction on $l \leqslant \gamma^i$. For above Z, we have $\gamma - le_i \in \mathbb{N}^I$, and by the proof of [1, Pro. 1.18], there exists a unique $Z' \in \mathrm{Irr}(\mathbf{M}^{sp}_{\Pi_Q, \gamma - le_i})$ and $Z_{i,c} \in \mathrm{Irr}(\mathbf{M}^{sp}_{\Pi_Q, le_i})$ such that $|\varepsilon_i(Z')| = 0$ and $[Z'][Z_{i,c}] = Z + \sum_{|\varepsilon_i(\widetilde{Z})| > l} a_{\widetilde{Z}}[\widetilde{Z}]$ for some $a_{\widetilde{Z}} \in \mathbb{Q}$. By applying the induction hypothesis to $\widetilde{Z}$ we have that the subspace spanned by $\{[Z] | \exists i, s.t. |\varepsilon_i(Z)| \geqslant d_i\}$ is contained in $\sum_{i \in I, |c| = d_i} \mathcal{H}^0[Z_{i,c}]$. Thus the two subspaces coincide. □

The dual of representations of Π_Q induces a bijection $* : \mathrm{Irr}(\mathbf{M}^{sp}_{\Pi_Q, \gamma}) \to \mathrm{Irr}(\mathbf{M}^{sp}_{\Pi_Q, \gamma}), Z \mapsto Z^*$, thus an antiautomorphism of $\mathcal{H}^0$. Then the dual of the above theorem holds:

Theorem 2.20. *The subspace spanned by $\{[Z] | \exists i, s.t. |\varepsilon_i(Z^*)| \geqslant d_i\}$ coincides with* $\sum_{i \in I, |c| = d_i} [Z_{i,c}]\mathcal{H}^0$.

3. 2*CY* categories and Donaldson–Thomas series

In Section 2 we discussed the semicanonical basis obtained as a result of the dimensional reduction from 3*CY* category to the 2-dimensional category. In this section we are going to discuss DT-series for 2*CY* categories.

3.1. Motivic stack functions and motivic Hall algebras: reminder

Let X be a constructible set over a field $\mathbf{k}$ of characteristic zero, G an affine algebraic group acting on X. In this section we are going to recall the definition of the abelian group of stack functions $\mathrm{Mot}_{st}((X,G))$ following [12, Section 4] (see also [9] for a different exposition).

Let us consider the following 2-category of constructible stacks over $\mathbf{k}$. Objects are pairs (X,G), where X is a constructible set, and G is an affine algebraic group acting on it. The category of 1-morphisms $\mathrm{Hom}((X_1,G_1),(X_2,G_2))$ consists of pairs (Z,f), where Z is a $G_1\times G_2$-constructible set such that $\{e\}\times G_2$ acts freely on Z in such a way that we have the induced G_1-equivariant isomorphism $Z/G_2\simeq X_1$, and $f:Z\to X_2$ is a $G_1\times G_2$-equivariant map with trivial action of G_1 on X_2. Furthermore, objects of $\mathrm{Hom}((X_1,G_1),(X_2,G_2))$ form naturally a groupoid. The 2-category of constructible stacks carries a direct sum operation induced by disjoint union of stacks.

After the above preliminaries we define the group of motivic stack functions $\mathrm{Mot}_{st}((X,G))$ as the abelian group generated by isomorphism classes of 1-morphisms of stacks $[(Y,H)\to(X,G)]$ with the fixed target (X,G), subject to the relations

- $[((Y_1,H_1)\sqcup(Y_2,H_2))\to(X,G)]=[(Y_1,H_1)\to(X,G)]+[(Y_2,H_2)\to(X,G)]$,
- $[(Y_2,H)\to(X,G)]=[(Y_1\times\mathbb{A}^d_k,H)\to(X,G)]$ if $Y_2\to Y_1$ is an H-equivariant constructible vector bundle of rank d.

One can define pullbacks, pushforwards and fiber products of elements of $\mathrm{Mot}_{st}((X,G))$ in the natural way (see loc. cit.).

Let $\mathcal{C}$ be an ind-constructible locally regular (e.g., locally Artin) triangulated A_∞-category over a field $\mathbf{k}$ (see [12]). Then the stack of objects admits a countable decomposition into the union of quotient stacks $\mathcal{O}b(\mathcal{C})=\sqcup_{i\in I}(Y_i,GL(N_i))$, where Y_i is a reduced algebraic scheme acted by the group $GL(N_i)$.

Definition 3.1 (cf. [12]). The motivic Hall algebra $H(\mathcal{C})$ is the $\mathrm{Mot}(\mathrm{Spec}(\mathbf{k}))$-module

$$\bigoplus_{i\in I}\mathrm{Mot}_{st}(Y_i,GL(N_i))[\mathbb{L}^n,n<0]$$

(i.e., we extend the direct sum of the groups of motivic stack functions by adding negative powers of the Lefschetz motive $\mathbb{L}$), endowed with the product defined below.

The product is defined as follows. Let us denote $\dim\mathrm{Ext}^i(E,F)$ by $(E,F)_i$, and use the truncated Euler characteristic $(E,F)_{\le N}=\sum_{i\le N}(-1)^i(E,F)_i$. Let $[\pi_i:Y_i\to\mathcal{O}b(\mathcal{C})], i=1,2$ be two elements of $H(\mathcal{C})$, then for any $n\in\mathbb{Z}$ we have constructible sets

$$W_n=\{(y_1,y_2,\alpha)|y_i\in Y_i,\alpha\in\mathrm{Ext}^1(\pi_2(y_2),\pi_1(y_1)),(\pi_2(y_2),\pi_1(y_1))_{\le 0}=n\}.$$

Then $[\mathrm{tot}((\pi_1\times\pi_2)^*(\mathcal{EXT}^1))\to\mathcal{O}b(\mathcal{C})]=\sum_{n\in\mathbb{Z}}[W_n\to\mathcal{O}b(\mathcal{C})]$. Define the product $[Y_1\to\mathcal{O}b(\mathcal{C})]\cdot[Y_2\to\mathcal{O}b(\mathcal{C})]=\sum_{n\in\mathbb{Z}}[W_n\to\mathcal{O}b(\mathcal{C})]\mathbb{L}^{-n}$, where the map $W_n\to\mathcal{O}b(\mathcal{C})$ is given by $(y_1,y_2,\alpha)\mapsto\mathrm{Cone}(\alpha:\pi_2(y_2)[-1]\to\pi_1(y_1))$.

Theorem 3.2 (see [12, Prop. 10]). *The algebra $H(\mathcal{C})$ is associative.*

For a constructible stability condition on $\mathcal{C}$ with an ind-constructible class map $cl : K_0(\mathcal{C}) \to \Gamma$, a central charge $Z : \Gamma \to \mathbb{C}$, a strict sector $V \subset \mathbb{R}^2$ and a branch Log of the logarithm function on V, we have (see [12]) the category $\mathcal{C}_V := \mathcal{C}_{V,\mathrm{Log}}$ generated by semistables with the central charge in V. Then we define the corresponding completed motivic Hall algebra

$$\widehat{H}(\mathcal{C}_V) := \prod_{\gamma \in (\Gamma \cap C(V,Z,Q)) \cup \{0\}} H(\mathcal{C}_V \cap cl^{-1}(\gamma)).$$

It contains an invertible element $A_V^{Hall} = 1 + \cdots = \sum_{i \in I} \mathbf{1}_{(Ob(\mathcal{C}_V) \cap Y_i, GL(N_i))}$, where 1 comes from the zero object. The element A_V corresponds (roughly) to the sum over all isomorphism classes of objects of $\mathcal{C}_V$, each counted with the weight given by the inverse to the motive of the group of automorphisms.

Theorem 3.3 (see [12, Prop. 11]). *The elements A_V^{Hall} satisfy the Factorization Property:*

$$A_V^{\mathrm{Hall}} = A_{V_1}^{\mathrm{Hall}} \cdot A_{V_2}^{\mathrm{Hall}}$$

for a strict sector $V = V_1 \sqcup V_2$ (decomposition in the clockwise order).

Let us fix the following data:

(1) a triple $(\Gamma, \langle \bullet, \bullet \rangle, Q)$ consisting of a free abelian group Γ of finite rank endowed with a bilinear form $\langle \bullet, \bullet \rangle : \Gamma \otimes \Gamma \to \mathbb{Z}$, and a quadratic form Q on $\Gamma_{\mathbb{R}} = \Gamma \otimes \mathbb{R}$,
(2) an ind-constructible, $\mathrm{Gal}(\overline{\mathbf{k}}/\mathbf{k})$-equivariant homomorphism $cl_{\overline{\mathbf{k}}} : K_0(\mathcal{C}(\overline{\mathbf{k}})) \to \Gamma$ compatible with the Euler form of $\mathcal{C}$ and the bilinear form $\langle \bullet, \bullet \rangle$,
(3) a constructible stability condition $\sigma \in \mathrm{Stab}(\mathcal{C}, cl)$ compatible with the quadratic form Q in the sense that $Q|_{\mathrm{Ker}(Z)} < 0$ and $Q(cl_{\overline{\mathbf{k}}}(E)) \geq 0, \forall E \in \mathcal{C}^{ss}(\overline{\mathbf{k}})$.

We define the quantum torus $\mathcal{R}_{\Gamma,R}$ over a given commutative unital ring R containing an invertible symbol $\mathbb{L}^{\frac{1}{2}}$ as an R-linear associative algebra $\mathcal{R}_{\Gamma,R} := \bigoplus_{\gamma \in \Gamma} R \cdot \widehat{e}_\gamma$, where the generators $\widehat{e}_\gamma, \gamma \in \Gamma$ satisfy the relations

$$\widehat{e}_{\gamma_1} \widehat{e}_{\gamma_2} = \mathbb{L}^{\frac{1}{2}\langle \gamma_1, \gamma_2 \rangle} \widehat{e}_{\gamma_1 + \gamma_2}, \quad \widehat{e}_0 = 1.$$

For any strict sector $V \subset \mathbb{R}^2$, we define the quantum torus associated with V by $\mathcal{R}_{V,R} := \prod_{\gamma \in \Gamma \cap C_0(V,Z,Q)} R \cdot \widehat{e}_\gamma$, where $C_0(V,Z,Q) := C(V,Z,Q) \cup \{0\}$, and $C(V,Z,Q)$ is the convex cone generated by $S(V,Z,Q) = \{x \in \Gamma_{\mathbb{R}} \setminus \{0\} | Z(x) \in V, Q(x) \geq 0\}$.

In the case when $\mathcal{C}$ is a $3CY$ category, one can define a homomorphism from the algebra $\widehat{H}(\mathcal{C}_V)$ to an appropriate motivic quantum torus (the word "motivic" here means that the coefficient ring R is a certain ring of motivic functions). This homomorphism was defined in [12] via the motivic Milnor fiber of the potential of the $3CY$ category. The notion of motivic DT-series was also introduced in the loc. cit.

It was later shown in [13] that in the case of quivers with potential one can define motivic DT-series differently, using equivariant critical cohomology (cf. our Section 2). In that case instead of the motivic Hall algebra one uses COHA.

3.2. A class of $2CY$ categories

Let us consider a class of 2-dimensional Calabi–Yau categories $\mathcal{C}$ which are:

1) Ind-constructible and locally ind-Artin in the sense of [12].
2) Endowed with a constructible homomorphism of abelian groups (class map) $cl : K_0(\mathcal{C}) \to \Gamma$, where $\Gamma \simeq \mathbb{Z}^I$ carries a symmetric integer bilinear form $\langle \bullet, \bullet \rangle$, and the class map cl satisfies
$$\langle cl(E), cl(F) \rangle = \chi(E,F) := \sum_{i\in\mathbb{Z}} (-1)^i \dim \mathrm{Ext}^i(E,F).$$
3) Generated by a spherical collection $\mathcal{E} = (E_i)_{i\in I}$ in the sense of loc. cit. such that $cl(E_i) \in \Gamma_+ \simeq \mathbb{Z}^I_{\geq 0}$. This means that $\mathrm{Ext}^\bullet(E_i, E_i) \simeq H^\bullet(S^2)$, and that $\mathrm{Ext}^m(E_i, E_j)$ can be non-trivial for $m = 1$ only as long as $i \neq j$.
4) For any $\gamma \in \Gamma_+$ the stack $\mathcal{C}_\gamma(\mathcal{E})$ of objects F of the heart of the t- structure corresponding to $(E_i)_{i\in I}$ such that $cl(F) = \gamma$ is a countable disjoint union of Artin stacks of dimensions less or equal than $-\frac{1}{2}\langle \gamma, \gamma \rangle$.
5) For any strict sector $V \subset \mathbb{R}^2$ with the vertex at $(0,0)$, and a constructible stability central charge $Z : \Gamma \to \mathbb{C}$ such that $\mathrm{Im}(Z(E_i)) := Z(cl(E_i)) \in V, i \in I$, the stack of objects of the category $\mathcal{C}_V$ generated by semistable objects with central charges in V is a finite union of Artin stacks satisfying the inequality of 4) above.

With the category from our class one can associate a symmetric quiver (vertices correspond to spherical objects E_i and arrows correspond to a basis in $\mathrm{Ext}^1(E_i, E_j)$). Similarly to [12], Section 8 one can prove a classification theorem for our categories in terms of Ginzburg algebras associated with quivers. Many $2CY$ categories which appear in "nature" belong to our class. For example, if Q is not an ADE quiver, then the derived category of finite-dimensional representations of Π_Q belongs to our class. Without any restrictions on Q one can construct a $2CY$ category as the category of dg-modules over the corresponding Ginzburg algebra.

3.3. Stability conditions and braid group action

Assume that $\mathcal{C}$ is a $2CY$ category from our class. We consider an open subset of the space $\mathrm{Stab}(\mathcal{C})$ of stability conditions which is defined as $U := \prod_{i\in I}(\mathrm{Im}\, z_i > 0)$, i.e., it is a product of upper half-planes. A point $Z = (z_i)_{i\in I} \in U$ defines the central charge $Z : \Gamma := \mathbb{Z}^I \to \mathbb{C}$ which maps classes of spherical generators to the open upper half-plane (hence the stability condition is determined by Z and the t-structure in $\mathcal{C}$ generated by $(E_i)_{i\in I}$).

Recall that with every $i_0 \in I$ we can associate an autoequivalence of $\mathcal{C}$ (called *reflection functor*) by the formula
$$R_{E_{i_0}} : F \mapsto \mathrm{Cone}(\mathrm{Ext}^\bullet(E_{i_0}, F) \otimes F \to F).$$

Then $R_{E_{i_0}}(E_{i_0}) = E_{i_0}[-1]$, and $R_{E_{i_0}}(E_j), j \neq i_0$ is determined as the middle term in the extension
$$0 \to E_j \to R_{E_{i_0}}(E_j) \to E_{i_0} \otimes \mathrm{Ext}^1(E_{i_0}, E_j) \to 0.$$

The inverse reflection functor $R_{E_{i_0}}^{-1}$ is given by

$$R_{E_{i_0}}^{-1}(E_{i_0}) = E_{i_{i_0}}[1],$$
$$0 \to E_{i_0} \otimes \mathrm{Ext}^1(E_{i_0}, E_j) \to R_{E_{i_0}}^{-1}(E_j) \to E_j \to 0.$$

Reflection functors $R_{E_i}, i \in I$ generate a subgroup $\mathrm{Braid}_{\mathcal{C}} \subset \mathrm{Aut}(\mathcal{C})$, which induces an action on $\mathrm{Stab}(\mathcal{C})$. The orbit $D := \mathrm{Braid}_{\mathcal{C}}(U) \subset \mathrm{Stab}(\mathcal{C})$ is the union of consecutive "chambers" obtained one from another one by reflection functor R_{E_j}. Such consecutive chambers have a common real codimension one boundary singled out by the condition $Im\, Z(E_j) = 0$.

Remark 3.4. The group $\mathrm{Braid}_{\mathcal{C}}$ plays a role of the braid group (or Weyl group) in the theory of Kac–Moody algebras. If we add also the group $\mathbb{Z}$ of shifts $F \mapsto F[n], n \in \mathbb{Z}$ then we obtain an affine version of the braid group $\mathrm{Braid}_{\mathcal{C}} \times \mathbb{Z}$. In some examples $\mathbb{Z} \subset \mathrm{Braid}_{\mathcal{C}}$.

3.4. Motivic DT-series for $2CY$ categories

Let $\mathcal{C}$ be an ind-constructible locally regular $2CY$ category over $\mathbf{k}$. Let us fix $R = \mathrm{Mot}(\mathrm{Spec}(\mathbf{k}))[\mathbb{L}^{\frac{1}{2}}, \mathbb{L}^{-1}, [GL(n)]_{n\geqslant 1}^{-1}]$ as the ground ring for the quantum torus $\mathcal{R}_{\Gamma,R}$. We will denote the latter by $\mathcal{R}_\Gamma$. It is a commutative algebra generated by elements $\widehat{e}_\gamma, \gamma \in \Gamma$ such that $\widehat{e}_{\gamma_1+\gamma_2} = \widehat{e}_{\gamma_1}\widehat{e}_{\gamma_2}, \widehat{e}_0 = 1$. Let us also fix a stability condition on $\mathcal{C}$ with the central charge $Z : \Gamma \to \mathbb{C}$.

Definition 3.5. The motivic weight $\omega \in \mathrm{Mot}(\mathcal{O}b(\mathcal{C}))$ is defined by

$$\omega(E) = \mathbb{L}^{\frac{1}{2}(\chi(E,E))}.$$

Then we have the following result.

Proposition 3.6. *The map $\Phi : H(\mathcal{C}) \to \mathcal{R}_\Gamma$ given by $\Phi(\nu) = (\nu, \omega)\widehat{e}_\gamma, \nu \in H(\mathcal{C})_\gamma$ satisfies the condition $\Phi(\nu_1 \cdot \nu_2) = \Phi(\nu_1)\Phi(\nu_2)$ for $\mathrm{Arg}(\gamma_1) > \mathrm{Arg}(\gamma_2)$, where $\nu_i \in H(\mathcal{C})_{\gamma_i}$. (Here $(\bullet, \bullet)$ is the pairing between motivic measures and motivic functions.)*

In other words, Φ can be written as

$$[\pi : Y \to \mathcal{O}b(\mathcal{C})] \mapsto \int_Y \mathbb{L}^{\frac{1}{2}\chi(\pi(y),\pi(y))}\widehat{e}_{cl(\pi(y))}.$$

Proof. It suffices to prove the theorem for $\nu_{E_i} = [\delta_{E_i} : pt \to \mathcal{O}b(\mathcal{C})]$, where $\delta_{E_i}(pt) = E_i \in \mathcal{O}b(\mathcal{C})$. Recall that we denote $\dim \mathrm{Ext}^i(E, F)$ by $(E, F)_i, i \in \mathbb{Z}$.

We have $\Phi(\nu_{E_i}) = \mathbb{L}^{\frac{1}{2}\chi(E_i,E_i)}\widehat{e}_{\gamma_i}$, which implies that

$$\Phi(\nu_{E_1})\Phi(\nu_{E_2}) = \mathbb{L}^{\frac{1}{2}(\chi(E_1,E_1)+\chi(E_2,E_2))}\widehat{e}_{\gamma_1+\gamma_2}.$$

On the other hand,

$$\nu_{E_1} \cdot \nu_{E_2} = \mathbb{L}^{-(E_2,E_1)_{\leq 0}}[\pi_{21} : \mathrm{Ext}^1(E_2, E_1) \to \mathcal{O}b(\mathcal{C})].$$

Then

$$\begin{aligned}
&\Phi(\nu_{E_1}\cdot\nu_{E_2})\\
&=\mathbb{L}^{-(E_2,E_1)_{\le 0}}\int_{\alpha\in \mathrm{Ext}^1(E_2,E_1)}\mathbb{L}^{\frac12\chi(E_\alpha,E_\alpha)}\widehat{e}_{\gamma_1+\gamma_2}\\
&=\mathbb{L}^{-(E_2,E_1)_{\le 0}}\mathbb{L}^{\frac12(\chi(E_1,E_1)+\chi(E_2,E_2)+\chi(E_1,E_2)+\chi(E_2,E_1))}\int_{\alpha\in \mathrm{Ext}^1(E_2,E_1)}\widehat{e}_{\gamma_1+\gamma_2}\\
&=\mathbb{L}^{-(E_2,E_1)_{\le 0}+\frac12(\chi(E_1,E_1)+\chi(E_2,E_2))+\chi(E_2,E_1)}\mathbb{L}^{(E_2,E_1)_1}\widehat{e}_{\gamma_1+\gamma_2}\\
&=\mathbb{L}^{\frac12(\chi(E_1,E_1)+\chi(E_2,E_2))+(E_2,E_1)_2}\widehat{e}_{\gamma_1+\gamma_2}
\end{aligned}$$

If $\mathrm{Arg}(\gamma_1)>\mathrm{Arg}(\gamma_2)$, then $(E_2,E_1)_2=(E_1,E_2)_0=0$. Thus

$$\Phi(\nu_{E_1}\cdot\nu_{E_2})=\Phi(\nu_{E_1})\Phi(\nu_{E_2}).$$ □

Recall the categories $\mathcal{C}_V$ and set $V=l$ be a ray. For a generic central charge Z let us consider the generating function

$$\begin{aligned}
A_l^{\mathrm{mot}}&=\sum_{[E],E\in Ob(\mathcal{C}_l)}\frac{\omega(E)\widehat{e}_{cl(E)}}{[\mathrm{Aut}(E)]}\\
&=\sum_{[E],E\in Ob(\mathcal{C}_l)}\mathbb{L}^{\frac12(\chi(E,E))}\frac{t^{cl(E)}}{[\mathrm{Aut}(E)]},
\end{aligned}$$

where $t=\widehat{e}_{\gamma_0}$ for a primitive γ_0 such that $Z(\gamma_0)\in l$ generates $Z(\Gamma)\cap l$ and $[\mathrm{Aut}(E)]$ denotes the motive of the group of automorphisms of E. More invariantly, $A_l^{\mathrm{mot}}=\Phi(A_l^{\mathrm{Hall}})$ where $A_l^{\mathrm{Hall}}\in H(\mathcal{C}_l)$ corresponds to the characteristic function of the stack of objects of the full subcategory $\mathcal{C}_l\subset\mathcal{C}$ generated by semistables E such that $Z(E)\in l$ (cf. loc. cit.).

Definition 3.7. We call A_l^{mot} the motivic DT-series of $\mathcal{C}$ corresponding to the ray l.

Suppose that $\mathcal{C}$ is associated with the preprojective algebra Π_Q. One can show that A_l^{mot} can be obtained from the motivic DT-series for the $3CY$ category associated with $(\widehat{Q},W)$ by the reduction to $\mathcal{C}$. Similarly to A_l^{mot} we define A_V^{mot} for any strict sector V.

Proposition 3.6 implies that the series A_V^{mot} is the (clockwise) product of A_l^{mot} over all rays $l\subset V$. This can be also derived from the dimensional reduction and the results of [12].

Corollary 3.8. *The collections of elements $A_V^{\mathrm{mot}}=\Phi(A_V^{\mathrm{Hall}})$ parametrized by strict sectors $V\subset\mathbf{R}^2$ with the vertex in the origin satisfies the Factorization Property: if a strict sector V is decomposed into a disjoint union $V=V_1\sqcup V_2$ in the clockwise order, then $A_V^{\mathrm{mot}}=A_{V_1}^{\mathrm{mot}}A_{V_2}^{\mathrm{mot}}$.*

Proposition 3.9. *Motivic DT-series A_V^{mot} is constant on each connected component of the space of stability conditions.*

Proof. Similarly to the case of $3CY$ categories, each element A_V^{mot} does not change when we move in the space of stability conditions on $\mathcal{C}$ in such a way that central charges of semistable object neither enter nor leave the sector V. But in the case of $2CY$ categories the Euler form is symmetric, hence the motivic quantum torus is commutative. It follows that the wall-crossing formulas from [12] are trivial. This implies the result. □

For a $2CY$ category form our class one can construct the corresponding $3CY$ category (see Introduction). We expect that the motivic DT-series arising in this situation are quantum admissible in the sense of [13] and can be described in terms of the corresponding COHA (the latter is expected to exist for quite general $3CY$ categories, see [19]).

Therefore, by analogy with the case of $3CY$ categories, we can define DT-invariants $\Omega(\gamma)$ in $2CY$ case using (quantum) admissibility (see [13], Section 6) of our DT-series by the formula:

$$A_V^{\mathrm{mot}} = \mathrm{Sym}\left(\sum_{n\geq 0}\mathbb{L}^n \sum_{\gamma\neq 0, Z(\gamma)\in V}\Omega(\gamma)\widehat{e}_\gamma\right)$$

$$= \mathrm{Sym}\left(\frac{\sum_{\gamma\neq 0, Z(\gamma)\in V}\Omega(\gamma)\widehat{e}_\gamma}{1-\mathbb{L}}\right).$$

By Proposition 3.9 our motivic DT-invariants $\Omega(\gamma)$ depend only on the connected component of $\mathrm{Stab}(\mathcal{C})$ which contains Z. The Conjecture 3.10 (see the next subsection) says that $\Omega(\gamma)$ is (essentially) the same as Kac polynomial $a_\gamma(\mathbb{L})$ (or the motivic DT-invariant of the corresponding $3CY$ category, see Introduction).

Let us fix the connected component in $\mathrm{Stab}(\mathcal{C})$ which contains such central charge Z that for each spherical generator E_i of $\mathcal{C}$ we have $Z(E_i) = (0, \dots, 1, \dots, 0)$ (the only nontrivial element 1 at the ith place). We will call the corresponding t-structure *standard.* We denote the corresponding motivic DT-invariants by $\Omega_{\mathcal{C}}^{\mathrm{mot}}(\gamma)$.

3.5. Kac polynomial of a $2CY$ category

We can now introduce an analog of the Kac polynomial in the case of a $2CY$ category from our class following the ideas of [16].

Notice that the coefficient ring $\mathrm{Mot}(\mathrm{Spec}(\mathbf{k}))[\mathbb{L}^{\frac{1}{2}}, \mathbb{L}^{-1}, [GL(n)]^{-1}_{n\geqslant 1}]$ of the quantum torus $\mathcal{R}_\Gamma$ has a λ-ring structure, which can be lifted to the quantum torus (which is commutative in the case of $2CY$ categories). Recall that for a λ-ring we can introduce the operation of symmetrization by the formula:

$$\mathrm{Sym}(r) = \sum_{n\geqslant 0}\mathrm{Sym}^n(r) = \sum_{n\geqslant 0}(-1)^n\lambda^n(-r) = \Big(\sum_{n\geqslant 0}(-1)^n\lambda^n(r)\Big)^{-1}.$$

For any ray $l \subset \mathbb{H}_+$, where $\mathbb{H}_+$ is the upper half-plane, we have the (quantum) admissible element A_l^{mot}.

Let $\mathcal{C}$ be a $2CY$ category from our class. We fix the standard t-structure. Recall the motivic DT-series A_l^{mot}.

Conjecture 3.10. *There exist elements*

$$a_\gamma^{\mathrm{mot}}(\mathbb{L}) \in \mathrm{Mot}(\mathrm{Spec}(\mathbf{k}))[\mathbb{L}^{\frac{1}{2}}, \mathbb{L}^{-1}, [GL(n)]_{n\geqslant 1}^{-1}]$$

which are polynomials in $\mathbb{L}$ *and such that the following formula holds in the (commutative) motivic quantum torus:*

$$A_l^{\mathrm{mot}} = Sym\left(\frac{\sum_{\gamma, Z(\gamma)\in l}(-a_\gamma^{\mathrm{mot}}(\mathbb{L})\cdot \mathbb{L})\widehat{e}_\gamma}{1-\mathbb{L}}\right).$$

Furthermore, there exists a $3CY$ *category* $\mathcal{B}$ *such that the elements* $a^{\mathrm{mot}}(\mathbb{L})$ *coincide with motivic* DT*-invariants with respect to some stability condition on* $\mathcal{B}$.

Some related results can be found in [3], [6] [8], and especially in [16]. In fact Theorem 5.1 from [16] establishes the conjecture in the framework of quivers. More precisely, if $\mathcal{C}$ is the $2CY$ category associated with the preprojective algebra of a quiver, then for its standard t-structure the element $a_\gamma^{\mathrm{mot}}(\mathbb{L})$ coincides with the Kac polynomial $a_\gamma(\mathbb{L})$ of the Kac–Moody algebra corresponding to the quiver.

We plan to discuss the general case in the future work.

4. Appendix by Ben Davison: Comparison of Schiffmann–Vasserot product with Kontsevich–Soibelman product

Let Q be a quiver with the set of vertices $Q_0 := I$, and the set of arrows $Q_1 = \Omega$. We can construct the double quiver $\overline{Q}$ and the triple quiver with potential $(\widehat{Q}, W)$ as in Section 2.2. Recall the representation spaces $\mathbf{M}_{\overline{Q},\gamma}$ and $\mathbf{M}_{\widehat{Q},\gamma}$. Since $\mathbf{M}_{\overline{Q},\gamma} \cong T^*(\mathbf{M}_{Q,\gamma})$, there is a natural action of the group $\mathbf{G}_\gamma = \prod_{i\in I} GL(\gamma^i, \mathbb{C})$ on the cotangent space. Let L be the subquiver of $\widehat{Q}$ containing only the loops l_i, and denote its representation space $\mathbf{M}_{L,\gamma}$.

Let μ_γ denote the moment map $\mathbf{M}_{\overline{Q},\gamma} \to \mathfrak{gl}_\gamma$ or the map $\mathbf{M}_{\widehat{Q},\gamma} \to \mathfrak{gl}_\gamma$ both given by evaluating at $\sum_{a\in Q_1}[a, a^*]$.

In [18] the following special case was considered. Let Q be the Jordan quiver (i.e., the quiver with one vertex and one loop). We denote by C_n the commuting variety, i.e., the variety of pairs of commuting $n\times n$ matrices. Then $C_n = \mu_n^{-1}(0)$, where $\gamma = n$ is the dimension vector.

4.1. Schiffmann–Vasserot product

Notice that the affine algebraic variety $\mathbf{M}_{\overline{Q},\gamma}$ carries the action of the 2-torus $T := (\mathbb{C}^*)^2$ given by

$$(z_1, z_2)\cdot \rho(a) = z_1\rho(a)$$
$$(z_1, z_2)\cdot \rho(a^*) = z_2\rho(a^*),$$

where $\rho : \mathbb{C}Q \to Gl(V_\gamma)$ is a representation of $\mathbb{C}Q$. Define $\tilde{\mathbf{G}}_\gamma := \mathbf{G}_\gamma \times T$. For $\gamma_1, \gamma_2 \in \mathbb{N}^I$ with $\gamma_1 + \gamma_2 = \gamma$ we define $\tilde{\mathbf{G}}_{\gamma_1,\gamma_2} = \mathbf{G}_{\gamma_1,\gamma_2} \times T$, and $\tilde{\mathbf{G}}_{\gamma_1 \times \gamma_2} = \mathbf{G}_{\gamma_1} \times \mathbf{G}_{\gamma_2} \times T$. Following Schiffmann and Vasserot [18] let us consider the space

$$\mathrm{H}_{c,\tilde{\mathbf{G}}_\gamma}(\mu_\gamma^{-1}(0))^\vee,$$

the vector dual to the compactly supported equivariant cohomology (equivalently, the equivariant Borel–Moore homology).

For $\gamma = \gamma_1 + \gamma_2$ denote by $\mathfrak{gl}_{\gamma_1,\gamma_2} \subset \mathfrak{gl}_\gamma$ the subspace of parabolic matrices. Denote by Z_{γ_1,γ_2} the space of triples $(a, b, c) \in \mathbf{M}_{\overline{Q},\gamma_1} \times \mathbf{M}_{\overline{Q},\gamma_2} \times \mathfrak{gl}_{\gamma_1,\gamma_2}$ such that $c|_{\mathfrak{gl}_{\gamma_1} \times \mathfrak{gl}_{\gamma_2}} = \mu_{\gamma_1}(a) \times \mu_{\gamma_2}(b)$.

Then we have the following diagram

$$\begin{array}{ccccc}
\mathbf{M}_{\overline{Q},\gamma} & \xleftarrow{\;h\;} & \mathbf{M}_{\overline{Q},\gamma_1,\gamma_2} & \xrightarrow{\;f'\;} & Z_{\gamma_1,\gamma_2} \\
\uparrow & & \uparrow & & \uparrow g \\
\mu_\gamma^{-1}(0) & \xleftarrow[h|_{\mu^{-1}_{\gamma_1,\gamma_2}(0)}]{} & \mu^{-1}_{\gamma_1,\gamma_2}(0) & \longrightarrow & \mu_{\gamma_1}^{-1}(0) \times \mu_{\gamma_2}^{-1}(0)
\end{array}$$

where f' is the map

$$(\alpha) \mapsto (\alpha|_{\mathbf{M}_{\overline{Q},\gamma_1}}, \alpha|_{\mathbf{M}_{\overline{Q},\gamma_2}}, \mu_\gamma(\alpha))$$

and g is the map

$$(\alpha, \beta) \mapsto (\alpha, \beta, 0)$$

and we have set $\mu^{-1}_{\gamma_1,\gamma_2}(0) := \mu_\gamma^{-1}(0) \cap \mathbf{M}_{\overline{Q},\gamma_1,\gamma_2}$.

The rightmost square is Cartesian, meaning that when we restrict the natural morphism of sheaves

$$\left(f_! \mathbb{Q} \to \mathbb{Q}[\sum_{i \in I} \gamma_1^i \gamma_2^i - \sum_{a \in \overline{Q}_1} \gamma_1^{s(a)} \gamma_2^{t(a)}] \right) |_{\mu_{\gamma_1}^{-1}(0) \times \mu_{\gamma_2}^{-1}(0)}$$

we obtain a map

$$f'_* : \mathrm{H}_{c,\tilde{\mathbf{G}}_{\gamma_1,\gamma_2}}(\mu^{-1}_{\gamma_1,\gamma_2}(0)) \to \mathrm{H}_{c,\tilde{\mathbf{G}}_{\gamma_1,\gamma_2}}(\mu_{\gamma_1}^{-1}(0) \times \mu_{\gamma_2}^{-1}(0)).$$

We have also the map of sheaves on $\mathbf{M}_{\overline{Q},\gamma}$

$$\mathbb{Q}_{\mu_\gamma^{-1}(0)} \to h_* \mathbb{Q}_{\mu^{-1}_{\gamma_1,\gamma_2}(0)}$$

and since h is proper we may replace h_* with $h_!$ and after taking compactly supported cohomology we obtain a map

$$h^* : \mathrm{H}_{c,\tilde{\mathbf{G}}_{\gamma_1,\gamma_2}}(\mu_\gamma^{-1}(0)) \to \mathrm{H}_{c,\tilde{\mathbf{G}}_{\gamma_1,\gamma_2}}(\mu^{-1}_{\gamma_1,\gamma_2}(0)).$$

Via the inclusion $\mathbf{G}_{\gamma_1,\gamma_2} \hookrightarrow \mathbf{G}_\gamma$ we obtain a map

$$r : \mathrm{H}_{c,\tilde{\mathbf{G}}_\gamma}(\mu_\gamma^{-1}(0)) \to \mathrm{H}_{c,\tilde{\mathbf{G}}_{\gamma_1,\gamma_2}}(\mu_\gamma^{-1}(0)) \tag{5}$$

as in [13]. Finally the Schiffmann–Vasserot multiplication is given by setting

$$m_{SV} = (f'_* h^* r)^\vee$$

where we identify

$$\mathrm{H}_{c,\tilde{\mathbf{G}}_{\gamma_1,\gamma_2}}(\mu_{\gamma_1}^{-1}(0)\times\mu_{\gamma_2}^{-1}(0))^\vee \cong \mathrm{H}_{c,\tilde{\mathbf{G}}_{\gamma_1}}(\mu_{\gamma_1}^{-1}(0))^\vee \otimes_{\mathrm{H}_T(\mathrm{pt})} \mathrm{H}_{c,\tilde{\mathbf{G}}_{\gamma_2}}(\mu_{\gamma_2}^{-1}(0))^\vee$$

via the affine fibration $\mathbf{G}_{\gamma_1,\gamma_2}\to\mathbf{G}_{\gamma_1}\times\mathbf{G}_{\gamma_2}$ and the T-equivariant Künneth isomorphism.

4.2. Critical COHA product

Consider the quiver with potential $(\widehat{Q}, W=\sum_{a\in\Omega}[a,a^*]l)$ constructed in Section 2.2. The torus T acts on $\mathbf{M}_{\widehat{Q},\gamma}$ via

$$(z_1,z_2)\cdot\rho(a)=z_1\rho(a)$$
$$(z_1,z_2)\cdot\rho(a^*)=z_2\rho(a^*)$$
$$(z_1,z_2)\cdot\rho(l_i)=z_1^{-1}z_2^{-1}\rho(l_i).$$

Note that $Tr(W)_\gamma$ is invariant with respect to the T-action, and the natural projection $\mathbf{M}_{\widehat{Q},\gamma}\to\mathbf{M}_{\overline{Q},\gamma}$ is $\tilde{\mathbf{G}}_\gamma$-equivariant.

We consider the space

$$\bigoplus_\gamma \mathrm{H}^{\mathrm{crit}}_{c,\tilde{\mathbf{G}}_\gamma}(\mathbf{M}_{\widehat{Q},\gamma},W_\gamma)^\vee.$$

Consider the correspondence diagram

$$\mathbf{M}_{\widehat{Q},\gamma} \xleftarrow{\tau} \mathbf{M}_{\widehat{Q},\gamma_1,\gamma_2} \xrightarrow{\zeta} \mathbf{M}_{\widehat{Q},\gamma_1}\times\mathbf{M}_{\widehat{Q},\gamma_2}.$$

Applying $\varphi_{Tr(W)_\gamma}$ to the map of sheaves $\mathbb{Q}\to\tau_*\mathbb{Q}$ we obtain a map $\varphi_{Tr(W)_\gamma}\mathbb{Q}\to\tau_*\varphi_{Tr(W)_{\gamma_1,\gamma_2}}\mathbb{Q}$ by properness of τ, and a map

$$\tau^*:\mathrm{H}^{\mathrm{crit}}_{c,\tilde{\mathbf{G}}_{\gamma_1,\gamma_2}}(\mathbf{M}_{\widehat{Q},\gamma},W_\gamma)\to\mathrm{H}^{\mathrm{crit}}_{c,\tilde{\mathbf{G}}_{\gamma_1,\gamma_2}}(\mathbf{M}_{\widehat{Q},\gamma_1,\gamma_2},W_{\gamma_1,\gamma_2}).$$

Since ζ is an affine fibration, and $Tr(W)_{\gamma_1,\gamma_2}=(Tr(W)_{\gamma_1}\boxplus Tr(W)_{\gamma_2})\circ\zeta$, there is a natural shifted isomorphism

$$\mathrm{H}^{\mathrm{crit}}_{c,\tilde{\mathbf{G}}_{\gamma_1,\gamma_2}}(\mathbf{M}_{\widehat{Q},\gamma_1,\gamma_2},W_{\gamma_1,\gamma_2})\to\mathrm{H}^{\mathrm{crit}}_{c,\tilde{\mathbf{G}}_{\gamma_1,\gamma_2}}(\mathbf{M}_{\widehat{Q},\gamma_1}\times\mathbf{M}_{\widehat{Q},\gamma_2},W_{\gamma_1}\boxplus W_{\gamma_2}).$$

Changing the group as in (5), and composing the duals of the above two maps, we obtain the $H_T(\mathrm{pt})$-linear Kontsevich–Soibelman product m_{KS} (here we use the Thom–Sebastiani isomorphism rather than Künneth – via dimensional reduction this becomes a special case of the equivariant Künneth isomorphism):

$$\mathrm{H}^{\mathrm{crit}}_{c,\tilde{\mathbf{G}}_{\gamma_1}}(\mathbf{M}_{\widehat{Q},\gamma_1},W_{\gamma_1})^\vee\otimes_{\mathrm{H}_T(\mathrm{pt})}\mathrm{H}^{\mathrm{crit}}_{c,\tilde{\mathbf{G}}_{\gamma_2}}(\mathbf{M}_{\widehat{Q},\gamma_2},W_{\gamma_2})^\vee\xrightarrow{m_{KS,\gamma_1,\gamma_2}}\mathrm{H}^{\mathrm{crit}}_{c,\tilde{\mathbf{G}}_\gamma}(\mathbf{M}_{\widehat{Q},\gamma},W_\gamma)^\vee.$$

4.3. Dimensional reduction

We give $\mathbf{M}_{\widehat{Q},\gamma}$ a further $\mathbb{C}^*$-action as follows

$$z \cdot \rho(a) = \rho(a)$$
$$z \cdot \rho(a^*) = \rho(a^*)$$
$$z \cdot \rho(l_i) = z\rho(l_i).$$

Then $Tr(W)_\gamma$ is an equivariant function with respect to this action, where $\mathbb{C}^*$ acts on the target $\mathbb{C}$ via the weight one (scaling) action. There is a projection $\pi : \mathbf{M}_{\widehat{Q},\gamma} \to \mathbf{M}_{\overline{Q},\gamma}$ given by forgetting the action on the loops l_i. We write $\mathbf{M}_{\widehat{Q},\gamma} = \mathbf{M}_{\overline{Q},\gamma} \times \mathbf{M}_{L,\gamma}$. In this way we consider $\mu_\gamma^{-1}(0) \times \mathbf{M}_{L,\gamma}$ as a subspace of $\mathbf{M}_{\widehat{Q},\gamma}$. Note that $\mu_\gamma^{-1}(0) \times \mathbf{M}_{L,\gamma} \subset Tr(W)_\gamma^{-1}(0)$.

There is a canonical map $\varphi_{Tr(W)_\gamma} \mathbb{Q}_{\mathbf{M}_{\widehat{Q},\gamma}}[-1] \to \mathbb{Q}_{Tr(W)_\gamma^{-1}(0)}$. The dimensional reduction theorem of the appendix of [5] (see also [13] in the case of rapid decay cohomology) states that the induced map

$$\pi_!\big((\varphi_{Tr(W)_\gamma} \mathbb{Q}_{\mathbf{M}_{\widehat{Q},\gamma}}[-1] \to \mathbb{Q}_{Tr(W)_\gamma^{-1}(0)}\big)_{\mu_\gamma^{-1}(0)\times \mathbf{M}_{L,\gamma}}\big)$$

is an isomorphism, and so taking compactly supported cohomology we obtain a (shifted) isomorphism

$$\mathrm{H}^{\mathrm{crit}}_{c,\tilde{\mathbf{G}}_\gamma}(\mathbf{M}_{\widehat{Q},\gamma}, W_\gamma) \to \mathrm{H}_{c,\tilde{\mathbf{G}}_\gamma}(\mu_\gamma^{-1}(0)). \tag{6}$$

This establishes an isomorphism between the underlying vector spaces of the two COHAs (in fact it is an isomorphism of mixed Hodge structures, up to the already-mentioned cohomological shift and a Tate twist). The purpose of this appendix is to compare the two algebra structures m_{SV} and m_{KS}.

4.4. Proof of (almost) preservation of product

The following lemma accounts for the sign that appears in the comparison of m_{SV} and m_{KS}.

Lemma 4.1. *Let*

$$X = \mathrm{Spec}(\mathbb{C}[x_1, \dots, x_m]) \times \mathrm{Spec}(\mathbb{C}[y_1, \dots, y_m]) \times \mathrm{Spec}(\mathbb{C}[z_1, \dots, z_n])$$

and let

$$f = \sum_{i \le m} x_i y_i.$$

In the following diagram

$$\begin{array}{ccc}
\mathrm{H}_c(\mathrm{pt}, \mathbb{Q}) & \xleftarrow{\;a\;} & \mathrm{H}_c(Z(x_1, \dots, x_m), \mathbb{Q}) \\
\uparrow{\scriptstyle b} & & \uparrow{\scriptstyle c} \\
\mathrm{H}_c(Z(y_1, \dots, y_m), \mathbb{Q}) & \xleftarrow{\;d\;} & \mathrm{H}_c^{\mathrm{crit}}(X, f)
\end{array}$$

the maps a and b are the canonical shifted isomorphisms associated to affine fibrations, while the maps c and d are dimensional reduction isomorphisms. Then we have the equality

$$ac = bd \cdot (-1)^m.$$

Proof. By the Thom–Sebastiani isomorphism it is enough to prove the same result in the case $m = 1$ and $n = 0$, i.e., consider the maps

$$\begin{array}{ccc}
\mathrm{H}_c(\mathrm{pt}, \mathbb{Q}) & \xleftarrow[\pi_{x,*}]{} & \mathrm{H}_c(Z(x), \mathbb{Q}) \\
{\scriptstyle \pi_{y,*}} \uparrow & & \uparrow {\scriptstyle (i_x)^*_\varphi} \\
\mathrm{H}_c(Z(y), \mathbb{Q}) & \xleftarrow[(i_y)^*_\varphi]{} & \mathrm{H}_c^{\mathrm{crit}}(\mathbb{A}^2, xy)
\end{array}$$

where $i_x : Z(x) \hookrightarrow \mathbb{A}^2$ and $i_y : Z(y) \hookrightarrow \mathbb{A}^2$ are the obvious inclusions, and π_x, π_y are the projections to a point. From now on we adopt the notation that when we pull back or push forward vanishing cycle cohomology, we insert a φ as a subscript or superscript, as we will be comparing ordinary compactly supported cohomology with vanishing cycle cohomology and operations on the two should be distinguished by the notation. Now $\mathrm{H}_c^{\mathrm{crit}}(\mathbb{A}^2, xy)$ is concentrated in degree 2, and is explicitly described as follows. There is a distinguished triangle

$$\mathrm{H}_c^{\mathrm{crit}}(\mathbb{A}^2, xy) \to \mathrm{H}_c(Z(xy), \mathbb{Q}) \to \mathrm{H}_c(Z(xy-1), \mathbb{Q}).$$

The variety $Z(xy)$ has two components, so that $\mathrm{H}_c^2(Z(xy), \mathbb{Q}) \cong \mathbb{Q}^{\oplus 2}$, with $(1, 0)$ representing the fundamental class of $Z(x)$ and $(0, 1)$ representing the fundamental class of $Z(y)$. On the other hand, $Z(xy-1)$ has only one component, and we may write the map $\mathrm{H}_c^2(Z(xy), \mathbb{Q}) \to \mathrm{H}_c^2(Z(xy-1), \mathbb{Q})$ as follows

$$\mathbb{Q}^{\oplus 2} \xrightarrow{(a,b) \mapsto a+b} \mathbb{Q}$$

and deduce that $\mathrm{H}_c^{2,\mathrm{crit}}(\mathbb{A}^2, xy)$ is represented by the class $\alpha = (1, -1) \in \mathrm{H}_c(\{xy = 0\})$. The claim is then clear; α is sent to $1 \in \mathrm{H}_c(\mathrm{pt}, \mathbb{Q}) \cong \mathbb{Q}$ by $\pi_{x,*}(i_x)^*_\varphi$ and to -1 by $\pi_{y,*}(i_y)^*_\varphi$. □

Throughout the rest of the appendix we will use the following diagram.

$$\begin{array}{ccccc}
\mathbf{M}_{\widehat{Q},\gamma_1} \times \mathbf{M}_{\widehat{Q},\gamma_2} & \xleftarrow{a} & Z_{\gamma_1,\gamma_2} \times \mathbf{M}_{L,\gamma_1,\gamma_2} & \xleftarrow{b} & \mathbf{M}_{\widehat{Q},\gamma_1,\gamma_2} \\
{\scriptstyle c} \uparrow & & \downarrow {\scriptstyle d} & & \downarrow {\scriptstyle e} \\
\mathbf{M}_{\overline{Q},\gamma_1} \times \mathbf{M}_{\overline{Q},\gamma_2} \times \mathbf{M}_{L,\gamma} & \xrightarrow[f]{} & Z_{\gamma_1,\gamma_2} \times \mathbf{M}_{L,\gamma} & \xleftarrow{g} & \mathbf{M}_{\overline{Q},\gamma_1,\gamma_2} \times \mathbf{M}_{L,\gamma} \\
 & & & & \downarrow {\scriptstyle h} \\
 & & & & \mathbf{M}_{\widehat{Q},\gamma}
\end{array} \tag{7}$$

Remark 4.2. Each of these spaces carries a function induced by W, and regardless of the space we will call this function $Tr(W)$. For all the spaces apart from those in the middle column, there is a natural inclusion of the space inside $\mathbf{M}_{\widehat{Q},\gamma}$ and we just define the function $Tr(W)$ to be the restriction of the function $Tr(W)$ on the ambient space. In the case of $(\rho', \rho'', p) \in Z_{\gamma_1,\gamma_2}$ and $\xi \in \mathbf{M}_{L,\gamma}$ we define $Tr(W)(\rho', \rho'', p, \xi) = Tr(p\xi)$ to be the function $Tr(W)$ on $Z_{\gamma_1,\gamma_2} \times \mathbf{M}_{L,\gamma}$, and we restrict this same function to define the function $Tr(W)$ on $Z_{\gamma_1,\gamma_2} \times \mathbf{M}_{L,\gamma_1,\gamma_2}$.

Next we will define the constituent maps of diagram (7).

- We define a as follows. If $(\rho', \rho'', p) \in Z_{\gamma_1,\gamma_2}$ and $\xi \in \mathbf{M}_{L,\gamma_1,\gamma_2}$ we forget p, and restrict ξ to the block diagonals to get elements ξ' of $\mathbf{M}_{L,\gamma_1}$ and $\xi'' \in \mathbf{M}_{L,\gamma_2}$. Via the natural isomorphism $\mathbf{M}_{\widehat{Q},\gamma} \cong \mathbf{M}_{\overline{Q},\gamma} \times \mathbf{M}_{L,\gamma}$ we get an element $(\rho', \xi', \rho'', \xi'') \in \mathbf{M}_{\widehat{Q},\gamma_1} \times \mathbf{M}_{\widehat{Q},\gamma_2}$.
- b is defined via the composition
$$\mathbf{M}_{\widehat{Q},\gamma_1,\gamma_2} \cong \mathbf{M}_{\overline{Q},\gamma_1,\gamma_2} \times \mathbf{M}_{L,\gamma_1,\gamma_2} \xrightarrow{f' \times \mathrm{id}} Z_{\gamma_1,\gamma_2} \times \mathbf{M}_{L,\gamma_1,\gamma_2}$$
where f' is as in the definition of m_{SV}.
- Via forgetting the offdiagonal blocks of a representation there is a $\mathbf{G}_{\gamma_1} \times \mathbf{G}_{\gamma_2}$-equivariant affine fibration $\varpi : \mathbf{M}_{L,\gamma} \to \mathbf{M}_{L,\gamma_1} \times \mathbf{M}_{L,\gamma_2}$, and for c we take the composition
$$\begin{aligned}&\mathbf{M}_{\overline{Q},\gamma_1} \times \mathbf{M}_{\overline{Q},\gamma_2} \times \mathbf{M}_{L,\gamma}\\ &\xrightarrow{\mathrm{id} \times \varpi} \mathbf{M}_{\overline{Q},\gamma_1} \times \mathbf{M}_{\overline{Q},\gamma_2} \times \mathbf{M}_{L,\gamma_1} \times \mathbf{M}_{L,\gamma_2} \cong \mathbf{M}_{\widehat{Q},\gamma_1} \times \mathbf{M}_{\widehat{Q},\gamma_2}.\end{aligned}$$
- d is the product of the identity on Z_{γ_1,γ_2} with the obvious inclusion i : $\mathbf{M}_{L,\gamma_1,\gamma_2} \hookrightarrow \mathbf{M}_{L,\gamma}$.
- e is the composition
$$\mathbf{M}_{\widehat{Q},\gamma_1,\gamma_2} \cong \mathbf{M}_{\overline{Q},\gamma_1,\gamma_2} \times \mathbf{M}_{L,\gamma_1,\gamma_2} \xrightarrow{\mathrm{id} \times i} \mathbf{M}_{\overline{Q},\gamma_1,\gamma_2} \times \mathbf{M}_{L,\gamma}.$$
- f is given by $f(\rho', \rho'', \xi) = (\rho', \rho'', \mu_{\gamma_1}(\rho') \otimes \mu_{\gamma_2}(\rho''), \xi)$.
- $g = f' \times \mathrm{id}$ where f' is as in the definition of m_{SV}.
- h is the obvious inclusion.

Remark 4.3. The functions $Tr(W)$ commute with all of the maps in (7).

Denote by $\cdot : \mathbb{Z}^I \otimes \mathbb{Z}^I \to \mathbb{Z}$ the usual dot product
$$\gamma_1 \cdot \gamma_2 := \sum_{i \in I} \gamma_1^i \gamma_2^i.$$

Theorem 4.4. *Let* $\nu : \mathrm{H}^{\mathrm{crit}}_{c,\tilde{\mathbf{G}}_\gamma}(\mathbf{M}_{\widehat{Q},\gamma}, W_\gamma) \to \mathrm{H}_{c,\tilde{\mathbf{G}}_\gamma}(\mu_\gamma^{-1}(0), \mathbb{Q})$ *be the dimensional reduction isomorphism. Let* $\gamma_1, \gamma_2 \in \mathbb{N}^I$ *with* $\gamma = \gamma_1 + \gamma_2$. *Then*
$$m_{SV}^\vee = (\nu_{\gamma_1} \otimes \nu_{\gamma_2}) \circ m_{KS}^\vee \circ \nu_\gamma^{-1} \cdot (-1)^{\gamma_1 \cdot \gamma_2}.$$

Proof. Firstly we observe that, with the notation of diagram (7)
$$ab = \zeta$$
$$he = \tau$$

where these are the maps appearing in the definition of the KS product, and so m_{KS} is obtained by the (dual of the operation) $a_*^{\varphi} b_*^{\varphi} e_{\varphi}^* h_{\varphi}^* = \zeta_*^{\varphi} \tau_{\varphi}^*$ in vanishing cycle cohomology.

The rightmost square of (7) is Cartesian, d is the inclusion of a vector bundle over Z_{γ_1,γ_2}, and e is the pullback of this inclusion along the map $\mathbf{M}_{\overline{Q},\gamma_1,\gamma_2} \to Z_{\gamma_1,\gamma_2}$. Then we obtain commutativity of the square of sheaves on $Z_{\gamma_1,\gamma_2} \times \mathbf{M}_{L,\gamma}$

$$\begin{array}{ccc} d_!\mathbb{Q} & \longleftarrow & g_!e_!\mathbb{Q} = d_!b_!\mathbb{Q} \\ \uparrow & & \uparrow \\ \mathbb{Q} & \longleftarrow & g_!\mathbb{Q} \end{array}$$

and so the commutativity of the square

$$\begin{array}{ccc} \mathrm{H}^{\mathrm{crit}}_{c,\tilde{\mathbf{G}}_{\gamma_1,\gamma_2}}(Z_{\gamma_1,\gamma_2} \times \mathbf{M}_{L,\gamma_1,\gamma_2}, W) & \xleftarrow{\;b_*^{\varphi}\;} & \mathrm{H}^{\mathrm{crit}}_{c,\tilde{\mathbf{G}}_{\gamma_1,\gamma_2}}(\mathbf{M}_{\widehat{Q},\gamma_1,\gamma_2}, W) \\ \uparrow{\scriptstyle d_{\varphi}^*} & & \uparrow{\scriptstyle e_{\varphi}^*} \\ \mathrm{H}^{\mathrm{crit}}_{c,\tilde{\mathbf{G}}_{\gamma_1,\gamma_2}}(Z_{\gamma_1,\gamma_2} \times \mathbf{M}_{L,\gamma}, W) & \xleftarrow{\;g_*^{\varphi}\;} & \mathrm{H}^{\mathrm{crit}}_{c,\tilde{\mathbf{G}}_{\gamma_1,\gamma_2}}(\mathbf{M}_{\overline{Q},\gamma_1,\gamma_2} \times \mathbf{M}_{L,\gamma}, W). \end{array}$$

Since the dual of KS product is given by $a_*^{\varphi} b_*^{\varphi} e_{\varphi}^* h_{\varphi}^*$, we may thus rewrite it as

$$m_{KS}^{\vee} = a_*^{\varphi} d_{\varphi}^* g_*^{\varphi} h_{\varphi}^*. \tag{8}$$

The following diagrams of sheaves commute:

$$\begin{array}{ccc} \varphi_{Tr(W)}\mathbb{Q} & \longrightarrow & \varphi_{Tr(W)}h_*\mathbb{Q} \\ \downarrow & & \downarrow \\ \mathbb{Q} & \longrightarrow & h_*\mathbb{Q} \end{array}$$

on $\mathbf{M}_{\widehat{Q},\gamma}$ and

$$\begin{array}{ccc} \varphi_{Tr(W)}g_!\mathbb{Q} & \longrightarrow & \varphi_{Tr(W)}\mathbb{Q} \\ \downarrow & & \downarrow \\ g_!\mathbb{Q} & \longrightarrow & \mathbb{Q} \end{array}$$

on $Z_{\gamma_1,\gamma_2} \times \mathbf{M}_{L,\gamma}$, from which we deduce the commutativity of the following squares

$$\begin{array}{ccc} \mathrm{H}^{\mathrm{crit}}_{c,\tilde{\mathbf{G}}_{\gamma_1,\gamma_2}}(\mathbf{M}_{\widehat{Q},\gamma}, W) & \xrightarrow{\;h_{\varphi}^*\;} & \mathrm{H}^{\mathrm{crit}}_{c,\tilde{\mathbf{G}}_{\gamma_1,\gamma_2}}(\mathbf{M}_{\overline{Q},\gamma_1,\gamma_2} \times \mathbf{M}_{L,\gamma}, W) \\ \downarrow{\scriptstyle \cong} & & \downarrow{\scriptstyle \cong} \\ \mathrm{H}_{c,\tilde{\mathbf{G}}_{\gamma_1,\gamma_2}}(\mathbf{M}_{\widehat{Q},\gamma} \cap \mu_{\gamma}^{-1}(0)) & \xrightarrow{\;h^*\;} & \mathrm{H}_{c,\tilde{\mathbf{G}}_{\gamma_1,\gamma_2}}\left(\left(\mathbf{M}_{\overline{Q},\gamma} \times \mathbf{M}_{L,\gamma}\right) \cap \mu_{\gamma}^{-1}(0)\right) \end{array}$$

and

$$
\begin{array}{ccc}
\mathrm{H}^{\mathrm{crit}}_{c,\tilde{\mathbf{G}}_{\gamma_1,\gamma_2}}((\mathbf{M}_{\overline{Q},\gamma_1,\gamma_2} \times \mathbf{M}_{L,\gamma}, W) & \xrightarrow{\;g^{\varphi}_*\;} & \mathrm{H}^{\mathrm{crit}}_{c,\tilde{\mathbf{G}}_{\gamma_1,\gamma_2}}((Z_{\gamma_1,\gamma_2} \times \mathbf{M}_{L,\gamma}, W) \\
\downarrow \cong & & \downarrow \cong \\
\mathrm{H}_{c,\tilde{\mathbf{G}}_{\gamma_1,\gamma_2}}((\mathbf{M}_{\overline{Q},\gamma} \times \mathbf{M}_{L,\gamma}) \cap \mu_\gamma^{-1}(0)) & \xrightarrow{\;g_*\;} & \mathrm{H}_{c,\tilde{\mathbf{G}}_{\gamma_1,\gamma_2}}((Z_{\gamma_1,\gamma_2} \times \mathbf{M}_{L,\gamma}) \cap \mu_\gamma^{-1}(0))
\end{array}
$$

where in both cases the vertical isomorphisms are given by dimensional reduction.

Note that dimensionally reducing along the fibre $\mathbf{M}_{L,\gamma}$ does indeed impose the full moment map relations, as in each case the derivative of $Tr(W)$ with respect to the function keeping track of the i,jth entry of l_k imposes the condition that $\mu_\gamma(\rho)_{ji} = 0$.

Similarly, we have the commuting diagram

$$
\begin{array}{ccc}
\mathrm{H}^{\mathrm{crit}}_{c,\tilde{\mathbf{G}}_{\gamma_1,\gamma_2}}(Z_{\gamma_1,\gamma_2} \times \mathbf{M}_{L,\gamma}, W) & \xrightarrow{\;f^*_{\varphi}\;} & \mathrm{H}^{\mathrm{crit}}_{c,\tilde{\mathbf{G}}_{\gamma_1,\gamma_2}}(\mathbf{M}_{\overline{Q},\gamma_1} \times \mathbf{M}_{\overline{Q},\gamma_2} \times \mathbf{M}_{L,\gamma}, W) \\
\downarrow \cong & & \downarrow \cong \\
\mathrm{H}_{c,\tilde{\mathbf{G}}_{\gamma_1,\gamma_2}}((Z_{\gamma_1,\gamma_2} \times \mathbf{M}_{L,\gamma}) \cap \mu_\gamma^{-1}(0)) & \xrightarrow{\;f^*\;} & \mathrm{H}_{c,\tilde{\mathbf{G}}_{\gamma_1,\gamma_2}}((\mathbf{M}_{\overline{Q},\gamma_1} \times \mathbf{M}_{\overline{Q},\gamma_2} \times \mathbf{M}_{L,\gamma}) \cap \mu_\gamma^{-1}(0))
\end{array}
$$

where $\mu : Z_{\gamma_1,\gamma_2} \times \mathbf{M}_{L,\gamma} \to \mathfrak{g}_{\gamma_1,\gamma_2}$ is defined by $((\rho', \rho'', p), \xi) \mapsto p$. Putting the previous three diagrams together, we conclude that the following diagram commutes

$$
\begin{array}{ccc}
\mathrm{H}^{\mathrm{crit}}_{c,\tilde{\mathbf{G}}_{\gamma_1,\gamma_2}}(\mathbf{M}_{\hat{Q},\gamma}, W) & \xrightarrow{\;f^*_{\varphi} g^{\varphi}_* h^*_{\varphi}\;} & \mathrm{H}^{\mathrm{crit}}_{c,\tilde{\mathbf{G}}_{\gamma_1,\gamma_2}}(\mathbf{M}_{\overline{Q},\gamma_1} \times \mathbf{M}_{\overline{Q},\gamma_2} \times \mathbf{M}_{L,\gamma}, W) \\
\downarrow & & \downarrow \\
\mathrm{H}_{c,\tilde{\mathbf{G}}_{\gamma_1,\gamma_2}}(\mathbf{M}_{\hat{Q},\gamma} \cap \mu_\gamma^{-1}(0)) & \xrightarrow{\;m^{\vee}_{SV} \otimes \mathrm{id}_{\mathrm{H}_c(\mathbf{M}_{L,\gamma})}\;} & \mathrm{H}_{c,\tilde{\mathbf{G}}_{\gamma_1,\gamma_2}}((\mathbf{M}_{\overline{Q},\gamma_1} \times \mathbf{M}_{\overline{Q},\gamma_2} \times \mathbf{M}_{L,\gamma}) \cap \mu_\gamma^{-1}(0))
\end{array}
$$

Note that the function $Tr(W)$ on $\mathbf{M}_{\overline{Q},\gamma_1} \times \mathbf{M}_{\overline{Q},\gamma_2} \times \mathbf{M}_{L,\gamma}$ is the box sum of the zero function on the off block-diagonal entries of $\mathbf{M}_{L,\gamma}$ and the function $Tr(W)$ on $\mathbf{M}_{\hat{Q},\gamma_1} \times \mathbf{M}_{\hat{Q},\gamma_2}$, so that we have a commuting square

$$
\begin{array}{ccc}
\mathrm{H}^{\mathrm{crit}}_{c,\tilde{\mathbf{G}}_{\gamma_1 \times \gamma_2}}(\mathbf{M}_{\overline{Q},\gamma_1} \times \mathbf{M}_{\overline{Q},\gamma_2} \times \mathbf{M}_{L,\gamma}, W) & \xrightarrow{\;c^{\varphi}_*\;} & \mathrm{H}^{\mathrm{crit}}_{c,\tilde{\mathbf{G}}_{\gamma_1 \times \gamma_2}}(\mathbf{M}_{\hat{Q},\gamma_1} \times \mathbf{M}_{\hat{Q},\gamma_2}, W) \\
\downarrow \cong^{\text{dim red}} & & \downarrow \nu_{\gamma_1}^{-1} \otimes \nu_{\gamma_2}^{-1} \\
\mathrm{H}_{c,\tilde{\mathbf{G}}_{\gamma_1 \times \gamma_2}}((\mathbf{M}_{\overline{Q},\gamma_1} \times \mathbf{M}_{\overline{Q},\gamma_2} \times \mathbf{M}_{L,\gamma}) \cap \mu^{-1}(0)) & \rightarrow & \mathrm{H}_{c,\tilde{\mathbf{G}}_{\gamma_1 \times \gamma_2}}((\mathbf{M}_{\hat{Q},\gamma_1} \times \mathbf{M}_{\hat{Q},\gamma_2}) \cap \mu^{-1}(0))
\end{array}
$$

where the bottom (shifted) isomorphism is the usual isomorphism associated to an affine fibration (note that the function μ is insensitive to the $\mathbf{M}_L$ component).

So putting everything together, we have that

$$
m^{\vee}_{SV,\gamma_1,\gamma_2} = (\nu_{\gamma_1} \otimes \nu_{\gamma_2}) c^{\varphi}_* f^*_{\varphi} g^{\varphi}_* h^*_{\varphi} \nu_\gamma^{-1} \tag{9}
$$

and

$$m^{\vee}_{KS,\gamma_1,\gamma_2} = a^{\varphi}_{*} b^{\varphi}_{*} e^{*}_{\varphi} h^{*}_{\varphi} \tag{10}$$

$$= a^{\varphi}_{*} d^{*}_{\varphi} g^{\varphi}_{*} h^{*}_{\varphi} \quad \text{using equation (8)} \tag{11}$$

while what we want to show is that

$$m^{\vee}_{SV,\gamma_1,\gamma_2} \stackrel{?}{=} (\nu_{\gamma_1} \otimes \nu_{\gamma_2}) m^{\vee}_{KS,\gamma_1,\gamma_2} \nu_{\gamma}^{-1}. \tag{12}$$

We will see that we do not actually have a strict equality. By comparing (9) and (11) with the left and right-hand side of (12) we see that the left- and right-hand side are the same, except we have substituted $c^{\varphi}_{*} f^{*}_{\varphi}$ for $a^{\varphi}_{*} d^{*}_{\varphi}$, and so compatibility of the KS product with the SV product is settled by (almost) commutativity of the following diagram:

$$\begin{array}{ccc} \mathrm{H}^{\mathrm{crit}}_{c,\tilde{\mathbf{G}}_{\gamma_1\times\gamma_2}}(\mathbf{M}_{\widehat{Q},\gamma_1} \times \mathbf{M}_{\widehat{Q},\gamma_2}) & \xleftarrow{a^{\varphi}_{*}} & \mathrm{H}^{\mathrm{crit}}_{c,\tilde{\mathbf{G}}_{\gamma_1\times\gamma_2}}(Z_{\gamma_1,\gamma_2} \times \mathbf{M}_{L,\gamma_1,\gamma_2}) \\ \uparrow c^{\varphi}_{*} & & \uparrow d^{*}_{\varphi} \\ \mathrm{H}^{\mathrm{crit}}_{c,\tilde{\mathbf{G}}_{\gamma_1\times\gamma_2}}(\mathbf{M}_{\overline{Q},\gamma_1} \times \mathbf{M}_{\overline{Q},\gamma_2} \times \mathbf{M}_{L,\gamma}) & \xleftarrow[f^{*}_{\varphi}]{} & \mathrm{H}^{\mathrm{crit}}_{c,\tilde{\mathbf{G}}_{\gamma_1\times\gamma_2}}(Z_{\gamma_1,\gamma_2} \times \mathbf{M}_{L,\gamma}). \end{array} \tag{13}$$

Consider $Z_{\gamma_1,\gamma_2} \times \mathbf{M}_{L,\gamma}$ as a $\tilde{\mathbf{G}}_{\gamma_1\times\gamma_2}$-equivariant bundle over $\mathbf{M}_{\widehat{Q},\gamma_1} \times \mathbf{M}_{\widehat{Q},\gamma_2}$. The function $Tr(W)_{\gamma}$ splits as a box sum $\tau \boxplus \zeta$, where ζ is a quadratic function on the fibres and τ is the function $Tr(W)$ on the base. Consider a point x of $(\mathbf{M}_{\widehat{Q},\gamma_1} \times \mathbf{M}_{\widehat{Q},\gamma_2}, \tilde{\mathbf{G}}_{\gamma_1\times\gamma_2})_N$, and the fibre of $(Z_{\gamma_1,\gamma_2} \times \mathbf{M}_{L,\gamma}, \tilde{\mathbf{G}}_{\gamma_1\times\gamma_2})_N$ over it. This fibre splits as a direct sum $A \oplus B \oplus C$, where $A \cong \mathfrak{p}_{\gamma_1,\gamma_2}/(\mathfrak{g}_{\gamma_1} \times \mathfrak{g}_{\gamma_2})$, B is given by the top right-hand block of the $\rho(l_i)$, and C is given by the bottom left-hand block of $\rho(l_i)$. The factor A comes from the fibre of the projection $Z_{\gamma_1,\gamma_2} \to \mathbf{M}_{\overline{Q},\gamma_1} \times \mathbf{M}_{\overline{Q},\gamma_2}$. Via the Killing form, we have the isomorphism $k : A \cong C^{\vee}$, and ζ is the function $(a, b, c) \mapsto k(a)(c)$. Applying the Thom–Sebastiani isomorphism, pushing forward to our point x in the base $(\mathbf{M}_{\widehat{Q},\gamma_1} \times \mathbf{M}_{\widehat{Q},\gamma_2}, \tilde{\mathbf{G}}_{\gamma_1\times\gamma_2})_N$, the underlying diagram of sheaves of diagram (13) is given by tensoring the constant map on $\varphi_{\tau_N}[-1]\mathbb{Q}$ with

$$\begin{array}{ccc} \mathrm{H}_c(\mathrm{pt}, \mathbb{Q}) & \longleftarrow & \mathrm{H}_c(A \oplus B, \mathbb{Q}) \\ \uparrow & & \uparrow \\ \mathrm{H}_c(B \oplus C, \mathbb{Q}) & \longleftarrow & \mathrm{H}^{\mathrm{crit}}_c(A \oplus B \oplus C, \zeta). \end{array}$$

We claim that the diagram commutes, up to the sign $(-1)^{\dim(A)}$. Explicitly, we may choose $x_1, \ldots, x_n$ coordinates for A, $y_1, \ldots, y_m$ coordinates for B, and $z_1, \ldots, z_n$ coordinates for C such that $\zeta = \sum_{i=1}^{n} x_i z_i$. Note that after taking cohomology, the bottom map and the rightmost map are just dimensional reduction isomorphisms, while the other two maps are the isomorphisms in compactly supported cohomology induced by affine fibrations. Then the claim is just Lemma 4.1, and the result follows. □

Define the modified dimensional reduction map $\nu_\gamma^\circ : \mathrm{H}^{\mathrm{crit}}_{c,\mathbf{G}_\gamma}(\mathbf{M}_{\widehat{Q},\gamma}, W) \to \mathrm{H}_{c,\mathbf{G}_\gamma}(\mu_\gamma^{-1}(0), \mathbb{Q})$ by

$$\nu_\gamma^\circ = \nu_\gamma \cdot \sqrt{-1}^{\gamma\cdot\gamma}.$$

Corollary 4.5. *The map*

$$\nu^{\circ\vee} : \bigoplus_{\gamma\in\mathbb{N}^I} \mathrm{H}_{c,\mathbf{G}_\gamma}(\mu_\gamma^{-1}(0), \mathbb{Q})^\vee \otimes \mathbb{C} \to \bigoplus_{\gamma\in\mathbb{N}^I} \mathrm{H}^{\mathrm{crit}}_{c,\mathbf{G}_\gamma}(\mathbf{M}_{\widehat{Q},\gamma}, W_\gamma)^\vee \otimes \mathbb{C}$$

is an isomorphism of algebras for all $\gamma_1, \gamma_2 \in \mathbb{N}^I$ with $\gamma_1 + \gamma_2 = \gamma$ there is an equality of maps

$$m_{KS,\gamma_1,\gamma_2} \circ (\nu_{\gamma_1}^{\circ\vee} \otimes \nu_{\gamma_2}^{\circ\vee}) = \nu_\gamma^{\circ\vee} \circ m_{SV,\gamma_1,\gamma_2}.$$

Proof. We have

$$\begin{aligned}
&m_{KS,\gamma_1,\gamma_2} \circ (\nu_{\gamma_1}^{\circ\vee} \otimes \nu_{\gamma_2}^{\circ\vee}) \\
&= \sqrt{-1}^{\gamma_1\cdot\gamma_1+\gamma_2\cdot\gamma_2} m_{KS,\gamma_1,\gamma_2} \circ (\nu_{\gamma_1}^{\vee} \otimes \nu_{\gamma_2}^{\vee}) \\
&= \sqrt{-1}^{\gamma_1\cdot\gamma_1+\gamma_2\cdot\gamma_2} \nu_\gamma^\vee \circ m_{SV,\gamma_1,\gamma_2} \cdot (-1)^{\gamma_1\cdot\gamma_2} \qquad \text{(Theorem 4.4)} \\
&= \nu_\gamma^\vee \circ m_{SV,\gamma_1,\gamma_2} \cdot \sqrt{-1}^{\gamma\cdot\gamma} \\
&= \nu_\gamma^{\circ\vee} \circ m_{SV,\gamma_1,\gamma_2}.
\end{aligned}$$

□

As we can see from the proof, the isomorphism holds over the subfield $\mathbb{Q}[i] \subset \mathbb{C}$, where $i = \sqrt{-1}$.

Acknowledgement

The idea of the paper goes back to the discussions of the second author (Y.S.) with Maxim Kontsevich during the conference on DT-invariants in Budapest in 2012 (the conference was sponsored by AIM). We are grateful to Maxim Kontsevich for sharing with us his ideas on the subject. We also thank to George Lusztig for encouraging us to continue the study of the relationship between COHA and semicanonical bases. Y.S. thanks to IHES for excellent research conditions. His work was partially supported by an NSF grant.

References

1. T. Bozec, *Quivers with loops and generalized crystals*, arXiv: 1403.0846.
2. W. Crawley-Boevey, *Preprojective algebras, differential operators and Conze embedding for deformations of Kleinian singularities*, Comment. Math. Helv., 74, (1999), 548–574.
3. W. Crawley-Boevey, M. Van den Bergh, *Absolutely indecomposable representations and Kac-Moody Lie algebras (with an appendix by Hiraku Nakajima)*, arXiv:math/0106009.
4. N. Chriss, V. Ginzburg, *Representation Theory and Complex Geometry*, Birkhäuser, Boston-Basel-Berlin, 1997.

5. B. Davison, *The critical COHA of a quiver with potential*, arXiv: 1311.7172.
6. B. Davison, *Purity of critical cohomology and Kac's conjecture*, arXiv: 1311.6989.
7. A. Efimov, *Non-commutative Hitchin systems*, preprint, 2014.
8. T. Hausel, E. Letellier, F. Rodriguez-Villegas, *Positivity of Kac polynomials and DT-invariants for quivers*, arXiv:1204.2375.
9. D. Joyce, *Motivic invariants of Artin stacks and 'stack functions'*, arXiv:math/0509722.
10. M. Kashiwara, P. Schapira, *Sheaves on Manifolds*, Springer-Verlag, 1990.
11. M. Kontsevich, Y. Soibelman, *Notes on A_∞-algebras, A_∞-categories and non-commutative geometry. I*, arXiv: 0606241v2.
12. M. Kontsevich, Y. Soibelman, *Stability structures, motivic Donaldson–Thomas invariants and cluster transformations*, arXiv: 0811.2435.
13. M. Kontsevich, Y. Soibelman, *Cohomological Hall algebra, exponential Hodge structures and motivic Donaldson–Thomas invariants*, Comm. Num. Th. and Phys. **5** (2011), no. **2**, 231–252.
14. G. Lusztig, *Quivers, perverse sheaves, and quantized enveloping algebras*, Journal of the AMS, Vol. **4**, No. **2** (1991), 365–421.
15. G. Lusztig, *Semicanonical bases arising from enveloping algebras*, Adv. Math. Vol. **151**, Iss. **2** (2000), 129–139.
16. S. Mozgovoy, *Motivic Donaldson–Thomas invariants and McKay correspondence*, arXiv:1107.6044.
17. O. Schiffmann, *Lectures on canonical and crystal bases of Hall algebras*, arXiv: 0910.4460v2.
18. O. Schiffmann, E. Vasserot, *Cherednik algebras, W-algebras and the equivariant cohomology of the moduli space of instantons on $\mathbb{A}^2$*, Pub. Mat. de l'IHES (2012).
19. Y. Soibelman, *Remarks on Cohomological Hall algebras and their representations*, arXiv:1404.1606.
20. Y. Yang, G. Zhao, *Cohomological Hall algebra of a preprojective algebra*, arXiv:1407.7994.

Jie Ren and Yan Soibelman
Department of Mathematics
Kansas State University
Manhattan, KS 66506, USA
e-mail: `jren@ksu.edu`
`soibel@math.ksu.edu`

Progress in Mathematics, Vol. 324, 295–364

Fukaya A_∞-structures Associated to Lefschetz Fibrations. II

Paul Seidel

To Maxim Kontsevich, on his 50th birthday

Abstract. Consider the Fukaya category associated to a Lefschetz fibration. It turns out that the Floer cohomology of the monodromy around ∞ gives rise to natural transformations from the Serre functor to the identity functor, in that category. We pay particular attention to the implications of that idea for Lefschetz pencils.

Mathematics Subject Classification (2010). primary 53D40; secondary 53D12, 32S50.

Keywords. Floer homology, Lefschetz fibrations, Fukaya categories.

1. Introduction

This paper is part of an investigation of the Floer-theoretic structures arising from Lefschetz fibrations. Its purpose is to introduce a new piece of that puzzle; and also, to set up a wider algebraic framework into which that piece should (conjecturally) fit, centered on a notion of noncommutative pencil.

1a. Symplectic geometry

It is well known that there is a version of the Fukaya category tailored to a Lefschetz fibration (the idea is originally due to Kontsevich). In [44], it was pointed out that these categories always come with some added structure: a distinguished natural transformation from the Serre functor to the identity functor (see [41, 8, 45, 2] for related theoretical developments, and [31, 32, 3] for applications). Concretely, let us fix a symplectic Lefschetz fibration

$$\pi : E^{2n} \longrightarrow \mathbb{C} \tag{1.1}$$

with fibre M^{2n-2}. For technical simplicity, we will impose an exactness condition on the symplectic form, hence assume that the fibre is a Liouville domain. Choose a basis of Lefschetz thimbles, and let $\mathcal{A}$ be the associated (directed) Fukaya A_∞-algebra, defined over a coefficient field $\mathbb{K}$. In general, $\mathcal{A}$ is $\mathbb{Z}/2$-graded, and this lifts to a $\mathbb{Z}$-grading if $c_1(E) = 0$. Natural transformations of degree n from the Serre functor to the identity functor (in the triangulated envelope, or equivalently derived A_∞-category, of $\mathcal{A}$) can be described as maps of A_∞-bimodules

$$\mathcal{A}^\vee[-n] \longrightarrow \mathcal{A}. \tag{1.2}$$

Here $\mathcal{A}$ is the diagonal bimodule; and $\mathcal{A}^\vee$ is its dual, to which we have applied an upwards shift by n. To be more precise, A_∞-bimodules form a dg category, denoted here by $[\mathcal{A},\mathcal{A}]$. By a bimodule map (1.2), we mean an element of

$$H^0(\hom_{[\mathcal{A},\mathcal{A}]}(\mathcal{A}^\vee[-n],\mathcal{A})) \cong H^n(\hom_{[\mathcal{A},\mathcal{A}]}(\mathcal{A}^\vee,\mathcal{A})). \tag{1.3}$$

In these terms, the story so far can be summarized as:

Lemma 1.1. *$\mathcal{A}$ always comes with a distinguished element of* (1.3), *denoted here by ρ.*

The definition in [45] uses the geometry and Floer cohomology of Lefschetz thimbles in E. However, there is another description (equivalent up to automorphisms of $\mathcal{A}^\vee$), in terms of the functor from $\mathcal{A}$ to the standard Fukaya category $\mathcal{F}(M)$ of the fibre (this second description was the one originally proposed in [44]; the relation between the two is [45, Corollary 7.1]).

The specific contribution of this paper is to introduce another geometric source of maps (1.2). This uses fixed point Floer cohomology for symplectic automorphisms ϕ of M (which are exact and equal the identity near the boundary). To define it, we make an auxiliary choice of perturbation, by the time ϵ map of the Reeb flow near the boundary, and write the outcome as $HF^*(\phi,\epsilon)$ (for small $|\epsilon|$, this appears in [39, 33]; for the general case, see [54] or the exposition in the body of this paper).

Theorem 1.2. *Let $\mu : M \to M$ be the monodromy of the Lefschetz fibration around a large circle. For sufficiently small $\epsilon > 0$, there is a canonical map*

$$HF^{*+2}(\mu,\epsilon) \longrightarrow H^*(\hom_{[\mathcal{A},\mathcal{A}]}(\mathcal{A}^\vee[-n],\mathcal{A})). \tag{1.4}$$

This is the outcome of a Floer cohomology computation in E which follows [33], together with an application of a suitable open-closed string map, similar to those in [1]. Generally, both sides of (1.4) are $\mathbb{Z}/2$-graded; when $c_1(E) = 0$, this lifts to $\mathbb{Z}$-gradings, and the map between them will respect those gradings.

An important special class of Lefschetz fibrations comes from anticanonical Lefschetz pencils: by which we mean, on a monotone closed symplectic manifold E^{cl}, a symplectic Lefschetz pencil of hypersurfaces representing the first Chern class. One obtains E from E^{cl} by removing a suitable neighbourhood of one such hypersurface; hence, $c_1(E) = 0$ always holds in this case. The Reeb flow on ∂M is periodic (say with period 1), and the monodromy is the boundary twist associated to that periodic flow:

$$\mu = \tau_{\partial M}. \tag{1.5}$$

This satisfies

$$\mathrm{HF}^{*+2}(\tau_{\partial M},\epsilon) \cong \begin{cases} \mathrm{H}^*(M,\partial M) & \epsilon \in (0,1), \\ \mathrm{H}^*(M) & \epsilon \in (1,2). \end{cases} \tag{1.6}$$

Theorem 1.2 applies only to the first case of (1.6). However, by a more precise analysis, one can show that in the second case, an analogue of the map (1.4) can be defined in the lowest degree $* = 0$. Hence, the element of $\mathrm{HF}^2(\tau_{\partial M},\epsilon)$, $\epsilon \in (1,2)$, corresponding to $1 \in \mathrm{H}^0(M)$ still gives rise to a map (1.3), which we denote by σ. We summarize the consequence:

Theorem 1.3. *For a Lefschetz fibration arising from an anticanonical Lefschetz pencil, there is a distinguished pair (ρ,σ) of maps* (1.2) *(obtained, respectively, by Lemma* 1.1, *and a modified version of the argument from Theorem* 1.2*).*

Remark 1.4. More generally, one can consider a Lefschetz pencil such that $c_1(E^{\mathrm{cl}})$ is $(1+m)$ times the class of the hypersurfaces in the pencil, for some $m \in \mathbb{Z}$. One then still has $c_1(E) = 0$, but the appropriate generalization of (1.6) is

$$\mathrm{HF}^{*+2}(\tau_{\partial M},\epsilon) \cong \begin{cases} \mathrm{H}^{*+2m}(M,\partial M) & \epsilon \in (0,1), \\ \mathrm{H}^{*+2m}(M) & \epsilon \in (1,2). \end{cases} \tag{1.7}$$

While ρ still has degree 0, our construction now leads to a σ which has degree $-2m$.

1b. Noncommutative geometry

The starting point for our algebraic framework is the notion of *noncommutative divisor*. While the terminology is new, the concept already appeared in [41], and a version was considered in [45, 27]. The last two references include a cyclic symmetry condition, which corresponds more specifically to a *noncommutative anticanonical divisor* (see Remark 2.11); we have chosen not to impose that condition here, for the sake of simplicity.

While precise definitions will be given later on, it makes sense to give an outline now. Let $\mathcal{A}$ be an A_∞-algebra. A noncommutative divisor on $\mathcal{A}$ consists of an A_∞-bimodule $\mathcal{P}$ which is invertible (with respect to tensor product), together with an A_∞-algebra structure on

$$\mathcal{B} = \mathcal{A} \oplus \mathcal{P}[1], \tag{1.8}$$

which extends the given one on $\mathcal{A}$ as well as the $\mathcal{A}$-bimodule structure of $\mathcal{P}$. If one then considers $\mathcal{B}$ as an $\mathcal{A}$-bimodule, it fits into a short exact sequence

$$0 \rightarrow \mathcal{A} \longrightarrow \mathcal{B} \longrightarrow \mathcal{P}[1] \rightarrow 0. \tag{1.9}$$

As boundary homomorphism of that sequence, one gets a bimodule map $\theta : \mathcal{P} \rightarrow \mathcal{A}$, which we call the *section* associated to the noncommutative divisor. We extend the algebro-geometric language to related structures:

(1.10) (i) The A_∞-algebra $\mathcal{B}$ is thought of as describing the *divisor by itself* (independently of its relationship with the *ambient space* $\mathcal{A}$).

(1.10) (ii) There is a localization process, in which one makes θ invertible by passing to a quotient A_∞-algebra. We consider this to be the abstract analogue of taking the *complement of the divisor* (see [41, Section 1] for an explanation), and write the outcome accordingly as $\mathcal{A} \setminus \mathcal{B}$. This depends only on the section associated to the divisor.

Generalizing the previous concept, we will introduce *noncommutative pencils.* As in classical algebraic geometry, such a pencil has an associated pair of bimodule maps (sections, in our geometrically inspired terminology)

$$\rho, \sigma : \mathcal{P} \longrightarrow \mathcal{A}. \tag{1.11}$$

This leads to enhancements of the previously described constructions:

(1.12) (i) A noncommutative pencil determines a family of noncommutative divisors, parametrized by $z \in \mathbb{P}^1 = \mathbb{K} \cup \{\infty\}$. The sections associated to those divisors are (at least up to a scalar multiple)

$$\theta_z = \sigma + z\rho \quad \text{for } z \in \mathbb{K},$$
$$\theta_\infty = \rho.$$

In particular, one gets an A_∞-algebra $\mathcal{B}_\infty$. More interestingly, one can work with varying z. For instance, taking $z = 1/q$, where q is a formal variable, leads to a formal deformation of $\mathcal{B}_\infty$, which we denote by $\hat{\mathcal{B}}_\infty$.

(1.12) (ii) $\mathcal{A} \setminus \mathcal{B}_\infty$ also acquires a distinguished (curved) formal deformation, whose deformation parameter has degree 2. Let us call this the *noncommutative Landau–Ginzburg model*, and denote it by $\mathcal{LG}$.

Remark 1.5. This by no means exhausts the noncommutative geometry structures which arise in this context. For one thing, from a purely algebraic perspective, there is no particular reason to single out the point $z = \infty$, which means that there are more general versions of the constructions listed above. Two other notions, the *graph* and *base locus* of a noncommutative pencil, will be mentioned briefly in Remark 2.15. Different kinds of algebraic structures, notably ones involving Hochschild and cyclic homology (such as, the Gauß–Manin connection for the periodic cyclic homology of the fibres of the noncommutative pencil), are entirely beyond the scope of our discussion.

1c. Speculations

The A_∞-algebra $\mathcal{A}$ associated to a Lefschetz fibration comes with a natural structure of a noncommutative divisor, with associated bimodule

$$\mathcal{P} = \mathcal{A}^\vee[-n]. \tag{1.13}$$

The algebraic constructions introduced above have the following meaning:

(1.14) (i) The divisor by itself, $\mathcal{B}$, is the full subcategory of $\mathcal{F}(M)$ consisting of the vanishing cycles in our basis. In fact, this is how the noncommutative divisor was constructed in [44, 41], by identifying $\mathcal{A}$ with the directed A_∞-subcategory of $\mathcal{B}$.

(1.14) (ii) The complement of the divisor, $\mathcal{A} \setminus \mathcal{B}$, yields a full subcategory of the wrapped Fukaya category of the total space, $\mathcal{W}(E)$. This is the main result of [2].

We will now propose a continuation of this line of thought, to the point where it can include Theorem 1.3. In spite of its tentative nature, this direction seems worth while exploring, because of the potential implications for symplectic geometry and mirror symmetry.

Conjecture 1.6. *Take an exact symplectic Lefschetz fibration coming from an anticanonical Lefschetz pencil. Then, the A_∞-algebra $\mathcal{A}$ carries a canonical structure of a noncommutative pencil, satisfying* (1.13)*, whose associated sections are those described in Theorem* 1.3.

The appeal of the noncommutative pencil structure in this context is that it can serve as an intermediate object between various other Fukaya A_∞-structures. Consider:

(1.15) (i) $\mathcal{B}_\infty$ should be a full subcategory of $\mathcal{F}(M)$ (formally, this is the same statement as in (1.14)(i); but the level of difficulty of proving it depends on how one would go about constructing the noncommutative pencil structure). Now, M admits a natural closure M^{cl}, obtained by gluing in a disc bundle over the base locus of the original pencil. Associated to that is the relative Fukaya category [40, 48, 50], which is a formal deformation of $\mathcal{F}(M)$. More generally, one can modify the deformation by including a suitable bulk term, as in [20]. The conjecture is that, for a specific choice of bulk term, and after a change in the deformation parameter, this deformation is $\hat{\mathcal{B}}_\infty$. This is made more precise in [47].

(1.15) (ii) E itself embeds into a closed manifold E^{cl}, the manifold on which the original Lefschetz pencil was defined. This leads to another relative Fukaya category, but where now (because of monotonicity) the deformation parameter has degree 2. At least on the basis of formal analogy, one expects this to be related (by a cohomologically full embedding) to the Landau–Ginzburg category $\mathcal{LG}$ from (1.12)(ii).

Remark 1.7. We have limited ourselves to anticanonical Lefschetz pencils (the setup of Theorem 1.3) for the sake of concreteness, but other cases are also of interest. For instance, along the same lines as in Conjecture 1.6, the situation from Remark 1.4 should lead to a $\mathbb{Z}/2m$-graded noncommutative pencil. The residual information given by the $\mathbb{Z}$-grading of $\mathcal{A}$ means that this pencil is homogeneous with respect to a circle action on $\mathbb{P}^1$ of weight $2m$.

1d. Homological Mirror Symmetry

The mirror (B-model) counterpart of Conjecture 1.6 is far more straightforward, and amounts to embedding ordinary algebraic geometry into its noncommutative cousin. Suppose that we have a smooth algebraic variety A, a line bundle P on it, and a section $t \in \Gamma(A, P^{-1})$ of its dual. Let $\mathcal{A}$ be an A_∞-algebra underlying the bounded derived category $D^b\mathrm{Coh}(A)$. Then, P gives rise to an invertible bimodule $\mathcal{P}$, and t to a bimodule map $\theta : \mathcal{P} \to \mathcal{A}$. In fact, there is a noncommutative divisor structure on $\mathcal{A}$ of which θ is part. This fits in with mirror symmetry in the formulation of [5], where the crucial ingredient is a choice of anticanonical divisor (see [41, Section 6], where the connection is made explicit in a simple example). Similarly, suppose that for A and P as before, we have two elements of $\Gamma(A, P^{-1})$; this yields the structure of a noncommutative pencil on $\mathcal{A}$.

Homological Mirror Symmetry can now be thought of as operating on the level of pencils, schematically like this:

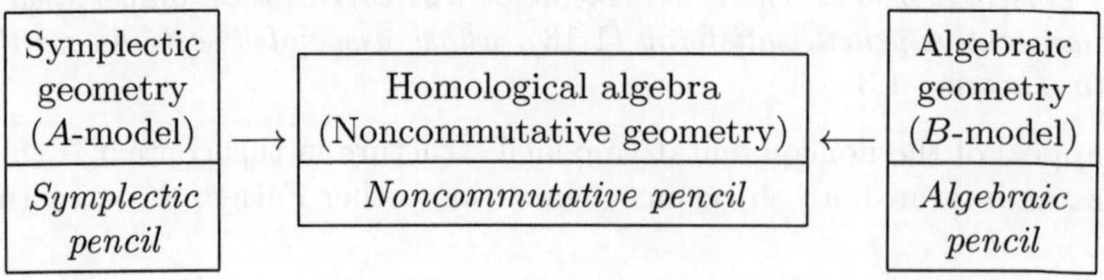

(1.16)

As a consequence, one gets a unified approach to the associated equivalences of categories. By (1.15)(ii), this picture includes counts of holomorphic discs, following the model of [5].

Acknowledgment. Conversations with Anatoly Preygel were useful at an early stage of this project. Mohammed Abouzaid and Maxim Kontsevich provided very valuable comments about (1.15)(i), which are reflected in its current formulation. The definition (2.75) was suggested by Ludmil Katzarkov. I am grateful to the anonymous referee for keeping me on the straight and narrow. Partial support was provided by: NSF grant DMS-1005288; the Simons Foundation, through a Simons Investigator award; and the Radcliffe Institute for Advanced Study at Harvard University, through a Radcliffe Fellowship. I would also like to thank Boston College, where the first version of this paper was written, for its hospitality.

2. Homological algebra

This section is of an elementary algebraic nature. To make the exposition more self-contained, we recall the relation between Hochschild homology and bimodule maps. We then discuss noncommutative divisors, largely following [41, Section 3] except for the terminology. Finally, we adapt the same ideas to noncommutative pencils.

2a. A_∞-bimodules

Let $\mathcal{A}$ be a graded vector space over a field $\mathbb{K}$. We denote by $T(\mathcal{A}[1])$ the tensor algebra over the (downwards) shifted space $\mathcal{A}[1]$. The structure of an A_∞-algebra on $\mathcal{A}$ is given by a map

$$\mu_{\mathcal{A}} : T(\mathcal{A}[1]) \longrightarrow \mathcal{A}[2], \tag{2.1}$$

which vanishes on the constants $\mathbb{K} \subset T(\mathcal{A}[1])$. The components of (2.1) are operations $\mu_{\mathcal{A}}^d : \mathcal{A}^{\otimes d} \to \mathcal{A}[2-d]$ (for $d \geq 1$). We will assume that $\mathcal{A}$ is strictly unital, denoting the unit by $e_{\mathcal{A}}$, and write $\bar{\mathcal{A}} = \mathcal{A}/\mathbb{K}\, e_{\mathcal{A}}$.

An A_∞-bimodule over $\mathcal{A}$ consists of a graded vector space $\mathcal{P}$ and a map

$$\mu_{\mathcal{P}} : T(\mathcal{A}[1]) \otimes \mathcal{P} \otimes T(\mathcal{A}[1]) \longrightarrow \mathcal{P}[1], \tag{2.2}$$

whose components we write as $\mu_{\mathcal{P}}^{s;1;r} : \mathcal{A}^{\otimes s} \otimes \mathcal{P} \otimes \mathcal{A}^{\otimes r} \to \mathcal{P}[1-r-s]$ (for $r, s \geq 0$). All A_∞-bimodules will be assumed to be strictly unital. A basic example is the diagonal bimodule $\mathcal{P} = \mathcal{A}$, which has

$$\begin{aligned} &\mu_{\mathcal{A}}^{s;1;r}(a'_s, \dots, a'_1; a; a_r, \dots, a_1) \\ &\quad = (-1)^{\|a_1\| + \dots + \|a_r\| + 1} \mu_{\mathcal{A}}^{r+1+s}(a'_s, \dots, a'_1, a, a_r, \dots, a_1). \end{aligned} \tag{2.3}$$

Here and later on, $\|a\| = |a| - 1$ stands for the reduced degree. A_∞-bimodules over $\mathcal{A}$ form a dg category, which we denote by $[\mathcal{A}, \mathcal{A}]$. An element of $\hom_{[\mathcal{A},\mathcal{A}]}(\mathcal{P}, \mathcal{Q})$ of degree d is given by a map

$$\phi : T(\mathcal{A}[1]) \otimes \mathcal{P} \otimes T(\mathcal{A}[1]) \longrightarrow \mathcal{Q}[d], \tag{2.4}$$

which factors through the projection to $T(\bar{\mathcal{A}}[1])$ on both sides. We denote the components of (2.4) again by $\phi^{s;1;r}$. The cocycles in $\hom_{[\mathcal{A},\mathcal{A}]}(\mathcal{P}, \mathcal{Q})$ are solutions of

$$\begin{aligned} &\sum_{i,j} (-1)^{|\phi|(\|a_1\| + \dots + \|a_i\|)} \\ &\qquad \mu_{\mathcal{Q}}^{s-j;1;i}(a'_s, \dots; \phi^{j;1;r-i}(a'_j, \dots, a'_1; p; a_r, \dots, a_{i+1}); \dots, a_1) \\ = \; &\sum_{i,j} (-1)^{|\phi| + \|a_1\| + \dots + \|a_i\|} \\ &\qquad \phi^{s-j;1;i}(a'_s, \dots; \mu_{\mathcal{P}}^{j;1;r-i}(a'_j, \dots, a'_1; p; a_r, \dots, a_{i+1}); \dots, a_1) \\ + &\sum_{i,j} (-1)^{|\phi| + \|a_1\| + \dots + \|a_i\|} \\ &\qquad \phi^{s;1;r-j+1}(a'_s, \dots; p; a_r, \dots, \mu_{\mathcal{A}}^j(a_{i+j}, \dots, a_{i+1}), \dots, a_1) \\ + &\sum_{i,j} (-1)^{|\phi| + \|a_1\| + \dots + \|a_r\| + |p| + \|a'_1\| + \dots + \|a'_i\|} \\ &\qquad \phi^{s-j+1;1;r}(a'_s, \dots, \mu_{\mathcal{A}}^j(a'_{i+j}, \dots, a'_{i+1}), \dots; p; a_r, \dots, a_1). \end{aligned} \tag{2.5}$$

For instance, the identity endomorphism $\mathrm{id}_{\mathcal{P}}$ has only one nonzero component, $\mathrm{id}_{\mathcal{P}}^{0;1;0}(p) = p$.

To relate these structures to their (classical) cohomology level counterparts, on equips $\mathrm{H}(\mathcal{A})$ with the graded associative algebra structure given by

$$[a_2] \cdot [a_1] = (-1)^{|a_1|} [\mu_{\mathcal{A}}^2(a_2, a_1)], \tag{2.6}$$

and makes $\mathrm{H}(\mathcal{P})$ into a graded bimodule over this algebra by setting

$$\begin{aligned} [a'] \cdot [p] &= -(-1)^{|p|} [\mu_{\mathcal{P}}^{1;1;0}(a'; p)], \\ [p] \cdot [a] &= [\mu_{\mathcal{P}}^{0;1;1}(p; a)]. \end{aligned} \tag{2.7}$$

If $\phi : \mathcal{P} \to \mathcal{Q}$ is as in (2.5), the associated cohomology level bimodule map is

$$\begin{aligned} H(\phi) : H^*(\mathcal{P}) &\longrightarrow H^{*+|\phi|}(\mathcal{Q}), \\ [p] &\longmapsto (-1)^{|\phi|\,|p|}[\phi^{0;1;0}(p)]. \end{aligned} \tag{2.8}$$

Conventions 2.1. *Our sign conventions for A_∞-algebras follow* [43]. *The sign conventions for A_∞-bimodules follow* [41] *except that, for greater compatibility with* [43], *we reverse the ordering of the entries. This applies in particular to* (2.5), *which agrees with* [41] *up to ordering (but differs from the convention for A_∞-modules in* [43]).

The following observation may help address some sign issues. Given any A_∞-bimodule structure, the sign change

$$\begin{aligned} &\mu_{\mathcal{P}}^{s;1;r}(a_s', \dots, a_1'; p; a_r, \dots, a_1) \\ &\quad \longmapsto (-1)^{\|a_1\|+\cdots+\|a_r\|+\|a_1'\|+\cdots+\|a_s'\|+1} \mu_{\mathcal{P}}^{s;1;r}(a_s', \dots, a_1'; p; a_r, \dots, a_1) \end{aligned} \tag{2.9}$$

yields another A_∞-bimodule structure. The two structures are generally distinct, but isomorphic (by an isomorphism that acts by ± 1 on each graded piece). Hence, whenever we define some construction of A_∞-bimodules, there are potentially at least two equivalent versions, which differ by (2.9).

We need to recall a few specific operations on A_∞-bimodules:

- The shifted space $\mathcal{P}[1]$ becomes an A_∞-bimodule, with

$$\begin{aligned} &\mu_{\mathcal{P}[1]}^{s;1;r}(a_s', \dots, a_1'; p; a_r, \dots, a_1) \\ &\quad = (-1)^{\|a_1\|+\cdots+\|a_r\|+1} \mu_{\mathcal{P}}^{s;1;r}(a_s', \dots, a_1'; p; a_r, \dots, a_1). \end{aligned} \tag{2.10}$$

- The dual is $\mathcal{P}^\vee = \mathrm{Hom}(\mathcal{P}, \mathbb{K})$, with

$$\begin{aligned} &\langle \mu_{\mathcal{P}^\vee}^{s;1;r}(a_s, \dots, a_1; \pi; a_r', \dots, a_1'), p \rangle \\ &\quad = (-1)^{|p|+1} \langle \pi, \mu_{\mathcal{P}}^{r;1;s}(a_r', \dots, a_1'; p; a_s, \dots, a_1) \rangle. \end{aligned} \tag{2.11}$$

- Given two A_∞-bimodules $\mathcal{Q}$ and $\mathcal{P}$, one defines the tensor product $\mathcal{Q} \otimes_{\mathcal{A}} \mathcal{P}$ to be the graded vector space $\mathcal{Q} \otimes T(\bar{\mathcal{A}}[1]) \otimes \mathcal{P}$, with the differential

$$\begin{aligned} &\mu_{\mathcal{Q}\otimes_{\mathcal{A}}\mathcal{P}}^{0;1;0}(q \otimes a_t' \otimes \cdots \otimes a_1' \otimes p) \\ &\quad = \sum_i (-1)^{|p|+\|a_1'\|+\cdots+\|a_i'\|} \mu_{\mathcal{Q}}^{0;1;t-i}(q; a_t', \dots, a_{i+1}') \otimes \cdots \otimes a_1' \otimes p \\ &\quad + \sum_{i,j} (-1)^{|p|+\|a_1'\|+\cdots+\|a_i'\|} q \otimes a_t' \otimes \cdots \otimes \mu_{\mathcal{A}}^j(a_{i+j}', \dots, a_{i+1}') \otimes \cdots \otimes a_1' \otimes p \\ &\quad + \sum_i q \otimes a_t' \otimes \cdots \otimes \mu_{\mathcal{P}}^{i;1;0}(a_i', \dots, a_1'; p); \end{aligned} \tag{2.12}$$

the operations (for $r > 0$ or $s > 0$)

$$\mu^{s;1;0}_{\mathcal{Q}\otimes_{\mathcal{A}}\mathcal{P}}(a''_s,\dots,a''_1;q\otimes a'_t\otimes\cdots\otimes a'_1\otimes p)$$
$$=\sum_i(-1)^{|p|+\|a'_1\|+\cdots+\|a'_i\|}\mu^{s;1;t-i}_{\mathcal{Q}}(a''_s,\dots,a''_1;q;a'_t,\dots,a'_{i+1})\otimes\cdots\otimes a'_1\otimes p,$$
$$\mu^{0;1;r}_{\mathcal{Q}\otimes_{\mathcal{A}}\mathcal{P}}(q\otimes a'_t\otimes\cdots\otimes a'_1\otimes p;a_r,\dots,a_1)$$
$$=\sum_i q\otimes a'_t\otimes\cdots\otimes\mu^{i;1;r}_{\mathcal{P}}(a'_i,\dots,a'_1;p;a_r,\dots,a_1); \tag{2.13}$$

and with $\mu^{s;1;r}_{\mathcal{Q}\otimes_{\mathcal{A}}\mathcal{P}}=0$ if both r and s are positive.

Tensor product with the diagonal bimodule is essentially a trivial operation. More precisely, there are canonical quasi-isomorphisms [41, Eqn. (2.21)–(2.24)]

$$\mathcal{A}\otimes_{\mathcal{A}}\mathcal{P}\simeq\mathcal{P}\simeq\mathcal{P}\otimes_{\mathcal{A}}\mathcal{A}. \tag{2.14}$$

For arbitrary $\mathcal{A}$-bimodules $\mathcal{P}$, $\mathcal{Q}$, $\mathcal{R}$, there are canonical isomorphisms

$$\hom_{[\mathcal{A},\mathcal{A}]}(\mathcal{P},\mathcal{Q}^\vee)\cong\hom_{[\mathcal{A},\mathcal{A}]}(\mathcal{Q},\mathcal{P}^\vee), \tag{2.15}$$
$$\begin{aligned}\hom_{[\mathcal{A},\mathcal{A}]}(\mathcal{P}\otimes_{\mathcal{A}}\mathcal{Q},\mathcal{R}^\vee)&\cong\hom_{[\mathcal{A},\mathcal{A}]}(\mathcal{Q}\otimes_{\mathcal{A}}\mathcal{R},\mathcal{P}^\vee)\\&\cong\hom_{[\mathcal{A},\mathcal{A}]}(\mathcal{R}\otimes_{\mathcal{A}}\mathcal{P},\mathcal{Q}^\vee).\end{aligned} \tag{2.16}$$

As an application of (2.14) and (2.16), one gets

$$\hom_{[\mathcal{A},\mathcal{A}]}(\mathcal{P}^\vee,\mathcal{P}^\vee)\simeq\hom_{[\mathcal{A},\mathcal{A}]}(\mathcal{A}\otimes_{\mathcal{A}}\mathcal{P}^\vee,\mathcal{P}^\vee)\cong\hom_{[\mathcal{A},\mathcal{A}]}(\mathcal{P}^\vee\otimes_{\mathcal{A}}\mathcal{P},\mathcal{A}^\vee), \tag{2.17}$$
$$\hom_{[\mathcal{A},\mathcal{A}]}(\mathcal{P}^\vee,\mathcal{P}^\vee)\simeq\hom_{[\mathcal{A},\mathcal{A}]}(\mathcal{P}^\vee\otimes_{\mathcal{A}}\mathcal{A},\mathcal{P}^\vee)\cong\hom_{[\mathcal{A},\mathcal{A}]}(\mathcal{P}\otimes_{\mathcal{A}}\mathcal{P}^\vee,\mathcal{A}^\vee). \tag{2.18}$$

By starting with $\mathrm{id}_{\mathcal{P}^\vee}$ and going through (2.17), (2.18), one obtains canonical maps

$$\lambda_{\mathrm{left}}:\mathcal{P}^\vee\otimes_{\mathcal{A}}\mathcal{P}\longrightarrow\mathcal{A}^\vee, \tag{2.19}$$
$$\lambda_{\mathrm{right}}:\mathcal{P}\otimes_{\mathcal{A}}\mathcal{P}^\vee\longrightarrow\mathcal{A}^\vee. \tag{2.20}$$

There are explicit formulae for (2.14)–(2.20), but we prefer to omit them. For more foundational material on A_∞-bimodules, see [53, 28, 26, 29, 41].

2b. Invertible bimodules

A bimodule $\mathcal{P}$ is called invertible if there is another bimodule $\mathcal{P}^{-1}$ and quasi-isomorphisms

$$\mathcal{P}^{-1}\otimes_{\mathcal{A}}\mathcal{P}\simeq\mathcal{A}\simeq\mathcal{P}\otimes_{\mathcal{A}}\mathcal{P}^{-1}. \tag{2.21}$$

In that case, taking the tensor product with $\mathcal{P}$ (on objects, and the tensor product with $\mathrm{id}_{\mathcal{P}}$ on morphisms) is an automorphism, or more rigorously a quasi-equivalence, of the dg category $[\mathcal{A},\mathcal{A}]$.

Lemma 2.2. *If $\mathcal{P}$ is invertible, λ_{left} and λ_{right} are quasi-isomorphisms.*

Proof. Consider the chain of quasi-isomorphisms

$$\begin{aligned}\hom_{[\mathcal{A},\mathcal{A}]}(\mathcal{Q},\mathcal{P}^\vee)&\simeq\hom_{[\mathcal{A},\mathcal{A}]}(\mathcal{A}\otimes_{\mathcal{A}}\mathcal{Q},\mathcal{P}^\vee)\\&\cong\hom_{[\mathcal{A},\mathcal{A}]}(\mathcal{Q}\otimes_{\mathcal{A}}\mathcal{P},\mathcal{A}^\vee)\simeq\hom_{[\mathcal{A},\mathcal{A}]}(\mathcal{Q},\mathcal{A}^\vee\otimes_{\mathcal{A}}\mathcal{P}^{-1}).\end{aligned} \tag{2.22}$$

This is functorial in $\mathcal{Q}$, hence (by the Yoneda Lemma) arises from a quasi-isomorphism

$$\mathcal{P}^\vee \simeq \mathcal{A}^\vee \otimes_{\mathcal{A}} \mathcal{P}^{-1}. \tag{2.23}$$

More concretely, one obtains that quasi-isomorphism by setting $\mathcal{Q} = \mathcal{P}^\vee$ in (2.22), and then taking the identity in the leftmost morphism group. By comparing this with (2.17), one sees that the quasi-isomorphism is in fact $\lambda_{\mathrm{left}} \otimes_{\mathcal{A}} \mathrm{id}_{\mathcal{P}^{-1}}$. By tensoring with $\mathrm{id}_{\mathcal{P}}$, it follows that λ_{left} itself is a quasi-isomorphism. The other case is similar. □

Note that, by combining (2.23) with its counterpart for $\mathcal{P}^{-1} \otimes_{\mathcal{A}} \mathcal{A}^\vee$, we can get a canonical quasi-isomorphism (for invertible $\mathcal{P}$)

$$\mathcal{A}^\vee \otimes_{\mathcal{A}} \mathcal{P}^{-1} \simeq \mathcal{P}^{-1} \otimes_{\mathcal{A}} \mathcal{A}^\vee. \tag{2.24}$$

2c. Hochschild homology

The Hochschild homology of $\mathcal{A}$ with coefficients in a bimodule $\mathcal{P}$, written as $\mathrm{HH}_*(\mathcal{A}, \mathcal{P})$, is the homology of $\mathrm{CC}_*(\mathcal{A}, \mathcal{P}) = T(\bar{\mathcal{A}}[1]) \otimes \mathcal{P}$, with differential

$$\begin{aligned}
&\partial(a_d \otimes \cdots \otimes a_1 \otimes p) \\
&\quad = \sum_{i,j} (-1)^{|p|+\|a_1\|+\cdots+\|a_i\|} a_d \otimes \cdots \otimes \mu^j_{\mathcal{A}}(a_{i+j}, \dots, a_{i+1}) \otimes \cdots \otimes a_1 \otimes p \\
&\qquad + \sum_{i,j} (-1)^* a_{d-i} \otimes \cdots \otimes \mu^{j;1;i}_{\mathcal{P}}(a_j, \dots, a_1; p; a_d, \dots, a_{d-i+1}),
\end{aligned} \tag{2.25}$$

where $* = (\|a_{d-i+1}\| + \cdots + \|a_d\|)(|p| + \|a_1\| + \cdots + \|a_{d-i}\|)$. In spite of the subscript notation, the grading on $\mathrm{CC}_*(\mathcal{A}, \mathcal{P})$ is cohomological, meaning that ∂ has degree 1.

There is a canonical isomorphism

$$\mathrm{CC}_*(\mathcal{A}, \mathcal{Q} \otimes_{\mathcal{A}} \mathcal{P})^\vee \cong \hom_{[\mathcal{A},\mathcal{A}]}(\mathcal{P}, \mathcal{Q}^\vee). \tag{2.26}$$

In particular,

$$\mathrm{CC}_*(\mathcal{A}, \mathcal{P})^\vee \simeq \hom_{[\mathcal{A},\mathcal{A}]}(\mathcal{A}, \mathcal{P}^\vee). \tag{2.27}$$

By spelling out (2.27), one obtains the following:

Lemma 2.3. *Suppose that we have a cocycle $\xi \in \mathrm{CC}_0(\mathcal{A}, \mathcal{P})^\vee$, whose constant term $\xi^0 \in \mathcal{P}^\vee$ has the property that*

$$\begin{aligned}
\mathrm{H}^*(\mathcal{A}) &\longrightarrow \mathrm{H}^*(\mathcal{P})^\vee, \\
[a] &\longmapsto [\xi^0(\mu^{1;1;0}_{\mathcal{P}}(a; \cdot))]
\end{aligned} \tag{2.28}$$

is an isomorphism. Then, by (2.27), ξ gives rise to a quasi-isomorphism $\mathcal{A} \simeq \mathcal{P}^\vee$.

If $\mathcal{A}$ is proper (has finite-dimensional cohomology), the natural map $\mathcal{A} \to (\mathcal{A}^\vee)^\vee$ is a quasi-isomorphism. In that case, (2.26) implies that

$$\hom_{[\mathcal{A},\mathcal{A}]}(\mathcal{P}, \mathcal{A}) \simeq \mathrm{CC}_*(\mathcal{A}, \mathcal{A}^\vee \otimes_{\mathcal{A}} \mathcal{P})^\vee. \tag{2.29}$$

Hochschild homology is cyclically invariant, which means that there is a canonical isomorphism of chain complexes

$$\mathrm{CC}_*(\mathcal{A}, \mathcal{Q} \otimes_{\mathcal{A}} \mathcal{P}) \cong \mathrm{CC}_*(\mathcal{A}, \mathcal{P} \otimes_{\mathcal{A}} \mathcal{Q}). \tag{2.30}$$

The dual of this isomorphism, using (2.26), is (2.15). Similarly, (2.30) implies that

$$\mathrm{CC}_*(\mathcal{A}, \mathcal{R} \otimes_{\mathcal{A}} \mathcal{Q} \otimes_{\mathcal{A}} \mathcal{P}) \cong \mathrm{CC}_*(\mathcal{A}, \mathcal{P} \otimes_{\mathcal{A}} \mathcal{R} \otimes_{\mathcal{A}} \mathcal{Q}), \tag{2.31}$$

and the dual of that is (2.16). One can apply the same idea to the tensor powers of a single bimodule, leading to an action of the cyclic group $\mathbb{Z}/i$ on $\mathrm{CC}_*(\mathcal{A}, \mathcal{P}^{\otimes_{\mathcal{A}} i})$. By applying (2.29), one gets a $\mathbb{Z}/i$-action on $\mathrm{H}^*(hom_{[\mathcal{A},\mathcal{A}]}((\mathcal{A}^\vee)^{\otimes_{\mathcal{A}} i-1}, \mathcal{A}))$, assuming that $\mathcal{A}$ is proper.

2d. Hochschild cohomology

Even though we have emphasized Hochschild homology, it makes sense to also mention the Hochschild cohomology $\mathrm{HH}^*(\mathcal{A}, \mathcal{P})$, whose underlying chain complex is $\mathrm{CC}^*(\mathcal{A}, \mathcal{P}) = Hom(T(\bar{\mathcal{A}}[1]), \mathcal{P})$. It is well known that

$$\mathrm{CC}^*(\mathcal{A}, \mathcal{P}) \simeq hom_{[\mathcal{A},\mathcal{A}]}(\mathcal{A}, \mathcal{P}). \tag{2.32}$$

One can interpret (2.27) and Lemma 2.3 in those terms, given that

$$\mathrm{CC}_*(\mathcal{A}, \mathcal{P})^\vee = \mathrm{CC}^*(\mathcal{A}, \mathcal{P}^\vee). \tag{2.33}$$

There is another, less straightforward, connection between the two Hochschild theories. Namely, suppose that $\mathcal{A}$ is proper and homologically smooth, in which case $\mathcal{A}^\vee$ is always invertible (see, e.g., [51, Theorem 4.5]). Then (see, e.g., [26, Remark 8.2.4])

$$\mathrm{CC}_*(\mathcal{A}, \mathcal{P}) \simeq hom_{[\mathcal{A},\mathcal{A}]}((\mathcal{A}^\vee)^{-1}, \mathcal{P}) \simeq \mathrm{CC}^*(\mathcal{A}, \mathcal{A}^\vee \otimes_{\mathcal{A}} \mathcal{P}). \tag{2.34}$$

If one specializes this to $\mathcal{P} = \mathcal{A}$, the outcome is a quasi-isomorphism between $\mathrm{CC}_*(\mathcal{A}, \mathcal{A})$ and its dual, which gives rise to a nondegenerate pairing on $\mathrm{HH}_*(\mathcal{A}, \mathcal{A})$ (see [51] for an extended discussion).

2e. Self-conjugation

Fix an invertible bimodule $\mathcal{P}$, and some $i \geq 1$. We define an endomorphism of $hom_{[\mathcal{A},\mathcal{A}]}(\mathcal{P}^{\otimes_{\mathcal{A}} i}, \mathcal{A})$, unique up to chain homotopy, through the homotopy commutative diagram

$$\begin{array}{ccc}
hom_{[\mathcal{A},\mathcal{A}]}(\mathcal{P}^{\otimes_{\mathcal{A}} i}, \mathcal{A}) & \dashrightarrow & hom_{[\mathcal{A},\mathcal{A}]}(\mathcal{P}^{\otimes_{\mathcal{A}} i}, \mathcal{A}) \\
{\scriptstyle id_{\mathcal{P}} \otimes_{\mathcal{A}} \cdot}\downarrow\simeq & & \simeq\downarrow{\scriptstyle \cdot \otimes_{\mathcal{A}} id_{\mathcal{P}}} \\
hom_{[\mathcal{A},\mathcal{A}]}(\mathcal{P}^{\otimes_{\mathcal{A}} i+1}, \mathcal{P} \otimes_{\mathcal{A}} \mathcal{A}) & & hom_{[\mathcal{A},\mathcal{A}]}(\mathcal{P}^{\otimes_{\mathcal{A}} i+1}, \mathcal{A} \otimes_{\mathcal{A}} \mathcal{P}). \\
 & \searrow^{\simeq} \quad {}^{\simeq}\swarrow & \\
 & hom_{[\mathcal{A},\mathcal{A}]}(\mathcal{P}^{\otimes_{\mathcal{A}} i+1}, \mathcal{P}) &
\end{array} \tag{2.35}$$

Again up to homotopy, these maps are compatible with the ring structure induced by the tensor product

$$\begin{aligned} hom_{[\mathcal{A},\mathcal{A}]}(\mathcal{P}^{\otimes_{\mathcal{A}} j}, \mathcal{A}) \otimes hom_{[\mathcal{A},\mathcal{A}]}(\mathcal{P}^{\otimes_{\mathcal{A}} i}, \mathcal{A}) &\longrightarrow hom_{[\mathcal{A},\mathcal{A}]}(\mathcal{P}^{\otimes_{\mathcal{A}} i+j}, \mathcal{A} \otimes_{\mathcal{A}} \mathcal{A}) \\ &\xrightarrow{\simeq} hom_{[\mathcal{A},\mathcal{A}]}(\mathcal{P}^{\otimes_{\mathcal{A}} i+j}, \mathcal{A}). \end{aligned} \tag{2.36}$$

We call the induced cohomology automorphisms *self-conjugation maps*, and denote them by

$$K_i \circlearrowright H^*(\hom_{[\mathcal{A},\mathcal{A}]}(\mathcal{P}^{\otimes_{\mathcal{A}} i}, \mathcal{A})). \tag{2.37}$$

Example 2.4. Let $A = \Lambda^*(\mathbb{K})$ be the exterior algebra in one variable (denoted by x, and given degree 1). Given any $\lambda \in \mathbb{K}^\times$, one has an invertible A-bimodule P_λ, whose underlying graded vector space is the same as A, but where the action of x on the right is multiplied by λ (while that on the left remains the same). Consider the bimodule map (of degree 1) $t : P_\lambda \to A$ which sends 1 to x. Then

$$\begin{cases} P_\lambda \otimes_A P_\lambda \xrightarrow{t \otimes_A \mathrm{id}_{P_\lambda}} A \otimes_A P_\lambda \cong P_\lambda & \text{maps } 1 \otimes 1 \mapsto x, \\ P_\lambda \otimes_A P_\lambda \xrightarrow{\mathrm{id}_{P_\lambda} \otimes_A t} P_\lambda \otimes_A A \cong P_\lambda & \text{maps } 1 \otimes 1 \mapsto \lambda x. \end{cases} \tag{2.38}$$

This translates straightforwardly into the A_∞-world, and provides an example where (2.37), for $i = 1$, is not the identity.

For any invertible bimodule $\mathcal{P}$, there is a homotopy commutative diagram

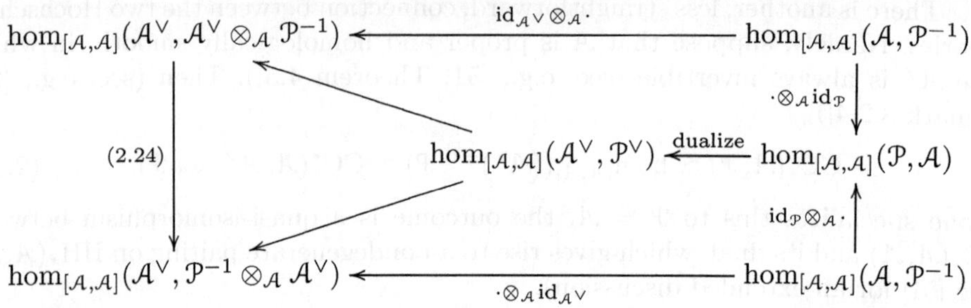

(2.39)

The diagonal arrows are (2.23) and its counterpart; hence, the left triangle in the diagram commutes by definition of (2.24). Proving the commutativity of the remaining parts requires one to go back to (2.22), and we will not explain the details here. Going vertically down the right column of (2.39) gives an automorphism of $H^*(\hom_{[\mathcal{A},\mathcal{A}]}(\mathcal{A}, \mathcal{P}^{-1}))$, which is a form of K_1. Suppose that $\mathcal{A}^\vee$ is an invertible bimodule. Then, by going around the diagram the other way, we get another description of that automorphism, which now involves the interaction of $\mathcal{P}$ and $\mathcal{A}^\vee$.

Example 2.5. Suppose that $\mathcal{A}$ is weakly Calabi–Yau of dimension n, by which we mean that it comes with a quasi-isomorphism of bimodules

$$\mathcal{A}^\vee[-n] \simeq \mathcal{A}. \tag{2.40}$$

By inserting that quasi-isomorphism on both sides of (2.24), one gets a distinguished automorphism of $\mathcal{P}^{-1}$ (and correspondingly of $\mathcal{P}$), which describes the failure of the tensor product with that bimodule to be compatible with the Calabi–Yau structure. It then follows from (2.39) that K_1 is the composition with that automorphism. This agrees with Example 2.4.

Lemma 2.6. *Suppose that $\mathcal{A}$ is proper. Suppose also that $\mathcal{A}^\vee$ is invertible (which holds if $\mathcal{A}$ is smooth). Then, the generator of the $\mathbb{Z}/(i+1)$-action on*

$$\mathrm{H}^*(\hom_{[\mathcal{A},\mathcal{A}]}((\mathcal{A}^\vee)^{\otimes_{\mathcal{A}} i}, \mathcal{A}))$$

equals the self-conjugation map K_i, for the bimodule $\mathcal{A}^\vee$.

We will only explain the counterpart of this in classical algebra. Given a finite-dimensional algebra A, a bimodule map

$$(A^\vee)^{\otimes_A i} \longrightarrow A \tag{2.41}$$

is given by an element

$$\sum_j a_{i+1}^j \otimes \cdots \otimes a_1^j \in A^{\otimes_{\mathbb{K}} i+1} \tag{2.42}$$

satisfying the equation

$$\sum_j a_{i+1}^j a \otimes a_i^j \otimes \cdots \otimes a_1^j = \sum_j a_{i+1}^j \otimes a a_i^j \otimes \cdots \otimes a_1^j \tag{2.43}$$

for $a \in A$, as well as all its cyclic permutations. The bimodule map associated to (2.42) is

$$\alpha_1 \otimes \cdots \otimes \alpha_i \longmapsto \sum_j \alpha_1(a_1^j) \cdots \alpha_i(a_i^j) a_{i+1}^j \in A. \tag{2.44}$$

If one tensors this with the identity map of $A^\vee$ on the right or left, the outcome are bimodule maps $(A^\vee)^{\otimes_A i+1} \to A^\vee$ given by, respectively,

$$\alpha_1 \otimes \cdots \otimes \alpha_{i+1} \longmapsto \sum_j \alpha_1(a_1^j) \cdots \alpha_i(a_i^j) \alpha_{i+1}(\cdot\, a_{i+1}^j), \tag{2.45}$$

$$\alpha_1 \otimes \cdots \otimes \alpha_{i+1} \longmapsto \sum_j \alpha_2(a_1^j) \cdots \alpha_{i+1}(a_i^j) \alpha_1(a_{i+1}^j\, \cdot). \tag{2.46}$$

Again using (2.43), one can write the right-hand side of (2.46) as

$$\sum_j \alpha_2(\cdot\, a_1^j) \alpha_3(a_2^j) \cdots \alpha_1(a_{i+1}^j), \tag{2.47}$$

which is indeed obtained from (2.45) by cyclically permuting the tensor factors in (2.42).

Example 2.7. For some $i \geq 1$, consider a graded algebra $A = \bigoplus_{k \in \mathbb{Z}/(i+1)} \mathbb{K}e_k \oplus \mathbb{K}x_k$. We draw it in quiver form as follows:

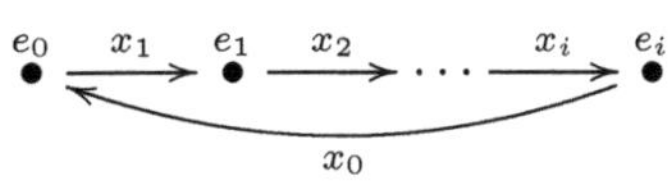

(2.48)

The vertices give rise to mutually orthogonal idempotents e_k, and the x_k satisfy

$$\begin{aligned} e_{k+1} x_k e_k &= x_k, \\ x_{k+1} x_k &= 0. \end{aligned} \tag{2.49}$$

The e_k have degree zero, and we choose the degrees of the x_k so that they add up to $i-1$. Following (2.42), each of the expressions

$$x_k \otimes x_{k-1} \otimes \cdots \otimes x_{k-i} \tag{2.50}$$

gives rise to a bimodule map (2.41) of degree $i-1$.

If one thinks of A (in the straightforward way) as an A_∞-algebra $\mathcal{A}$, the elements (2.50) induce A_∞-bimodule maps $(\mathcal{A}^\vee)^{\otimes_{\mathcal{A}} i} \to \mathcal{A}[i-1]$, which are cyclically exchanged by K^i (up to signs). It is a familiar fact that $\mathcal{A}$ is "virtually Calabi–Yau of dimension $(i-1)/(i+1)$", meaning that

$$(\mathcal{A}^\vee)^{\otimes_{\mathcal{A}} i+1} \simeq \mathcal{A}[i-1]. \tag{2.51}$$

Hence,

$$\mathrm{H}^{i-1}(\hom_{[\mathcal{A},\mathcal{A}]}((\mathcal{A}^\vee)^{\otimes_{\mathcal{A}} i}, \mathcal{A})) \cong \mathrm{H}^0(\hom_{[\mathcal{A},\mathcal{A}]}(\mathcal{A}, \mathcal{A}^\vee)) \cong \mathrm{HH}_0(\mathcal{A},\mathcal{A})^\vee. \tag{2.52}$$

One easily computes that

$$\mathrm{HH}_*(\mathcal{A},\mathcal{A}) = \begin{cases} \mathbb{K}^{i+1} & * = 0, \\ \mathbb{K} & * < 0, \\ 0 & * > 0. \end{cases} \tag{2.53}$$

The elements of (2.52) which we have constructed are a dual basis of $\mathrm{HH}_0(\mathcal{A},\mathcal{A})$.

Example 2.8. Start with the graded algebra from the previous example (with $i > 1$), but now let $\mathcal{A}$ be an A_∞-deformation of it, in which $\mu_{\mathcal{A}}^{i+1}(x_k, \dots, x_{k-i})$ is a nonzero multiple of e_k (if this is the case for one k, it must be true for all k, since these operations are related to each other by the A_∞-associativity equations; in fact, $\mathcal{A}$ is unique up to A_∞-isomorphism). The counterpart of (2.53) is that

$$\mathrm{HH}_*(\mathcal{A},\mathcal{A}) = \begin{cases} \mathbb{K}^i & * = 0, \\ 0 & * \neq 0. \end{cases} \tag{2.54}$$

This deformation still satisfies (2.51), and hence (2.52). The $\mathbb{Z}/(i+1)$-action on (2.52) generated by K_i is the action on $\mathbb{K}^i$ obtained by taking the standard cyclic permutation representation, and dividing by the one-dimensional diagonal subspace.

2f. Noncommutative divisors

We now have all the ingredients needed to flesh out the outline previously given in Section 1b.

Definition 2.9. A noncommutative divisor on $\mathcal{A}$, with underlying invertible bimodule $\mathcal{P}$, is an A_∞-algebra structure $\mu_{\mathcal{B}}$ on the graded vector space (1.8), with the following properties:

(i) $\mathcal{A} \subset \mathcal{B}$ is an A_∞-subalgebra.

(ii) Consider $\mathcal{B}$ as an $\mathcal{A}$-bimodule (by restriction of the diagonal bimodule $\mathcal{B}$ to the subalgebra $\mathcal{A}$). Then, the induced $\mathcal{A}$-bimodule structure on $\mathcal{B}/\mathcal{A} = \mathcal{P}[1]$ agrees with the previously given one.

Two noncommutative divisors (with the same underlying bimodule) are isomorphic if there is an isomorphism of the associated A_∞-algebras $\mathcal{B}$ which restricts to the identity map on $\mathcal{A}$, and induces the identity map on the $\mathcal{A}$-bimodule $\mathcal{B}/\mathcal{A}$.

Example 2.10. There is a trivial special case, denoted by $\mathcal{B}^{\mathrm{triv}}$, which is the trivial extension algebra obtained from $\mathcal{A}$ and $\mathcal{P}$. In that case, the only nonzero components of $\mu_{\mathcal{B}}$ are those dictated by (i) and (ii) above.

What does this mean concretely? If we restrict $\mu_{\mathcal{B}}$ to $T(\mathcal{A}[1]) \subset T(\mathcal{B}[1])$, it takes values in $\mathcal{A}$ and agrees with $\mu_{\mathcal{A}}$. Next, if we restrict $\mu_{\mathcal{B}}$ to $T(\mathcal{A}[1]) \otimes \mathcal{P}[2] \otimes T(\mathcal{A}[1]) \subset T(\mathcal{B}[1])$, it has the form

$$\begin{aligned} \mu_{\mathcal{B}}^{r+s+1}(a'_s, \dots, a'_1, p, a_r, \dots, a_1) = \mu_{\mathcal{P}}^{s;1;r}(a'_s, \dots, a'_1; p; a_r, \dots, a_1) \\ + \theta^{s;1;r}(a'_s, \dots, a'_1; p; a_r, \dots, a_1) \end{aligned} \tag{2.55}$$

for some

$$\theta : T(\mathcal{A}[1]) \otimes \mathcal{P} \otimes T(\mathcal{A}[1]) \longrightarrow \mathcal{A}, \tag{2.56}$$

which is the first new piece of information (not described by $\mathcal{A}$ and $\mathcal{P}$) in the noncommutative divisor. The A_∞-associativity equation for $\mu_{\mathcal{B}}$ implies that θ is a bimodule map $\mathcal{P} \to \mathcal{A}$, meaning that it satisfies (2.5). We call its cohomology class

$$[\theta] \in \mathrm{H}^0(\hom_{[\mathcal{A},\mathcal{A}]}(\mathcal{P}, \mathcal{A})) \tag{2.57}$$

the section associated to the noncommutative divisor. It is an isomorphism invariant, which classifies the bimodule extension (1.9). For instance, $\mathcal{B}^{\mathrm{triv}}$ has $\theta = 0$. Hence, (2.57) is an obstruction (not the only one, in general) to the triviality of a noncommutative divisor.

The next piece of the structure of a noncommutative divisor is obtained by restricting $\mu_{\mathcal{B}}$ to $T(\mathcal{A}[1]) \otimes \mathcal{P}[2] \otimes T(\mathcal{A}[1]) \otimes \mathcal{P}[2] \otimes T(\mathcal{A}[1])$, and then projecting the outcome to $\mathcal{P}[3]$. This can be viewed as an element of $\hom_{[\mathcal{A},\mathcal{A}]}(\mathcal{P} \otimes_{\mathcal{A}} \mathcal{P}, \mathcal{P})$ of degree -1. It is not a cocycle; instead, as shown in [41, Lemma 3.2], it provides a homotopy (in the dg category of bimodules) between the two sides of the diagram

$$\begin{array}{ccccc} & & \mathcal{P} \otimes_{\mathcal{A}} \mathcal{P} & & \\ & \swarrow^{\mathrm{id}_{\mathcal{P}} \otimes_{\mathcal{A}} \theta} & & \searrow^{\theta \otimes_{\mathcal{A}} \mathrm{id}_{\mathcal{P}}} & \\ \mathcal{P} \otimes_{\mathcal{A}} \mathcal{A} & & & & \mathcal{A} \otimes_{\mathcal{A}} \mathcal{P} \\ & \searrow_{\simeq} & & \swarrow_{\simeq} & \\ & & \mathcal{P}. & & \end{array} \tag{2.58}$$

In terms of (2.37), this means that

$$K_1([\theta]) = [\theta]. \tag{2.59}$$

One can put the previous discussion on a more systematic footing, as follows. Consider the Hochschild cochain complex of $\mathcal{B}^{\mathrm{triv}}$. We recall (from Section 2d) that this is

$$CC^*(\mathcal{B}^{\mathrm{triv}}, \mathcal{B}^{\mathrm{triv}})[1] = \mathrm{Hom}\big(T(\bar{\mathcal{B}}^{\mathrm{triv}}[1]), \mathcal{B}^{\mathrm{triv}}[1]\big), \tag{2.60}$$

with a differential given by $\mu_{\mathcal{B}^{\mathrm{triv}}}$, which means by $\mu_{\mathcal{A}}$ and $\mu_{\mathcal{P}}$. Let us equip $\mathcal{B}^{\mathrm{triv}}$ with an additional grading (called *weight grading* to distinguish it from the usual one), in which $\mathcal{A}$ has weight 0 and $\mathcal{P}[1]$ has weight -1. For any $i > 0$, let $F^i\mathfrak{g}$ be the subspace of (2.60) consisting of maps which increase the weight by (exactly) i. Write $\mathfrak{g}$ for the (bigraded) space which combines all the $F^i\mathfrak{g}$. The Hochschild differential maps each $F^i\mathfrak{g}$ to itself. Moreover, its cohomology forms a long exact sequence

$$\begin{aligned} \cdots \to H^{*-2i}(\hom_{[\mathcal{A},\mathcal{A}]}(\mathcal{P}^{\otimes_{\mathcal{A}} i+1}, \mathcal{P})) \longrightarrow H^*(F^i\mathfrak{g}) \\ \longrightarrow H^{*-2i+1}(\hom_{[\mathcal{A},\mathcal{A}]}(\mathcal{P}^{\otimes_{\mathcal{A}} i}, \mathcal{A})) \longrightarrow H^{*-2i+1}(\hom_{[\mathcal{A},\mathcal{A}]}(\mathcal{P}^{\otimes_{\mathcal{A}} i+1}, \mathcal{P})) \to \cdots \end{aligned} \tag{2.61}$$

where the last map is the difference between $\phi \mapsto \mathrm{id}_{\mathcal{P}} \otimes_{\mathcal{A}} \phi$ and $\phi \mapsto \phi \otimes_{\mathcal{A}} \mathrm{id}_{\mathcal{P}}$. Since either map is an isomorphism, one can identify $H^*(\hom_{[\mathcal{A},\mathcal{A}]}(\mathcal{P}^{\otimes_{\mathcal{A}} i+1}, \mathcal{P})) \cong H^*(\hom_{[\mathcal{A},\mathcal{A}]}(\mathcal{P}^{\otimes_{\mathcal{A}} i}, \mathcal{A}))$, and then (2.61) becomes

$$\begin{aligned} \cdots \to H^{*-2i}(\hom_{[\mathcal{A},\mathcal{A}]}(\mathcal{P}^{\otimes_{\mathcal{A}} i}, \mathcal{A})) \longrightarrow H^*(F^i\mathfrak{g}) \\ \longrightarrow H^{*-2i+1}(\hom_{[\mathcal{A},\mathcal{A}]}(\mathcal{P}^{\otimes_{\mathcal{A}} i}, \mathcal{A})) \xrightarrow{K_i - \mathrm{id}} H^{*-2i+1}(\hom_{[\mathcal{A},\mathcal{A}]}(\mathcal{P}^{\otimes_{\mathcal{A}} i}, \mathcal{A})) \to \cdots \end{aligned} \tag{2.62}$$

where K_i is (2.37). From this point of view, (2.59) holds because the structure of a noncommutative divisor specifies a lift of $[\theta]$ to $H^1(F^1\mathfrak{g})$.

Remark 2.11. In our applications, $\mathcal{A}$ is proper and smooth (hence $\mathcal{A}^\vee$ is invertible), and

$$\mathcal{P} = \mathcal{A}^\vee[-n]. \tag{2.63}$$

By Lemma 2.6, the kernel of the boundary map in (2.61) is

$$H^*(\hom_{[\mathcal{A},\mathcal{A}]}((\mathcal{A}^\vee)^{\otimes_{\mathcal{A}} i}, \mathcal{A}))^{\mathbb{Z}/(i+1)}.$$

This hints at the existence of a more refined notion, considered in [45, 27]. Let us suppose for simplicity that $\mathcal{A}$ is finite-dimensional. More importantly, we assume that our coefficient field $\mathbb{K}$ has characteristic 0. Then, a *noncommutative anticanonical divisor* is one with $\mathcal{P}$ as in (2.63), and such that the A_∞-structure $\mu_{\mathcal{B}}$ is cyclic with respect to the obvious inner product on $\mathcal{B} = \mathcal{A} \oplus \mathcal{A}^\vee[1-n]$. The counterparts $F^i\mathfrak{g}^{\mathrm{cyc}}$ of the spaces $F^i\mathfrak{g}$ satisfy

$$H^*(F^i\mathfrak{g}^{\mathrm{cyc}}) \cong H^{*+(n-2)i+1}(\hom_{[\mathcal{A},\mathcal{A}]}((\mathcal{A}^\vee)^{\otimes_{\mathcal{A}} i}, \mathcal{A})^{\mathbb{Z}/(i+1)}). \tag{2.64}$$

Let us keep the assumption that the coefficient field $\mathbb{K}$ has characteristic 0. The complex (2.60) carries the structure of dg Lie algebra, and since the bracket respects weight gradings, we get the structure of a (bigraded) dg Lie algebra on $\mathfrak{g}$. A noncommutative divisor is given by a solution of the Maurer–Cartan equation [23, 30]

$$\beta \in \mathfrak{g}^1, \quad d_{\mathfrak{g}}\beta + \tfrac{1}{2}[\beta, \beta] = 0 \tag{2.65}$$

(the trivial solution $\beta = 0$ corresponds to $\mathcal{B}^{\mathrm{triv}}$). One can decompose (2.65) into a series of equations for the weight components $(\beta^1, \beta^2, \dots)$, which are

$$d_{\mathfrak{g}}\beta^i + \tfrac{1}{2}\sum_j [\beta^j, \beta^{i-j}] = 0. \tag{2.66}$$

The notion of isomorphism for noncommutative divisors reduces to the usual gauge equivalence for solutions of (2.65). Standard obstruction theory yields:

Lemma 2.12. *Suppose that*

$$H^*(\hom_{[\mathcal{A},\mathcal{A}]}(\mathcal{P}^{\otimes_{\mathcal{A}} i}, \mathcal{A})) = 0 \quad \textit{for all } i \geq 1 \textit{ and } * < 0. \tag{2.67}$$

Then, the structure of a noncommutative divisor is determined (up to isomorphism) by (2.57), *which moreover must satisfy* (2.59). □

Here is another equivalent formulation, which will be convenient when considering generalizations. Take the one-dimensional vector space $\mathbb{K}[-1]$, placed in degree 1 and weight 1, and considered as a dg Lie algebra with trivial differential and bracket. A solution of (2.66) is the same as an L_∞-homomorphism

$$\mathbb{K}[-1] \longrightarrow \mathfrak{g} \tag{2.68}$$

which preserves the weight grading (more precisely, the image of $1^{\otimes i}$ under the i-linear part of (2.68) should be divided by $i!$ to get β^i; this is the usual action of L_∞-homomorphisms on Maurer–Cartan elements [25, Section 4.3]).

Following up on (1.10), we have:

(2.69) (i) Of course, $\mathcal{B}$ itself is an A_∞-algebra.

(2.69) (ii) Take $\mathcal{B}[[u]]$, where the formal variable u has degree 2. Let

$$\mathcal{D} = \mathcal{A} \oplus \prod\nolimits_{i \geq 1} u^i \mathcal{B} = \mathcal{A}[[u]] \oplus u(\mathcal{P}[1])[[u]] \subset \mathcal{B}[[u]]$$

be the space of $\mathcal{B}$-valued formal series in u whose constant term lies in $\mathcal{A}$. Following [44], we make $\mathcal{D}$ into a curved A_∞-algebra by extending $\mu_{\mathcal{B}}$ u-linearly, and then adding a curvature term $\mu^0_{\mathcal{D}} = u\, e_{\mathcal{A}}$. Then, $\mathcal{A}$ is a (right) A_∞-module over $\mathcal{D}$, by pullback under the projection $\mathcal{D} \to \mathcal{A}$. One can then define $\mathcal{A} \setminus \mathcal{B}$ as the endomorphism ring of that module (it is important that we consider $\mathcal{D}$ to be defined over $\mathbb{K}$, but equipped with the u-adic topology). Explicitly, the underlying vector space [41, Equation (4.7)] is

$$\mathcal{A} \setminus \mathcal{B} = \bigoplus_{\substack{k \geq 0 \\ i_1, \dots, i_k > 0}} \begin{aligned}[t] \mathrm{Hom}\big(\mathcal{A} \otimes T(\bar{\mathcal{A}}[1]) \otimes u^{i_k}\mathcal{B}[1] \otimes T(\bar{\mathcal{A}}[1]) \otimes \cdots \\ \cdots \otimes u^{i_1}\mathcal{B}[1] \otimes T(\bar{\mathcal{A}}[1]), \mathcal{A}\big). \end{aligned}$$

The identification of this construction with a suitable categorical localization (or quotient) is provided by [41, Theorem 4.1].

2g. Noncommutative pencils

We continue in the same framework as before, including the assumption that $\mathbb{K}$ has characteristic 0. Fix a two-dimensional vector space V, which will be given weight grading -1 (and ordinary grading 0). Denote the symmetric algebra by $\mathrm{Sym}^*(V)$, and the dual vector space by $W = V^\vee$.

Definition 2.13. A noncommutative pencil on $\mathcal{A}$, with underlying bimodule $\mathcal{P}$, is a map

$$\wp : T(\mathcal{B}[1]) \longrightarrow \mathcal{B}[2] \otimes \mathrm{Sym}^*(V) \tag{2.70}$$

which preserves both the ordinary grading and the one by weight, and with the following property: for any $w \in W$, we can use the associated evaluation map $\mathrm{Sym}^*(V) \longrightarrow \mathbb{K}$ to specialize (2.70) to $\mu_{\mathcal{B},w} : T(\mathcal{B}[1]) \longrightarrow \mathcal{B}[2]$; and all these should be noncommutative divisors.

As before, this definition lends itself to piecewise analysis. The first term not determined by $\mathcal{A}$ and $\mathcal{P}$ has the form (after dualizing)

$$W \otimes T(\mathcal{A}[1]) \otimes \mathcal{P} \otimes T(\mathcal{A}[1]) \longrightarrow \mathcal{A}. \tag{2.71}$$

This is a family of bimodule maps $\mathcal{P} \to \mathcal{A}$, depending linearly on W. To make things even more concrete, suppose that we identify V and W with $\mathbb{K}^2$ by choosing dual bases. Then, (2.71) consists of two bimodule maps ρ and σ, corresponding to $w = (1, 0)$ resp. $w = (0, 1)$, which we have called the sections associated to the noncommutative pencil.

Let us return to the bigraded dg Lie algebra $\mathfrak{g}$. Take $W[-1]$, as a vector space placed in degree 1 and weight 1, and equip it with the trivial dg Lie structure. A noncommutative pencil is the same as an L_∞-homomorphism, preserving the weight grading,

$$W[-1] \longrightarrow \mathfrak{g}. \tag{2.72}$$

There is a standard obstruction theory for such homomorphisms, which in particular yields an analogue of Lemma 2.12:

Lemma 2.14. *If* (2.67) *holds, the structure of a noncommutative pencil is determined up to isomorphism by the cohomology classes* $[\rho]$, $[\sigma]$ *(each of which must be fixed by* K_1*).* □

By definition, a noncommutative pencil yields a noncommutative divisor for each $w \in W$. For $w = 0$, this reduces to the trivial extension algebra $\mathcal{B}^{\mathrm{triv}}$. Assuming as before that $W = \mathbb{K}^2$, we write $\mathcal{B}_\infty$ for the noncommutative divisor associated to $w = (1, 0)$, and $\mathcal{B}_z$ for that associated to $w = (z, 1)$. The associated sections are those given in (1.12). This essentially exhausts all the possibilities, since the noncommutative divisors associated to w and λw, for $\lambda \in \mathbb{K}^\times$, are related by an automorphism of $\mathcal{B}$ (which acts trivially on $\mathcal{A}$ and rescales $\mathcal{P}$ by λ).

Let us spell out the other structures mentioned in (1.12):

(2.73) (i) Set $w = (1, q)$, where q is a formal variable. The corresponding specialization of (2.70) is a map $T(\mathcal{B}[1]) \longrightarrow (\mathcal{B}[2])[[q]]$. This defines a $\mathbb{C}[[q]]$-linear A_∞-structure on $\mathcal{B}[[q]]$, which is a deformation of $\mathcal{B}_\infty$. This is what we denoted by $\hat{\mathcal{B}}_\infty$.

(2.73) (ii) Using the basis $r = (1,0)$ and $s = (0,1)$ of V, write $\wp^d = \sum_{i,j} \wp^d_{i,j} \otimes r^i s^j$. Because $\wp^d_{i,j}$ increases weights by $i+j$, it is nontrivial only if $d \geq i+j$. Moreover, if we extend it u-linearly to a map $\mathcal{D}^{\otimes d} \to \mathcal{D}$, then it is divisible by u^{i+j}. Hence, the expression

$$\sum_{i,j} \wp^d_{i,j} u^{-j} h^j$$

makes sense, as a family of maps $\mathcal{D}^{\otimes d} \to \mathcal{D}$ depending on an auxiliary formal variable h of degree 2. We add to it a curvature term as in (2.69), and get a formal deformation of our previous curved A_∞-structure on $\mathcal{D}$, with parameter h. This induces a deformation of $\mathcal{A} \setminus \mathcal{B}_\infty$ (with a curvature term, which vanishes if we set $h = 0$). We write it as $\mathcal{LG}$.

Remark 2.15. (i) A pencil in ordinary algebraic geometry has a (compactified) graph, see, e.g., [52, p. 154]. Here is one possible approach towards defining its noncommutative analogue: over the projective line $\mathbb{P}^1 = \mathbb{P}(W)$, consider the graded quasi-coherent sheaf

$$\mathcal{A} = \mathcal{A} \otimes \mathcal{O}_{\mathbb{P}^1} \oplus \mathcal{P}[1] \otimes \mathcal{O}_{\mathbb{P}^1}(-1). \tag{2.74}$$

Then, (2.70) makes this into a sheaf of A_∞-algebras. Using, e.g., Čech resolutions, one can associate to this a single A_∞-algebra, whose underlying cohomology is $H^*(\mathbb{P}^1, \mathcal{A} \otimes \mathrm{End}(\mathcal{E}))$ for some generator $\mathcal{E}$ of the derived category of $\mathbb{P}^1$, for instance $\mathcal{E} = \mathcal{O}_{\mathbb{P}^1} \oplus \mathcal{O}_{\mathbb{P}^1}(-1)$.

(ii) A pencil in classical algebraic geometry also has a base locus (or axis). A first idea for its noncommutative analogue is

$$\mathcal{B}_0 \otimes_{\mathcal{A}} \mathcal{B}_\infty, \tag{2.75}$$

but this is not a priori an A_∞-algebra (only a bimodule over $\mathcal{A}$, and as such, does not use the full structure of the noncommutative pencil). However, even at this level, the notion has geometric significance: we conjecture that, for the noncommutative pencils obtained from anticanonical Lefschetz pencils as in Conjecture 1.6, (2.75) is trivial, meaning quasi-isomorphic to zero (this property would distinguish them from Lefschetz fibrations with singular fibre at ∞).

In classical algebraic geometry, (i) and (ii) are related, since (under suitable smoothness assumptions) the graph is obtained by blowing up the base locus. Noncommutative counterparts of this relation would be of interest, because they arise when considering the relation between the Fukaya category of an ample hypersurface and the wrapped Fukaya category of its complement. Concretely, in the case where (2.75) is trivial, the Fukaya category of the ample hypersurface should appear as a formal completion of the wrapped Fukaya category of the complement.

2h. Discussion of the assumptions

Throughout this section, we have required that all A_∞-structures should be $\mathbb{Z}$-graded. One can actually work with $\mathbb{Z}/2$-gradings throughout, with the same results, except for Lemmas 2.12 and 2.14.

Remark 2.16. This is not quite the end of this topic, as illustrated by the intermediate situation mentioned in Remark 1.7. There, $\mathcal{A}$ and $\mathcal{P}$ are $\mathbb{Z}$-graded, and so is $\wp$ provided that we take $V = \mathbb{K} \oplus \mathbb{K}[2m]$. The associated sections ρ and σ have degrees 0 and $2m$, respectively. While $\mathcal{B}_\infty$ is still $\mathbb{Z}$-graded, the other $\mathcal{B}_z$, $z \in \mathbb{K}$, are only $\mathbb{Z}/2$-graded. The original $\mathbb{Z}$-grading leads to an isomorphism $\mathcal{B}_z \to \mathcal{B}_{\lambda^{2m}z}$, for any $\lambda \in \mathbb{K}^\times$. Similarly, the formal deformation $\hat{\mathcal{B}}_\infty$ is $\mathbb{Z}$-graded if we assign degree $-2m$ to the deformation variable q.

The other assumption we have imposed, starting with (2.65), is that the coefficient field $\mathbb{K}$ has characteristic 0. This can be dropped as well, at the cost of making some formulations a little more complicated. Maurer–Cartan theory, which means the classification theory of solutions of (2.65), does not apply in positive characteristic. However, a simpler obstruction theory argument suffices to prove Lemma 2.12 (and its variant, Lemma 2.14), and that does not require any assumption on $\mathbb{K}$. The second case where we have used the assumption on $\mathbb{K}$ is in the definition of noncommutative pencil, which was formulated by specializing two-variable polynomials, appearing as elements of $\mathrm{Sym}(V)$, to one-dimensional linear subspaces. In positive characteristic, such specializations may fail to recover the original polynomial ($f(x,y) = x^p y - xy^p$ vanishes on every line, for $\mathbb{K} = \mathbb{F}_p$). Hence, the definition of noncommutative pencil would have to be reformulated as a condition on (2.70) itself.

Finally, there is a variation on our general setup which will be important for applications. Namely, one can replace the ground field $\mathbb{K}$ by the semisimple ring

$$R = \mathbb{K}^m = \mathbb{K}e_1 \oplus \cdots \oplus \mathbb{K}e_m. \tag{2.76}$$

An A_∞-algebra over R consists of a graded R-bimodule $\mathcal{A}$, together with R-bimodule maps

$$\mu^d_{\mathcal{A}} : \mathcal{A} \otimes_R \mathcal{A} \otimes_R \cdots \otimes_R \mathcal{A} \longrightarrow \mathcal{A}[2-d]. \tag{2.77}$$

This is in fact the same as an A_∞-category with objects labeled by $\{1,\dots,m\}$. The strict unit $e_{\mathcal{A}}$ must lie in the diagonal part of the R-bimodule $\mathcal{A}$, which means that it can be written as $e_{\mathcal{A}} = e_{\mathcal{A},1} + \cdots + e_{\mathcal{A},m}$ with $e_{\mathcal{A},i} \in e_i \mathcal{A} e_i$. Similarly, an $\mathcal{A}$-bimodule consists of a graded R-bimodule $\mathcal{P}$, with structure maps that are linear over R as in (2.77). The same applies to the definition of morphisms in $[\mathcal{A},\mathcal{A}]$ and of Hochschild homology. All definitions and results in this section then carry over without any other significant modifications.

3. Hamiltonian Floer cohomology

This section recalls the most familiar version of Floer cohomology [17], adapted to Liouville domains following [55]. The only non-standard aspect of the exposition is our extensive use of the BV (Batalin–Vilkovisky or loop rotation) operator [42, 49, 12]. As a specific example, we will consider Liouville domains whose boundary is a contact circle bundle. Their Floer cohomology is very well understood in relation to Symplectic Field Theory, thanks to [10, 13] and ongoing work of Diogo–Lisi; but our limited needs can be met by a more elementary approach.

3a. Basic definitions

We will work in the following setting, which benefits from strong exactness assumptions:

Setup 3.1.

(i) Let M^{2n} be a Liouville domain. This means that it is a compact manifold with boundary, equipped with an exact symplectic structure $\omega_M = d\theta_M$, such that the Liouville vector field Z_M dual to θ_M points strictly outwards along the boundary. The Liouville flow provides a canonical collar embedding $(-\infty, 0] \times \partial M \hookrightarrow M$. The exponential of the time variable then gives a function ρ_M (defined on the image of the embedding in M) such that

$$\begin{cases} \rho_M|\partial M = 1, \\ Z_M.\rho_M = \rho_M. \end{cases} \tag{3.1}$$

We denote by R_M the Hamiltonian vector field of ρ_M. This is the natural extension of the Reeb vector field $R_{\partial M}$ associated to the contact one-form $\theta_M|\partial M$.

(ii) During part of our argument, we will additionally assume that M comes with a symplectic Calabi–Yau structure (a distinguished homotopy class of trivializations of its canonical bundle $K_M = \Lambda^n_{\mathbb{C}}(TM)^\vee$, for some compatible almost complex structure).

(iii) We consider functions $H \in C^\infty(M, \mathbb{R})$ with

$$H = \epsilon\rho_M \quad \text{near } \partial M \tag{3.2}$$

for some $\epsilon \in \mathbb{R}$. Throughout, it is assumed that

$$\epsilon R_{\partial M} \text{ has no 1-periodic orbits;} \tag{3.3}$$

equivalently, $\epsilon \neq 0$ and there are no Reeb orbits on ∂M whose period is $|\epsilon|/k$, for any positive integer k.

(iv) We use compatible almost complex structures J such that

$$\theta_M \circ J = d\rho_M \quad \text{near } \partial M; \tag{3.4}$$

equivalently, $JZ_M = R_M$.

Example 3.2.

(i) The most important case for us is when ∂M is a *contact circle bundle*, by which we mean that the Reeb flow is a free S^1-action (for concreteness, we assume that this has period 1). In that case, M can be embedded into a closed symplectic manifold M^{cl} by attaching a disc bundle over $\partial M/S^1$. The zero-section $\partial M/S^1 \hookrightarrow M^{\mathrm{cl}}$ represents a positive multiple of the symplectic class $[\omega_{M^{\mathrm{cl}}}] \in \mathrm{H}^2(M^{\mathrm{cl}};\mathbb{R})$.

(ii) In the same situation as before, suppose additionally that

$$c_1(M^{\mathrm{cl}}) = m\,[\partial M/S^1] \in \mathrm{H}^2(M^{\mathrm{cl}};\mathbb{Z}), \quad \text{for some } m \in \mathbb{Z}. \tag{3.5}$$

For $m = 0$, this means that $K_{M^{\mathrm{cl}}}$ is trivial; one can choose a symplectic Calabi–Yau structure on M^{cl}, which then obviously restricts to one on M. For general m, one can still find a section of $K_{M^{\mathrm{cl}}}$ which has an order m "pole" along $\partial M/S^1$, and is nonzero elsewhere. By restricting that to M, one again obtains a symplectic Calabi–Yau structure.

In its most basic form, the Floer cohomology $\mathrm{HF}^*(M,\epsilon)$ is a $\mathbb{Z}/2$-graded vector space over a field $\mathbb{K}$ of characteristic 2, which depends on M and a choice of constant ϵ satisfying (3.3). To define it, take a family of functions $H = (H_t)$ as in (3.2), as well as a family of almost complex structures $J = (J_t)$ as in (3.4), both parametrized by $t \in S^1 = \mathbb{R}/\mathbb{Z}$. On the free loop space $\mathcal{L} = C^\infty(S^1, M)$, consider the H-perturbed action functional:

$$A_H(x) = \int_{S^1} -x^*\theta_M + H_t(x(t))\,dt. \tag{3.6}$$

Its critical points are solutions $x \in \mathcal{L}$ of

$$\dot{x} = X_{H,t}, \tag{3.7}$$

where X_H is the time-dependent Hamiltonian vector field of H. Such x correspond bijectively to fixed points $x(1)$ of ϕ^1_{H}, where (ϕ^t_{H}) is the Hamiltonian isotopy generated by H. By assumption on ϵ, all such x are contained in the interior of M. For a generic choice of H, all x are nondegenerate; assume from now on that this is the case. One defines the Floer cochain space $\mathrm{CF}^*(M,H)$ to be the $\mathbb{Z}/2$-graded vector space with one summand $\mathbb{K}$ for each x (the $\mathbb{Z}/2$-grading encodes the local Lefschetz number). Consider solutions of Floer's gradient flow equation for (3.6). These "Floer trajectories" are maps $u : \mathbb{R} \times S^1 \to M$, satisfying

$$\partial_s u + J_t(\partial_t u - X_{H,t}) = 0, \tag{3.8}$$

with limits

$$\lim_{s\to\pm\infty} u(s,\cdot) = x_\pm \tag{3.9}$$

as in (3.7). While M is not closed, a standard maximum principle argument shows that Floer trajectories can never reach ∂M. We consider the moduli space of solutions to (3.8), (3.9) (more precisely, solutions that are not constant in s, modulo translation in that direction). For generic choice of J, these moduli spaces are regular. One counts isolated points in them to obtain the coefficients of the differential d on $\mathrm{CF}^*(M,H)$, of which $\mathrm{HF}^*(M,\epsilon)$ is the cohomology.

The free loop space admits an obvious circle action (loop rotation). In classical cohomology, a circle action gives rise to an operator of degree -1, dual to moving cycles around orbits. The analogue for Floer cohomology is the BV operator

$$\Delta : \mathrm{HF}^*(M,\epsilon) \longrightarrow \mathrm{HF}^{*-1}(M,\epsilon). \tag{3.10}$$

To define it, suppose that we have fixed (H, J) as necessary to define the Floer cochain complex. The rotated versions, for $r \in S^1$, are

$$(\mathrm{H}_t^{(r)}, J_t^{(r)}) = (H_{t-r}, J_{t-r}). \tag{3.11}$$

Similarly, if x is as in (3.7), we write $x^{(r)}(t) = x(t-r)$. Choose a family of functions and almost complex structures $(H_{r,s,t}, J_{r,s,t})$, depending on $(r, s, t) \in S^1 \times \mathbb{R} \times S^1$, which lie in the same class as before and satisfy

$$(H_{r,s,t}, J_{r,s,t}) = \begin{cases} (H_t^{(r)}, J_t^{(r)}) & s \leq -1, \\ (H_t, J_t) & s \geq 1. \end{cases} \tag{3.12}$$

One considers the parametrized moduli space of pairs (r, u), consisting of $r \in S^1$ and a solution $u : \mathbb{R} \times S^1 \to M$ of the r-dependent equation

$$\partial_s u + J_{r,s,t}(\partial_t u - X_{H,r,s,t}) = 0, \tag{3.13}$$

with limits

$$\begin{cases} \lim_{s \to -\infty} u(s, \cdot) = x_-^{(r)}, \\ \lim_{s \to +\infty} u(s, \cdot) = x_+. \end{cases} \tag{3.14}$$

The same maximum principle argument as before applies to solutions of (3.13), (3.14), ensuring that they can never reach ∂M. For suitably generic choices, counting isolated points in the parametrized moduli space yields a chain map

$$\delta : \mathrm{CF}^*(M, H) \longrightarrow \mathrm{CF}^{*-1}(M, H), \tag{3.15}$$

which induces (3.10) on cohomology. Unlike the differential d, the operator δ is not always compatible with the action filtration. More precisely, given a solution of (3.13), (3.14), one has

$$E(u) = \int_{\mathbb{R} \times S^1} |\partial_s u|^2 \, ds \wedge dt \leq A_H(x_-) - A_H(x_+) + \int_{\mathbb{R} \times S^1} \|\partial_s H_{r,s,t}\|_\infty \, ds \wedge dt. \tag{3.16}$$

If H was autonomous (independent of t) one could choose $H_{r,s,t} = H$ for all (r, s, t), and then the last term in (3.16) would vanish. In general, this interferes with the nondegeneracy of 1-periodic orbits; but one can always choose H to be a small perturbation of an autonomous function, and thereby make that term as small as needed. For such choices, δ will decrease actions only by a small amount.

Remark 3.3. By counting Floer trajectories with appropriate signs, one can define Floer cohomology over an arbitrary coefficient field, as already pointed out in [17]. A symplectic Calabi–Yau structure on M determines a lift of the $\mathbb{Z}/2$-grading on $\mathrm{HF}^*(M, \epsilon)$ to a $\mathbb{Z}$-grading [37]. Since these are standard additions to the basic construction, we will continue our exposition in the simplest setup ($\mathrm{char}(\mathbb{K}) = 2$ and $\mathbb{Z}/2$-gradings), but take care that all formulae remain correct when signs and $\mathbb{Z}$-gradings are added. For instance, in the $\mathbb{Z}$-graded case, the BV operator has degree -1, as indicated in (3.10).

Floer cohomology has a Poincaré type duality, obtained by reversing loops in $\mathcal{L}$:

$$\mathrm{HF}^*(M,-\epsilon) \cong \mathrm{HF}^{2n-*}(M,\epsilon)^\vee. \tag{3.17}$$

This comes from an isomorphism of the underlying chain complexes, assuming that the choices of (H,J) have been suitably coordinated. For equally straightforward reasons, it is compatible with BV operators.

As is implicit in the notation, Floer cohomology is independent (up to canonical isomorphism) of all the auxiliary choices made in its construction, and the same holds for the BV operator. It is maybe useful to elaborate on this slightly. Suppose that we have two possible choices, giving rise to Floer complexes $\mathrm{CF}^*(M,H)$ and $\mathrm{CF}^*(M,\tilde{H})$, with their differentials d, $\tilde{d}$ and chain level BV operators δ, $\tilde{\delta}$. The continuation map method [37] provides maps

$$\begin{aligned} k &: \mathrm{CF}^*(M,H) \longrightarrow \mathrm{CF}^*(M,\tilde{H}),\\ \kappa &: \mathrm{CF}^*(M,H) \longrightarrow \mathrm{CF}^{*-2}(M,\tilde{H}). \end{aligned} \tag{3.18}$$

The first one is a chain map (and actually a quasi-isomorphism), and the second one satisfies

$$\tilde{d}\kappa - \kappa d = \tilde{\delta}k - k\delta. \tag{3.19}$$

Moreover, the maps (3.18) are themselves unique in a suitable homotopical sense.

3b. Reeb orbits

We now consider the dependence on ϵ. This usually appears as part of the construction of symplectic cohomology (in the sense of [55]).

Lemma 3.4. *Suppose that all $\epsilon \in [\epsilon_-, \epsilon_+]$ satisfy* (3.3). *Then we have an isomorphism, compatible with BV operators,*

$$\mathrm{HF}^*(M,\epsilon_-) \cong \mathrm{HF}^*(M,\epsilon_+). \tag{3.20}$$

Proof. Even though that is not absolutely necessary, we find it convenient to introduce finite enlargements of our Liouville domain. Such an enlargement, by an amount $C > 1$, is

$$\hat{M} = M \cup_{\partial M} ([1,C] \times \partial M). \tag{3.21}$$

The conical part which we have added carries the one-form $\theta_{\hat{M}} = r(\theta_M|\partial M)$ (r being the coordinate in $[1,C]$). One extends ρ_M to $\hat{M}$ by setting $\rho_{\hat{M}}(r,x) = r$. Choose a function

$$\begin{cases} h_+ : (0,1] \longrightarrow \mathbb{R}, \\ h_+(a) = \epsilon_- a \quad \text{for } a \text{ sufficiently small}, \\ h_+'(a) = \epsilon_+ \quad \text{for } a \text{ close to } 1, \\ h_+''(a) \geq 0 \quad \text{everywhere}, \\ h_+''(a) > 0 \quad \text{for all } a \text{ such that } \epsilon_- < h_+'(a) < \epsilon_+. \end{cases} \tag{3.22}$$

Given the H_- used to define $\mathrm{HF}^*(M,\epsilon_-)$, we extend it to a time-dependent function $\hat{H}_+$ on $\hat{M}$ by setting

$$\hat{H}_{+,t}(r,x) = C\,h_+(C^{-1}r). \tag{3.23}$$

This makes sense provided that C is large (so that $h_+(a) = \epsilon_- a$ near $a = C^{-1}$). This extension does not quite belong to the class (3.2) for the manifold $\hat{M}$ and constant ϵ_+, but that could be remedied by adding a constant, which has no effect on our construction. Hence, $\hat{H}_+$ can be used to define $\mathrm{HF}^*(\hat{M},\epsilon_+)$.

Note that on $[1,C] \times \partial M$,

$$X_{\hat{H}_{+,t}} = h'_+(C^{-1}r)\,R_{\partial M}. \tag{3.24}$$

Because of the assumption, it follows that all 1-periodic orbits of (3.23) are actually contained in M. A general almost complex structure on M does not naturally extend to $\hat{M}$, so there we proceed in the other direction: we choose $\hat{J} = (\hat{J}_t)$ on $\hat{M}$ which satisfy the analogue of (3.4) in a neighbourhood of $[1,C] \times \partial M$, and then take $J = (J_t)$ to be the restriction to M. A maximum principle argument shows that Floer trajectories in $\hat{M}$ are in fact all contained in M. As a consequence, we get an isomorphism (of complexes, and hence of Floer cohomology groups)

$$\mathrm{HF}^*(\hat{M},\epsilon_+) \cong \mathrm{HF}^*(M,\epsilon_-). \tag{3.25}$$

For similar reasons, this is compatible with the BV operators. A repetition of the same argument, with a linear function instead of h_+, shows that

$$\mathrm{HF}^*(\hat{M},\epsilon_+) \cong \mathrm{HF}^*(M,\epsilon_+). \tag{3.26}$$

Combining the two isomorphisms yields the desired result. □

Lemma 3.5. *Take $\epsilon_- < \epsilon_+$, both of which satisfy* (3.3)*, and suppose that there is exactly one $\epsilon \in (\epsilon_-,\epsilon_+)$ for which* (3.3) *fails. Then there is a long exact sequence*

$$\cdots \to \mathrm{HF}^*(M,\epsilon_-) \longrightarrow \mathrm{HF}^*(M,\epsilon_+) \longrightarrow \mathrm{H}^*(\mathrm{Q}) \to \cdots \tag{3.27}$$

where $\mathrm{H}^*(\mathrm{Q})$ depends only on the local geometry near the 1-periodic orbits of $\epsilon R_{\partial M}$. Moreover, $\mathrm{H}^*(\mathrm{Q})$ carries an endomorphism of degree -1, with the same locality property, and which fits in with (3.27) *and the BV operators on $\mathrm{HF}^*(M,\epsilon_\pm)$.*

Proof. Let us use the same extension $\hat{H}_+$ as in the proof of Lemma 3.4. This time, there are additional 1-periodic orbits in the conical part. Their actions are

$$C(h_+(a_*) - a h'_+(a_*)), \quad \text{for the unique } a_* \text{ such that } h'_+(a_*) = \epsilon. \tag{3.28}$$

The convexity properties of h_+ ensure that $h_+(a_*) - a_* h'_+(a_*) < 0$. Therefore, (3.28) goes to $-\infty$ as we make C large. In particular, we can choose C so that (3.28) is smaller than the actions of any 1-periodic orbits of H_-. The 1-periodic orbits lying in the conical part will be degenerate, but one can remedy that by a small time-dependent perturbation of $\hat{H}_+$ (concentrated near those orbits), and the statement about actions will continue to hold. Hence, the 1-periodic orbits lying in M form a subcomplex of $\mathrm{CF}^*(\hat{M},\hat{H}_+)$. Using the integrated maximum principle (see [4]), one can show that the differential on this subcomplex agrees with the Floer differential in M (alternatively, one can arrive at the same conclusion by taking C very large, and using the Monotonicity Lemma; or by the standard maximum principle and a Gromov compactness argument, applied to the limit where the perturbation of $\hat{H}_+$ goes to 0). In other words, one has an inclusion of chain complexes

$$\mathrm{CF}^*(M, H_-) \subset \mathrm{CF}^*(\hat{M}, \hat{H}_+). \tag{3.29}$$

Defining Q^* to be the quotient, one obviously gets a long exact sequence

$$\cdots \to \mathrm{HF}^*(M, \epsilon_-) \longrightarrow \mathrm{HF}^*(\hat{M}, \epsilon_+) \longrightarrow \mathrm{H}^*(\mathrm{Q}) \to \cdots \tag{3.30}$$

In the original situation (3.28), all 1-periodic orbits in the conical part had the same action. The perturbation will destroy that, but only by a small amount. Hence, $\mathrm{H}^*(\mathrm{Q})$ is still a local Floer cohomology group, in the sense of [36, 22]. As in Lemma 3.4, one can replace $\mathrm{HF}^*(\hat{M}, \epsilon_+)$ by $\mathrm{HF}^*(M, \epsilon_+)$ in (3.30), which yields (3.27).

The corresponding statements concerning BV operators are slightly more tricky. Let us choose all our Hamiltonians to be small perturbations of autonomous ones. In that case, it follows from (3.28) and (3.16) that the chain level BV map on $\mathrm{CF}^*(\hat{M}, \hat{H}_+)$ will preserve the subcomplex (3.29). A Gromov compactness argument (where one decreases the size of the perturbations, as briefly mentioned before) implies that the induced endomorphism of Q^* is again local in nature. To show that its restriction to $\mathrm{CF}^*(M, H_-)$ counts only solutions lying in M, one combines the same idea of Gromov compactness with the methods we've used for the differential (integrated maximum principle, or Monotonicity Lemma). □

Lemma 3.6. *In the situation of Lemma* 3.5, *assume additionally that the* 1*-periodic orbits of* $\epsilon R_{\partial M}$ *form a Morse–Bott nondegenerate connected manifold* F. *Then,*

$$\mathrm{H}^*(\mathrm{Q}) \cong \mathrm{H}^{*+k}(F; \xi), \tag{3.31}$$

for some k *and local* $\mathbb{K}$*-coefficient system* $\xi \to F$ *(with holonomy* ± 1*). That coefficient system is equivariant with respect to the* S^1*-action given by loop rotation on* F. *Hence,* $\mathrm{H}^*(F; \xi)$ *comes with an endomorphism of degree* -1, *which fits in with the BV operators and* (3.27).

The connectedness assumption is for notational convenience only (otherwise, each component of F comes with its own index offset k, and one has to take the direct sum of their contributions).

Example 3.7. Suppose that $\epsilon R_{\partial M}$ has exactly one 1-periodic orbit (up to loop rotation), which moreover is transversally nondegenerate. This means that $F \cong S^1$. The local coefficient system ξ is trivial if the orbit is "good", and has holonomy -1 if it is "bad" (in the terminology from Symplectic Field Theory; for an explanation in a framework close to ours, see [9, §5] or [11, §4.4]).

Lemma 3.6 can be proved by using the Morse–Bott formalism for Floer cohomology [6, 7]. While we will not give a complete proof, it makes sense to describe that formalism in the appropriate form, since we'll return to it later on.

We start with a given H_- and extend that to $\hat{H}_+$ as in (3.23), but do not perturb this extension further. As usual, C is assumed to be large, to get the necessary action inequalities. Fix a Morse–Smale pair on F, consisting of a Morse function f and a metric g. The Morse–Bott–Floer complex $\mathrm{CF}^*(\hat{M}, \hat{H}_+, f)$ has two kinds of generators: ones coming from the 1-periodic orbits in M, and ones coming from the critical points of f (the constant k and local coefficient system ξ appear when we take gradings and signs into account). Let us write this as

$$\mathrm{CF}^*(\hat{M}, \hat{H}_+, f) = \mathrm{CF}^*(M, H_-) \oplus \mathrm{CM}^{*+k}(f; \xi). \tag{3.32}$$

The differential on (3.32) is of the form

$$d = \begin{pmatrix} d_{\mathrm{Floer}} & d_{\mathrm{mixed}} \\ 0 & d_{\mathrm{Morse}} \end{pmatrix}, \tag{3.33}$$

where the various parts are defined as follows:

(i) If x_- and x_+ are 1-periodic orbits in M, one counts Floer trajectories as before, and this yields d_{Floer}.

(ii) If x_- and x_+ are both critical points of f, one counts trajectories of $-\nabla_g f$ going from x_- to x_+, which yields d_{Morse}.

(iii) Suppose that x_- lies in M, and x_+ is a critical point of f. In that case, one considers a mixed moduli space consisting of pairs (u, v), where u is a Floer trajectory, and $v : [0, \infty) \to F$ is a positive half flow line of $-\nabla_g f$, with

$$\begin{cases} \lim_{s \to -\infty} u(s, \cdot) = x_-, \\ \lim_{s \to +\infty} u(s, \cdot) = v(0), \\ \lim_{s \to +\infty} v(s) = x_+. \end{cases} \tag{3.34}$$

Counting solutions of this equation yields d_{mixed}.

Assume that the original H_- was a small perturbation of an autonomous Hamiltonian. We then choose a family of Hamiltonians $H_{-,r,s,t}$ on M as in (3.12), which are perturbations of the same autonomous Hamiltonian, and extend them to $\hat{H}_{+,r,s,t}$ on $\hat{M}$ as in (3.23). The main property, established by similar arguments as in the proof of Lemma 3.5, is that any solution of (3.13) whose left-hand limit $x_-^{(r)}$ lies in F must be stationary, which means $u(s,t) = x_-^{(r)}(t) = x_+(t)$. Then, the Morse–Bott version of (3.15) takes on a form parallel to (3.33):

$$\delta = \begin{pmatrix} \delta_{\mathrm{Floer}} & \delta_{\mathrm{left}} + \delta_{\mathrm{right}} \\ 0 & \delta_{\mathrm{Morse}} \end{pmatrix}. \tag{3.35}$$

The definition is a more elaborate version of the previous one:

(i) δ_{Floer} is defined by considering solutions of (3.13) whose limits both lie in M.

(ii) Loop rotation restricts to an S^1-action on F. Denote by $f^{(r)}$ and $g^{(r)}$ the images of f and g under that action, for time r. Choose generic families $f_{r,s}$

and $g_{r,s}$, depending on $(r,s) \in S^1 \times \mathbb{R}$, which satisfy the analogue of (3.12):

$$(f_{r,s}, g_{r,s}) = \begin{cases} (f^{(r)}, g^{(r)}) & s \leq -1, \\ (f, g) & s \geq 1. \end{cases} \tag{3.36}$$

The counterpart of (3.13), (3.14) is an equation for $r \in S^1$ and $v : \mathbb{R} \to F$:

$$\begin{cases} \partial_s v + \nabla_{g_{r,s}} f_{r,s} = 0, \\ \lim_{s \to -\infty} v(s) = x_-^{(r)}, \\ \lim_{s \to +\infty} v(s) = x_+, \end{cases} \tag{3.37}$$

where $x_\pm$ are critical points of f. By counting isolated solutions of this equation, one defines δ_{Morse}.

(iii) Suppose that x_- lies in M, and x_+ is a critical point of f. Consider triples (r,u,v), where (r,u) is a solution of the same parametrized equation as in (i), and v is a half flow line of $-\nabla_g f$, with asymptotic conditions

$$\begin{cases} \lim_{s \to -\infty} u(s, \cdot) = x_-^{(r)}, \\ \lim_{s \to +\infty} u(s, \cdot) = v(0), \\ \lim_{s \to +\infty} v(s) = x_+. \end{cases} \tag{3.38}$$

This defines δ_{left}, which satisfies

$$\delta_{\mathrm{left}} d_{\mathrm{Morse}} + \delta_{\mathrm{Floer}} d_{\mathrm{mixed}} + d_{\mathrm{Floer}} \delta_{\mathrm{left}} + b = 0. \tag{3.39}$$

The two last terms in (3.39) correspond to limits where a Floer trajectory "breaks off" from a sequence of solutions u on the left-hand side $s \ll 0$. If the limit of the u themselves is non-stationary, such broken solutions are accounted for by $d_{\mathrm{Floer}} \delta_{\mathrm{left}}$ in the standard way. The other term b counts the remaining limiting configurations. It is defined using a parametrized moduli space of triples (r,u,v), where: u is a solution of Floer's equation; v is a half flow line; and

$$\begin{cases} \lim_{s \to -\infty} u(s, \cdot) = x_-, \\ \lim_{s \to +\infty} u(s, \cdot) = v(0)^{(-r)}, \\ \lim_{s \to +\infty} v(s) = x_+. \end{cases} \tag{3.40}$$

(iv) For $x_\pm$ as in (iii), we consider an equation for pairs (u,v) with additional parameters $(q,r) \in [0,\infty) \times S^1$. Asymptotic conditions are as in (3.40), and u is always a solution of Floer's equation. The q-dependence lies entirely in the equation satisfied by v. For $q \gg 0$, that equation should be that which defines δ_{Morse}, but applied to the positive half-line only, and with the s-parameter shifted by q:

$$\partial_s v + \nabla_{g_{r,s-q}} f_{r,s-q} = 0. \tag{3.41}$$

In contrast, for $q = 0$ the equation should be the gradient flow equation for $-\nabla_g f$ (and one connects those two behaviours by choosing some intermediate data). The outcome is that

$$d_{\mathrm{Floer}}\delta_{\mathrm{right}} + \delta_{\mathrm{right}}d_{\mathrm{Morse}} + d_{\mathrm{mixed}}\delta_{\mathrm{Morse}} - b = 0, \qquad (3.42)$$

where the last two terms correspond to $q \to \infty$ and $q = 0$, respectively.

Lemma 3.6 follows directly once one establishes that the Morse–Bott approach yields the same Floer cohomology (and BV operator) as the original definition. There are several ways of doing that, the most natural one being a suitable generalization of continuation maps; but we will not explain the details.

3c. Autonomous Hamiltonians

The relation between Floer cohomology and ordinary cohomology is established by the following classical result:

Proposition 3.8. *For sufficiently small $\epsilon > 0$, we have $\mathrm{HF}^*(M, \epsilon) \cong \mathrm{H}^*(M)$. Moreover, the BV operator Δ vanishes.*

The most natural proof is a version of [34], but here we choose instead to follow [18, 24], which is more elementary. Take a time-independent H and J, such that H is Morse and the metric associated to J is Morse–Smale. Then, after possibly multiplying H with a small positive constant, we have:

(3.43) (i) all one-periodic orbits of X_H are constant (at the critical points of H);

(3.43) (ii) any solution of Floer's equation is constant in t, hence a negative gradient flow line for H [24, Lemma 7.1];

(3.43) (iii) for the linearization of Floer's equation (at a gradient flow line), all solutions are also constant in t [37, Proposition 4.2]; together with an easy index computation, this shows that the moduli spaces are all regular.

Hence, $\mathrm{CF}^*(M, H)$ is isomorphic to the Morse complex of H (this continues to hold when signs are taken into account). Moreover, when defining the BV operator, one may choose $(H_{r,s,t}, J_{r,s,t}) = (H, J)$, in which case there are no isolated solutions of (3.13), since r can be changed freely. Hence, δ itself vanishes.

We have proved that the BV operator is zero on the chain level, for a specific kind of Hamiltonian and almost complex structure. One can be slightly more precise about the meaning of this vanishing result; for that, it is convenient to introduce some simple algebraic language.

Setup 3.9.

(i) Suppose that we have a chain complex (C^*, d) together with a chain endomorphism δ of degree -1. Recall that a nullhomotopy for δ is a map $\chi : \mathrm{C}^* \to \mathrm{C}^{*-2}$ satisfying $d\chi - \chi d = \delta$. Two nullhomotopies are called equivalent if the difference between them is a nullhomotopic chain map of degree -2. Equivalence classes are called *homotopy trivializations* of δ.

(ii) Take two chain complexes (C, d) and $(\tilde{\mathrm{C}}, \tilde{d})$, with endomorphisms δ and $\tilde{\delta}$. Suppose that we have maps

$$\begin{aligned} k &: \mathrm{C}^* \longrightarrow \tilde{\mathrm{C}}^*, \\ \kappa &: \mathrm{C}^* \longrightarrow \tilde{\mathrm{C}}^{*-2}, \end{aligned} \tag{3.44}$$

of which the first one is a chain map, and the second satisfies (3.19). Given homotopy trivializations $[\chi]$ and $[\tilde{\chi}]$ on our complexes, we say that they are *compatible with* (k, κ) if there is a map

$$g : \mathrm{C}^* \longrightarrow \tilde{\mathrm{C}}^{*-3} \tag{3.45}$$

such that

$$\tilde{\chi}k - k\chi - \kappa = \tilde{d}g + gd. \tag{3.46}$$

It is easy to see that compatibility depends only on the equivalence classes of χ and $\tilde{\chi}$. Moreover, if k is a quasi-isomorphism, every homotopy trivialization on one complex determines a unique compatible homotopy trivialization on the other one.

In this terminology, the desired statement is that the chain complex underlying $\mathrm{HF}^*(M, \epsilon)$ (still in the situation of Proposition 3.8, but now with (H, J) arbitrary) comes with a canonical homotopy trivialization of the BV operator, which is compatible with (3.18). To prove that, one takes two choices (H, J) and $(\tilde{H}, \tilde{J})$ to which the proof of Proposition 3.8 applies. The associated BV operators vanish on the chain level, hence have trivial nullhomotopies. From (3.19), one gets a degree -2 chain map between these Floer complexes (itself canonical up to chain homotopy). One can show that, within a slightly more precisely defined class of Hamiltonians and almost complex structures, those chain maps are always nullhomotopic. Nullhomotopies for them provide the desired maps (3.45), which show compatibility of the homotopy trivializations. After that, one uses continuation maps to extend to arbitrary (H, J). We omit the details.

Proposition 3.10. *Suppose that ∂M is a contact circle bundle. Take $\epsilon \in (1, 2)$. Then there is a long exact sequence (of $\mathbb{Z}/2$-graded spaces)*

$$\cdots \to \mathrm{H}^*(M) \longrightarrow \mathrm{HF}^*(M, \epsilon) \longrightarrow \mathrm{H}^*(\partial M) \to \cdots \tag{3.47}$$

The connecting map in (3.47), $\mathrm{H}^*(\partial M) \to \mathrm{H}^{*+1}(M)$, vanishes on $\mathrm{H}^0(\partial M)$. Moreover, the BV operator vanishes on the preimage of $\mathrm{H}^0(\partial M)$ inside $\mathrm{HF}^*(M, \epsilon)$. Finally, suppose that M carries a symplectic Calabi–Yau structure constructed as in Example 3.2(ii). Then, (3.47) becomes $\mathbb{Z}$-graded if we take the rightmost term to be $\mathrm{H}^{*+2m-2}(\partial M)$.

The existence of such an exact sequence is a special case of Lemma 3.6, combined with Proposition 3.8; what is new are the additional properties (of the connecting map and BV operator).

Proof. Choose $\epsilon_- > 0$ small, and define $\mathrm{HF}^*(M, \epsilon_-)$ using time-independent H_- (and the same for the almost complex structure), as in Proposition 3.8. For $\epsilon_+ = \epsilon \in (1,2)$, extend H_- to $\hat{H}_+$ as in (3.23). This enlargement produces a Morse–Bott-nondegenerate manifold of 1-periodic orbits F, which is a copy of ∂M (and one can check that it carries a trivial local system ξ). Floer trajectories whose limits $x_\pm$ lie in M remain entirely within that subset, and have the same description as in Proposition 3.8. The remaining point is to ensure that Floer trajectories with mixed limits (x_- lies in M, and x_+ in F) can be made regular. Because x_+ is a simple periodic orbit, such trajectories are themselves simple, in the sense of [18, p. 279]. Then, [18, Theorem 7.4] ensures that one can choose J (by varying it near x_-) so that regularity holds generically. Three technical remarks are appropriate. First, transversality theory needs to be carried out in an analytic formalism suitable for the Morse–Bott case. Secondly, while the results in [18] are stated for trajectories with both limits being constant orbits, only one such limit is actually necessary for the argument to go through (and, since the constant orbits are still nondegenerate in our situation, their treatment does not need to modified). Finally, we need a technical condition on the Hessian of H_- at its critical points [18, Definition 7.1], but it is unproblematic to arrange that it holds.

Given that, we can use this same almost complex structure everywhere in the definition of (3.33) and (3.35). Then, as in Proposition 3.8 (which means using the fact that r is a free parameter), we get

$$\delta_{\mathrm{Floer}} = 0, \tag{3.48}$$

$$\delta_{\mathrm{left}} = 0. \tag{3.49}$$

Now let x_+ be a local minimum (Morse index 0 critical point) of f, so that for degree reasons,

$$\delta_{\mathrm{Morse}}(x_+) = 0. \tag{3.50}$$

There is an open subset of F consisting of points $v(0)$ which flow to x_+ under $-\nabla_g f$. The Floer trajectories u such that $\lim_{s \to +\infty} u(s, \cdot)$ lies in that open subset form a space of dimension ≥ 1 (since one can rotate them a little in the t-variable). The same applies if one replaces the gradient flow equation by (3.41). As a consequence,

$$d_{\mathrm{mixed}}(x_+) = 0, \tag{3.51}$$

$$\delta_{\mathrm{right}}(x_+) = 0. \tag{3.52}$$

The desired properties of (3.47) and Δ follow directly from this. The observation about $\mathbb{Z}$-gradings is a standard Conley–Zehnder index computation. □

As before, it can be useful to encode this observation in a more abstract framework, which also allows one to formulate a uniqueness property.

Setup 3.11. (i) In the situation of Setup 3.9(i), fix some integer j. Let $\mathrm{C}^{\leq j} \subset \mathrm{C}^*$ be the subcomplex consisting of all cocycles of degree j, together with all cochains of degree $< j$. Note that this is automatically preserved by δ. A *homotopy* *trivialization of* δ *in degrees* $\leq j$ is a homotopy trivialization of $\delta^{\leq j} = \delta|\mathrm{C}^{\leq j}$, in the previously defined sense. The existence of such a trivialization implies that $[\delta] : \mathrm{H}^*(\mathrm{C}) \to \mathrm{H}^{*-1}(\mathrm{C})$ vanishes in degrees $\leq j$, since the map $\mathrm{H}^*(\mathrm{C}^{\leq j}) \to \mathrm{H}^*(\mathrm{C})$ is an isomorphism in those degrees.

(ii) In the situation of Setup 3.9(ii), suppose that we have homotopy trivializations of C and $\tilde{\mathrm{C}}$ in degrees $\leq j$. We say that these are compatible if there exists a map (3.45) defined on $\mathrm{C}^{\leq j}$, with the same properties as before.

Suppose that we are in the situation of Proposition 3.10, with a symplectic Calabi–Yau structure constructed as in Example 3.2(ii). Then, the proof of that proposition yields (for a very special choice of Morse–Bott complex underlying Floer cohomology) a homotopy trivialization of the BV operator in degrees $\leq 2m - 2$. The relevant uniqueness result would say that this is compatible with continuation maps. Proving this requires an analogue of Proposition 3.10 for continuation maps, which we will not explain here.

Remark 3.12. The statement of Proposition 3.10 is actually not optimal. One can show that the connecting map vanishes on $\mathrm{im}(\mathrm{H}^*(\partial M/S^1) \to \mathrm{H}^*(\partial M))$, and that the BV operator is zero on the preimage of that subspace inside $\mathrm{HF}^*(M, \epsilon)$. To do that, one has to use a chain level model for $\mathrm{H}^*(\partial M)$ which is more closely adapted to the circle action than what we have done (one possibility is to perturb the Reeb flow on ∂M so that the one-periodic orbits become transversally nondegenerate).

Alternatively, one can use the S^1-equivariant version of Hamiltonian Floer cohomology (see, e.g., [42]), which is a $\mathbb{K}[[u]]$-module $\mathrm{HF}^*_{S^1}(M, \epsilon)$. In the situation of Proposition 3.10, it fits into a long exact sequence of such modules,

$$\cdots \to \mathrm{H}^*(M)[[u]] \longrightarrow \mathrm{HF}^*_{S^1}(M, \epsilon) \longrightarrow \mathrm{H}^*(\partial M/S^1) \to \cdots \tag{3.53}$$

where u acts on $\mathrm{H}^*(\partial M/S^1)$ by cup product with the first Chern class of the circle bundle $\partial M \to \partial M/S^1$. Since that action is nilpotent, the boundary operator of (3.53) is necessarily zero. Now, (3.53) and (3.47) fit into a commutative diagram (3.54) whose rows and columns are long exact sequences.

The right-hand column of (3.54) is the standard Gysin sequence; and the BV operator is the composition of two vertical arrows (going from $\mathrm{HF}^*(M, \epsilon)$ to $\mathrm{HF}^*_{S^1}(M, \epsilon)$ and then back). Diagram-chasing yields the desired result: if a class in $\mathrm{HF}^*(M, \epsilon)$ has the property that its image in $\mathrm{H}^*(\partial M)$ is the pullback of a class in $\mathrm{H}^*(\partial M/S^1)$, then it necessarily comes from a class in $\mathrm{HF}^*_{S^1}(M, \epsilon)$, hence is killed by the BV operator.

$$
\begin{array}{ccccccccc}
 & & \vdots & & \vdots & & \vdots & & \\
 & & \big\downarrow{\scriptstyle \text{zero}} & & \big\downarrow & & \big\downarrow & & \\
\cdots & \xrightarrow{\text{zero}} & \mathrm{H}^*(M)[[u]] & \longrightarrow & \mathrm{HF}^*_{S^1}(M,\epsilon) & \longrightarrow & \mathrm{H}^*(\partial M/S^1) & \xrightarrow{\text{zero}} & \cdots \\
 & & \big\downarrow{\scriptstyle u} & & \big\downarrow{\scriptstyle u} & & \big\downarrow{\scriptstyle u} & & \\
\cdots & \xrightarrow{\text{zero}} & \mathrm{H}^*(M)[[u]] & \longrightarrow & \mathrm{HF}^*_{S^1}(M,\epsilon) & \longrightarrow & \mathrm{H}^*(\partial M/S^1) & \xrightarrow{\text{zero}} & \cdots \\
 & & \big\downarrow & & \big\downarrow & & \big\downarrow & & \\
\cdots & \longrightarrow & \mathrm{H}^*(M) & \longrightarrow & \mathrm{HF}^*(M,\epsilon) & \longrightarrow & \mathrm{H}^*(\partial M) & \longrightarrow & \cdots \\
 & & \big\downarrow{\scriptstyle \text{zero}} & & \big\downarrow & & \big\downarrow & & \\
 & & \vdots & & \vdots & & \vdots & &
\end{array}
\tag{3.54}
$$

4. Fixed point Floer cohomology

The generalization of Floer cohomology from Hamiltonian to general symplectic automorphisms was introduced in [14, 15]. Here, we use a version for Liouville domains, as in [39, Section 4], [33], or [54].

4a. Definition

For M as before, we will consider the following situation.

Setup 4.1.

(i) Let ϕ be a symplectic automorphism of M which is equal to the identity near ∂M and exact. The latter condition means that there is a function G_ϕ (necessarily locally constant near ∂M) such that $\phi^*\theta_M - \theta_M = dG_\phi$.

(ii) If M carries a symplectic Calabi–Yau structure, we will assume that ϕ is compatible with it, and in fact comes with a choice of grading (making it a graded symplectic automorphism [38]).

Example 4.2. Suppose that ∂M is a contact circle bundle. Fix a function F on M which agrees with ρ_M near the boundary, and let (ϕ^t_F) be its flow. By construction, ϕ_F^{-1} is the identity near the boundary. We call it the *boundary twist* of M (see [38, Section 4] for a general discussion of such symplectic automorphisms), and denote it by $\tau_{\partial M}$.

Let us now assume that M carries a symplectic Calabi–Yau structure as in Example 3.2(ii). Then, $\tau_{\partial M}$ can be equipped with the structure of a graded symplectic automorphism. In fact, there are two reasonable choices (which coincide for $m = 1$):

(i) One can take the trivial grading of the identity, and extend it continuously over the isotopy (ϕ_F^t) to get a grading of $\tau_{\partial M}$. Near the boundary, that grading is a shift $[2-2m]$.
(ii) Alternatively, by changing the previous grading by a constant $2-2m$, one can get the unique grading of $\tau_{\partial M}$ which is trivial near the boundary.

Take families of functions $H=(H_t)$ and $J=(J_t)$, parametrized by $t \in \mathbb{R}$. Each of those should be as in (3.2) and (3.4), but now with ϕ-twisted periodicity properties:

$$\begin{cases} H_{t+1}(x) = H_t(\phi(x)), \\ J_{t+1} = \phi^* J_t. \end{cases} \tag{4.1}$$

Take the space $\mathcal{L}_\phi = \{x : \mathbb{R} \to M \, : \, x(t) = \phi(x(t+1))\}$, with action functional

$$A_{\phi,H}(x) = \Big(\int_0^1 -x^*\theta_M + H_t(x(t))\,dt\Big) - G_\phi(x(1)). \tag{4.2}$$

Its critical points are solutions $x \in \mathcal{L}_\phi$ of (3.7), and correspond bijectively to fixed points $x(1)$ of $\phi_{\mathrm{H}}^1 \circ \phi$. For a generic choice of H, these will be nondegenerate, and we use them as generators of a $\mathbb{Z}/2$-graded $\mathbb{K}$-vector space $\mathrm{CF}^*(\phi, H)$. To define the differential, one considers solutions $u : \mathbb{R}^2 \to M$ of (3.8), but now satisfying

$$u(s,t) = \phi(u(s,t+1)). \tag{4.3}$$

The resulting Floer cohomology is denoted by $\mathrm{HF}^*(\phi,\epsilon)$, with the previous $\mathrm{HF}^*(M,\epsilon)$ being the special case $\phi = \mathrm{id}_M$. Remark 3.3 carries over to this more general situation; except that to make $\mathrm{HF}^*(\phi,\epsilon)$ $\mathbb{Z}$-graded, one needs to impose conditions on ϕ as well as on M, as in Setup 4.1(ii).

Remark 4.3. In general, $\mathrm{HF}^*(\phi,\epsilon)$ does not carry a BV operator. The exception is when ϕ can be written as the time-one map of a Hamiltonian flow (ϕ^t). Here, the general ϕ^t do not have to be equal to the identity near ∂M, but they have to preserve θ_M near ∂M (one example of this would be the flow that leads to $\phi = \tau_{\partial M}$). One then defines a circle action on $\mathcal{L}_\phi$ by mapping a twisted loop x to

$$x^{(r)}(t) = \phi^t(x(t-r)). \tag{4.4}$$

There is a related situation where fixed point Floer cohomology admits a discrete symmetry, which was extensively studied in [46, 35]. Namely, suppose that ϕ admits a kth root $\phi^{1/k}$ (such that $(\phi^{1/k})^*\theta_M = \theta_M$ near ∂M). This gives rise to an action of $\mathbb{Z}/k$ on $\mathcal{L}_\phi$, whose generator maps x to

$$x^{(1/k)}(t) = \phi^{1/k}(x(t-1/k)). \tag{4.5}$$

One gets an induced action of $\mathbb{Z}/k$ on $\mathrm{HF}^*(\phi)$ (in the previously considered case of Hamiltonian flows, ϕ has kth roots for any k; but the induced $\mathbb{Z}/k$-actions on Floer cohomology are trivial, which is intuitively clear since they embed into a continuous symmetry).

Floer cohomology is invariant under isotopies of ϕ (within the class of symplectic automorphisms that are allowed). One has a straightforward counterpart of (3.17):

$$\mathrm{HF}^*(\phi^{-1}, -\epsilon) \cong \mathrm{HF}^{2n-*}(\phi, \epsilon)^\vee. \tag{4.6}$$

As for the dependence on ϵ, Lemmas 3.4–3.6, with the parts about BV operators omitted, carry over with essentially the same proofs.

Lemma 4.4. *Suppose that ∂M is a contact circle bundle, and let $\tau_{\partial M}$ be the boundary twist. For any ϕ and ϵ,*

$$\mathrm{HF}^*(\tau_{\partial M}^{-1} \circ \phi, \epsilon - 1) \cong \mathrm{HF}^*(\phi, \epsilon). \tag{4.7}$$

This isomorphism is compatible with $\mathbb{Z}$-gradings if we use option (i) *from Example 4.2 as a grading for $\tau_{\partial M}$; if instead we use option* (ii)*, the right-hand side of* (4.7) *should be replaced with $\mathrm{HF}^{*+2-2m}(\phi, \epsilon)$.*

Proof. When defining $\tau_{\partial M}$, one can choose the flow (ϕ_F^t) to be supported arbitrarily close to ∂M, so that it commutes with ϕ. Consider the diffeomorphism

$$\begin{aligned} &\mathcal{F} : \mathcal{L}_{\tau_{\partial M}^{-1} \circ \phi} \longrightarrow \mathcal{L}_\phi, \\ &(\mathcal{F}x)(t) = \phi_F^t(x(t)), \end{aligned} \tag{4.8}$$

which satisfies

$$\begin{aligned} &\mathcal{F}^* A_{\phi,H} = A_{\tau_{\partial M}^{-1} \circ \phi, \tilde{H}}, \\ &\tilde{H}_t(x) = H_t(\phi_F^t(x)) - F(x). \end{aligned} \tag{4.9}$$

Note that close to ∂M, $\tilde{H}_t = \epsilon \rho_M - \rho_M$. Given suitable choices of almost complex structures, (4.8) gives rise to an isomorphism of Floer cochain complexes, which induces (4.7). □

Example 4.5. Assuming choice (ii) for the grading,

$$\mathrm{HF}^{*+2-2m}(\tau_{\partial M}, \epsilon) \cong \mathrm{HF}^*(M, \epsilon - 1).$$

In particular, using Proposition 3.8 and (3.17), one obtains (1.6) (for $m = 0$) and its generalization (1.7) (for arbitrary m).

4b. Symplectic mapping tori

As usual, the mapping torus construction relates discrete dynamics (symplectic automorphisms) and its continuous counterpart (Hamiltonian flows).

Setup 4.6.

(i) Let M be a Liouville domain, and μ an exact symplectic automorphism (as in Setup 4.1, with associated function G_μ). Its symplectic mapping torus is the fibration

$$E = \frac{\mathbb{R}^2 \times M}{(p, q, x) \sim (p, q - 1, \mu(x))} \xrightarrow{\pi(p,q,x)=(p,q)} \mathbb{R} \times S^1. \tag{4.10}$$

The symplectic form is $\omega_E = dp \wedge dq + \omega_M$. To obtain a primitive, one chooses a function $G = G(q,x)$ such that $G_\mu(x) = G(q,x) - G(q-1,\mu(x))$, and sets $\theta_E = p\,dq + \theta_M + dG$.

(ii) If M has a symplectic Calabi–Yau structure, and μ is a graded symplectic automorphism, the mapping torus inherits a symplectic Calabi–Yau structure.

(iii) We use the class of functions H on E of the following form. Fix constants ϵ, as in (3.3), and $\gamma_\pm \in \mathbb{R} \setminus \mathbb{Z}$. Let H_M be a function on the fibre, which equals $\epsilon \rho_M$ near the boundary, and which is invariant under μ (such functions can easily be constructed as cutoffs of $\epsilon\rho_M$). Let $H_{\mathbb{R}\times S^1}$ be a function on the base, such that $H_{\mathbb{R}\times S^1}(p,q) = \gamma_\pm p$ if $\pm p \gg 0$. Then, there should be an open subset $U \subset E$, which contains ∂E and has compact complement, such that

$$H(p,q,x) = H_{\mathbb{R}\times S^1}(p,q) + H_M(x) \quad \text{for } (p,q,x) \in U. \tag{4.11}$$

We stress that if we have several different H, the associated $H_{\mathbb{R}\times S^1}$ and H_M can also be different (they do not have to be fixed once and for all).

(iv) We will use almost complex structures J on E such that: outside a compact subset, π is J-holomorphic (with respect to the standard complex structure i on the base); and near the boundary, $J = J_{\mathbb{R}\times S^1} \times J_{M,p,q}$, where $J_{\mathbb{R}\times S^1}$ is an almost complex structure on the base which is standard outside a compact subset, and the $J_{M,p,q}$ are as in Setup 3.1.

Given time-dependent $H = (H_t)$ and $J = (J_t)$, one can build a Floer complex $\mathrm{CF}^*(E,H)$ as before. Of course, one has to check that solutions of Floer's equation $u : \mathbb{R} \times S^1 \to E$ remain inside a compact subset of $E \setminus \partial E$. To see that this is the case, note that on the subset where the $|p|$-coordinate of u is large, $v = \pi(u)$ will itself be a solution of

$$\partial_s v + i(\partial_t v - \gamma_\pm \partial_q) = 0, \tag{4.12}$$

(∂_q stands for the unit vector field in q-direction), hence its p-component is harmonic. On the other hand, where u is close to ∂E, one can consider its fibre component alone, and then apply the maximum principle in the same way as when constructing Floer cohomology inside M. The resulting groups $\mathrm{HF}^*(E,\gamma_-,\gamma_+,\epsilon)$ depend only on the constants involved (and we can change those, without affecting Floer cohomology, as long as no "forbidden values" are crossed, in parallel with Lemma 3.4). They also carry a BV operator.

Lemma 4.7. *Suppose that $\gamma_- \in (0,1)$ and $\gamma_+ \in (1,2)$. Then*

$$\mathrm{HF}^*(E,\gamma_-,\gamma_+,\epsilon) \cong \mathrm{H}^*(S^1) \otimes \mathrm{HF}^*(\mu,\epsilon). \tag{4.13}$$

The BV operator is given by rotation on S^1, tensored with the identity map on $\mathrm{HF}^*(\mu,\epsilon)$.

The Floer cohomology computation (4.13) is [33, Theorem 1.3], which can be proved relatively straightforwardly by a Morse–Bott approach. A similar argument determines the BV operator.

5. Rotations at infinity

The next step is to apply Hamiltonian Floer cohomology to total spaces of Lefschetz fibrations. Besides making the necessary adjustments to the construction of Floer cohomology, we will consider some specific computations. Those follow [33] fairly closely, but with the BV operator as an added ingredient.

5a. Target spaces

The Lefschetz (complex nondegeneracy) condition is not important yet, and we will allow considerably more freedom for the local geometry. On the other hand, we impose quite strict conditions on the behaviour near infinity.

Setup 5.1. (i) An *exact symplectic fibration with singularities* is a $2n$-dimensional manifold with boundary E, together with an exact symplectic form $\omega_E = d\theta_E$, and a proper map

$$\pi : E \longrightarrow \mathbb{C}, \tag{5.1}$$

subject to the following conditions.

At any $x \in E$, define the horizontal subspace $TE^h_x \subset TE$ as the ω_E-orthogonal complement of $TE^v_x = \ker(D\pi_x)$. We ask that there should be an open subset $U \subset E$ which contains ∂E and has compact complement, such that at each $x \in U$, one has: TE^v_x is a codimension 2 symplectic subspace of TE (hence, x is a regular point of π, and $D\pi_x|TE^h_x$ is an isomorphism); and

$$(\omega_E - \pi^*\omega_{\mathbb{C}})\,|\,TE^h_x = 0, \tag{5.2}$$

where $\omega_{\mathbb{C}} = d\mathrm{re}(y) \wedge d\mathrm{im}(y)$ is the standard symplectic form on the base.

Next, at each point $x \in \partial E$, TE^h_x should lie inside $T_x(\partial E)$.

Finally, there should be a $y_* > 0$ such that: all fibres $E_y = \pi^{-1}(y)$ with $|y| \geq y_*$ lie inside the previously introduced subset U; and $M = E_{y_*}$, with its induced exact symplectic structure, is a Liouville domain (in all subsequent developments, we will assume that such a y_* has been fixed).

(ii) For part of our considerations, we will assume that E comes with a symplectic Calabi–Yau structure. This induces the same kind of structure on M.

Let us consider the implications of these conditions. We have symplectic parallel transport maps defined in a neighbourhood of ∂E. More precisely, let $W \subset M$ be a small open neighbourhood of ∂M. Then, parallel transport yields a canonical embedding, whose image is an open neighbourhood of $\partial E \subset E$:

$$\begin{cases} \Theta : \mathbb{C} \times W \longrightarrow E, \\ \Theta(\mathbb{C} \times \partial M) = \partial E, \\ \pi(\Theta(y,x)) = y, \\ \Theta(y_*, x) = x, \\ \Theta^*\omega_E = \omega_{\mathbb{C}} + \omega_M. \end{cases} \tag{5.3}$$

To see why that is the case, note that (5.2) implies that the symplectic connection is flat near ∂E. Hence, the relevant part of parallel transport is independent of the choice of path, and $\Theta^*\omega_E - \omega_M$ must be locally the pullback of some two-form on $\mathbb{C}$; to determine that two-form, one again appeals to (5.2).

We also have the same flatness property outside a compact subset, with the following consequence. Let μ be the monodromy around the circle of radius y_*, which is an exact symplectic automorphism of M (and restricts to the identity on W). Then, for $p_* = \log(y_*)$, there is a unique covering map

$$\begin{cases} \Psi : [p_*, \infty) \times \mathbb{R} \times M \longrightarrow \{|\pi(x)| \geq y_*\} \subset E, \\ \pi(\Psi(p,q,x)) = e^{p+iq}, \\ \Psi(p_*, 0, x) = x, \\ \Psi(p, q+2\pi, x) = \Psi(p, q, \mu(x)), \\ \Psi^* \omega_E = e^{2p} dp \wedge dq + \omega_M. \end{cases} \tag{5.4}$$

The partial trivializations (5.3) and (5.4) are compatible, in the sense that

$$\Theta(e^{p+iq}, x) = \Psi(p,q,x) \quad \text{for } p \geq p_* \text{ and } x \in W. \tag{5.5}$$

In words, what we have observed is that (5.1) is symplectically trivial near ∂E, whereas outside the preimage of a disc, it is like a symplectic mapping torus for μ (but with a different symplectic structure on the base than before). Finally, in the situation of Setup 5.1(ii), μ is a graded symplectic automorphism (in a preferred way, such that the grading is trivial near ∂M).

Example 5.2. The most important examples for us are those obtained from anticanonical Lefschetz pencils, or more generally, from Lefschetz pencils satisfying the condition from Remark 1.4. In that case, the boundary of the fibre is a contact circle bundle; E carries a symplectic Calabi–Yau structure, such that the induced structure on the fibre is as in Example 3.2(ii); and as already stated in (1.5), the monodromy is the associated boundary twist.

Setup 5.3.

(i) We consider functions $H \in C^\infty(E, \mathbb{R})$ such that

$$\begin{cases} H(\Theta(y,x)) = H_{\mathbb{C}}(y) + \epsilon \rho_M(x) & \text{for } x \text{ near } \partial M, \\ H(\Psi(p,q,x)) = \gamma e^{2p}/2 + H_M(x) & \text{for } p \gg 0. \end{cases} \tag{5.6}$$

Here, the constant ϵ is as in (3.3), and $\gamma \in \mathbb{R} \setminus 2\pi\mathbb{Z}$. $H_{\mathbb{C}}$ is a function on the base such that $H_{\mathbb{C}}(y) = \gamma|y|^2/2$ for $|y| \gg 0$, and H_M is as in Setup 4.6.

(ii) We use compatible almost complex structures J on E satisfying the following conditions. Outside a compact subset, π is J-holomorphic, with respect to the standard complex structure i on the base; and at points $(y,x) \in \mathbb{C} \times M$ with x sufficiently close to ∂M,

$$\Theta^* J = J_{\mathbb{C}} \times J_{M,y}. \tag{5.7}$$

Here, $J_{\mathbb{C}}$ is an almost complex structure on $\mathbb{C}$ which is standard outside a compact subset, and each $J_{M,y}$ is an almost complex structure on M as in Setup 3.1.

5b. Floer cohomology and its properties

The construction of Hamiltonian Floer cohomology in this context proceeds pretty much as in Section 4b. One gets chain complexes $\mathrm{CF}^*(E,H)$ and cohomology groups $\mathrm{HF}^*(E,\gamma,\epsilon)$, together with a BV operator. A suitable analogue of Lemma 3.4 holds. There are also counterparts of Lemma 3.5 for changing either γ or ϵ. We will not consider them in full generality, but one important special case is this:

Lemma 5.4. *For $\gamma_- \in (0,2\pi)$, $\gamma_+ \in (2\pi,4\pi)$, and any ϵ, we have a long exact sequence*

$$\cdots \rightarrow \mathrm{HF}^*(E,\gamma_-,\epsilon) \longrightarrow \mathrm{HF}^*(E,\gamma_+,\epsilon) \longrightarrow \mathrm{H}^*(S^1)\otimes \mathrm{HF}^{*+2}(\mu,\epsilon) \rightarrow \cdots \tag{5.8}$$

where μ is the monodromy. This is compatible with BV operators, where the operator on the rightmost group is as in Lemma 4.7. Moreover, if E has a symplectic Calabi–Yau structure, (5.8) *is compatible with $\mathbb{Z}$-gradings.*

Proof. To simplify the notation, we assume that one can take $y_* = 1$. Fix a function H_M which is supported in a small neighbourhood of ∂M (hence is invariant under μ), and which equals $\epsilon\rho_M$ close to ∂M. Via (5.3), there is an obvious extension of this function to all of E, which we again denote by H_M. Choose a function

$$\begin{cases} h_+ : [0,\infty) \longrightarrow \mathbb{R}, \\ h_+'(a) = 0 \quad \text{if and only if } a \leq 1/2, \\ h_+'(a) = \gamma_- \quad \text{for } a \text{ close to } 2, \\ h_+(a) = \gamma_+ a \quad \text{for } a \gg 0, \\ h_+''(a) \geq 0 \quad \text{everywhere}, \\ h_+''(a) > 0 \quad \text{whenever } h_+'(a) \in (\gamma_-,\gamma_+). \end{cases} \tag{5.9}$$

Define $H_+ : E \to \mathbb{R}$ by

$$H_+(x) = h_+(|\pi(x)|^2/2) + H_M(x). \tag{5.10}$$

This has two kinds of 1-periodic orbits:

(i) The first kind are contained in a fibre $\pi^{-1}(y)$, $|y| \leq 1$. They are either constant, or else nontrivial 1-periodic orbits of H_M (lying close to ∂M).

(ii) The second kind of 1-periodic orbit is fibered over the circle $|y|^2/2 = a_*$, where $a_* > 2$ is the unique value such that $h_+'(a_*) = 2\pi$. In each fibre, they correspond to fixed points of $\phi^1_{H_M} \circ \mu$.

Let us perturb H_+ slightly (in a time-dependent way, and without changing the notation), so as to make the 1-periodic points nondegenerate; we will assume that the perturbation is trivial near ∂E, as well as outside a compact subset. The outcome lies in the class of Hamiltonians which can be used to construct $\mathrm{HF}^*(E,\gamma_+,\epsilon)$. As long as the perturbation is small, one can still maintain the distinction between type (i) and (ii) orbits, even though they no longer have exactly the same properties as before.

We now consider Floer trajectories. When perturbing H_+, we can assume that the perturbation is trivial near the preimage of the circle $Z = \{|y| = 2\}$. We will also use almost complex structures $J = (J_t)$ such that π becomes pseudo-holomorphic near that same preimage. As a consequence, if u is a Floer trajectory, then $v = \pi(u)$ satisfies

$$\partial_s v + i(\partial_t v - \gamma_- iv) = 0 \quad \text{whenever } v \text{ is close to } Z. \tag{5.11}$$

We exploit that by introducing a degenerate version of the action functional (this is essentially what [45] called a "barrier argument"). On the base, choose a two-form $\bar{\omega}_{\mathbb{C}}$ which is: rotationally invariant; everywhere nonnegative; supported in a small neighbourhood of Z; and positive at every point of Z. Then,

$$0 \leq \int_{\mathbb{R}\times S^1} \bar{\omega}_{\mathbb{C}}(\partial_s v, \partial_t v - \gamma_- iv) = \bar{A}(y_-) - \bar{A}(y_+), \tag{5.12}$$

where $\bar{A}(y_\pm)$ depends only on the limits $y_\pm$ of v. To compute it explicitly, fix a primitive $\bar{\omega}_{\mathbb{C}} = d\bar{\theta}_{\mathbb{C}}$. Also, consider the function $\bar{H}_{\mathbb{C}}$ such that $d\bar{H}_{\mathbb{C}} = \bar{\omega}_{\mathbb{C}}(\cdot, \gamma_- iy)$, normalized by asking that it should vanish near the origin. Then,

$$\bar{A}(y) = \int_{S^1} -y^*\bar{\theta}_{\mathbb{C}} + \bar{H}_{\mathbb{C}}(y(t))\, dt. \tag{5.13}$$

Concretely,

$$\bar{A}(y) = \begin{cases} 0 & \text{if } y = \pi(x), \text{ with } x \text{ of type (i) as listed above,} \\ (\frac{\gamma_-}{2\pi} - 1)\int_{\mathbb{C}} \bar{\omega}_{\mathbb{C}} < 0 & \text{if } x \text{ is of type (ii).} \end{cases} \tag{5.14}$$

Finally, note that equality in (5.12) holds only if $v^{-1}(Z) = \emptyset$. It follows that there are only three kinds of Floer trajectories:

(i, i) If both limits $x_\pm$ are of type (i), the Floer trajectory remains entirely on the inside of $\pi^{-1}(Z)$;

(ii, ii) If both limits $x_\pm$ are of type (ii), the Floer trajectoy remains entirely on the outside of $\pi^{-1}(Z)$, since $\bar{A}(y_-) = \bar{A}(y_+)$ (the same purpose was achieved by the "minimum principle argument" in [33, Lemma 6.1]);

(i, ii) Finally, x_- could be of type (i), and x_+ of type (ii).

Take a function $h_- : [0, \infty) \to \mathbb{R}$ which agrees with h on $[0, 2]$, and satisfies $h_-'(a) = \gamma_-$ for $a \geq 2$. Let H_- be a (time-dependent) function which agrees with H_+ over the preimage of $\{|y| \leq 2\}$, and satisfies $H_-(x) = h_-(|\pi(x)|^2/2) + H_M(x)$ elsewhere. This has only the type (i) periodic orbits, and the associated chain complex $\mathrm{CF}^*(E, H_-)$ has cohomology $\mathrm{HF}^*(E, \gamma_-, \epsilon)$. From the argument above, we get (assuming suitable choices of almost complex structures) a short exact sequence of chain complexes

$$0 \to \mathrm{CF}^*(E, H_-) \longrightarrow \mathrm{CF}^*(E, H_+) \longrightarrow \mathrm{Q}^* \to 0. \tag{5.15}$$

The differential on the quotient Q^* counts type (ii, ii) trajectories only, hence defines a (slightly modified) version of Floer cohomology inside the mapping torus of μ, which is as computed in Lemma 4.7. This shows that (5.15) induces the desired long exact sequence (5.8). The argument about BV operators is parallel, since one can arrange that the relevant equation (3.13) has the same property (5.11). The statement about gradings is straightforward (the shift by 2 reflects the Conley–Zehnder index of the circle $|y|^2/2 = a_*$, as a 1-periodic orbit of the Hamiltonian $h_+(|y|^2/2)$ on $\mathbb{C}$). □

Proposition 5.5. *For $\gamma \in (0, 2\pi)$ and sufficiently small $\epsilon > 0$,*

$$\mathrm{HF}^*(E, \gamma, \epsilon) \cong \mathrm{H}^*(E). \tag{5.16}$$

Moreover, the BV operator vanishes.

This is an analogue of Proposition 3.8, and has the same proof. We will now dig a little deeper into the consequences. In the situation of Lemma 5.4, choose a class $z \in \mathrm{HF}^{*+2}(\mu, \epsilon)$. Using notation as in (5.15), take $[\text{point}] \otimes z \in \mathrm{H}^1(S^1) \otimes \mathrm{HF}^{*+2}(\mu, \epsilon) \subset \mathrm{H}^{*+1}(\mathrm{Q})$, represent it by a cocycle in Q^*, and lift that cocycle to a cochain

$$x \in \mathrm{CF}^{*+1}(E, H_+), \quad dx \in \mathrm{CF}^{*+2}(E, H_-). \tag{5.17}$$

Assume that γ_- and ϵ are small. Using the proof of Proposition 3.8 (as applied to our situation in Lemma 5.5), one can arrange that the chain level BV operator vanishes on $\mathrm{CF}^*(E, H_-) \subset \mathrm{CF}^*(E, H_+)$. As a result, we get a cocycle

$$\bar{x} = \delta x \in \mathrm{CF}^*(E, H_+), \tag{5.18}$$

since $d\bar{x} = -\delta(dx) = 0$. Clearly, $\bar{x}$ is independent of the choice of lift x. Moreover, if the original cohomology class was trivial, we would have $x - dy \in \mathrm{CF}^{*+1}(E, H_-)$ for some y, and hence

$$\bar{x} = \delta(dy) = -d(\delta y). \tag{5.19}$$

This shows that $[\bar{x}] \in \mathrm{HF}^*(E, \gamma, \epsilon_+)$ depends only on z. Finally, if we project $\bar{x}$ to Q^*, it represents $1 \otimes z \in \mathrm{H}^0(S^1) \otimes \mathrm{HF}^{*+2}(\mu, \epsilon) \subset \mathrm{H}^*(\mathrm{Q})$ (by the description of the BV operator in Lemmas 4.7 and 5.4). The outcome of this argument can be summarized as follows:

Proposition 5.6. *Take the situation from Lemma 5.4, with $\epsilon > 0$ small. Then the map*

$$\mathrm{HF}^*(E, \gamma_+, \epsilon) \longrightarrow \mathrm{H}^*(S^1) \otimes \mathrm{HF}^{*+2}(\mu, \epsilon) \tag{5.20}$$

from (5.8) has a partial splitting, defined on $\mathrm{H}^0(S^1) \otimes \mathrm{HF}^{*+2}(\mu, \epsilon)$. □

For the benefit of readers concerned by the apparent ad hoc nature of the previous argument, we can explain how it fits into an appropriate formal framework.

Setup 5.7.

(i) Take a chain complex C^*_+ together with an endomorphism δ of degree -1. Let C^*_- be a subcomplex such that $\delta(\mathrm{C}^*_-) \subset \mathrm{C}^*_-$. Then, a homotopy trivialization

$[\chi]$ of $\delta|C^*_-$ induces a lift

$$\begin{array}{ccc} & & H^{*-1}(C_+) \\ & \nearrow & \downarrow \\ H^*(Q) & \longrightarrow & H^{*-1}(Q) \end{array} \tag{5.21}$$

where the horizontal arrow is the operation induced by δ on the quotient complex $Q^* = C^*_+/C^*_-$, and the vertical arrow is the projection map. The lift in (5.21) is defined as follows: given a cocycle in Q^*, lift it to a cochain x in C^*_+, and then map that to the cocycle

$$\bar{x} = \delta x + \chi(dx). \tag{5.22}$$

One sees immediately that this induces a map on cohomology, which depends only on the equivalence class $[\chi]$.

(ii) Take two such complexes C^*_+ and $\tilde{C}^*_+$, each with the same structure as in (i) (including subcomplexes C^*_- and $\tilde{C}^*_-$). Suppose that we have maps between them as in Setup 3.9(ii), which preserve the subcomplexes. If we are given mutually compatible homotopy trivializations $[\chi]$ and $[\tilde{\chi}]$ on those subcomplexes, which are compatible in the sense of Setup 3.9(ii), then (by a straightforward computation) the lifts from (5.21) fit into a commutative diagram

$$\begin{array}{ccc} H^*(C_+) & \longrightarrow & H^*(\tilde{C}_+) \\ \uparrow & & \uparrow \\ H^*(Q) & \longrightarrow & H^*(\tilde{Q}). \end{array} \tag{5.23}$$

Mapping this abstract framework onto our specific situation is fairly straightforward. We had $H^*(Q) \cong H^*(S^1) \otimes HF^{*+2}(\mu, \epsilon)$, with the BV operator on the quotient as described in Lemma 4.7. We arranged that $\delta|C^*_-$ vanished, and made the trivial choice $\chi = 0$. The partial splitting of (5.20) is the map which fits into the commutative diagram

$$\begin{array}{ccc} HF^*(E, \gamma_+, \epsilon) & & \\ \uparrow & \nwarrow & \\ H^1(S^1) \otimes HF^*(\mu, \epsilon) & \xrightarrow{\cong} & H^0(S^1) \otimes HF^*(\mu, \epsilon) \end{array} \tag{5.24}$$

where the vertical arrow is the restriction of the lift (5.21), and the horizontal arrow is the BV operator. Inspection shows that this indeed recovers the formula (5.18). As in the discussion following Setup 3.9, one can use the fact that the homotopy trivialization is canonical to show that the partial splitting is independent of all auxiliary choices.

We will also need a modified version of Proposition 5.6, whose proof combines much of the technical material about Floer cohomology that has been mentioned so far:

Proposition 5.8. *Take the situation from Lemma* 5.4. *Assume that ∂M is a contact circle bundle, and that $\epsilon \in (1,2)$. Additionally, assume that E has a symplectic Calabi–Yau structure, such that the induced symplectic Calabi–Yau structure on M is as in Example* 3.2(ii). *Then* (5.20) *has a partial splitting, defined on the subspace of* $\mathrm{H}^0(S^1) \otimes \mathrm{HF}^{*+2}(\mu, \epsilon)$ *where* $* \leq -2m$.

Proof. One can set up a Morse–Bott chain complex which computes $\mathrm{HF}^*(E, \gamma_-, \epsilon)$ exactly as in Proposition 3.10. Namely, generators consist of constant orbits, together with a Morse–Bott nondegenerate submanifold F which is a copy of ∂M (located in some fibre of π). Moreover, the Hamiltonian and almost complex structure can be taken to be time-independent. Then, the argument used there shows that the Morse–Bott analogue of the BV operator vanishes on all cochains of degree $\leq 2 - 2m$.

What does this mean for our original complex $\mathrm{CF}^*(E, H_-)$? It is related to the Morse–Bott version by a chain homotopy equivalence, which is a version of a continuation map, hence comes with a secondary map as in (3.18), expressing its compatibility with BV operators. Using Setup 3.11(ii), we obtain a nullhomotopy for the BV operator on the correspondingly truncated subcomplex:

$$\chi : \mathrm{CF}^*(E, H_-)^{\leq 2-2m} \longrightarrow \mathrm{CF}^{*-2}(E, H_-)^{\leq 2-2m}. \tag{5.25}$$

Now take (5.17), where $* \leq -2m$, and replace (5.18) with (5.22), which provides the desired splitting; this makes sense because dx is a cocycle in $\mathrm{CF}^*(E, H_-)$ of degree $* + 2 \leq 2 - 2m$, hence lies in the domain of definition of (5.25). □

As before, one can interpret this as an instance of a more general setup:

Setup 5.9.

(i) Consider a variation on the situation of Setup 5.7(i), where the homotopy trivialization exists only on $(\mathrm{C}^*_-)^{\leq j}$, for some j. Then, (5.22) defines a partial lift, as in (5.21) but defined only on $\mathrm{H}^*(Q)$ for $* \leq j - 1$.

(ii) There is a corresponding variation of Setup 5.7(ii), with two homotopy trivializations defined on $(\mathrm{C}^*_-)^{\leq j}$ respectively $(\tilde{\mathrm{C}}^*_-)^{\leq j}$. Then, one gets a diagram (5.23) in degrees $* \leq j - 1$.

To apply this to our case, we first used a Morse–Bott formalism to put ourselves in a situation where part of the BV operator can be explicitly computed (allowing an obvious choice of partial homotopy trivialization), and then used continuation maps to carry over that trivialization to the general case. In principle, part (ii) of Setup 5.9 can be used to show that the partial splitting from Proposition 5.8 is canonical, but that would require additional arguments not carried out here.

Example 5.10. In the situation of Example 5.2, we can use (1.7) for $\epsilon \in (1,2)$. The splitting from Proposition 5.8 then takes $1 \in H^0(M) \cong HF^{2-2m}(\mu,\epsilon)$ to an element of $HF^{-2m}(E,\gamma_+,\epsilon)$.

6. Translations at infinity

We remain in the same class of manifolds (Setup 5.1), but now approach Floer cohomology differently. First of all, we follow the version for symplectic automorphisms, rather than the Hamiltonian one. More importantly, the perturbations involved have different behaviour at infinity, modeled on [45].

6a. Geometric data

As usual, we begin by assembling the basic geometric ingredients, this time starting with the class of symplectic automorphisms that are allowed.

Setup 6.1. (i) We consider exact symplectic automorphisms $\phi : E \to E$ with the following properties. There is an open subset $U \subset E$ of the same kind as in Setup 5.1, such that

$$\pi(\phi(x)) = \phi_{\mathbb{C}}(\pi(x)) \quad \text{for } x \in U. \tag{6.1}$$

Here, $\phi_{\mathbb{C}}$ is a symplectic automorphism of the base, which is the identity outside a compact subset. Moreover, if we restrict ϕ to a fibre E_y, $|y| \gg 0$, then it should be the identity near ∂E_y.

(ii) As usual, if E has a symplectic Calabi–Yau structure, we assume that ϕ is a graded symplectic automorphism (then, the same holds for its restriction to any fibre E_y, $|y| \gg 0$).

(iii) We use functions $H \in C^\infty(E,\mathbb{R})$ such that

$$\begin{cases} H(\Theta(y,x)) = H_{\mathbb{C}}(y) + \epsilon\rho_M(x) & \text{for } x \text{ near } \partial M, \\ H(\Psi(p,q,x)) = \delta\,\mathrm{re}(e^{p+iq}) + H_M(x) & \text{for } p \gg 0. \end{cases} \tag{6.2}$$

Here, ϵ is as in (3.3), and $\delta \in \mathbb{R} \setminus \{0\}$. $H_{\mathbb{C}}$ is a function on the base such that $H_{\mathbb{C}}(y) = \delta\,\mathrm{re}(y)$ outside a compact subset, and H_M is as in Setup 4.6.

(iv) We use compatible almost complex structures J on E of the following kind. Near the boundary, we impose the same condition (5.7) as before. At points (p,q,x), where $p \gg 0$ and $-\pi \leq q \leq \pi$,

$$\Psi^* J = \begin{cases} i \times J^+_{M,\mathrm{re}(e^{p+iq})} & \text{for } 0 \leq q \leq \pi, \\ i \times J^-_{M,\mathrm{re}(e^{p+iq})} & \text{for } -\pi \leq q \leq 0. \end{cases} \tag{6.3}$$

Here, $J^{\pm}_{M,r}$ are two auxiliary families of compatible almost complex structures on the fibre, both parametrized by $r \in \mathbb{R}$ and belonging to the class (3.4), and additionally satisfying

$$\begin{cases} J^-_{M,r} = J^+_{M,r} & \text{if } r \gg 0, \\ J^-_{M,r} = \mu_* J^+_{M,r} & \text{if } r \ll 0. \end{cases} \tag{6.4}$$

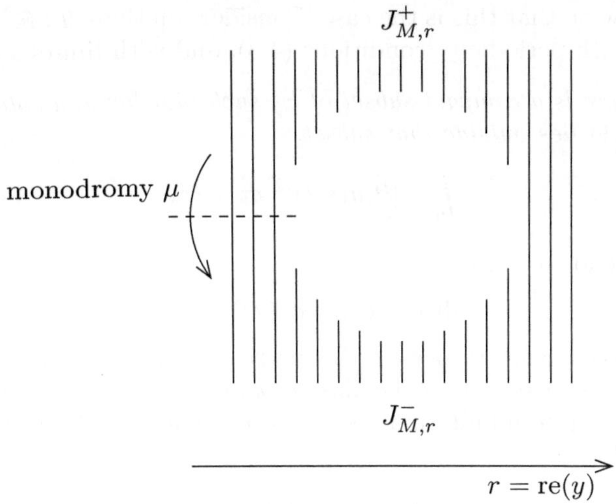

FIGURE 1

The assumption (6.1) is quite restrictive, since it implies that ϕ must commute with symplectic parallel transport between the fibres, inside U. In combination with the other condition, this implies that

$$\phi(\Theta(y,x)) = \Theta(\phi_{\mathbb{C}}(y),x) \quad \text{for } x \text{ close to } \partial M. \tag{6.5}$$

Similarly, we have

$$\phi(\Psi(p,q,x)) = \Psi(p,q,\phi_M(x)) \quad \text{if } p \gg 0, \tag{6.6}$$

where ϕ_M is an exact symplectic automorphism of M, which is the identity near ∂M, and which must commute with μ.

Example 6.2. Fix a function on $\mathbb{C}$, which equals $\pi|y|^2$ outside a compact subset, and pull it back to E. Then, its flow for time 1 gives a symplectic automorphism within the class defined above. We denote it by ν, and call it the *global monodromy*. The associated automorphism of the fibre is the monodromy μ.

The class of almost complex structures introduced above has the property that π is J-holomorphic outside a compact subset. Moreover, because of (6.3), the induced almost complex structures on the fibres E_y, $|y| \gg 0$, are locally constant under parallel transport in imaginary direction (this is sketched in Figure 1).

6b. The compactness argument

Given ϕ, choose $H = (H_t)$ and $J = (J_t)$, with the same periodicity condition as in (4.1) (this makes sense because the relevant classes of Hamiltonians and almost complex structures are invariant under ϕ). The solutions of (3.7) in $\mathcal{L}_\phi$, or equivalently the fixed points of $\phi^1_H \circ \phi$, will always be contained in a compact subset of $E \setminus \partial E$. For generic choice of (H_t), these fixed points will also be nondegenerate;

assume from now on that this is the case. Consider solutions $u : \mathbb{R}^2 \to E$ of Floer's equation (3.8) with periodicity conditions (4.3), and with limits $x_\pm$ as usual.

Lemma 6.3. *There is a compact subset of E, such that for any solution u and any s such that $u(s,1)$ lies outside that subset,*

$$\int_{[0,1]} |\partial_s u(s,t)|^2 \, dt \geq \delta^2. \tag{6.7}$$

Proof. For x outside a compact subset,

$$\mathrm{dist}(x, (\phi_{\mathrm{H}}^1 \circ \phi)(x)) \geq \delta, \tag{6.8}$$

where the distance is with respect to the metric associated to any almost complex structure in our class (recall that the fibres of π are compact, so "outside a compact subset" means going to infinity in base direction). For essentially the same reason, $\int_{[0,1]} |\partial_s u(s,t)| \, dt \geq \delta$. □

Lemma 6.4. *There is a constant $B > 0$, such that $|\mathrm{re}(\pi(u))| \leq B$ for all solutions u.*

Proof. Write $v = \pi(u)$. On the subset where $\mathrm{re}(v)$ is large, it satisfies the equation

$$\begin{cases} \partial_s v + i(\partial_t v - i\delta) = 0, \\ v(s,t+1) = v(s,t) + i\delta. \end{cases} \tag{6.9}$$

Hence, $\mathrm{re}(v)$ is harmonic, so that we can apply the maximum principle. □

Lemma 6.5. *There is a neighbourhood $V \subset M$ of ∂M, such that the image of any solution u is disjoint from $\Theta(\mathbb{C} \times V)$.*

This is again a maximum principle argument, but now used in fibre direction.

Lemma 6.6. *There is a constant C such that $\|du\|_\infty \leq C$ for any Floer trajectory.*

Proof. This is a simple Gromov compactness argument, following [45, Proposition 5.1], and we will only give limited details. Suppose that the result is false. After passing to a subsequence, $\|\partial_s u_k\|$ goes to infinity. Take points $z_k = (s_k, t_k)$ where $|\partial_s u_k|$ reaches its maximum.

If the $u_k(z_k)$ remain inside a compact subset of E, one can rescale locally near z_k, and get a non-constant pseudo-holomorphic plane as a limit. Since we have a priori bounds on the energy, this plane has finite energy, hence extends to a pseudo-holomorphic sphere, in contradiction to exactness. *Suppose on the other hand that the $u_k(z_k)$ do not remain inside a compact subset of E.* In view of Lemma 6.4, one can pass to a subsequence and then assume that $\mathrm{im}(\pi(u_k(z_k)))$ converges to either $+\infty$ or $-\infty$. Then, after a translation in imaginary direction over the base, the same argument as before applies, except that the limit lies in $\mathbb{C} \times M$, and is pseudo-holomorphic for an almost complex structure $i \times J^\pm_{M,\mathrm{re}(y)}$, where y is the coordinate on $\mathbb{C}$ (it is here that we use the specific property of our almost complex structures indicated in Figure 1). □

Proposition 6.7. *There is a compact subset of $E \setminus \partial E$ which contains all Floer trajectories.*

Proof. Assume that this is not true. By Lemmas 6.4 and 6.5, there must be a sequence (u_k) of solutions, such that the maximal value of $|\mathrm{im}(\pi(u_k))|$ goes to infinity. Because we have an absolute bound on $|\partial_s u_k|$ from Lemma 6.6, $u_k(s,1)$ has to spend an increasingly large interval (in s) inside the region where Lemma 6.3 applies. But that contradicts the a priori bound on the energy. □

6c. Floer cohomology and its properties

Having obtained Proposition 6.7, it is now a familiar process to set up Floer complexes $\mathrm{CF}^*(\phi, H)$, whose cohomology we denote by $\mathrm{HF}^*(\phi, \delta, \epsilon)$. The appropriate version of (4.6) is

$$\mathrm{HF}^*(\phi^{-1}, -\delta, -\epsilon) \cong \mathrm{HF}^{2n-*}(\phi, \delta, \epsilon)^\vee. \tag{6.10}$$

Lemma 6.8. *For small $\delta > 0$ and $\epsilon > 0$, $\mathrm{HF}^*(\mathrm{id}, \delta, \epsilon) \cong \mathrm{H}^*(E, \{\mathrm{re}(\pi) \ll 0\})$.*

We will not explain the proof of this, which can be done by reduction to Morse theory as in Proposition 3.8 (if π is a Lefschetz fibration, $\mathrm{H}^*(E, \{\mathrm{re}(\pi) \ll 0\})$ is concentrated in degree n, and has one generator for each critical point; one can in fact arrange that the underlying chain complex has the same property, and thereby give an elementary proof of Lemma 6.8 for that special case).

The dependence of Floer cohomology on the parameters (δ, ϵ) is a more interesting issue than before. One can show (for instance, using parametrized moduli spaces) that only the sign of δ matters. In fact, one also has isomorphisms

$$\mathrm{HF}^*(\phi, -\delta, \epsilon) \cong \mathrm{HF}^*(\phi, \delta, \epsilon), \tag{6.11}$$

but not canonical ones. To see that, note that the sign of δ depends on our identification of the base with the standard complex plane $\mathbb{C}$. Reversing that identification $(y \mapsto -y)$ takes δ to $-\delta$. To construct (6.11), one has to rotate the plane by some amount in $\pi + 2\pi\mathbb{Z}$, and different choices yield different maps (6.11).

Remark 6.9. One way to think of the ambiguity in (6.11) is as follows. $\mathrm{HF}^*(\phi, \delta, \epsilon)$ carries a canonical automorphism, induced by a full rotation of the plane (whose angle is thought of as an additional parameter), or equivalently by conjugation with the global monodromy. Then, (6.11) is unique up to composition with powers of that automorphism.

Another version of the same explanation goes as follows: on can allow translations over the base in any direction, corresponding to a parameter $\delta \in \mathbb{C}^*$ (for compatibility with our previous notation, the translation would have to be by $i\delta$). For fixed ϵ, these more general Floer cohomology groups $\mathrm{HF}^*(\phi, \delta, \epsilon)$ would be canonical locally trivial in δ, which means that they would form a local system over $\mathbb{C}^*$. The previously mentioned automorphism is just the holonomy of the local system, and (6.11) would be a parallel transport map between two different fibres.

Lemma 6.10. *For any $\epsilon_- < \epsilon_+$ which satisfy* (3.3)*, one has* $\mathrm{HF}^*(\phi, \delta, \epsilon_-) \cong \mathrm{HF}^*(\phi, \delta, \epsilon_+)$.

Proof. Our strategy follows that of Lemma 3.4. We enlarge the given E by attaching a conical piece to the boundary of each fibre, as in (3.21):

$$\hat{E} = E \cup_{\partial E} (\mathbb{C} \times [1, C] \times \partial M), \tag{6.12}$$

where ∂E is identified with $\mathbb{C} \times \partial M$ using parallel transport, which means (5.3). Extend the given ϕ to an automorphism $\hat{\phi}$ of $\hat{E}$ by setting it equal to $\phi_{\mathbb{C}} \times \mathrm{id}_{[1,C]\times\partial M}$ on the conical part, where $\phi_{\mathbb{C}}$ is as in (6.5).

Suppose that H_- is the function used to define the chain complex $CF^*(\phi, H_-)$ underlying $HF^*(\phi, \delta, \epsilon_-)$. We extend it to a function $\hat{H}_+$ on $\hat{E}$ by a fibrewise version of (3.23):

$$\hat{H}_{t,+}(r, y, x) = H_{\mathbb{C},t}(y) + Ch(C^{-1}r) \tag{6.13}$$

where h is as in (3.22), and the $H_{\mathbb{C},t}$ are functions as in (6.2). It is unproblematic to show that $CF^*(\hat{\phi}, \hat{H}_+)$ computes $HF^*(\phi, \delta, \epsilon_+)$. At this point, we want to be more specific about the choice of function for the original Floer cohomology group. Using the fact that the group of compactly supported symplectic automorphisms of $\mathbb{C}$ is connected, one can find a time-dependent $H_{\mathbb{C}}$ such that

$$(\phi^1_{H_{\mathbb{C}}} \circ \phi_{\mathbb{C}})(y) = y + i\delta \tag{6.14}$$

is simply a translation. In that case, $CF^*(\hat{\phi}, \hat{H}_+)$ has the same generators as $CF^*(\phi, H_-)$, and an easy maximum principle argument (in fibre direction) shows that the differentials also coincide. □

Remark 6.11. Alternatively, one can avoid the use of specific choices of $H_{\mathbb{C}}$, and argue as in Lemma 3.5. Namely, suppose that there is only one $\epsilon \in (\epsilon_-, \epsilon_+)$ such that $\epsilon R_{\partial M}$ has 1-periodic orbits. Then, there is a long exact sequence

$$\cdots \to HF^*(\phi, \delta, \epsilon_-) \longrightarrow HF^*(\phi, \delta, \epsilon_+) \longrightarrow H(Q^*) \to \cdots \tag{6.15}$$

One can arrange that $Q^* \cong Q^*_{\mathbb{C}} \otimes Q^*_M$, where the first factor is a version of the Floer chain complex for $\phi^1_{H_{\mathbb{C}}} \circ \phi_{\mathbb{C}}$, which is then shown to be acyclic (this last step would again use the connectedness of the group of compactly supported symplectic automorphisms of $\mathbb{C}$).

Our final topic is the relation between the two versions of Floer cohomology on E. Because of the different classes of perturbations used, this is not quite straightforward. We only need a partial result:

Lemma 6.12. *Take $\gamma_- \in (2\pi(k-1), 2\pi k)$, and arbitrary δ, ϵ. Let ν be the global monodromy. Then there is a canonical map*

$$HF^*(E, \gamma_-, \epsilon) \longrightarrow HF^*(\nu^k, \delta, \epsilon). \tag{6.16}$$

To clarify the grading conventions: the source in (6.16) carries its natural $\mathbb{Z}$-grading (as a form of Hamiltonian Floer cohomology, on a manifold with a symplectic Calabi–Yau structure). For the target, we use the structure of the global monodromy ν as a graded symplectic automorphism which comes from its original construction by a flow (see Example 6.2; this is parallel to (i) in Example 4.2).

Proof. The argument follows Lemma 5.4 closely, and we will only give a few details. Set $\gamma_+ = 2\pi k$, and choose a function h_+ as in (5.9). We can assume that ν^k is defined as the time-one map of the function $F(x) = h_+(|\pi(x)|^2/2)$. When choosing a perturbation H_+ to be used to define $\mathrm{CF}^*(\nu^k, H_+)$, we can assume that $H_{t,+}(x) = H_M(x)$ close to $Z = \{|\pi(x)| = 2\}$.

A suitable "barrier" argument shows that (for small $|\delta|$) the generators corresponding to fixed points lying on the inside of Z form a subcomplex of $\mathrm{CF}^*(\nu^k, H_+)$. Moreover, that subcomplex can be identified with $\mathrm{CF}^*(E, H_-)$, where

$$H_{t,-}(x) = \begin{cases} H_{t,+}(\phi_F^{-t}(x)) + F(x) & |\pi(x)| \leq 2, \\ H_M(x) + h_-(|\pi(x)|^2/2) & |\pi(x)| \geq 2. \end{cases} \tag{6.17}$$

Here, h_- is the function that agrees with h_+ on $[0, 2]$, and satisfies $h'_-(a) = \gamma_-$ for all $a \geq 2$. But (6.17) defines $\mathrm{HF}^*(E, \gamma_-, \epsilon)$. □

6d. Additional remarks

One can show that the map (6.16) fits into a long exact sequence

$$\cdots \to \mathrm{HF}^*(E, \gamma_-, \epsilon) \longrightarrow \mathrm{HF}^*(\nu^k, \delta, \epsilon) \longrightarrow \mathrm{HF}^{*+2k}(\mu^{k+1}, \epsilon) \to \cdots \tag{6.18}$$

It seems likely (but we have not checked the details) that there is a similar long exact sequence

$$\cdots \to \mathrm{HF}^*(\nu^k, \delta, \epsilon) \longrightarrow \mathrm{HF}^*(E, \gamma_+, \epsilon) \longrightarrow \mathrm{HF}^{*+2k+1}(\mu^{k+1}, \epsilon) \to \cdots \tag{6.19}$$

where $\gamma_+ \in (2\pi k, 2\pi(k+1))$. One piece of supporting evidence is that the combination of the two sequences above (in the appropriate order) is compatible with (5.8). Repeated use of those two sequences gives a step-by-step "decomposition" of all the Floer cohomology groups $\mathrm{HF}^*(E, \gamma, \epsilon)$ and $\mathrm{HF}^*(\nu^k, \delta, \epsilon)$. Alternatively, one can approach the same idea through spectral sequences:

Lemma 6.13. *For any ϵ and any $\gamma \in (2\pi k, 2\pi(k+1))$, $k \geq 0$, there is a spectral sequence converging to $\mathrm{HF}^*(E, \gamma, \epsilon)$, whose starting page is*

$$\mathrm{E}_1^{pq} = \begin{cases} \mathrm{H}^{q+1}(E, \{\mathrm{re}(\pi) \ll 0\}) & p = 1, \\ \mathrm{HF}^q(\mu^{-\lfloor p/2 \rfloor}, \epsilon) & -2k \leq p \leq 0, \\ 0 & \textit{otherwise}. \end{cases} \tag{6.20}$$

Lemma 6.14. *For any $k \geq 0$, $\delta > 0$, and ϵ, there is a spectral sequence converging to $\mathrm{HF}^*(\nu^k, \delta, \epsilon)$, whose starting page is*

$$\mathrm{E}_1^{pq} = \begin{cases} \mathrm{H}^{q+1}(E, \{\mathrm{re}(\pi) \ll 0\}) & p = 1, \\ \mathrm{HF}^q(\mu^{-\lfloor p/2 \rfloor}, \epsilon) & -2k+1 \leq p \leq 0, \\ 0 & \textit{otherwise}. \end{cases} \tag{6.21}$$

In fact, these spectral sequences are more straightforward to prove than (6.18) or (6.19). The first one is a weaker version of that in [33], and the second one is not substantially different. We will not explain them further, but we do want to show one application.

Example 6.15. Take the situation arising from an anticanonical Lefschetz pencil (see Examples 3.2(ii) and 5.2, but where $m = 0$). By repeatedly applying Lemma 4.4, we get

$$\mathrm{HF}^*(\mu^j, \epsilon) \cong \mathrm{HF}^{*-2j}(M, \epsilon - j). \tag{6.22}$$

Moreover, for any $\epsilon > 0$, there is a Morse–Bott spectral sequence converging to $\mathrm{HF}^*(M, \epsilon)$ (generalizing Proposition 3.10), with

$$\mathrm{E}_1^{pq} = \begin{cases} \mathrm{H}^{p+q}(M) & p = 0, \\ \mathrm{H}^{q-p}(\partial M) & -\lfloor \epsilon \rfloor \leq p < 0, \\ 0 & \text{otherwise.} \end{cases} \tag{6.23}$$

In particular, $\mathrm{HF}^*(M, \epsilon)$ is always concentrated in degrees $* \geq 0$.

Choose some $k > 0$, and take $\epsilon > k$. From (6.22) and (6.23), it follows that

$$\begin{cases} \mathrm{HF}^*(M, \epsilon) \text{ is concentrated in degrees } * \geq 0, \\ \mathrm{HF}^*(\mu, \epsilon) \cong \mathrm{HF}^{*-2}(M, \epsilon - 1) \text{ is concentrated in degrees } * \geq 2, \\ \dots \\ \mathrm{HF}^*(\mu^k, \epsilon) \cong \mathrm{HF}^{*-2k}(M, \epsilon - k) \text{ is concentrated in degrees } * \geq 2k. \end{cases} \tag{6.24}$$

By feeding that into (6.20) and (6.21), it follows that $\mathrm{HF}^*(E, \gamma, \epsilon)$ and $\mathrm{HF}^*(\nu^k, \delta, \epsilon)$ are all concentrated in nonnegative degrees (moreover, all the contributions coming from closed Reeb orbits on ∂M land in degrees ≥ 2).

To summarize, we have two infinite sequences of "closed string" Floer cohomology groups associated to any Lefschetz fibration. The first of these sequences, $\mathrm{HF}^*(E, \gamma, \epsilon)$ for $\gamma \in (2\pi k, 2\pi(k + 1))$, comes with an additional dependence on ϵ (one can remove that dependence by passing to the direct limit $\epsilon \to \infty$, as in the definition of symplectic cohomology; the price to pay is that the resulting groups will typically be infinite-dimensional). Moreover, these groups carry BV operators. The second sequence is $\mathrm{HF}^*(\nu^k, \delta, \epsilon)$. These groups are independent of ϵ, do not have BV operators, but come with canonical automorphisms (mentioned in Remark 6.9).

Remark 6.16. It is possible to interpret this situation in terms of mirror symmetry. Consider a smooth projective variety A, together with a section r of its anticanonical bundle, which gives rise to a smooth divisor $B = r^{-1}(0)$. We'll discuss the analogues of the two Floer cohomology groups mentioned above, but in reverse order.

Take the sheaves $\Omega^i_A(kB)$ of algebraic i-forms with poles of order at most k along B, and form

$$\bigoplus_i \mathrm{H}^{*+i}(A, \Omega^i_A(kB)) \tag{6.25}$$

For $k = 0$, this is the Hochschild homology of A. It admits a BV type operator, induced by the de Rham differential, but that is known to vanish. Derived autoequivalences act on Hochschild homology, and in particular, one gets a distinguished automorphism from the action of the Serre functor. More concretely, this automorphism is given by multiplying with the exponential of the class of the canonical bundle in $\mathrm{H}^1(A, \Omega^1_A)$. For $k > 0$, (6.25) does not carry a natural BV type operator, since the de Rham differential increases pole order (but one can still define a canonical automorphism, as before).

Instead, consider the subsheaves

$$\Omega^i_A((k-1)B + \log B) \subset \Omega^i_A(kB) \tag{6.26}$$

of those differential forms α such that both α and $d\alpha$ have poles of order $\leq k$ along B. This gives rise to another sequence of graded groups,

$$\bigoplus_i \mathrm{H}^{*+i}(A, \Omega^i_A((k-1)B + \log B)), \tag{6.27}$$

which do carry BV operators. To be more precise, the situation we have just considered (with a smooth B) is mirror to working with a Lefschetz fibration which has closed fibres; hence, there is no parameter corresponding to our ϵ here.

We also would like to consider a variant of Lemma 6.14, where some of the columns in the E_1 page have already been combined, as in (6.18). Again, the proof is omitted.

Lemma 6.17. *There is a spectral sequence converging to* $\mathrm{HF}^*(\nu^2, \delta, \epsilon)$, *with*

$$E_1^{pq} = \begin{cases} \mathrm{HF}^q(E, \gamma, \epsilon) & p = 0, \ \textit{where } \gamma > 0 \textit{ is small,} \\ \mathrm{HF}^{q+1}(\mu, \epsilon) \oplus \mathrm{HF}^{q+2}(\mu, \epsilon) & p = -1, \\ \mathrm{HF}^{q+1}(\mu^2, \epsilon) & p = -2, \\ 0 & \textit{otherwise.} \end{cases} \tag{6.28}$$

This spectral sequence is compatible with the $\mathbb{Z}/2$*-action on* $\mathrm{HF}^*(\nu^2, \delta, \epsilon)$ *from Remark* 4.3*: the induced* $\mathbb{Z}/2$*-action on* (6.28) *is trivial except in the* $p = -2$ *column, where it is the corresponding action on* $\mathrm{HF}^*(\mu^2, \epsilon)$.

Example 6.18. Take the situation from Example 6.15 with $\epsilon > 2$. Then the $p = -2$ column contributes only in positive degrees. Hence, the $\mathbb{Z}/2$-action on $\mathrm{HF}^0(\nu^2, \delta, \epsilon)$ is trivial, at least if we assume that $\mathrm{char}(\mathbb{K}) \neq 2$.

7. Open-closed string maps

We now combine fixed point Floer cohomology, in the version considered in Section 6, with its counterpart for Lagrangian submanifolds. The relation between the two theories, together with the preceding Floer cohomology computations, leads directly to our main results (Theorems 1.2 and 1.3).

7a. Lagrangian Floer cohomology

We will continue to work in the situation of Setups 5.1 and 6.1, with the following additional geometric ingredient.

Setup 7.1. (i) We consider oriented exact Lagrangian submanifolds $L \subset E \setminus \partial E$ such that $\pi|L$ is proper, and

$$\pi(L) = \{\text{compact subset}\} \cup \{\text{re}(y) \gg 0,\ \ \text{im}(y) = o\} \subset \mathbb{C} \quad \text{for some } o \in \mathbb{R}. \tag{7.1}$$

(ii) If E comes with a Calabi–Yau structure, we will assume that L is a graded Lagrangian submanifold.

(iii) Independently, one may want to assume that L comes with a *Spin* structure.

At any point $x \in L$ outside a compact subset, we know that TE^v_x is a symplectic subspace, and hence that

$$\text{rank}(D\pi_x|TL_x) = n - \dim(TL_x \cap TE^v_x) \geq 1. \tag{7.2}$$

By combining this with (7.1), one sees that equality holds in (7.2), hence that $D\pi_x|TL_x : TL_x \to \mathbb{R}$ is onto. This implies that there is a unique closed Lagrangian submanifold $L_M \subset M\setminus\partial M$ such that, if $y = e^{p+iq}$ with $q \in (-\pi/2, \pi/2)$, $\text{re}(y) \gg 0$ and $\text{im}(y) = o$, then

$$L \cap \pi^{-1}(y) = \Psi((p, q) \times L_M). \tag{7.3}$$

In other words, at infinity L is fibered over a horizontal half-infinite path, with each fibre being equal to L_M. Note that L_M is exact, and inherits an orientation (as well as the other structure mentioned in (ii) and (iii) above, whenever that exists on L).

Given two such submanifolds L_0, L_1, such that the corresponding numbers (7.1) satisfy $o_1 - o_0 \neq 0$, there is a well-defined Floer cohomology $HF^*(L_0, L_1)$. To make the setup formally parallel to fixed point Floer cohomology, we will also introduce a perturbed version $HF^*(L_0, L_1, \delta, \epsilon)$, which reduces to the previous one for $\delta = \epsilon = 0$, and is defined under the assumption that

$$o_1 - o_0 - \delta \neq 0. \tag{7.4}$$

Floer cohomology is invariant under automorphisms of E,

$$HF^*(\phi(L_0), \phi(L_1), \delta, \epsilon) \cong HF^*(L_0, L_1, \delta, \epsilon). \tag{7.5}$$

The analogue of (4.6) says that

$$HF^*(L_1, L_0, -\delta, -\epsilon) \cong HF^{n-*}(L_0, L_1, \delta, \epsilon)^\vee. \tag{7.6}$$

Lemma 7.2. *Suppose that $\delta_\pm$ are such that $o_1 - o_0 - \delta_\pm$ have the same sign. Then* $HF^*(L_0, L_1, \delta_-, \epsilon) \cong HF^*(L_0, L_1, \delta_+, \epsilon)$.

Lemma 7.3. *For any $\epsilon_\pm$,* $HF^*(L_0, L_1, \delta, \epsilon_-) \cong HF^*(L_0, L_1, \delta, \epsilon_+)$.

The two lemmas above are analogues of Lemma 3.4. Note that this time, there are no "forbidden values" of ϵ. The passage through the unique "forbidden value" of δ is described by the following result, whose proof we omit:

Lemma 7.4. *Take $\delta_- < o_1 - o_0 < \delta_+$. Then there is a long exact sequence*

$$\cdots \to \mathrm{HF}^*(L_0, L_1, \delta_-, \epsilon) \longrightarrow \mathrm{HF}^*(L_0, L_1, \delta_+, \epsilon) \longrightarrow \mathrm{HF}^*(L_{0,M}, L_{1,M}) \to \cdots \tag{7.7}$$

where the third group is the Floer cohomology of the associated closed Lagrangian submanifolds (7.3) *in the fibre* M.

Finally, we have an analogue of Proposition 3.8, also given here without proof:

Lemma 7.5. *For* $\delta > 0$, $\mathrm{HF}^*(L, L, \delta, \epsilon) \cong \mathrm{H}^*(L)$.

Remark 7.6. In general, $\mathrm{HF}^*(L_0, L_1, \delta, \epsilon)$ is a $\mathbb{Z}/2$-graded space over a coefficient field $\mathbb{K}$ of characteristic 2. In the situation from Setup 7.1(ii), one can obtain $\mathbb{Z}$-graded Floer cohomology groups; and in that of Setup 7.1(iii), an arbitrary coefficient field $\mathbb{K}$ can be allowed (see [19, Chapter 8] or [43, Section 12]).

To define the chain complex $\mathrm{CF}^*(L_0, L_1, H)$ underlying $\mathrm{HF}^*(L_0, L_1, \delta, \epsilon)$, one chooses functions $H = (H_t)$ and almost complex structures $J = (J_t)$ as in Setup 6.1, but this time parametrized by $t \in [0, 1]$. On the path space

$$\mathcal{L}_{L_0,L_1} = \{x : [0,1] \to E \, : \, x(0) \in L_0, \; x(1) \in L_1\}, \tag{7.8}$$

one has a counterpart of (4.2):

$$A_{L_0,L_1,H}(x) = \Big(\int_0^1 -x^*\theta_E + H_t(x(t))\, dt \Big) + G_{L_1}(x(1)) - G_{L_0}(x(0)), \tag{7.9}$$

where the G_{L_k} are functions such that $dG_{L_k} = \theta_E | L_k$. The critical points are solutions x of (3.7) in (7.8), hence correspond to points of $\phi^1_{\mathrm{H}}(L_0) \cap L_1$. The condition (7.4) implies that all such x are contained in a compact subset of E. Moreover, for generic choice of H, the x will be nondegenerate. Assuming this to be the case, one considers solutions $u : \mathbb{R} \times [0, 1] \to E$ of (3.8), with boundary conditions

$$u(s, 0) \in L_0, \;\; u(s, 1) \in L_1. \tag{7.10}$$

There is an analogue of Proposition 6.7 in this context, with similar strategy of proof. The rest of the construction of $\mathrm{HF}^*(L_0, L_1, \delta, \epsilon)$ follows the classical theory for closed Lagrangian submanifolds [16].

Remark 7.7. It may make sense to mention the one point where the situation here differs from that in Proposition 6.7 (by being easier, in fact). There is no analogue of (6.8) in the present context, but that is in fact unnecessary: because of (7.10), a bound on $\|du_\infty\|$ (obtained as in Lemma 6.6) implies that the image of u is contained in a region $\{\mathrm{im}(\pi) \leq D\}$. (The reader may also want to consult the proof of [45, Proposition 5.1], even though that result itself does not apply here, because of the different choice of inhomogeneous terms.)

7b. The formalism

We will now introduce the additional structures on Floer cohomology which underlie our main argument (for the moment, only the formal aspects will be considered; discussion of their actual construction will take place later on, in Section 7c).

Given Lagrangian submanifolds (L_0, L_1, L_2) and constants (δ_0, δ_1), (ϵ_0, ϵ_1) such that all Floer cohomology groups involved are well-defined, the triangle product (due to Donaldson) is a map

$$HF^*(L_1, L_2, \delta_1, \epsilon_1) \otimes HF^*(L_0, L_1, \delta_0, \epsilon_0) \longrightarrow HF^*(L_0, L_2, \delta_0 + \delta_1, \epsilon_0 + \epsilon_1). \quad (7.11)$$

This satisfies an appropriate associativity condition. As an application, fix Lagrangian submanifolds $L_1, \dots, L_m$ with pairwise different constants $o_1, \dots, o_m \in \mathbb{R}$. To these, one can associate an algebra A over (2.76), namely

$$A = R \oplus \bigoplus_{i<j} HF^*(L_i, L_j), \quad (7.12)$$

or equivalently

$$e_j A e_i = \begin{cases} HF^*(L_i, L_j) & i < j, \\ \mathbb{K} & i = j, \\ 0 & i > j. \end{cases} \quad (7.13)$$

Recall that here, the Floer cohomology groups under discussion are

$$HF^*(L_i, L_j, \delta, \epsilon) \quad \text{with} \quad \delta = \epsilon = 0$$

(in view of Lemmas 7.2 and 7.3, one could equivalently use any δ which lies on the same side of $o_j - o_i$ as the origin, and any ϵ, but that is not helpful for thinking about the products). The nontrivial part of the algebra structure of A consists of the products $HF^*(L_j, L_k) \otimes HF^*(L_i, L_j) \to HF^*(L_i, L_k)$ for $i < j < k$, which are special cases of (7.11). Next, take an automorphism ϕ as in Setup 6.1, and constants δ, ϵ satisfying

$$\delta \neq o_i - o_j \quad \text{for all } i, j \in \{1, \dots, m\} \text{ (including } i = j \text{, which means } \delta \neq 0). \quad (7.14)$$

Generalizing [45, Section 6.3], one associates to this a bimodule $P_{\phi,\delta,\epsilon}$ over A, namely:

$$P_{\phi,\delta,\epsilon} = \bigoplus_{i,j} HF^*(\phi(L_i), L_j, \delta, \epsilon). \quad (7.15)$$

These structures have cochain level refinements: an A_∞-algebra $\mathcal{A}$ (following Fukaya), and A_∞-bimodules $\mathcal{P}_{\phi,\delta,\epsilon}$ over $\mathcal{A}$. There are corresponding refinements of the properties of Lagrangian Floer cohomology mentioned above. Namely, there is a quasi-isomorphism (in fact, if the choices are suitably coordinated, an isomorphism)

$$\mathcal{P}_{\phi,\delta,\epsilon} \simeq \mathcal{P}^\vee_{\phi^{-1},-\delta,-\epsilon}[-n]. \quad (7.16)$$

Lemma 7.8. *The quasi-isomorphism type of $\mathcal{P}_{\phi,\delta,\epsilon}$ is independent of ϵ. It is also locally constant in δ, within the allowed range* (7.14).

This is a refinement of the previously mentioned invariance properties of Lagrangian Floer cohomology (Lemmas 7.2, 7.3). We omit the proof.

Example 7.9. In our eventual application, $(L_1, \dots, L_m)$ will be a basis of Lefschetz thimbles. In that case, if one takes $\delta \gg 0$, then $\mathcal{P}_{\mathrm{id},\delta,\epsilon}$ is quasi-isomorphic to the diagonal bimodule $\mathcal{A}$ [45, Corollary 6.1]; and if one takes $\delta \ll 0$, $\mathcal{P}_{\mathrm{id},\delta,\epsilon}$ is quasi-isomorphic to the shifted dual diagonal bimodule $\mathcal{A}^\vee[-n]$ [45, Corollary 6.2].

For ϕ as before, take (δ, ϵ) so that $\mathrm{HF}^*(\phi, \delta, \epsilon)$ is well-defined. Given any L_i, one has a canonical map relating the two kinds of Floer cohomology groups:

$$\mathrm{HF}^*(\phi(L_i), L_i, \delta, \epsilon) \longrightarrow \mathrm{HF}^{*+n}(\phi, \delta, \epsilon). \tag{7.17}$$

After applying (6.10), (7.6) and changing notation from (ϕ, δ, ϵ) to $(\phi^{-1}, -\delta, \epsilon)$, the map dual to (7.17) can be written as

$$\mathrm{HF}^*(\phi, \delta, \epsilon) \longrightarrow \mathrm{HF}^*(\phi(L_i), L_i, \delta, \epsilon). \tag{7.18}$$

Example 7.10. Let us specialize to $\phi = \mathrm{id}$ and $\delta > 0$. Applying Lemmas 7.5 and 6.8, one sees that (7.17) reduces to a map

$$\mathrm{H}^*(L_i) \longrightarrow \mathrm{H}^{*+n}(E, \{\mathrm{re}(\pi) \ll 0\}), \tag{7.19}$$

Let us instead take $\delta < 0$. Then, because of (7.6) and (6.10), the map (7.17) looks like this:

$$\mathrm{H}_*(L_i) \longrightarrow \mathrm{H}_*(E, \{\mathrm{re}(\pi) \ll 0\}), \tag{7.20}$$

and its dual (7.18) is correspondingly

$$\mathrm{H}^*(E, \{\mathrm{re}(\pi) \ll 0\}) \longrightarrow \mathrm{H}^*(L_i). \tag{7.21}$$

Unsurprisingly, (7.19) and (7.21) can be identified with the ordinary pushforward and restriction maps in cohomology; in particular, (7.21) factors through $\mathrm{H}^*(E)$.

As before, (7.17) is part of a more complicated structure, the *open-closed string map*

$$\mathrm{HH}_*(\mathcal{A}, \mathcal{P}_{\phi,\delta,\epsilon}) \longrightarrow \mathrm{HF}^{*+n}(\phi, \delta, \epsilon), \tag{7.22}$$

which is defined for all (δ, ϵ) such that both sides make sense. Let us use the same Floer-theoretic duality as before, as well as (2.27) and (7.16). After applying that, and suitably adjusting notation, one can write the dual of (7.22) as

$$\mathrm{HF}^*(\phi, \delta, \epsilon) \longrightarrow \mathrm{HH}^*(\mathcal{A}, \mathcal{P}_{\phi,\delta,\epsilon}) = \mathrm{H}^*(\hom_{[\mathcal{A},\mathcal{A}]}(\mathcal{A}, \mathcal{P}_{\phi,\delta,\epsilon})). \tag{7.23}$$

Remark 7.11. The composition of (7.22) and (7.23) yields a map from Hochschild homology to Hochschild cohomology (of degree n), for any bimodule $\mathcal{P}_{\phi,\delta,\epsilon}$. Taking into account (2.34) (since $\mathcal{A}$ is directed, it is homologically smooth), one can write that as

$$\mathrm{H}^*(\hom_{[\mathcal{A},\mathcal{A}]}((\mathcal{A}^\vee)^{-1}, \mathcal{P}_{\phi,\delta,\epsilon})) \longrightarrow \mathrm{H}^{*+n}(\hom_{[\mathcal{A},\mathcal{A}]}(\mathcal{A}, \mathcal{P}_{\phi,\delta,\epsilon})). \tag{7.24}$$

Generally speaking, one natural source of such homomorphisms are bimodule maps $\mathcal{A} \to (\mathcal{A}^\vee)^{-1}$ of degree n, or equivalently bimodule maps $\mathcal{A}^\vee[-n] \to \mathcal{A}$. At least in the case of Lefschetz fibrations, to be discussed later (and assuming

$\delta \gg 0$), it seems likely that (7.24) is induced by the bimodule map we have called ρ (Lemma 1.1).

We will also need a variation of (7.22) which involves two automorphisms ϕ_k ($k = 0, 1$), with their associated bimodules $\mathcal{P}_{\phi_k,\delta_k,\epsilon_k}$. For simplicity, we assume that $\epsilon_0/\delta_0 = \epsilon_1/\delta_1$. Set

$$\begin{cases} \phi = \phi_1\phi_0, \\ \delta = \delta_0 + \delta_1, \\ \epsilon = \epsilon_0 + \epsilon_1. \end{cases} \tag{7.25}$$

We suppose that $HF^*(\phi, \delta, \epsilon)$ is defined. Then, the new version of the open-closed string map has the form

$$HH_*(\mathcal{A}, \mathcal{P}_{\phi_1,\delta_1,\epsilon_1} \otimes_{\mathcal{A}} \mathcal{P}_{\phi_0,\delta_0,\epsilon_0}) \longrightarrow HF^{*+n}(\phi, \delta, \epsilon). \tag{7.26}$$

This is the same kind of construction as the "two-pointed open-closed string maps" of [21, Section 5.6]. In parallel with (7.23), but this time using (2.26), one can write the dual of (7.26) as

$$HF^*(\phi, \delta, \epsilon) \longrightarrow H^*\big(\hom_{[\mathcal{A},\mathcal{A}]}(\mathcal{P}_{\phi_0^{-1},-\delta_0,-\epsilon_0}, \mathcal{P}_{\phi_1,\delta_1,\epsilon_1})\big). \tag{7.27}$$

Remark 7.12. One can think of (7.26) as follows. Write $\mathcal{P}_k = \mathcal{P}_{\phi_k,\delta_k,\epsilon_k}$ and $\mathcal{P} = \mathcal{P}_{\phi,\delta,\epsilon}$, assuming that the latter is defined. The triangle products

$$\begin{aligned} &HF^*(\phi_1(L_j), L_k, \delta_1, \epsilon_1) \otimes HF^*(\phi_0(L_i), L_j, \delta_0, \epsilon_0) \\ &\quad \cong HF^*(\phi_1(L_j), L_k, \delta_1, \epsilon_1) \otimes HF^*(\phi(L_i), \phi_1(L_j), \delta_0, \epsilon_0) \longrightarrow HF^*(\phi(L_i), L_k, \delta, \epsilon) \end{aligned} \tag{7.28}$$

can be lifted to an A_∞-bimodule homomorphism

$$\mathcal{P}_1 \otimes_{\mathcal{A}} \mathcal{P}_0 \longrightarrow \mathcal{P}. \tag{7.29}$$

One can reduce (7.26) to (7.22) for ϕ, by writing it as the composition

$$HH_*(\mathcal{A}, \mathcal{P}_1 \otimes_{\mathcal{A}} \mathcal{P}_0) \longrightarrow HH_*(\mathcal{A}, \mathcal{P}) \longrightarrow HF^{*+n}(\phi, \delta, \epsilon), \tag{7.30}$$

where the first map is induced by (7.29) ([21] uses the same idea to relate "two-pointed open-closed string maps" to ordinary "one-pointed" ones). However, we will prefer a direct definition of (7.26), since that is a little simpler, and we have no use for the maps (7.29) by themselves.

Remark 7.13. Let us slightly change notation, and assume that (ϕ, δ, ϵ) is given, with the property that $\mathcal{P}_{\phi,\delta,\epsilon}$ and $HF^*(\phi^2, 2\delta, 2\epsilon)$ are well-defined. Then, (7.26) specializes to yield a map

$$HH_*(\mathcal{A}, \mathcal{P}_{\phi,\delta,\epsilon}^{\otimes_{\mathcal{A}} 2}) \longrightarrow HF^{*+n}(\phi^2, 2\delta, 2\epsilon). \tag{7.31}$$

The left-hand side carries a natural $\mathbb{Z}/2$-action, as discussed in Section 2c; so does the right-hand side, because it is the Floer cohomology of a square (Remark 4.3); and (7.31) is compatible with those actions (essentially, because Figure 4 is rotationally symmetric).

7c. Cauchy–Riemann equations

The previously mentioned algebraic structures belong to a kind of TCFT (Topological Conformal Field Theory), which means a general framework of algebraic operations parametrized by Riemann surfaces. We will now outline some of the ingredients of this construction. Compared to similar ideas in the literature, the main difference is that we allow nontrivial monodromy around interior punctures on the Riemann surfaces. In other respects, the exposition here is actually somewhat restrictive: for instance, unlike [4], we only allow closed one-forms on our Riemann surfaces.

Setup 7.14. (i) Fix constants (ϵ, δ). A *worldsheet* is a connected non-compact Riemann surface S, possibly with boundary, with the following properties and additional data.

We assume that there is a compactification $\bar{S} = S \sqcup \Sigma$, obtained by adding a finite set of points. We divide the points of this finite set Σ into closed string (interior) and open string (boundary) ones, and (independently, and arbitrarily) into inputs and outputs, denoting the respective subsets by $\Sigma^{\mathrm{cl/op,in/out}}$. These should come with local holomorphic coordinates (tubular and strip-like ends), of the form

$$\begin{cases} \epsilon_\zeta : (-\infty, 0] \times S^1 \longrightarrow S, & \zeta \in \Sigma^{\mathrm{cl,out}}, \\ \epsilon_\zeta : [0, \infty) \times S^1 \longrightarrow S, & \zeta \in \Sigma^{\mathrm{cl,in}}, \\ \epsilon_\zeta : (-\infty, 0] \times [0,1] \longrightarrow S, & \zeta \in \Sigma^{\mathrm{op,out}}, \\ \epsilon_\zeta : [0, \infty) \times [0,1] \longrightarrow S, & \zeta \in \Sigma^{\mathrm{op,in}}. \end{cases} \tag{7.32}$$

Let $\tilde{S} \to S$ be the universal covering. We fix lifts of (7.32),

$$\begin{cases} \tilde{\epsilon}_\zeta : (-\infty, 0] \times \mathbb{R} \longrightarrow \tilde{S}, & \zeta \in \Sigma^{\mathrm{cl,out}}, \\ \tilde{\epsilon}_\zeta : [0, \infty) \times \mathbb{R} \longrightarrow \tilde{S}, & \zeta \in \Sigma^{\mathrm{cl,in}}, \\ \tilde{\epsilon}_\zeta : (-\infty, 0] \times [0,1] \longrightarrow \tilde{S}, & \zeta \in \Sigma^{\mathrm{op,out}}, \\ \tilde{\epsilon}_\zeta : [0, \infty) \times [0,1] \longrightarrow \tilde{S}, & \zeta \in \Sigma^{\mathrm{op,in}}. \end{cases} \tag{7.33}$$

Write $\Gamma \cong \pi_1(S)$ for the covering group of $\tilde{S} \to S$. We want to have a homomorphism

$$\Phi : \Gamma \longrightarrow \mathrm{Symp}(E), \tag{7.34}$$

which takes values in the subgroup of those symplectic automorphisms of E described in Setup 6.1. In particular, given $\zeta \in \Sigma^{\mathrm{cl}}$, passing from $\tilde{\epsilon}_\zeta(s, t+1)$ to $\tilde{\epsilon}_\zeta(s,t)$ amounts to acting by an element of Γ, which determines an automorphism ϕ_ζ by (7.34).

To each point $\tilde{z} \in \partial\tilde{S}$ we want to associate a Lagrangian submanifold $L_{\tilde{z}}$ as in Setup 7.1, in a way which is locally constant in z, and compatible with (7.34):

$$L_{\gamma(\tilde{z})} = \Phi(\gamma)(L_{\tilde{z}}). \tag{7.35}$$

In particular, given $\zeta \in \Sigma^{\mathrm{op}}$, we have distinguished Lagrangian submanifolds $(L_{\zeta,0}, L_{\zeta,1})$ which are associated to the points $(\tilde{\epsilon}_\zeta(s,0), \tilde{\epsilon}_\zeta(s,1))$. By definition, these come with real numbers (7.1), which we denote by $(o_{\zeta,0}, o_{\zeta,1})$.

Finally, we want S to carry a closed one-form β_S with $\beta_S|\partial S = 0$, and which over each end (7.32) satisfies

$$\epsilon_\zeta^* \beta_S = \beta_\zeta \, dt, \tag{7.36}$$

where the β_ζ are constants. Set $\delta_\zeta = \beta_\zeta \delta$, $\epsilon_\zeta = \beta_\zeta \epsilon$. We require that for $\zeta \in \Sigma^{\mathrm{cl}}$, ϵ_ζ should satisfy (3.3), and $\delta_\zeta \neq 0$; while for $\zeta \in \Sigma^{\mathrm{op}}$, the pair $(L_{\zeta,0}, L_{\zeta,1})$ and the number δ_ζ must satisfy (7.4).

(ii) Given S, we consider families $J_S = (J_{S,\tilde{z}})$ of almost complex structures in the class from Setup 6.1, parametrized by $\tilde{z} \in \tilde{S}$. These should be equivariant with respect to (7.34), and over the strip-like ends they should be invariant under translation in the first variable. Concretely, this means that there are families $J_\zeta = (J_{\zeta,t})$ such that

$$J_{S,\tilde{\epsilon}_\zeta(s,t)} = J_{\zeta,t}; \tag{7.37}$$

and if $\zeta \in \Sigma^{\mathrm{cl}}$, the associated family J_ζ satisfies a periodicity condition as in (4.1) with respect to ϕ_ζ.

(iii) We will equip S with an *inhomogeneous term* K_S, which is a one-form on $\tilde{S}$ with values in the space of functions on E. More precisely, for any tangent vector ξ, $K_S(\xi) \in C^\infty(E, \mathbb{R})$ belongs to the class from Setup 6.1 for the constants $(\beta_S(\xi)\delta, \beta_S(\xi)\epsilon)$. This family must be equivariant with respect to (7.34), and if ξ is tangent to ∂S, then $K_S(\xi)|L_{\tilde{z}} = 0$. Finally, over each end, $\epsilon_\zeta^* K_S$ is invariant under translation in the first variable, and satisfies $(\epsilon_\zeta^* K_S)(\partial_s) = 0$. In analogy with (7.37), one can therefore write

$$\tilde{\epsilon}_\zeta^* K_S = H_{\zeta,t} \, dt; \tag{7.38}$$

and if $\zeta \in \Sigma^{\mathrm{cl}}$, H_ζ again satisfies (4.1).

We usually refer to a pair (J_S, K_S) as a *perturbation datum*, since it specifies a particular perturbed Cauchy–Riemann equation on S. Concretely, this is an equation for maps

$$\begin{cases} \tilde{u} : \tilde{S} \longrightarrow E, \\ \tilde{u}(\gamma(\tilde{z})) = \Phi(\gamma)(\tilde{u}(\tilde{z})), \\ \tilde{u}(\tilde{z}) \in L_{\tilde{z}} \quad \text{for } \tilde{z} \in \partial \tilde{S}, \end{cases} \tag{7.39}$$

namely:

$$(d\tilde{u} - Y_{K_S})^{0,1} = 0. \tag{7.40}$$

Here, Y_{K_S} is the one-form on $\tilde{S}$ with values in $C^\infty(TE)$ associated to K_S (by passing from functions to Hamiltonian vector fields). We use the complex structure on S, and the almost complex structures $J_{S,\tilde{z}}$, to form the $(0,1)$-part of the linear map $d\tilde{u} - Y_{K_S} : TS_{\tilde{z}} \longrightarrow TE_{\tilde{u}(\tilde{z})}$. One checks readily that (7.40) is invariant under the twisted Γ-periodicity condition from (7.39). Over the ends, which means for

$u_\zeta(s,t) = \tilde{u}(\tilde{\epsilon}_\zeta(s,t))$, (7.40) reduces to equations for suitable Floer trajectories:

$$\begin{cases} u_\zeta(s,k) \in L_{\zeta,k} \quad \text{for } k = 0,1, \text{ if } \zeta \in \Sigma^{\mathrm{op}}, \\ u_\zeta(s,t) = \phi_\zeta(u_\zeta(s,t+1)) \quad \text{if } \zeta \in \Sigma^{\mathrm{cl}}, \\ \partial_s u_\zeta + J_{\zeta,t}(\partial_t u_\zeta - X_{H_\zeta,t}) = 0. \end{cases} \tag{7.41}$$

It therefore makes sense to impose convergence conditions

$$\lim_{s\to\pm\infty} u_\zeta(s,\cdot) = x_\zeta. \tag{7.42}$$

The necessary compactness argument for the moduli space of solutions of (7.40) combines: an overall energy bound; a maximum principle argument in horizontal direction over the base (as in Lemma 6.4); and the following idea. Let us fix a metric on S which is standard over the ends. Then there is a constant c such that each point of S either (i) lies within distance c of ∂S, or else (ii) lies on a tubular end (the image of ϵ_ζ for some $\zeta \in \Sigma^{\mathrm{cl}}$). For preimages $\tilde{z}$ of points as in (i), one can bound $|\pi(\tilde{u}(\tilde{z}))|$ as in Lemma 6.6; whereas in case (ii), one uses an argument as in Lemma 6.3. Taken together, these ingredients yield an analogue of Proposition 6.7. Transversality is straightforward, given the freedom to choose J_S and K_S.

Counting solutions of (7.40), (7.42) for a single Riemann surface S yields a map between Floer cochain spaces, which induces a cohomology level map of degree $n(-\chi(\bar{S}) + |\Sigma^{\mathrm{out,op}}| + 2|\Sigma^{\mathrm{out,cl}}|)$:

$$\begin{array}{c} \bigotimes_{\zeta\in\Sigma^{\mathrm{in,op}}} HF^*(L_{\zeta,0}, L_{\zeta,1}, \delta_\zeta, \epsilon_\zeta) \otimes \bigotimes_{\zeta\in\Sigma^{\mathrm{in,cl}}} HF^*(\phi_\zeta, \delta_\zeta, \epsilon_\zeta) \\ \downarrow \\ \bigotimes_{\zeta\in\Sigma^{\mathrm{out,op}}} HF^*(L_{\zeta,0}, L_{\zeta,1}, \delta_\zeta, \epsilon_\zeta) \otimes \bigotimes_{\zeta\in\Sigma^{\mathrm{out,cl}}} HF^*(\phi_\zeta, \delta_\zeta, \epsilon_\zeta). \end{array} \tag{7.43}$$

Remark 7.15. A suitable analogue of Remarks 3.3 and 7.6 applies. In general, (7.43) will be a $\mathbb{Z}/2$-graded map, and defined over a coefficient field with $\mathrm{char}(\mathbb{K}) = 2$. However, if one assumes that: E carries a symplectic Calabi–Yau structure; that (7.34) comes with a lift to the graded symplectic automorphism group; and that all boundary conditions are graded Lagrangian submanifolds; then (7.43) will have the specified degree with respect to the $\mathbb{Z}$-gradings of the Floer cohomology groups involved. Moreover, if the Lagrangian submanifolds are *Spin*, arbitrary $\mathbb{K}$ are allowed.

So far, what we have explained is a version of the TQFT formalism for Floer theory. For the extension to a TCFT, which is relevant for us, one has to include families of Riemann surfaces. Strictly speaking, a TCFT framework would have to allow a class of such families which is large enough to form a chain level model for the homology of the relevant compactified moduli spaces of Riemann surfaces, and to allow appropriate composition (operad) structures. For our purpose, only certain specific families are needed; there, the underlying analysis remains the same as for the TQFT case (except that the gluing theory has to be carried out in a parametrized sense, which is something that can be regarded as well-understood).

7d. Moduli spaces

As before, we start with $(L_1, \dots, L_m)$ having pairwise different constants $(o_1, \dots, o_m)$. For $1 \leq i < j \leq m$, fix the additional data (functions H_{ij}, and almost complex structures J_{ij}) needed to define the Floer cochain complexes $\mathrm{CF}^*(L_i, L_j, H_{ij})$. To obtain an A_∞-algebra structure, consider worldsheets S which are discs with $d+1 \geq 3$ boundary punctures. We equip the components of ∂S with boundary conditions $L_{i_0}, \dots, L_{i_d}$, for $i_0 < \cdots < i_d$ (in positive order around $\partial \bar{S}$). The one-forms β_S are taken to be zero throughout. The perturbation data (J_S, K_S) need to be chosen so that their restriction to the strip-like ends, as in (7.37) and (7.38), reduces to the choices previously made to define Floer theory. Finally, the choices need to depend smoothly on the moduli of S, and there are conditions about their behaviour as this surface degenerates. We will not give any details, referring instead to [43, Section 9]. The outcome are maps

$$\mathrm{CF}^*(L_{i_{d-1}}, L_{i_d}, H_{i_{d-1}i_d}) \otimes \cdots \otimes \mathrm{CF}^*(L_{i_0}, L_{i_1}, H_{i_0 i_1}) \longrightarrow \mathrm{CF}^{*+2-d}(L_{i_0}, L_{i_d}, H_{i_0 i_d}), \tag{7.44}$$

which satisfy the A_∞-associativity equations. To define the cochain level structure underlying (7.12), one extends these operations in the unique way to a strictly unital A_∞-structure on

$$\mathcal{A} = R \oplus \bigoplus_{i<j} \mathrm{CF}^*(L_i, L_j, H_{ij}). \tag{7.45}$$

Fix an automorphism ϕ. To define $\mathcal{P} = \mathcal{P}_{\phi,\delta,\epsilon}$, one chooses Hamiltonians $H_{\phi,ij}$ and almost complex structures $J_{\phi,ij}$ for all (i,j), and then sets

$$\mathcal{P} = \bigoplus_{ij} \mathrm{CF}^*(\phi(L_i), L_j, H_{\phi,ij}), \tag{7.46}$$

with $\mu_{\mathcal{P}}^{0;1;0}$ the direct sum of Floer differentials. Fundamentally, the moduli spaces of Riemann surfaces which define the higher operations $\mu_{\mathcal{P}}^{s;1;r}$ are the same (Stasheff polyhedra) as for $\mu_{\mathcal{A}}^{r+1+s}$. However, the other data they carry are different, and we find it convenient to think of the Riemann surfaces themselves in a slightly different way, namely to write them as

$$S = T \setminus \{\zeta_1, \dots, \zeta_r, \zeta_1', \dots, \zeta_s'\}, \tag{7.47}$$

where: $T = \mathbb{R} \times [0,1]$; the $\zeta_i \in \mathbb{R} \times \{0\} \subset \partial T$ are increasing; and the $\zeta_i' \in \mathbb{R} \times \{1\} \subset \partial T$ are decreasing. The one-forms β_S should be equal to dt on the region where $|s| \gg 0$, and should vanish sufficiently close to the ζ_i and ζ_i'. The boundary conditions are

$$(\phi(L_{i_0}), \dots, \phi(L_{i_r})) \quad \text{along } \mathbb{R} \times \{0\} \tag{7.48}$$

(as s increases, and where $i_0 < \cdots < i_r$), respectively

$$(L_{i_0'}, \dots, L_{i_s'}) \quad \text{along } \mathbb{R} \times \{1\} \tag{7.49}$$

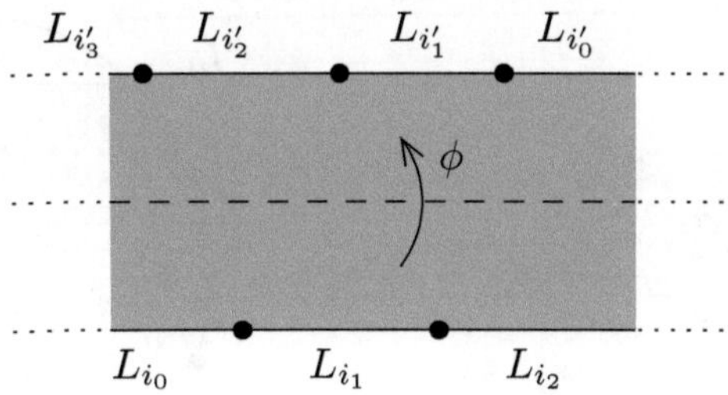

FIGURE 2

(as s decreases, and where again $i'_0 < \cdots < i'_s$). The inhomogeneous terms and almost complex structures are determined by those chosen for (7.46) over the ends where $|s| \gg 0$, and by the choices made for $\mathcal{A}$ on the remaining ends. As we vary over all possible S, the outcome are operations

$$\begin{gathered} CF^*(L_{i'_{s-1}}, L_{i'_s}, H_{i'_{s-1}i'_s}) \otimes \cdots \otimes CF^*(L_{i'_0}, L_{i'_1}, H_{i'_0 i'_1}) \\ \otimes CF^*(\phi(L_{i_r}), L_{i'_0}, H_{\phi, i_r i'_0}) \otimes \\ CF^*(L_{i_{r-1}}, L_{i_r}, H_{i_{r-1}i_r}) \otimes \cdots \otimes CF^*(L_{i_0}, L_{i_1}, H_{i_0 i_1}) \\ \downarrow \\ CF^{*+1-r-s}(\phi(L_{i_0}), L_{i'_s}, H_{\phi, i_0 i'_s}), \end{gathered} \tag{7.50}$$

which (extended over the units in $\mathcal{A}$ in the obvious way) constitute $\mu_{\mathcal{P}}^{s;1;r}$. It may be convenient to think of (7.47) as glued together from two half-strips $S_\pm = \{(s,t) \in S \,:\, \pm t \leq \frac{1}{2}\}$, each of which carries boundary conditions taken from the L_k, but where the gluing identifies the two target spaces E using ϕ. This amounts to rewriting the relevant equation (7.40) as two parts $u_\pm : S_\pm \to E$, joined by a "seam" $u_+(s, \frac{1}{2}) = \phi(u_-(s, \frac{1}{2}))$. Mathematically, this does not change anything, but it can be useful as an aid to the intuition, since it makes the connection with quilted Floer cohomology [56] (Figure 2).

Next, consider surfaces of the form

$$S = H \setminus \{\zeta_*, \zeta_1, \ldots, \zeta_d\} \tag{7.51}$$

where $H \subset \mathbb{C}$ is the closed upper half-plane, from which we remove a fixed interior point ζ_* (say $\zeta_* = i$) as well as boundary points $\zeta_1 < \cdots < \zeta_d$. We consider the ζ_k as inputs, and ζ_* as an output. The one-form β_S should vanish near the ζ_k, and its integral along a small (counterclockwise) loop around ζ_* should be equal to 1 (in other words, $\beta_{\zeta_k} = 0$ and $\beta_{\zeta_*} = 1$). The representation (7.34) maps that same loop to ϕ (Figure 3 shows two equivalent pictures of this; in the right-hand one, we have drawn H itself as a disc with one boundary point removed). Over $\partial H \subset H \setminus \{\zeta_*\}$, we can choose a section of the universal cover. Along that section we place the boundary conditions $L_{i_0}, \ldots, L_{i_d}$ with $i_0 < \cdots < i_d$, and then extend

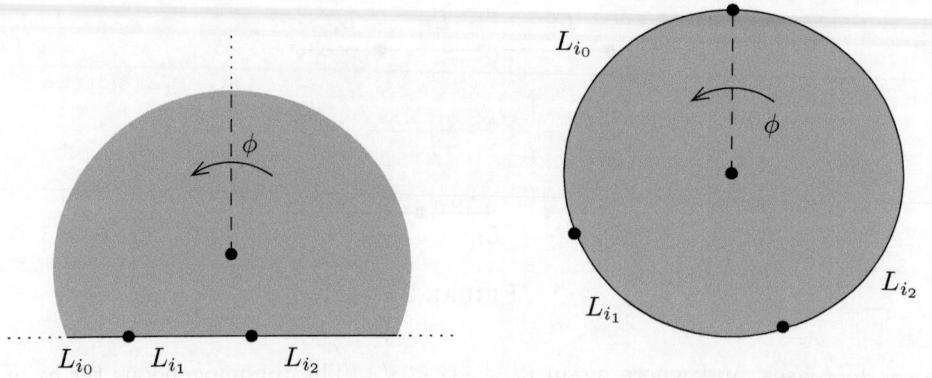

FIGURE 3

that to all of $\partial\tilde{S}$ in the unique way which satisfies (7.35). The outcome of this construction are maps

$$\begin{gathered}\mathrm{CF}^*(L_{i_{d-1}}, L_{i_d}, H_{i_{d-1}i_d}) \otimes \cdots \otimes \mathrm{CF}^*(L_{i_0}, L_{i_1}, H_{i_0 i_1}) \otimes \mathrm{CF}^*(\phi(L_{i_d}), L_{i_0}, H_{\phi, i_d i_0}) \\ \downarrow \\ \mathrm{CF}^{*+n-d}(\phi, H_\phi),\end{gathered} \tag{7.52}$$

where H_ϕ is a new Hamiltonian, chosen (together with a corresponding J_ϕ) to form $\mathrm{HF}^*(\phi, \delta, \epsilon)$. The (7.52) are the components of a chain map, which induces (7.22).

The construction of (7.26) is a mixture of the previous two. One considers surfaces

$$S = T \setminus \{\zeta_*, \zeta_1, \dots, \zeta_r, \zeta'_1, \dots, \zeta'_s\}, \tag{7.53}$$

where ζ_* is a fixed interior point of T, say $\zeta_* = (0, 1/2)$, and the other points are as in (7.47). Note however that this time, both ends $s \to \pm\infty$ will be considered as inputs. The map (7.34) has monodromy $\phi = \phi_1\phi_0$ around ζ_*. This structure, as well as the choice of boundary conditions, is indicated in Figure 4. One chooses one-forms β_S such that

$$\beta_S = \begin{cases} -(\epsilon_0/\epsilon)\; dt & s \ll 0, \\ (\epsilon_1/\epsilon)\; dt & s \gg 0, \end{cases} \tag{7.54}$$

and which vanish near the ζ_k, ζ'_k. As a consequence, the integral of β_S along a loop around ζ_* is necessarily $\epsilon_0/\epsilon + \epsilon_1/\epsilon = 1$.

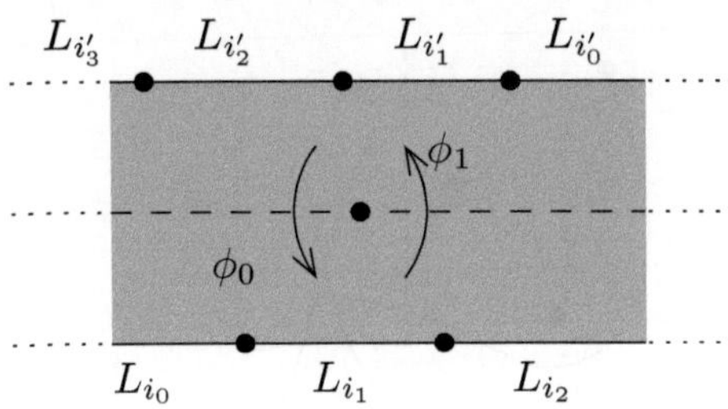

FIGURE 4

The outcome are maps

$$\begin{gathered} \mathrm{CF}^*(L_{i'_{s-1}}, L_{i'_s}, H_{i'_{s-1}i'_s}) \otimes \cdots \otimes \mathrm{CF}^*(L_{i'_0}, L_{i'_1}, H_{i'_0 i'_1}) \\ \otimes \mathrm{CF}^*(\phi_1(L_{i_r}), L_{i'_0}, H_{\phi_0, i_r i'_0}) \otimes \\ \mathrm{CF}^*(L_{i_{r-1}}, L_{i_r}, H_{i_{r-1}i_r}) \otimes \cdots \otimes \mathrm{CF}^*(L_{i_0}, L_{i_1}, H_{i_0 i_1}) \\ \otimes \mathrm{CF}^*(\phi_0(L_{i'_s}), L_{i_0}, H_{\phi_1, i_{s'} i_0}) \\ \downarrow \\ \mathrm{CF}^{*+n-r-s}(\phi, H_\phi). \end{gathered} \tag{7.55}$$

These are the components of a chain map, which induces (7.26).

7e. The Lefschetz condition

What has been missing so far is a natural source of Lagrangian submanifolds, which would make the construction of $\mathcal{A}$ meaningful. We will provide that now, in the form of Lefschetz thimbles.

Setup 7.16. (i) Take an exact symplectic fibration with singularities. We call it an *exact symplectic Lefschetz fibration* if it satisfies the following additional conditions. First of all, $TE^v_x = \ker(D\pi_x) \subset TE_x$ is a symplectic subspace at every regular point x. Secondly, near each critical point, there is an (integrable and ω_E-compatible) complex structure I_E, such that π is I_E-holomorphic, and the (complex) Hessian at the critical point is nondegenerate. Together with the previous conditions, this implies that there are only finitely many critical points. For practical bookkeeping purposes, we also impose the additional condition that there should be at most one critical point in each fibre.

(ii) Suppose that we have an exact symplectic Lefschetz fibration. A *basis of vanishing paths* (see Figure 5) is a collection of properly embedded half-infinite paths $l_1, \dots, l_m \subset \mathbb{C}$, whose endpoints are precisely the critical values of π, and with the following properties. There is a half-plane $\{\mathrm{re}(y) \leq C\}$ which contains all critical values of π, and such that the parts of the l_k lying in that half-plane are

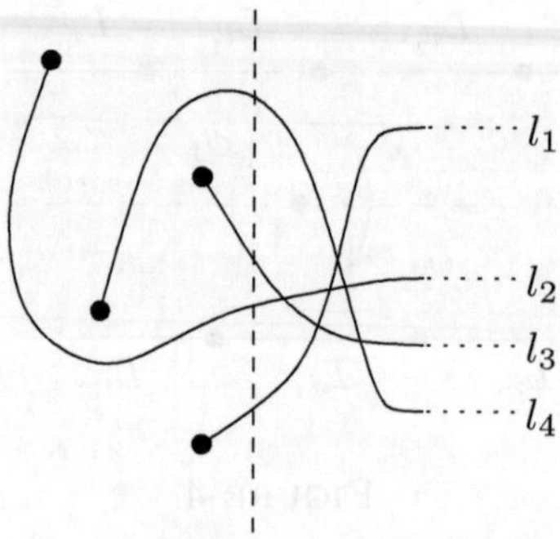

FIGURE 5

pairwise disjoint. Outside that half-plane, we have

$$l_k \cap \{\mathrm{re}(y) \geq C\} = \{\mathrm{im}(y) = \iota_k(\mathrm{re}(y))\}. \tag{7.56}$$

Here, the functions $\iota_1, \dots, \iota_m : [C, \infty] \to \mathbb{R}$ are constant near infinity, let us say $\iota_k(r) = o_k$ for $r \gg 0$, and

$$\begin{cases} \iota_1(C) < \cdots < \iota_m(C), \\ o_1 > \cdots > o_m. \end{cases} \tag{7.57}$$

The corresponding *basis of Lefschetz thimbles* consists of the unique Lagrangian submanifolds $L_1, \dots, L_m \subset E$ such that $\pi(L_k) = l_k$. We choose orientations arbitrarily (since the Lefschetz thimbles are diffeomorphic to $\mathbb{R}^n$, they carry unique *Spin* structures; and if E has a symplectic Calabi–Yau structure, they can be equipped with gradings).

Conjecture 7.17. *Let us use a basis of vanishing cycles to define the A_∞-algebra $\mathcal{A}$ and bimodules $\mathcal{P}_{\phi,\delta,\epsilon}$. Then, for any automorphism ϕ and $\delta \gg 0$, the open-closed string map* (7.22) *is an isomorphism.*

This is a kind of "decomposition of the diagonal" statement, which is of general interest since it would give a way of computing fixed point Floer cohomology in terms of open string (Fukaya category) data. It is plausible that it could be approached by the methods from [1], but we will not discuss that possibility further here.

Lemma 7.18. *Let ν be the global monodromy. For any $\delta < 0$, $\mathrm{HF}^*(\nu(L_i), L_i, \delta, \epsilon)$ is one-dimensional and concentrated in degree 0. Moreover, for any $i < j$ and any $\delta < o_j - o_i < 0$, the triangle product*

$$\mathrm{HF}^*(L_i, L_j) \otimes \mathrm{HF}^*(\nu(L_i), L_i, \delta, \epsilon) \longrightarrow \mathrm{HF}^*(\nu(L_i), L_j, \delta, \epsilon) \tag{7.58}$$

is an isomorphism.

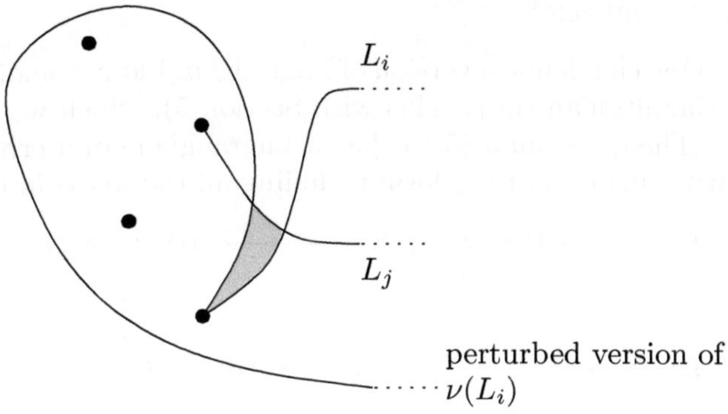

FIGURE 6

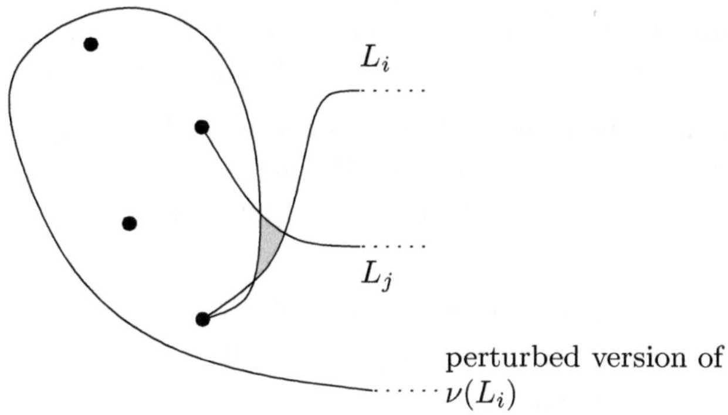

FIGURE 7

Sketch of proof. Let H be the Hamiltonian used to define $\mathrm{HF}^*(\nu(L_i), L_i, \delta, \epsilon)$. The generators of the underlying chain complex correspond to points of $(\phi^1_{\mathrm{H}} \circ \nu)(L_i) \cap L_i$. After a compactly supported isotopy (see Figure 6), these two Lagrangian submanifolds will intersect in a single point, which is a critical point of π; and one easily computes that its Maslov index is 0. For a suitable choice of auxiliary data, the product (7.58) is obtained by counting holomorphic triangles in E which project to the triangle in the base shaded in Figure 6. One can in principle determine those directly, but it is easier to apply a further isotopy as indicated in Figure 7, after which the triangle in the base can be shrunk to a point, making the computation straightforward (the same trick is used in [32, Figure 6]). □

Take $\gamma \in (0, 2\pi)$, $\delta > 0$, and a small $\epsilon > 0$. Consider the map

$$\mathrm{H}^*(E) \cong \mathrm{HF}^*(E, \gamma, \epsilon) \longrightarrow \mathrm{HF}^*(\nu, \delta, \epsilon) \longrightarrow \mathrm{HF}^*(\nu(L_i), L_i, \delta, \epsilon). \tag{7.59}$$

Here, the first isomorphism is Lemma 5.5; the map after that comes from Lemma 6.12; and the final one is (7.18). Let u_i be the image of $1 \in \mathrm{H}^0(E)$ under (7.59).

Lemma 7.19. *u_i is nontrivial.*

Sketch of proof. One can define a version of Lagrangian Floer homology perturbed by a rotational Hamiltonian (in parallel with Section 5), which we will denote by $\mathrm{HF}^*(L_i, L_i, \gamma, \epsilon)$. Then, the map (6.16) has a Lagrangian counterpart, which fits into a commutative diagram of the form including all the maps in (7.59):

$$\begin{array}{ccccc} \mathrm{H}^*(E) & \longrightarrow & \mathrm{HF}^*(E,\gamma,\epsilon) & \longrightarrow & \mathrm{HF}^*(\nu,\delta,\epsilon) \\ \downarrow & & \downarrow & & \downarrow \\ \mathrm{H}^*(L) & \longrightarrow & \mathrm{HF}^*(L_i,L_i,\gamma,\epsilon) & \longrightarrow & \mathrm{HF}^*(\nu(L_i),L_i,\delta,\epsilon). \end{array} \tag{7.60}$$

The leftmost $\downarrow$ is the standard restriction map; and all the $\rightarrow$ in the bottom row are isomorphisms (of one-dimensional vector spaces, concentrated in degree 0). □

Lemma 7.20. *Suppose that $\epsilon > 0$ is small, and that $\delta < o_m - o_1$. Take the element of $\mathrm{HF}^0(\nu, \delta, \epsilon)$ produced from $1 \in \mathrm{H}^0(E)$ as in (7.59). Applying the open-closed string map in its dual form (7.23) to that element yields a quasi-isomorphism $\mathcal{A} \rightarrow \mathcal{P}_{\nu,\delta,\epsilon}$.*

Proof. This is a direct application of Lemma 2.3 (with $\mathcal{P} = \mathcal{P}_{\nu^{-1},-\delta,-\epsilon}[n] \cong \mathcal{P}^{\vee}_{\nu,\delta,\epsilon}$). To use that criterion, one has to consider the maps

$$e_j \mathrm{H}^*(\mathcal{A}) e_i \longrightarrow e_j \mathrm{H}^*(\mathcal{P}_{\nu,\delta,\epsilon}) e_i = \mathrm{HF}^*(\nu(L_i), L_j, \delta, \epsilon) \tag{7.61}$$

given by

$$\begin{cases} \text{the product with } u_i, \text{ if } i < j; \\ \text{taking the unit } e_i \text{ to } u_i, \text{ if } i = j; \\ \text{zero, if } i > j. \end{cases} \tag{7.62}$$

The first two cases yield an isomorphism, by Lemmas 7.18 and 7.19. As for the last case, it is straightforward to show that $\mathrm{HF}^*(\nu(L_i), L_j, \delta, \epsilon) = 0$ for $i > j$. □

To summarize, we now know that

$$\text{for } \delta \ll 0, \quad \begin{cases} \mathcal{P}_{\nu,\delta,\epsilon} \simeq \mathcal{A}, \\ \mathcal{P}_{\nu^{-1},-\delta,-\epsilon} \simeq \mathcal{A}^{\vee}[-n], \end{cases} \tag{7.63}$$

where the second part of the statement follows from the first one by (7.16). Note that ϵ can be arbitrary, since the bimodules are independent of ϵ up to quasi-isomorphism. Applying (7.27), one therefore gets a map

$$\begin{aligned} \mathrm{HF}^*(\nu^2, \delta, \epsilon) \longrightarrow \mathrm{H}^*\big(\hom_{[\mathcal{A},\mathcal{A}]}(\mathcal{P}_{\nu^{-1},-\delta/2,-\epsilon/2}, \mathcal{P}_{\nu,\delta/2,\epsilon/2})\big) \\ \cong \mathrm{H}^*(\hom_{[\mathcal{A},\mathcal{A}]}(\mathcal{A}^{\vee}[-n], \mathcal{A})). \end{aligned} \tag{7.64}$$

Proof of Theorem 1.2. For small $\epsilon > 0$, Proposition 5.6 and Lemma 6.12 yield a map

$$\mathrm{HF}^{*+2}(\mu, \epsilon) \longrightarrow \mathrm{HF}^*(\nu^2, \delta, \epsilon). \tag{7.65}$$

One combines this with (7.64) to get the desired construction. □

Proof of Theorem 1.3. This is the same argument as before, but instead of Proposition 5.6, one uses Proposition 5.8 (and $1 < \epsilon < 2$). □

Remark 7.21. By comparing (2.61) and Conjecture 1.6, one sees that for the conjecture to hold, the bimodule map $\mathcal{A}^\vee \to \mathcal{A}$ constructed in the proof of Theorem 1.3 must necessarily be invariant under self-conjugation (2.37). Equivalently by Lemma 2.6, if one thinks of that map as an element of $HH_*(\mathcal{A}, \mathcal{A}^\vee \otimes_{\mathcal{A}} \mathcal{A}^\vee)^\vee$, it must be invariant under the $\mathbb{Z}/2$-action which exchanges the two tensor factors. Going back to the geometric definition, which means using (7.63), the desired statement is that the image of a specific element under the map

$$HF^0(\nu^2, \delta, \epsilon) \longrightarrow HH_0(\mathcal{A}, \mathcal{P}_{\nu^{-1},-\delta/2,-\epsilon/2} \otimes_{\mathcal{A}} \mathcal{P}_{\nu^{-1},-\delta/2,-\epsilon/2})^\vee \tag{7.66}$$

is $\mathbb{Z}/2$-invariant. Assuming $\mathrm{char}(\mathbb{K}) \neq 2$, it follows from Remark 7.13 and Example 6.18 that the entire image of (7.66) is $\mathbb{Z}/2$-invariant.

Remark 7.22. As a generalization of Conjecture 7.17, one could consider maps (7.26), but with an arbitrary number of bimodules involved. We will be interested only in one special case, written in dual form as in (7.27):

$$\begin{aligned} HF^*(\nu^k, \delta, \epsilon) &\longrightarrow H^*\big(\hom_{[\mathcal{A},\mathcal{A}]}(\mathcal{P}_{\nu^{-1},-\delta/k,-\epsilon/k}^{\otimes_{\mathcal{A}} k-1}, \mathcal{P}_{\nu,\delta/k,\epsilon/k})\big) \\ &\cong H^*(\hom_{[\mathcal{A},\mathcal{A}]}((\mathcal{A}^\vee)^{\otimes_{\mathcal{A}} k-1}, \mathcal{A})), \quad \delta \ll 0. \end{aligned} \tag{7.67}$$

Suppose that (7.67) is an isomorphism. Let us specialize to Lefschetz fibrations coming from anticanonical Lefschetz pencils. In that case, we have seen in Example 6.15 that, for geometric reasons, the Floer cohomology groups of ν^k are concentrated in nonnegative degrees. It would then follow that (2.67) is satisfied, so that Lemmas 2.12 and 2.14 would become applicable, leading to a proof of Conjecture 1.6 (however, this approach would not apply to other situations, such as that of Remark 1.7).

References

[1] M. Abouzaid and S. Ganatra. In preparation.

[2] M. Abouzaid and P. Seidel. Lefschetz fibration techniques in wrapped Floer cohomology. In preparation.

[3] M. Abouzaid and P. Seidel. Altering symplectic manifolds by homologous recombination. Preprint arXiv:1007.3281, 2010.

[4] M. Abouzaid and P. Seidel. An open string analogue of Viterbo functoriality. *Geom. Topol.*, 14:627–718, 2010.

[5] D. Auroux. Mirror symmetry and T-duality in the complement of an anticanonical divisor. *J. Gökova Geom. Topol.*, 1:51–91, 2007.

[6] D. Austin and P. Braam. Morse–Bott theory and equivariant cohomology. In *The Floer memorial volume*, volume 133 of *Progr. Math.*, pages 123–183. Birkhäuser, 1995.

[7] F. Bourgeois. *A Morse–Bott approach to contact homology.* PhD thesis, Stanford University, 2002.

[8] F. Bourgeois, T. Ekholm, and Ya. Eliashberg. Effect of Legendrian surgery (with an appendix by S. Ganatra and M. Maydanskiy). *Geom. Topol.*, 16:301–390, 2012.

[9] F. Bourgeois and K. Mohnke. Coherent orientations in symplectic field theory. *Math. Z.*, 248:123–146, 2004.

[10] F. Bourgeois and A. Oancea. An exact sequence for contact- and symplectic homology. *Invent. Math.*, 175:611–680, 2009.

[11] F. Bourgeois and A. Oancea. Symplectic homology, autonomous Hamiltonians, and Morse–Bott moduli spaces. *Duke Math. J.*, 146:71–174, 2009.

[12] F. Bourgeois and A. Oancea. The Gysin exact sequence for S^1-equivariant symplectic homology. *J. Topol. Anal.*, 5:361–407, 2013.

[13] L. Diogo. *Filtered Floer and symplectic homology via Gromov–Witten theory.* PhD thesis, Stanford Univ., 2012.

[14] S. Dostoglou and D. Salamon. Instanton homology and symplectic fixed points. In D. Salamon, editor, *Symplectic Geometry*, volume 192 of *LMS Lecture Note Series*, pages 57–94. Cambridge Univ. Press, 1993.

[15] S. Dostoglou and D. Salamon. Self dual instantons and holomorphic curves. *Annals of Math.*, 139:581–640, 1994.

[16] A. Floer. Morse theory for Lagrangian intersections. *J. Differential Geom.*, 28:513–547, 1988.

[17] A. Floer. Symplectic fixed points and holomorphic spheres. *Commun. Math. Phys.*, 120:575–611, 1989.

[18] A. Floer, H. Hofer, and D. Salamon. Transversality in elliptic Morse theory for the symplectic action. *Duke Math. J.*, 80:251–292, 1995.

[19] K. Fukaya, Y.-G. Oh, H. Ohta, and K. Ono. *Lagrangian intersection Floer theory – anomaly and obstruction.* Amer. Math. Soc., 2010.

[20] K. Fukaya, Y.-G. Oh, H. Ohta, and K. Ono. Lagrangian Floer theory on compact toric manifolds. II: Bulk deformations. *Selecta Math.*, 17:609–711, 2011.

[21] S. Ganatra. Symplectic cohomology and duality for the wrapped Fukaya category. Preprint arXiv:1304.7312, 2013.

[22] V. Ginzburg. The Conley conjecture. *Ann. of Math.* (2), 172:1127–1180, 2010.

[23] W. Goldman and J. Millson. The deformation theory of the fundamental group of compact Kähler manifolds. *IHES Publ. Math.*, 67:43–96, 1988.

[24] H. Hofer and D. Salamon. Floer homology and Novikov rings. In H. Hofer, C. Taubes, A. Weinstein, and E. Zehnder, editors, *The Floer memorial volume*, volume 133 of *Progress in Mathematics*, pages 483–524. Birkhäuser, 1995.

[25] M. Kontsevich. Deformation quantization of Poisson manifolds. *Lett. Math. Phys.*, 66:157–216, 2003.

[26] M. Kontsevich and Y. Soibelman. Notes on A_∞-algebras, A_∞-categories, and non-commutative geometry. In *Homological Mirror Symmetry: New Developments and Perspectives*, volume 757 of *Lecture Notes in Physics*, pages 153–219. Springer, 2008.

[27] M. Kontsevich and Y. Vlassopoulos. Pre-Calabi–Yau algebras and topological quantum field theories. In Preparation.

[28] K. Lefèvre-Hasegawa. *Sur les A_∞-catégories.* PhD thesis, Université Paris 7, 2002.

[29] V. Lyubashenko and O. Manzyuk. A_∞-bimodules and Serre A_∞-functors. In *Geometry and dynamics of groups and spaces*, volume 265 of *Progr. Math.*, pages 565–645, Basel, 2008. Birkhäuser.

[30] M. Manetti. Lectures on deformations of complex manifolds (deformations from differential graded viewpoint). *Rend. Mat. Appl.* (7), 24:1–183, 2004.

[31] M. Maydanskiy. *Exotic symplectic manifolds from Lefschetz fibrations.* PhD thesis, MIT, 2009.

[32] M. Maydanskiy and P. Seidel. Lefschetz fibrations and exotic symplectic structures on cotangent bundles of spheres. *J. Topology*, 3:157–180, 2010.

[33] M. McLean. Symplectic homology of Lefschetz fibrations and Floer homology of the monodromy map. *Selecta Math.*, 18:473–512, 2012.

[34] S. Piunikhin, D. Salamon, and M. Schwarz. Symplectic Floer–Donaldson theory and quantum cohomology. In C.B. Thomas, editor, *Contact and symplectic geometry*, pages 171–200. Cambridge Univ. Press, 1996.

[35] L. Polterovich and E. Shelukhin. Autonomous Hamiltonian flows, Hofer's geometry and persistence modules. Preprint arXiv:1412.8277, 2014.

[36] M. Poźniak. Floer homology, Novikov rings and clean intersections. In *Northern California Symplectic Geometry Seminar*, pages 119–181. Amer. Math. Soc., 1999.

[37] D. Salamon and E. Zehnder. Morse theory for periodic solutions of Hamiltonian systems and the Maslov index. *Comm. Pure Appl. Math.*, 45:1303–1360, 1992.

[38] P. Seidel. Graded Lagrangian submanifolds. *Bull. Soc. Math. France*, 128:103–146, 2000.

[39] P. Seidel. More about vanishing cycles and mutation. In K. Fukaya, Y.-G. Oh, K. Ono, and G. Tian, editors, *Symplectic Geometry and Mirror Symmetry* (*Proceedings of the* 4*th KIAS Annual International Conference*), pages 429–465. World Scientific, 2001.

[40] P. Seidel. Fukaya categories and deformations. In *Proceedings of the International Congress of Mathematicians* (*Beijing*), volume 2, pages 351–360. Higher Ed. Press, 2002.

[41] P. Seidel. A_∞-subalgebras and natural transformations. *Homology, Homotopy Appl.*, 10:83–114, 2008.

[42] P. Seidel. A biased survey of symplectic cohomology. In *Current Developments in Mathematics* (*Harvard,* 2006), pages 211–253. Intl. Press, 2008.

[43] P. Seidel. *Fukaya categories and Picard–Lefschetz theory.* European Math. Soc., 2008.

[44] P. Seidel. Symplectic homology as Hochschild homology. In *Algebraic Geometry: Seattle* 2005, volume 1, pages 415–434. Amer. Math. Soc., 2008.

[45] P. Seidel. Fukaya categories associated to Lefschetz fibrations. I. *J. Symplectic Geom.*, 10:325–388, 2012.

[46] P. Seidel. The equivariant pair-of-pants product in fixed point Floer cohomology. *Geom. Funct. Anal.*, 25:942–1007, 2015.

[47] P. Seidel. Fukaya A_∞-structures associated to Lefschetz fibrations. II 1/2. Preprint arXiv:1504.06317, 2015.

[48] P. Seidel. *Homological mirror symmetry for the quartic surface*, volume 1116 of *Memoirs of the Amer. Math. Soc.* Amer. Math. Soc., 2015.

[49] P. Seidel and J. Solomon. Symplectic cohomology and q-intersection numbers. *Geom. Funct. Anal.*, 22:443–477, 2012.

[50] N. Sheridan. Homological mirror symmetry for Calabi–Yau hypersurfaces in projective space. *Invent. Math.*, 199:1–186, 2015.

[51] D. Shklyarov. On Serre duality for compact homologically smooth DG algebras. Preprint arXiv:math/0702590, 2007.

[52] M. Tibar. *Polynomials and vanishing cycles.* Cambridge Univ. Press, 2007.

[53] T. Tradler. Infinity-inner-products on A-infinity-algebras. *J. Homotopy Related Struct.*, 3:245–271, 2008.

[54] I. Uljarevic. Symplectic homology for symplectomorphism and symplectic isotopy problem. Preprint arXiv:1404.2128, 2014.

[55] C. Viterbo. Functors and computations in Floer homology with applications, Part I. *Geom. Funct. Anal.*, 9:985–1033, 1999.

[56] K. Wehrheim and C. Woodward. Quilted Floer cohomology. *Geom. Topol.*, 14:833–902, 2010.

Paul Seidel
MIT 2-276
77 Massachusetts Ave
Cambridge, MA 02139, USA
e-mail: `pseidel@mit.edu`

MIX
Papier aus verantwortungsvollen Quellen
Paper from responsible sources
FSC® C105338
FSC
www.fsc.org

If you have any concerns about our products,
you can contact us on
ProductSafety@springernature.com

In case Publisher is established outside the EU,
the EU authorized representative is:
Springer Nature Customer Service Center GmbH
Europaplatz 3, 69115 Heidelberg, Germany

Printed by Libri Plureos GmbH
in Hamburg, Germany